A DICTIONARY OF ECOLOGY,
EVOLUTION AND SYSTEMATICS

A DICTIONARY OF ECOLOGY, EVOLUTION AND SYSTEMATICS

R. J. LINCOLN G. A. BOXSHALL P. F. CLARK

Department of Zoology, British Museum (Natural History)

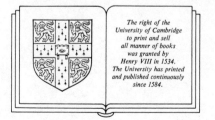

*The right of the
University of Cambridge
to print and sell
all manner of books
was granted by
Henry VIII in 1534.
The University has printed
and published continuously
since 1584.*

CAMBRIDGE UNIVERSITY PRESS

CAMBRIDGE

LONDON   NEW YORK   NEW ROCHELLE

MELBOURNE   SYDNEY

Published by the Press Syndicate of the University of Cambridge
The Pitt Building, Trumpington Street, Cambridge CB2 1RP
32 East 57th Street, New York, NY 10022, USA
10 Stamford Road, Oakleigh, Melbourne 3166, Australia

© Cambridge University Press 1982

First published 1982
First paperback edition 1983
Reprinted 1984, 1985, 1986

Printed in Great Britain at the University Press, Cambridge

Library of Congress Catalogue card number: 81–18013

British Library Cataloguing in Publication Data

Lincoln, R.J.
A dictionary of ecology.
1. Ecology—Dictionaries
I. Title   II. Boxshall, G.A.   III. Clark, P.F.
574.5'03   QH540.4

ISBN 0 521 23957 5 hard covers
ISBN 0 521 26902 4 paperback

CONTENTS

'The psychosemanticists will specialize in the havoc wrought by verbal artillery upon the fortress of reason. Their job will be to cope with the psychic trauma caused by linguistic meaninglessness, to prevent the language from degenerating into gibberish, and to save the sanity of persons threatened by the onset of polysyllabic monstrosititis'
James Thurber

'The Psychosemanticist Will See You Now, Mr Thurber.'

PREFACE

We selected the terms Ecology, Evolution and Systematics for our title because we felt they would indicate three focal points within the overall coverage of this dictionary to potential users. Biology, in common with the whole of Science is structured into a plethora of sub-disciplines exhibiting varying degrees of specialization, that often tend to inhibit or obscure a broadly based understanding. Our three major disciplines are in practice integrated by large areas of overlap into a modern biological science that has been called Evolutionary Biology, but might equally well be viewed as contemporary Natural History. The special emphasis is placed on the dynamics of the biosphere – change through time. Ecologists look especially at populations, communities, and their interrelationships with the physical environment. Evolutionists look at organisms and groups but explore in particular the mechanisms that produce or influence variation, adaptation and change. The trend towards more frequent intercourse between workers from traditionally separate disciplines in search of a better understanding of the causal explanations for observed biological phenomena has brought ecological and evolutionary studies into closer contact. The effect on the student or researcher is to open up a vast new literature with the inevitable plague of specialized terms which although relevant may be quite unfamiliar to the reader as they have their origins in other disciplines. Just such a situation exists in Natural History and provided the initial impetus for the compilation of an index of terms that led eventually to the preparation of this book. The ultimate choice of which terms to include and which to omit was our own, founded on personal research experience as well as on advice from colleagues at the British Museum (Natural History). There are few natural discontinuities in biology or biological terminology so our decisions concerning peripheral material were necessarily subjective. Our aim throughout has been to provide short working definitions of those terms that come within the routine reading matter of ecologists, taxonomists and the like.

The language of taxonomy, or systematics, has been given thorough coverage because, despite its reputation as a specialized cul-de-sac, it is a subject that underpins many ecological and evolutionary studies. Indeed, many ecologists at first disinterested in taxonomic practice have found themselves reluctantly recruited into the discipline when confounded by the twin obstacles of identification and species recognition. The application of statistics to ecology has also been covered – with biology becoming an ever more rigorous science a basic understanding of statistics is a necessity.

Special attention has been given to principles, processes and classifications. We have deliberately omitted basic morphological and anatomical terms since these can be tracked down fairly easily in relevant descriptive textbooks and, more to the point perhaps, any such list of structural terminology would be so immense as to swamp the present text and expand the work into several volumes. Individual entries are brief, with the emphasis on a working definition rather than an encyclopaedic or discursive essay. We aim to provide an indication of modern usage and have not attempted to impose a rigid definition where one does not exist as this can often be quite misleading. Where terms have been employed in several subtly different ways by different authors we have opted for a definition closest to the original use or to the etymological derivation of the word. The latter usually enables a term to be understood even when read in a context not covered directly by our source material. Where substantial divergences in the use of a term exist we have given separate definitions. We have no desire to stimulate the introduction of unnecessary new terms which although they may serve to classify or qualify particular events or assemblages can actually hinder efficient communication. As we have become increasingly aware the literature already carries a heavy burden of redundant words, synonyms, partial synonyms and cumbersome neologisms. Such terminological monstrosities as bathyplanktohyponeuston do little to facilitate communication.

The final section of the book is devoted to 21 Appendices in the form of maps, diagrams, tables and lists that have been used to summarize groups of associated terms or concepts.

We gratefully acknowledge the help of our colleagues in the Departments of Zoology, Botany, Entomology, Palaeontology and Central Services of the British Museum (Natural History). Without the benefit of their experience and expertise this work would be impoverished both in scope and interpretation.

September 1981   R.J.L., G.A.B., P.F.C.

A

a. Abbreviation of the Latin *annum*, meaning year.

a- Prefix meaning without; **an-**.

a posteriori Used of a method of reasoning, commencing with effect or experience to establish cause or general rule; *cf. a priori.*

a posteriori **weighting** The method of attaching different importance values to taxonomic characters already established as providing acceptable classifications; *cf. a priori* weighting.

a priori Used of a method of reasoning, commencing with theory or cause to establish fact or effect; *cf. a posteriori.*

a priori **weighting** The method of attaching different importance values to taxonomic characters on the basis of preconceived criteria; *cf. a posteriori* weighting.

ab. Abbreviation of the Latin *aberratio*, meaning aberration; aberrant individual.

ab- Prefix meaning from, away from.

abbreviation 1: Shortening, curtailment. 2: The successive shortening of ontogeny through loss of developmental stages.

abduct To move away from the midline; **abduction**; *cf.* adduct.

aberrant 1: Not conforming to type. 2: An individual variant exhibiting atypical characters due usually to external environmental influences rather than to genetic factors; **aberration**.

abience Withdrawal or retraction from a stimulus; an avoidance reaction; **abient**; *cf.* adience.

abiocoen The non-biotic component of an ecosystem or habitat.

abiogenesis Spontaneous generation; the concept that life can arise spontaneously from non-living matter by natural processes without the intervention of supernatural powers; archebiosis; archegenesis; autogenesis; nomogenesis; xenogenesis; *cf.* biogenesis.

abioseston The non-living component of the total particulate matter suspended in water; tripton; *cf.* seston.

abiotic Devoid of life; non-living; *cf.* biotic.

abiotic environment The non-living component of an ecosystem; the physical and chemical factors of the environment; *cf.* biotic environment.

abiotic factors Physical, chemical and other non-living environmental factors; *cf.* biotic factors.

ablation 1: The surgical removal of body tissue or organs. 2: The removal of a surface layer, as of ice by melting or evaporation.

abnormal soil Azonal soil *q.v.*

aboospore A spore produced from an unfertilized female gamete; azygospore; parthenospore.

aborigine The original or indigenous biota of a geographical region; **aboriginal**.

abort To arrest development; **abortion, abortive**.

abortive transduction Transduction *q.v.* in which the genetic material is not integrated or replicated but is otherwise functional.

abrasion The process of erosion by rubbing off or wearing away of surface material; attrition; **abrade, abrasive, abrasiveness**.

abscissa The horizontal axis, or *x*-axis of a graph; *cf.* ordinate.

abscission The natural process by which two parts of an organism separate; **abscise**.

absenteeism The behaviour shown by animals which nest away from their progeny but visit them from time to time with food, providing minimal parental care.

absolute abundance The precise number of individuals of a taxon in a given area, volume, population or community; *cf.* relative abundance.

absolute affinity The relationship established only when the suite of characters exhibited by a specimen matches to a predetermined degree that of only one other taxon; *cf.* relative affinity.

absolute age The precise geological age of a fossil or rock, usually calculated by means of radiometric dating; absolute date; *cf.* relative age.

absolute character species A characteristic species *q.v.* that has a geographical distribution more or less coincident with that of the vegetation unit to which it belongs.

absolute chronology Absolute dating *q.v.*

absolute dating A method of geological dating employing isotope decay, that gives a direct measure of the amount of time that has elapsed since formation of the rocks; absolute chronology; radiometric dating; *cf.* relative dating.

absolute growth rate The actual increase in size of an individual or population per unit time under known or specified conditions; *cf.* relative growth rate.

absolute humidity The actual amount of water vapour present in a unit mass of air, expressed in grams per cubic metre; *cf.* relative humidity.

absolute pollen frequency An estimate of the actual amount of pollen deposited in a given area per unit time; obtained by correcting the pollen content of a sediment for differences in the rate of sedimentation.

absolute ranking The assignment of formal rank to monophyletic groups, based on the criterion of age of origin rather than degree of divergence; *cf.* relative ranking.

absolute synonyms Homotypic synonyms *q.v.*

absolute tautonymy In nomenclature, the identical spelling of a generic name and epithet of one of the species or subspecies originally included in the genus; *cf.* virtual tautonymy.

absorption 1: The process by which one substance (the absorbate) is taken into and incorporated in another substance (the absorbent); **absorb, absorbency**. 2: In ecological energetics, that part of consumption *q.v.* not voided as egesta *q.v.*

abundance The total number of individuals of a taxon or taxa in an area, volume, population or community; absolute abundance; often measured as cover in plants; *cf.* relative abundance.

abundismus Incomplete melanism, produced by a relative increase in the number of black pigment spots within the non-black areas of the colour pattern.

abyssal Pertaining to great depths within the earth, or to zones of great depth in the oceans or lakes into which light does not penetrate; in oceanography occasionally restricted to depths below 2000 m but more usually used of depths between 4000 and 6000 m; hypobenthile; abysmal; **abyss**; see Appendix 7.

abyssal plain The more or less flat ocean floor below 4000 m, excluding ocean trenches, having a slope of less than 1 in 1000; see Appendix 7.

abyssobenthic Living on or in the ocean floor in the abyssal zone; abyssalbenthic; see Appendix 7.

abyssopelagic Living in the oceanic water column at depths between 4000 and 6000 m, away from the ocean floor; abyssal pelagic; see Appendix 7.

acarology The study of mites and ticks; **acarological**.

acarophilous Thriving in association with mites; **acarophile, acarophily**.

acarophytism Symbiosis between plants and mites; acarophytium.

accelerated erosion A marked increase in the natural rate of erosion as the result of the activities of animals or man, or to changes in vegetation cover or soil conditions.

acceleration 1: Increase in velocity. 2: Increase in the speed of ontogenetic development so that a character appears earlier in the ontogeny of a descendant than in that of the ancestor; *cf.* retardation.

acceleration, law of The theory that the sequence of development of organs and structures in ontogeny is directly related to their importance to the organism.

acceptance region In statistics, the region of sample space containing all outcomes for which the null hypothesis is accepted; *cf.* critical region.

accepted name A name adopted by an author as the correct name for a taxon.

accessibilité A phytosociological concept of the totality of conditions prevailing at a given site that may influence the possibility of a propagule reaching that site.

accessory chromosome Any chromosome differing from the normal A-chromosomes, including B-chromosome *q.v.*, m-chromosome *q.v.* and sex chromosome *q.v.*

accessory species A plant species with a moderate degree of fidelity *q.v.* in a given community or association.

accidental 1: Not normally occurring in a particular community or habitat. 2: Used of a plant species with a low degree of fidelity *q.v.* in a given community or association.

accidental parasite A parasite found associated with an organism which is not its normal host.

acclimation Phenotypic adaptation to environmental fluctuations; the gradual and reversible adjustment of physiology or morphology as a result of changing environmental conditions; often used with reference to an individual organism in an artificial or experimentally manipulated environment; habituation; *cf.* acclimatization, accommodation.

acclimatization Phenotypic adaptation to environmental fluctuations; the gradual and reversible adjustment of physiology and morphology to changing natural environmental conditions; often used to refer to the changes observed in a species over a number of generations; *cf.* acclimation, accommodation.

acclivous Having a gentle upward slope.

accommodation 1: The capacity of a plant to adapt to changes in the environment. 2: A decrease in response or sensation as a result of repetitive stimulation; *cf.* acclimation, acclimatization.

accrescent Increasing in size with age; used of plants that continue to grow after flowering.

accretion 1: Increase in size by the external addition of new material. 2: Deposition of material by sedimentation.

accumulation, degree of A measure of the biological concentration of heavy metals or minerals in plants, calculated as the ratio of the mineral concentration in plants on contaminated soils to that of plants on normal soils expressed as a percentage.

accumulation, zone of The B-horizon of the soil profile, the upper subsoil horizon.

accumulator organism Any organism that actively concentrates a particular element or compound in its tissues.

accuracy In statistics, the degree to which a measured quantity approaches the true value of what is being measured; *cf.* precision.

-aceae The ending of a name of a family in botanical nomenclature.

acellular Not composed of cells; *cf.* cellular.

acentric Used of a chromosome or chromosome fragment lacking a centromere; akinetic; *cf.* centric.

achiasmate meiosis Meiosis *q.v.* occurring without the formation of chiasmata; in species in which crossing over is limited to one sex, the achiasmate meiosis generally occurs in the heterogametic sex *q.v.*

achromatic Without colour; unpigmented; **achromic.**

A-chromosome Any of the normal chromosomes of a eukaryotic organism.

achroous Without colour, unpigmented; achromatic.

aciculilignosa Evergreen and deciduous needle leaved forest vegetation.

acidic Pertaining to habitat media having a pH less than 7; *cf.* alkaline.

acidic habitat A habitat of low pH, typically poor in nutrients.

acidobiontic Living in an acidic habitat.

acidophilic Thriving in an acidic environment; acidophilous; aciduric; oxyphilous; **acidophile, acidophily;** *cf.* acidophobic.

acidophobic Intolerant of acidic environments. **acidophobe, acidophoby;** *cf.* acidophilic.

acidotrophic Feeding on acidic food or acidic substrates.

aciduric Acidophilic *q.v.*

acme A period of maximum vigour; the highest point attained in phylogenetic or ontogenetic development; *cf.* epacme, paracme.

acme-zone A biostratigraphic unit *q.v.* characterized by the abundance of particular fossil form or forms.

acmic Pertaining to the acme *q.v.*, or to periods of seasonal change in populations.

acoustics The study of sound; **acoustic.**

acquired characters, inheritance of 1: Lamarckism; that changes in use or disuse of an organ result in changes in size and functional capacity and that these modified characters are transmitted to the offspring. 2: Neo-Lamarckism; that

characters acquired by organisms as a response to environmental factors are assimilated into the genome and transmitted to the offspring.

acquired trait A character or trait which is a result of direct environmental influences; acquired character.

acre A secondary fps unit of area equal to 4840 square yards or 4.046×10^3 square metres; see Appendix 13.

acre-foot The quantity of water sufficient to cover one acre to a depth of one foot; used as a measure of irrigation capacity or runoff volume.

acridophagous Feeding on grasshoppers; **acridophage, acridophagy.**

acrobryous Growing at the apex only.

acrocentric Used of a chromosome having the centromere located at one end; *cf.* metacentric.

acrodendrophilous Thriving in tree top habitats, **acrodendrophile, acrodendrophily.**

acrophytia Plant communities of alpine regions; **acrophyta.**

acropleustophyte A large aquatic plant floating freely at the water surface.

acrotropism An orientation response resulting in the continued growth of a plant in the direction in which growth originally commenced; **acrotropic.**

actic Pertaining to rocky shores; often used of the zone between high and low tides; littoral; intertidal.

actinobiology The study of the effects of radiation on living organisms; **actinology.**

action system A behaviour pattern in an organism.

actium A rocky shore community.

active chamaephyte A chamaephyte *q.v.* subtype in which the horizontal vegetative shoots persist through unfavourable seasons in a procumbent position.

active immunity Resistance to disease effected by the production of antibodies by an organism in response to antigens derived from a pathogen or disease organism.

active process Any process requiring the expenditure of metabolic energy; *cf.* passive process.

active space The space within which a pheromone *q.v.* is above threshold concentration.

active uptake The absorption of ions by processes requiring the expenditure of metabolic energy; *cf.* passive uptake.

actophilous Thriving on rocky seashores; **actophile, actophily.**

actualism The theory that seeks to explain the evolution of the earth in terms of relatively small scale natural fluctuations or events that have been operative throughout the entire geological history of the earth and have thus produced changes of great magnitude; uniformitarianism; *cf.* catastrophism.

actualistic palaeontology The study of living animals and present day environments as representing the source material for the fossils of the future; equivalent to neontology.

actuopalaeontology Palaeontology based on the philosophy of actualism.

acute Severe, often lethal; of short duration; undergoing rapid development; *cf.* chronic.

acyclic parthenogenesis Reproduction by parthenogenesis *q.v.* alone, in which the sexual phase of an alternation of generations cycle has been lost; anholocyclic parthenogenesis; *cf.* parthenogenesis.

ad- Prefix meaning to, on the side of.

ad int. Abbreviation of the Latin *ad interim*, meaning for the present, provisionally.

Adansonian taxonomy An early method of classification advocating the grouping of organisms on the basis of many equally-weighted attributes, a principle adopted in modern numerical taxonomy to avoid the subjectivity of classical taxonomic methodology; Adansonism.

adaptability The potential for adaptation *q.v.*; adaptable.

adaptation 1: The process of adjustment of an individual organism to environmental stress; adaptability. 2: Process of evolutionary modification which results in improved survival and reproductive efficiency. 3: Any morphological, physiological, developmental or behavioural character that enhances survival and reproductive success of an organism; **adaption**; **adaptive**.

adaptedness The extent of adaptation to environment.

adaptiogenesis The production of new adaptations.

adaption Adaptation *q.v.*

adaptive capacity The genetically determined physiological tolerance of an organism.

adaptive gap Nonadaptive zone *q.v.*

adaptive landscape The figurative representation of the fitness of organisms in the form of a topographical map, on which those fit genotypes (species) able to occupy particular ecological niches are depicted as adaptive peaks separated by adaptive valleys representing unfit gene combinations; adaptive surface; adaptive topography.

adaptive norm The more or less stable complex of genetic diversity within a well adapted population.

adaptive peak A peak on an adaptive landscape *q.v.*

adaptive radiation The evolutionary diversification of a taxon (adaptive type) into a number of different ecological roles or modes of life (adaptive zones), usually over a relatively short period of time and leading to the appearance of a variety of new forms.

adaptive selection Selection producing phenotypic similarity as a result of adaptation to environment.

adaptive strategy The sum total of the adaptations to environment of a species.

adaptive valley A valley in an adaptive landscape *q.v.*

adaptive value The comparative fitness *q.v.* of different genotypes in a given environment; the survival and reproductive value of one genotype relative to the other genotypes in the population.

adaptive zone A concept in evolutionary biology of the ecological pathways along which taxa evolve; the way of life and organizational level of a higher taxonomic group; at the species level, essentially similar to the fundamental niche *q.v.*; *cf.* nonadaptive zone, transitional adaptive zone.

additive factor Any of a group of non-allelic genes affecting the same phenotypic character, each enhancing the expression of the other in the phenotype.

additive hypothesis That there is a value corresponding to each link on any phylogenetic tree, and that the observed distance between each pair of contemporary taxonomic categories is equal to the sum of the values for the links connecting the pair of taxonomic categories.

adduct To move towards the midline; **adduction**; *cf.* abduct.

adeciduate Not falling off, or coming away.

adelfotype A topotype *q.v.* collected in the original collection of a species but not part of the original type series; isocotype.

adelphogamy Fertilization between two different individuals derived vegetatively from the same parent plant; sib mating; sib pollination.

adelphoparasite An organism parasitic on a closely related host organism; *cf.* alloparasite.

adelphophagy The fusion of two gametes of the same sex; **adelphophagia**.

adelphotaxy The mutual attraction between spores after extrusion.

adenotrophic viviparity A type of viviparity *q.v.* in some insects in which the larva is retained within the uterus until ready to pupate.

adhesion 1: Firm attachment. 2: The attraction of molecules in the walls of the interstices of a sediment, for water molecules.

adiabatic Pertaining to a thermodynamic process

in which heat is neither lost nor gained; *cf.* non-adiabatic.

adichogamy Simultaneous maturation of male and female reproductive organs of a flower or hermaphroditic organism; homogamy; **adichogamous;** *cf.* dichogamy.

adience Movement towards a stimulus; an approaching reaction; *cf.* abience.

adjacently sympatric Parapatric *q.v.*; used of populations that are geographically separated but whose boundaries are in contact at some point so that gene flow between them is possible; neighbouringly sympatric.

adjustment Acclimation *q.v.*; the functional response of an organism to stimuli.

admissible In taxonomy, the form of a name which can be validly published and the use of a name or epithet in accordance with the provisions of the Code *q.v.*

adnate Closely applied to; growing on; attached along the entire length.

adsere That stage of an ecological succession that precedes its transformation into the subclimax *q.v.*

adsorption The adhesion of molecules as an ultra-thin layer on the surface of solids or fluids; **adsorb.**

adspersed Widely distributed; scattered.

adtidal Living immediately below low tide level.

adultation The appearance, by acceleration *q.v.*, of adult ancestral characters in the larval stages of descendants; *cf.* recapitulation.

advection The process of transfer by virtue of motion; the transfer of heat or matter by horizontal movement of water masses.

advena (adv.) Latin, meaning alien, introduced.

adventitious Accidental; occurring at an unusual site; secondary; extraordinary.

adventitious embryony The production of an embryonic sporophyte by mitotic divisions from tissues of another sporophyte without an intervening gametophyte generation; seed production without a sexual process.

adventive Not native; an organism transported into a new habitat, whether by natural means or by the agency of man.

adynamandrous 1: Having non-functioning male reproductive organs; **adynamandry;** *cf.* adynamogynous. 2: Incapable of self-fertilization.

adynamogynous Having non-functioning female reproductive organs; apogynous; **adynamogyny;** *cf.* adynamandrous.

aedeotype The original specimen of a species to have its genitalia examined; also the actual slide preparation of the genitalia.

aegricorpus The single manifestation of the specific genetic interactions in and between a host and pathogen.

aeolation Erosion of a land surface by wind-blown sand and dust; **eolation.**

aeolian Pertaining to the action or effect of the wind; **eolian.**

aeolian deposit Wind borne soil deposit; *cf.* alluvial deposit, colluvial deposit.

aeon An indefinitely long period of time; **eon.**

aer- Prefix meaning air, atmosphere.

aerial Occurring in the air; airborne.

aerial water All forms of atmospheric water, as distinct from terrestrial water.

aerobic 1: Growing or occurring only in the presence of molecular oxygen; oxybiotic; **aerobe, aerobiosis;** *cf.* anaerobic. 2: Used of an environment in which the partial pressure of oxygen is similar to normal atmospheric levels; oxygenated.

aerobiology The study of airborne organisms.

aerochorous Disseminated by wind; anemochorous; aerophilous; **aerochore, aerochory.**

aerogenic Gas producing; **aerogenesis;** *cf.* anaerogenic.

aerohygrophilous Thriving in high atmospheric humidity; **aerohygrophile, aerohygrophily;** *cf.* aerohygrophobous.

aerohygrophobous Intolerant of high atmospheric humidity; **aerohygrophobe, aerohygrophoby;** *cf.* aerohygrophilous.

aeromorphosis A structural change resulting from exposure to air or wind; **aeromorphic.**

aerophilous 1: Pollinated by wind; fertilized by airborne pollen. 2: Thriving in exposed windy habitats; **aerophile, aerophily.** 3: Disseminated by wind; aerochorous.

aerophyte An epiphyte growing on a terrestrial plant and lacking direct contact with soil or water; air plant; aerial plant.

aerophytobiota The aerobic soil flora; **aerophytobiont;** *cf.* anaerophytobiota.

aeroplankton Those organisms freely suspended in the air and dispersed by wind; aerial plankton.

aeroscepsy The perception of airborne sound or chemical stimuli.

aerotaxis The directed movement of a motile organism towards (positive) or away from (negative) an air-liquid interface, or a concentration gradient of dissolved oxygen; **aerotactic.**

aerotolerant Used of anaerobic organisms having the capacity to grow to a limited extent under aerobic conditions.

aerotropism An orientation response to a gaseous stimulus; **aerotropic.**

aestatifruticeta A deciduous bush community; a subgroup of aestilignosa *q.v.*

aestatisilvae A deciduous woodland community; a subgroup of aestilignosa *q.v.*

aesthesia Sensibility; **aesthesis**; *cf.* anaesthesia.

aestidurilignosa A mixed evergreen deciduous hardwood forest community.

aestilignosa Broad leaved deciduous bush (aestatifruticeta) and woodland (aestatisilvae) vegetation of temperate regions that experiences alternating periods of mild, damp and cold, dry climate.

aestival Pertaining to the early summer season; *cf.* hibernal, vernal.

aestivation 1: Passing the summer or dry season in a dormant or torpid state; **estivation**; **aestivate**; *cf.* hibernation. 2: The manner in which plant structures are folded prior to expansion or opening.

aethogametism Asynethogametism *q.v.*

aetiologic agent Causal agent *q.v.*

aetiology 1: The branch of science dealing with the study of origins or causes. 2: The demonstrated cause of a disease or trait; causation; **etiology**; **aetiological**.

AFDW Abbreviation of ash-free dry weight.

affinis (aff.) Latin, meaning akin to, having an affinity but not identical.

affinity 1: The degree of relationship or causal connection; *cf.* absolute affinity, relative affinity. 2: Preference.

affinity index A measure of the similarity in species composition between communities A and B; calculated as $A = c/(a \times b)^{\frac{1}{2}}$, where a and b are the numbers of species occurring only in communities A and B respectively and c is the number of species common to both.

affix A sequence of letters forming part of a derived compound word.

afforestation The process of establishing a forest in a non-forested area; *cf.* reforestation.

African subkingdom A subdivision of the Palaeotropical kingdom; see Appendix 4.

Afro-Brasilian bridge The proposed land-bridge *q.v.* between Africa and South America following the breakup of Gondwanaland, persisting possibly until the Palaeocene; Atlanto–equatorial bridge; Inabresis.

Aftonian interglacial An interglacial period of the Quaternary Ice Age in North America; see Appendix 2.

agad A beach plant.

agamandroecious Used of a plant having male and neuter flowers in the same inflorescence; *cf.* agamogynoecious, agamohermaphrodite.

agameon A species comprising only non-sexually

reproducing individuals; agamospecies; binom.

agamete A mature reproductive cell which does not fuse with another to form a zygote; a non-copulating germ cell; *cf.* gamete.

agamic Without gametes; used of complexes of organisms in which all individuals reproduce asexually.

agamobium The asexual phase in an alternation of generations; *cf.* gamobium.

agamodeme A local interbreeding population of predominantly asexually reproducing (apomictic) individuals; *cf.* deme.

agamogenesis Asexual reproduction; agamogony.

agamogony Asexual reproduction by development of a new individual from a single cell, following binary or multiple fission, or budding.

agamogynoecious Used of a plant having female and neuter flowers in the same inflorescence; *cf.* agamandroecious, agamohermaphrodite.

agamohermaphrodite Used of a plant having hermaphrodite and neuter flowers in the same inflorescence; *cf.* agamandroecious, agamogynoecious.

agamont The asexual individual or generation producing agametes *q.v.*

agamospecies A species or population comprising only asexually reproducing (apomictic) individuals; agameon.

agamospermy Apomixis *q.v.* in which embryos and seeds are formed asexually, but not including vegetative reproduction; **agamospermous**.

agamotropic Used of flowers that do not close again once they have opened; *cf.* gamotropic, hemigamotropic.

agaricolous Living on mushrooms and toadstools; **agaricole**.

age 1: To become old; to attain maturity. 2: The period of time a group or organism has existed. 3: The length of geological time since the formation of a rock either by solidification from a molten state (igneous) or by sedimentation (sedimentary). 4: A period of geological history characterized by a dominant life form, such as the age of fishes. 5: The position of an event or organism in the geological time scale. 6: A geological time unit shorter than an epoch.

age and area hypothesis That the area occupied by a species is proportional to its evolutionary age.

age class A category comprising individuals of a given age within a population; cohort.

age distribution The number or percentage of individuals in each age class of a population; age composition; age structure.

age of amphibians That period of the Earth's

history dominated by amphibians; the Carboniferous and Permian periods.

age of cycads That period of the Earth's history dominated by cycads; the Jurassic period.

age of fishes That period of the Earth's history dominated by fishes; the Silurian and Devonian periods.

age of gymnosperms That period of the Earth's history dominated by gymnosperms; the Mesozoic era.

age of mammals That period of the Earth's history dominated by mammals; the Cenozoic era.

age of man That period of the Earth's history dominated by man; the Quaternary period.

age of marine invertebrates That period of the Earth's history dominated by marine invertebrates; the Ordovician and Cambrian periods.

age of reptiles That period of the Earth's history dominated by reptiles; the Mesozoic era.

age polyethism The adoption of different labour roles by members of a colony of social insects according to age; temporal polyethism; cf. caste polyethism.

age specific death rate The death rate for a given cohort or age class of a population; calculated as the number dying in age class x divided by the number that attain age class x.

age specific fecundity rate The average number of eggs or young produced per unit time by an individual of specified age.

age specific survival rate The number or percentage of individuals surviving to a specified age from an initial cohort of individuals.

age structure The number or percentage of individuals in each age class of a population; age distribution; age composition.

ageing The process of irreversible decline of bodily function and adaptibility with time or increasing age; the increasing force of mortality with time.

ageotropism 1: Apogeotropism q.v. 2: The absence of orientation movements in response to gravity; **ageotropic**.

agg. Abbreviation of **aggregate**; used to denote a group of species or hybrids which are morphologically similar and difficult to distinguish from one another; species aggregate.

agglomerative Used of a method of constructing hierarchies by the successive linking together of smaller units into larger, higher ranking units; agglomerative classification; cf. divisive.

agglutinate To clump together; often used with reference to blood cells, pollen grains or bacteria; **agglutination**.

aggradation The natural building up of a surface

or filling up of a channel by sediment deposition; **aggrade**.

aggregate 1: Species aggregate q.v. 2: Aggregation q.v.

aggregated distribution Contagious distribution q.v.

aggregation A society or group of conspecific organisms which have a social structure and consist of repeated members or modular units but with a low level of coordination, integration or genotypic relatedness.

aggression A hostile act or threat made to protect territory, the family group or offspring, or to establish dominance.

aggressive mimicry Mimicry q.v. in which a predator mimics a non-predatory model in order to deceive the operator, its prey; Peckhammian mimicry.

agium A beach community; aigialium.

agmatoploidy An increase in chromosome number due to fragmentation of holocentric chromosomes.

Agnotozoic Proterozoic q.v.

agonistic behaviour Social interaction between members of a species, involving aggression or threat and conciliation or retreat; **agonist**.

agonistic buffering The use of infants by adults to inhibit the aggressive behaviour of other adults.

agrad A cultivated plant.

agrarian Pertaining to cultivation or cultivated plants.

agrestal Growing on arable land.

agric A depositional B-horizon of clay and humus in a soil profile, formed as a result of cultivation.

agriotype The wild type; ancestral type.

agrium A community on cultivated land or land subject to influence by the activities of man; culture community.

agroclimatology The study of climate in relation to the productivity of plants and animals of agricultural importance.

agro-ecosystem An agricultural ecosystem.

agroecotype An edaphic ecotype adapted to cultivated soils.

agrology The branch of agriculture dealing with the study of soils.

agronomy The theory and practice of agricultural management, crop production and husbandry.

agrophilous Thriving in cultivated soils; **agrophile**, **agrophily**.

agrostology The study of grasses; graminology.

agrotype An agricultural variety or race.

Agulhas current A warm surface ocean current that flows south off the coast of South Africa, derived in part as an extension of the Mozambique current and Indian South Equatorial current; see Appendix 6.

aheliotropism Apheliotropism *q.v.*

ahermatypic 1: Pertaining to a non-colonial assemblage, or to an individual organism. 2: Used of a coral that lacks symbiotic algae; **ahermatype**.

A-horizon The dark coloured upper mineral horizon of a soil profile, immediately below the O-horizon and comprising some humified organic material as a result of biological activity or cultivation; see Appendix 12.

aigialium A beach community; agium.

aigialophilus Thriving in beach habitats; **aigialophile, aigialophily**.

aigialophyte A beach plant.

aigicolous Living in beach habitats; **aigicole**.

aiphyllophilus Thriving in evergreen woodland; **aiphyllophile, aiphyllophily**.

aiphyllus Evergreen; **aiophyllus**.

air capacity That volume of air remaining in a soil after saturation with water; air content.

air plant Aerophyte *q.v.*

air porosity The ratio of the volume of air in a given mass of soil to its total volume.

aithallium An evergreen thicket community.

aithalophilus Thriving in evergreen thickets; **aithalophile, aithalophily**.

aithalophyte A plant occurring in evergreen thickets.

aitiogenic Used of a movement or reaction induced by an external stimulus; paratonic; aitionomic; aitionomous; **aitiogenous**.

aitionomic Used of growth patterns and other phenomena imposed by the environment.

aitiotropism Any tropism resulting from an exogenous stimulus; **aitiotropic**.

akaryotic Lacking a discrete nucleus; non-nucleated; acaryotic; **akaryote**; *cf.* eukaryotic, prokaryotic.

akinesis Absence or cessation of movement.

akinetic Acentric *q.v.*

akineton Non-motile planktonic organisms or propagules.

aktological Pertaining to shallow inshore environments and communities.

aktology The study of shallow inshore ecosystems.

al. Abbreviation of the Latin *alii*, meaning others.

alarm call A sound produced by an animal when danger threatens, but is still a significant distance away; protective call.

alarm pheromone A chemical substance exchanged by members of a group, that induces a state of alarm or alertness.

alarm reaction The sum of all non-specific responses to the sudden exposure to stimuli to which the organism is not adapted.

Alaska current Aleutian current *q.v.*; see Appendix 6.

albedo A measure of surface reflectivity; that fraction of incident electromagnetic radiation that is reflected by a body or surface, usually expressed as a percentage.

albinism The absence or deficiency of pigmentation in animals; partial albinism in plants is termed variegation.

-ales The ending of a name of an order in botanical nomenclature.

aletophilous Thriving on roadside verges and beside railway tracks; **aletophile, aletophily**.

aletophyte A plant growing in a mesic habitat or on roadside verges.

Aleutian current A warm surface ocean current that flows north and west off the coast of Alaska, derived as a deflection of the North Pacific Gyre; Alaska current; see Appendix 6.

algal bloom An explosive increase in the density of phytoplankton within an area; bloom.

algal line The highest continuous line on the shore along which any particular algal species occurs; *cf.* tang line.

algal wash Shoreline drift, comprising mainly filamentous algae.

algeny Genetic engineering; the experimental manipulation of the genetic composition of an organism or cell.

algicolous Living on algae; **algicole**.

algology The study of algae; phycology; **algological**.

Algonkian Proterozoic *q.v.*

algophagous Feeding on algae; **algophage, algophagy**.

Algophytic Archaeophytic *q.v.*

algorithm A finite series of logical steps or instructions by which a particular type of numerical or algebraic problem can be solved; *cf.* branch swapping algorithm, Fitch algorithm.

alien Non-native; a species occurring in an area to which it is not native.

alien addition, monosomic A genotype comprising the characteristic complement of chromosomes plus a single chromosome from another species.

alienation, coefficient of A measure of the information loss during transcription of data from a matrix to a dendrogram.

alii (*al.*) Latin, meaning others.

alimentation 1: Feeding, taking-in nourishment. 2: Those processes, including precipitation, sublimation and refreezing, that serve to increase the mass of a glacier or snowfield.

aliquot An integral portion of a whole; an integral sample.

alkaline Pertaining to habitats or substances having a pH greater than 7; basic; *cf.* acidic.

alkalinity 1: The properties of an alkali. 2: A measure of the pH of sea water, calculated as the number of milliequivalents of hydrogen ion that is neutralized by one litre of sea water at 20° C.

alkaliphilic Thriving in alkaline habitats; **alkalophilic, alkaliphile, alkaliphily, alkalophile, alkalophily**; *cf.* alkaliphobic.

alkaliphobic Intolerant of alkaline habitats or conditions; **alkaliphobe, alkaliphoby**; *cf.* alkaliphilic.

alkaloduric Extremely tolerant of high pH (alkaline) conditions.

alkaloid A nitrogenous organic compound, produced especially by flowering plants; many alkaloids have a pronounced physiological activity in animals, such as nicotine, caffeine, morphine.

allaesthetic Used of characters effective only when perceived by another organism; **allesthetic**.

allautogamy Autoallogamy *q.v.*

Allee's principle That the density of a population varies according to the spatial distribution of the individuals (degree of aggregation) and that both overcrowding and undercrowding may be suboptimal; Allee's law.

Alleghany subregion A subdivision of the Nearctic region; Eastern subregion; see Appendix 3.

allelarkean society An independent, dense, non-nomadic and civilized human society; *cf.* autarkean society.

allele Any of the different forms of a gene occupying the same locus *q.v.* on homologous chromosomes, and which undergo meiotic pairing *q.v.* and can mutate one to another; allelomorph; allelomorphic series; **allelic, allelism**.

allele frequency Gene frequency *q.v.*

allelo- Prefix meaning one another.

allelochemic A secondary substance produced by an organism that has the effect of modifying the growth, behaviour or population dynamics of other species, often having an inhibitory or regulatory effect (allelopathic substance); **allelochemical**.

allelogenic Producing offspring in broods that are entirely of one sex; **allelogenous, allelogeny**; *cf.* amphogenic, arrhenogenic, monogenic, thelygenic.

allelomimetic behaviour Imitative behaviour exhibited by two or more animals as a result of mutual stimulation.

allelomorph Allele *q.v.*

allelopathic substance An allelochemic; a waste product, excretory product or metabolite having an inhibitory or regulatory effect on other organisms.

allelopathy Biogenic toxicity; the chemical inhibition of one organism by another; antibiosis.

Allen's law The generalization that the extremities (ears and tails for example) of mammals tend to be relatively shorter in colder climates than in warmer ones; proportion rule; Allen's rule.

alliance A ranked category in the classification of vegetation, comprising one or more closely related associations *q.v.*, equivalent in rank to a federation; the first word of an alliance name has the ending *-ion*.

alliogenesis Alloiogenesis *q.v.*

allo- Prefix meaning other.

allobiosphere That part of the eubiosphere *q.v.* in which heterotrophic organisms occur but into which organic food material must be transported as primary production does not take place; subdivided into hyperallobiosphere and hypoallobiosphere; *cf.* autobiosphere.

allocheiral Having reversed symmetry.

allochemic Any secondary compound produced by plants as part of their defence mechanism against herbivores; acting either as a toxin or digestibility reducer.

allochoric Occurring in two or more communities within a given geographical region; **allochorous, allochore**.

allochroic 1: Exhibiting colour variation. 2: Having the ability to change colour.

allochronic Not contemporary; existing at different times; used of populations or species living, growing or reproducing during different seasons of the year; *cf.* synchronic.

allochronic speciation 1: Speciation without geographical separation through the acquisition of different breeding seasons or patterns. 2: Speciation occurring by the sequential replacement of species through time.

allochronic species Species not occurring in the same time horizon.

allochthonous Exogenous; originating outside and transported into a given system or area; non-native; xenogenous; ectogenous; **allochthone, allochthony**; *cf.* autochthonous.

allochtone 1: Exogenous food material transported into a cave from outside. 2: A non-native organism; *cf.* autochtone.

allocryptic Used of organisms that conceal themselves under a covering of other material, living or non-living.

allodiploid A hybrid diploid in which one or more chromosome pairs are derived from different species.

allogamy Cross fertilization; **allogamous**.

allogene A recessive allele; *cf.* protogene.

allogenetic plankton Planktonic organisms transported into an area by movement of the medium, but normally living and reproducing elsewhere; expatriated plankton.

allogenic 1: Used of factors acting from outside the system, or of material transported into an area from outside; allochthonous. 2: Having different sets of genes; **allogeneic, allogeneous, allogenetic; allogenous**; *cf.* syngenic.

allogenic succession The replacement of one community by another as a result of extrinsic changes in the environment; *cf.* autogenic succession.

allogenous detritus Detritus carried into an area from outside.

allograft A tissue transplant from one individual to another of the same species; homoplastic graft.

allogrooming Grooming by one individual of another; allopreening; *cf.* self-grooming.

alloheteroploid An aneuploid *q.v.* hybrid derived from two different species; *cf.* autoheteroploid.

allohomoeotype A homoeotype of the same sex as the allotype *q.v.* or the allolectotype *q.v.*; allohomoiotype.

allohospitalic Used of two or more parasite species occurring on different host species only; *cf.* synhospitalic.

alloiogenesis Alternation of generations; an alternation between sexual and asexual phases in a life cycle; **alliogenesis**.

alloiometron A measurable change of proportion or intensity of development within a species or race.

allokinesis Passive or involuntary movement; drifting; planktonic transport; **allokinetic**.

allolectotype A type specimen of opposite sex to the lectotype and chosen from the type series subsequent to the original description; lectallotype.

allomaternal A female alloparent *q.v.*

allometric coefficient The ratio of relative growth rates.

allometric growth Differential growth of body parts (x and y), expressed by the equation $y = bx^a$, where a and b are fitted constants; change of shape or proportion with increase in size; heterauxesis, heterogony; **allometry**; *cf.* isometric growth.

allometric variability Variation in body proportions exhibited by individuals of different size in species which undergo allometric growth.

allometry Allometric growth *q.v.*

allomixis Cross fertilization; **allomictic**.

allomone A chemical substance produced and released by one species in order to communicate with another species; *cf.* pheromone.

allomorphosis 1: The condition of having individual, racial, specific or phylogenetic heterogeneity in allometric growth rates. 2: Evolution characterized by a rapid increase in structural specialization; evolutionary allometry; idio-adaptation; *cf.* aromorphosis.

alloparalectotype A paralectotype *q.v.* of opposite sex to the lectotype; *cf.* holoparalectotype.

allo-parapatric speciation Speciation in which initial segregation and differentiation of the diverging populations takes place in disjunction but complete reproductive isolation is attained after range adjustment so that the populations become separate but contiguous.

alloparasite An organism parasitic on an unrelated host organism; *cf.* adelphoparasite.

alloparatype A paratype *q.v.* of the same sex as the allotype; parallotype; *cf.* holoparatype.

alloparent An individual that assists a parent in the care of its young, either male (allopaternal) or female (allomaternal).

allopaternal A male alloparent *q.v.*

allopatric Used of populations, species or taxa occupying different and disjunct geographical areas; **allopatry** *cf.* dichopatric, parapatric, sympatric.

allopatric introgression The production of a new genotype by introgression *q.v.* which exists then in separate populations from the original genotype.

allopatric speciation The differentiation of, and attainment of complete reproductive isolation of, populations that are completely geographically separated.

allopatry Spatial separation; disjunction; **allopatric**.

allopelagic Used of organisms occurring at any depth in the pelagic zone; *cf.* autopelagic.

allophene An abnormal phenotype, usually of a particular tissue, not produced by mutation, which will develop a normal phenotype if transplanted to a wild-type host; *cf.* autophene.

allophilous Pollinated by non-adapted agencies; used of a plant that lacks morphological adaptations for attracting and guiding pollinators; **allophily**; *cf.* euphilous.

alloplesiotype A plesiotype *q.v.* of the same sex as the allotype; *cf.* holoplesiotype.

alloploideon A species derived by allopolyploidy *q.v.*

alloploidy Allopolyploidy *q.v.*; **alloploid**.

allopolyploid A polyploid hybrid having

chromosome sets derived from two different species or genera; **alloploid**; *cf*. autopolyploid.

allopolyploidy The state of being allopolyploid; **alloploidy**.

all-or-none character Binary character *q.v*.

allosematic Pertaining to coloration or markings that imitate warning patterns of other typically noxious or dangerous organisms.

allosomal inheritance Inheritance of characters controlled by genes located in an allosome *q.v*.

allosome Heterochromosome; any chromosome or chromosome fragment other than a normal A-chromosome.

allospecies Semispecies *q.v*.; the component species of a superspecies.

allosyndesis The pairing of homologous chromosomes in an allopolyploid *q.v*.; *cf*. autosyndesis.

allotetraploid Hybrid polyploid derived by the doubling of the chromosome complement of a diploid hybrid (allodiploid); amphidiploid; allopolyploid.

allotherm An organism having a body temperature determined largely by the ambient temperature; ectotherm; poikilotherm; *cf*. autotherm.

allotopic Used of populations or species that occupy different macrohabitats; **allotopy**; *cf*. syntopic.

allotopotype A type specimen from the original type locality, of the same sex as the allotype *q.v*.

allotriploid A hybrid triploid; a type of allopolyploid.

allotrophic 1: Obtaining nourishment from another organism; heterotrophic *q.v*. 2: Pertaining to the influx of nutrients into a water body or ecosystem from outside; **allotrophy**.

allotrophic lake A lake receiving organic material by drainage from the surrounding land; *cf*. autotrophic lake.

allotropism 1: The mutual attraction of cells, especially gametes; **allotropic**. 2: The condition of having an abnormal mitotic cycle. 3: The condition of a flower having a plentiful supply of readily available nectar.

allotropous Used of unspecialized insect species that are able to feed on a variety of kinds of flowers; *cf*. eutropous.

allotype A paratype of opposite sex to the holotype *q.v*. and originally designated by the author.

alloxenic Used of two or more different species of parasites occurring on different host species only; **alloxenia**; *cf*. homoxenic.

allozygous Used of a zygote homozygous at a given locus but with the two alleles not identical by descent, i.e. of independent origin; **allozygote**; *cf*. autozygote.

allozymes The various forms of an enzyme which have the same activity but which differ slightly in amino acid sequence, produced by different alleles at a single locus.

alluvial deposit A silty deposit transported by water; alluvial sediment; **alluvium**; *cf*. aeolian deposit, colluvial deposit.

Alluvial soil An azonal soil formed from a relatively unmodified recent alluvial deposit in flood plains and deltas.

alluviation The deposition of sediment by a river at any point along its course.

Alm's Fb coefficient The ratio of fish caught to total benthic biomass per hectare.

alpestrine Living at high altitude above the tree line *q.v*., commonly used as synonymous with alpine *q.v*.; occasionally used of plants found below the tree line (subalpine plants).

alpha The highest ranking individual in a dominance hierarchy.

alpha diversity The diversity of species within a particular habitat or community; *cf*. beta diversity, gamma diversity.

alpha karyology In the study of chromosomes, the determination of chromosome numbers and approximate sizes; *cf*. karyology.

alpha taxonomy Descriptive taxonomy; that aspect of taxonomy concerned with the description and designation of species, typically on the basis of morphological characters; *cf*. beta taxonomy, gamma taxonomy.

alphanumeric A code comprising both letters and numbers.

alpine 1: Pertaining to the alps. 2: Used of habitats and organisms found between the tree line and snow line in mountainous regions.

Alpine Meadow soil An intrazonal soil with a dark coloration, formed under meadow grass above the tree line in alpine regions; typically shallow and stony with thin litter and duff, A-horizon granular and strongly acidic, B-horizon absent, and C-horizon gleyed due to poor drainage.

Alpine Turf soil An intrazonal soil with a dark yellow-brown coloration formed under alpine turf; typically a well drained soil with thin litter and duff, A-horizon stony and strongly acidic, B-horizon stony and moderately acidic.

alsad A grove plant; alsophyte.

alsium A grove community; an area grown with trees and grasses.

alsocolous Living in woody groves; **alsocole**.

alsophilus Thriving in woody grove habitats; **alsophile, alsophily**.

alsophyte A grove plant; alsad.

alternating communities Alternes *q.v.*

alternation of generations The alternation of generations having different reproductive processes, typically of sexual (diploid) and asexual (haploid) phases, in the life cycle of an organism; alloiogenesis; metagenesis.

alternative hypothesis In statistics, the hypothesis accepted when the null hypothesis *q.v.* is rejected.

alternative names Two names for the same taxon, of the same rank, published simultaneously by an author.

alterne An ecological succession resulting from an abrupt change in environmental conditions; a disturbed zonation.

alternes Two or more communities alternating with each other in a particular area; alternating communities; twin communities.

altherbosa A tall herb community, often found in areas of forest clearance.

Altithermal period A postglacial interval (*ca.* 7500–4000 years B.P.) characterized by distinctly warmer conditions than at present; corresponds approximately to the Atlantic period *q.v.* or mid Hypsithermal period *q.v.*; Thermal Maximum; *cf.* Anathermal period, Medithermal period.

altitude The vertical distance between a given point and a datum surface, or chart datum; elevation.

altricial Used of offspring or of species that show a marked delay in the attainment of independent self-maintenance; **altrices**; *cf.* precocial.

altruism The situation in which one individual acts to promote or enhance the fitness of an unrelated individual or of other members of a group at the same time reducing its own fitness; the case of mutually beneficial behaviour is known as reciprocal altruism; **altruistic**; *cf.* selfishness, spite.

alvar Plant community dominated by mosses and herbs, occurring on shallow alkaline limestone soils.

amanthicolous Inhabiting sandy plains; amathicolous; **amanthicole, amathicole**.

amanthium A sandhill or plain community; amathium.

amanthophilus Thriving in sandy plains; **amathophilous, amanthophile, amanthophily, amathophile, amathophily**.

ambient Surrounding; background.

ambient temperature The temperature of the medium immediately surrounding an organism or object.

ambiguous Open to misinterpretation; of uncertain origin.

ambiguous genetic code A code in which a codon has more than one meaning during translation.

ambiguous name *Nomen ambiguum q.v.*

ambisexual 1: Having separate male and female flowers on the same plant; ambosexual; hermaphrodite; monoecious. 2: Pertaining to both sexes.

ambivalence 1: Behaviour resulting from two incompatible motivations, often taking the form of a mixture of the two motivational tendencies; displacement activity. 2: The property of a gene possessing both beneficial and detrimental actions; **ambivalent**.

ambivorous Feeding on grasses and broad leaved plants; **ambivore, ambivory**.

ambulatorial Adapted for walking.

ameiosis Nuclear division in which no reduction in chromosome number occurs.

ameiotic parthenogenesis Parthenogenesis in which meiosis has been suppressed so that neither chromosome reduction nor any corresponding phenomenon occurs.

ameiotic thelytoky Obligate ameiotic parthenogenesis in which females produce only female offspring.

amensalism An interspecific interaction in which one species population is inhibited, typically by toxin produced by the other, which is unaffected; amentalism.

amentalism Amensalism *q.v.*

ameristic Not divided into unitary parts; *cf.* meristic.

ametabolous development The ontogenetic pattern in which the offspring gradually assume adult characters through a series of moults; **ametabolic development, ametabolism**; *cf.* hemimetabolous development, holometabolous development.

ametoecious 1: Used of a parasite having only a single host during its life cycle; autoecious; autoxenous; *cf.* heteroecious. 2: Used of a parasite species that is highly host specific; **ametoecius**; *cf.* metoecious.

amictic 1: Pertaining to females that produce unfertilized eggs that develop into offspring of one sex only. 2: Used of a lake that has no period of overturn because it is perennially frozen; *cf.* mictic.

amitosis A process of nuclear division by simple constriction into two portions, often forming dissimilar daughter nuclei; holoschisis.

amixis 1: Absence or failure of interbreeding; **amixia**. 2: A process of pseudosexual reproduction in certain haploid fungi, in which both meiosis and karyogamy are absent.

ammochthad A plant living on sand banks; ammochthophyte.

ammochthium A sandbank community.

ammochthophilous Thriving on sand banks; **ammochthophile**, **ammochthophily**.

ammochthophyte A sand bank plant; ammochthad.

ammocolous Living or growing in sand; **ammocole**.

ammodyte A plant living in sand.

ammonification The release of ammonia from nitrogenous organic material by the action of microorganisms; ammonization.

ammonization Ammonification *q.v.*

ammophilous Thriving in sandy habitats; **ammophile**, **ammophily**.

amnicolous Living on sandy river banks; **amnicole**.

amorphic allele An inoperative allele that does not influence the phenotype; *cf.* hypermorphic allele, hypomorphic allele, neomorphic allele.

ampere (A) The standard SI base unit of electric current, defined as that constant current which will produce a force of 2×10^{-7} newtons per metre of two infinitely long straight parallel wires of negligible cross section placed one metre apart *in vacuo*.

ampheclexis Sexual selection *q.v.*

ampherotoky Amphitoky *q.v.*

amphi- Prefix meaning both; on both sides.

amphiatlantic Occurring on both sides of the Atlantic Ocean.

amphibiotic 1: Having aquatic larval stages and terrestrial adults. 2: Used of a microorganism participating in a symbiotic relationship with a given host, that can act either parasitically or mutualistically.

amphibious Adapted to life on land as well as in water; **amphibian**.

amphicarpogean Producing fruits above ground that are subsequently buried; **amphicarpogenous**; *cf.* hypocarpogean.

amphichromatism The occurrence of different coloured flowers on individual plants in different seasons; **amphichromatic**.

amphiclinous hybrid A hybrid that exhibits the phenotypic characters of one parent more than the other.

amphiclinous progeny Hybrid offspring, some of which resemble one parent and some the other parent.

amphicryptophyte A marsh plant with amphibious vegetative parts.

amphidiploid A polyploid formed of hybrid diploid parents by doubling of the chromosome set; allotetraploid.

amphidromic motion A large scale circular motion of a depression and elevation in the sea surface, caused by the deflection of tides by the Coriolis force.

amphidromic point A point in the sea having no vertical tidal movement.

amphigamy Normal cross fertilization; the fusion of male and female gametes.

amphigean Used of a plant having underground as well as aerial flowers.

amphigeic Pertaining to both the Old World and the New World; **amphigean**; *cf.* gerontogeic, neontogeic.

amphigenesis Sexual development; the fusion of two dissimilar gametes.

amphigony Sexual reproduction involving cross fertilization.

amphilepsis Inheritance of characters from both parents.

amphimict An obligate sexually reproducing organism; **amphimictic**.

amphimixis True sexual reproduction; the union of male and female gametes; may be either autogamy (inbreeding) or allogamy (outbreeding); **amphimict**, **amphimictic**; *cf.* apomixis.

amphinereid An amphibious plant.

amphioecious Used of a population or species showing broad variable tolerance of habitat and environmental conditions, reflected in the presence of clines and subspecific taxa; amphitopic; *cf.* euryoecious, stenoecious.

amphiphyte A plant able to live either rooted in damp soil above the water level or completely submerged.

amphiplexus A sexual embrace.

amphiploid A polyploid formed by the doubling of the chromosome set of a hybrid; allopolyploid.

amphithallism Heteromixis *q.v.*

amphitoky Parthenogenesis in which both male and female offspring are produced; ampherotoky; amphoterotoky; deuterotoky; gametotoky; *cf.* thelytoky.

amphitopic Used of a population or species showing broad, variable tolerance of habitat and environmental conditions, reflected in the existence of clines and subspecies; *cf.* eurytopic, stenotopic.

amphitrophic Used of an organism that lives as a phototroph during daylight and as a chemotroph in darkness; **amphitroph**.

amphogenic Producing male and female offspring in approximately equal numbers; **amphogenous**, **amphogeny**; *cf.* allelogenic, arrhenogenic, monogenic, thelygenic.

amphoterotoky Amphitoky *q.v.*

amplectant Clasping or twining for support.

amplexus Precopulation pairing; the grasping of the female by the male prior to copulation.

amplification Increased structural or functional complexity during ontogeny or phylogeny.

amplitude 1: The range of tolerance to environmental conditions of an organism or species; ecological amplitude. 2: The magnitude of the displacement of a wave from its mean value.

anabatic Used of winds caused by the upward movement of heated air.

anabiont A perennial plant producing flowers and fruits many times; **anabiontic**.

anabiosis A state of greatly reduced metabolic activity assumed during unfavourable environmental conditions; cryptobiosis; **anabiotic**.

anabolism That part of metabolism involving the manufacture of complex substances from simpler substrates with the consequent utilization of energy; **anabolic**; *cf.* catabolism.

anaboly Evolution by the addition of a new feature to the end of the embryonic period of morphogenesis *q.v.* and before the subsequent period in which growth proceeds without any accompanying change in shape; also used to refer to acceleration of the development of an organism; **anabolic**.

anachoric Living in fissures, holes and crevices; **anachoresis**.

anaclinotropism A tropism resulting in growth or movement in a horizontal plane; **anaclinotropic**.

anadromous Migrating from salt to fresh water, as in the case of a fish moving from the sea into a river to spawn; **anadromesis, anadromy**; *cf.* catadromous, oceanodromous, potamodromous.

anaerobic 1: Growing or occurring in the absence of molecular oxygen; **anaerobe, anaerobiosis**; *cf.* aerobic. 2: Used of an environment in which the partial pressure of oxygen is significantly below normal atmospheric levels; deoxygenated.

anaerobiotic Used of habitats devoid of, or significantly depleted in, molecular oxygen; **anaerobiosis**.

anaerogenic Not producing gas; used of microorganisms that do not produce visible gas during the breakdown of carbohydrate; *cf.* aerogenic.

anaerophytobiota The anaerobic soil flora; **anaerophytobiont**; *cf.* aerophytobiota.

anaesthesia Insensibility; **anaesthesis**; *cf.* aesthesia.

anagenesis 1: Progressive evolution towards higher levels of organization or specialization; **anagenetic**; *cf.* catagenesis, stasigenesis. 2: Phyletic evolution *q.v.*; *cf.* cladogenesis. 3: The regeneration of tissues.

anagenetic Pertaining to evolutionary change through time.

anagotoxic Having the potential to counteract the effects of a venom or toxin.

analogous 1: Pertaining to similarity of function, structure or behaviour due to convergent evolution rather than to common ancestry; **analogue, analogy**; *cf.* homologous. 2: Used of organisms having similar habitats or distributions.

analysis In botanical nomenclature, any illustration of a taxon showing sufficient detail for identification, with or without a caption.

analysis of covariance (ancova) A statistical method for determining whether the functional relationships described by two or more regression equations are the same (i.e. could have come from populations with the same slope); it comprises a combined application of linear regression *q.v.* and analysis of variance *q.v.* and is used when treatments are compared in the presence of accompanying variables which cannot be eliminated or regulated.

analysis of variance (anova) A statistical technique for partitioning the total variability affecting a set of observations between the possible and statistically independent causes of the variability; useful for examination of sources of uncontrolled variation in experimental situations.

anamnestic response The immune response to subsequent challenge, marked by a more rapid and stronger immune reaction than after the primary immunizing dose of the antigen.

anamorphosis 1: Evolution through a series of gradual changes. 2: Ametabolous development *q.v.*

anaphylaxis Hypersensitivity following re-exposure to an antigen subsequent to an initial sensitizing exposure.

anaplasia 1: The progressive phase in the ontogeny of an organism, prior to attaining maturity. 2: An evolutionary state characterized by increasing vigour and diversification of organisms; **anaplasis**; *cf.* cataplasia, metaplasia. 3: Dedifferentiation *q.v.*

anastomosis The union of branches to form a network.

anathermal Pertaining to a period of rising temperature; *cf.* catathermal.

Anathermal period A postglacial interval (*ca.* 10 000–7500 years B.P.) characterized by rising temperatures following the last glaciation; about equivalent to the Preboreal and Boreal periods *q.v.*; *cf.* Altithermal period, Medithermal period.

anatomical Pertaining to the structure of organisms, especially as revealed by dissection; **anatomy**.

anatropism The inversion of a seed or ovule so that the micropyle is bent downwards; **anatropic**.

anautogenous Used of a female insect that must

feed before it can produce mature eggs; non-autogenous; *cf.* autogenous.

anauxotrophic Prototrophic *q.v.*; used of nutritionally independent organisms; **anauxotroph**.

ancestor A progenitor of a more recent descendant taxon, group or individual; any preceding member of a lineage; archetype; **ancestral**.

ancestral Used of a primitive or plesiomorphic character state present or assumed to be present in an ancestor.

anchialine Maritime; used of coastal salt-water habitats having no surface connection to the sea.

ancium A canyon forest community.

ancocolous Living in canyons; **ancocole**.

ancophilus Thriving in canyon forests; **ancophilous, ancophile, ancophily**.

ancophyte A plant inhabiting canyon forests; **anchophyte**.

Andean region A subdivision of the Neotropical kingdom; see Appendix 4.

-andr- Combining affix meaning male, testis, stamen.

andric Male; *cf.* gynic.

androchorous Dispersed by the agency of man; **androchore, androchory**.

androcyclic parthenogenesis Parthenogenesis *q.v.*, shown by species which also exhibit cyclic parthenogenesis, in which a series of parthenogenetic generations is followed by the production, by a portion of the population only, of males as a sexual generation.

androdioecious Used of plant species having male and hermaphrodite flowers on separate plants; **androdioecy**; *cf.* gynodioecious.

androecious Used of a plant having male flowers only; *cf.* gynodioecious.

androgamy The impregnation of the male gamete by the female gamete.

androgenesis Male parthenogenesis; the development of an egg carrying paternal chromosomes only (patrogenesis) or the production of a haploid plant by the germination of a pollen grain within an anther; **androgenetic**; *cf.* gynogenesis.

androgenous Producing male offspring only; *cf.* gynogenous.

androgynous Having both male and female reproductive organs; hermaphrodite; **androgyne, androgynal**; androgynism.

andromonoecious Having male and hermaphrodite flowers on the same plant; *cf.* gynomonoecious.

andromorphic Having a morphological resemblance to males; **andromorphy**; *cf.* gynomorphic.

androphilous Thriving in proximity to man; **androphile, androphily**.

androphyte A male plant.

androsome Any chromosome found only in a male nucleus.

androspore 1: A male spore; pollen grain. 2: An asexual zoospore giving rise to a dwarf male plant.

androtype A male type *q.v.*; *cf.* gynetype.

anebous Prepubertal; immature.

anecdysis The absence of moulting, or a prolonged intermoult period.

anemo- Prefix meaning wind.

anemochorous Having seeds, spores or other propagules dispersed by wind; **anemochoric; anemochore, anemochory**.

anemodium A blowout *q.v.* community; anemium.

anemogamous Anemophilous *q.v.*; wind-pollinated; **anemogamy**.

anemoneuston Terrigenous organic and inorganic material transported by wind to the surface of water bodies.

anemophilous 1: Dispersed by wind; anemochorous. 2: Pollinated by wind-blown pollen; anemogamous; **anemophile, anemophily**. 3: Thriving in sand draws; **anemophilus**.

anemophobic Intolerant of exposure to wind; **anemophobe, anemophoby**.

anemophyte 1: A plant occurring in wind swept situations. 2: A plant fertilized by wind-blown pollen; anemophile.

anemoplankton Wind borne organisms; aerial plankton; aeroplankton.

anemotaxis A directed reaction of a motile organism towards (positive) or away from (negative) wind or air currents; **anemotactic**.

anemotropism Orientation response to an air current; **anemotropic**.

aneroid Containing or requiring no water or other liquid.

aneuploid Having a chromosome number that is more than or less than, but not an exact multiple of, the basic chromosome number; *cf.* euploid, monosomic, nullisomic, polysomic, tetrasomic, trisomic.

aneusomatic Pertaining to individuals containing both euploid and aneuploid cells.

aneuspory The formation of spores by irregular meiotic divisions; **aneusporous**.

Angara A continental landmass comprising most of central and northern Asia, formed from Eurasia *q.v.* after the break up of Pangaea *q.v.*; *cf.* Cathaysia, Euramerica.

Anglian glaciation A glaciation of the Quaternary Ice Age in the British Isles with an estimated duration of 90 thousand years; Lowestoft glaciation; see Appendix 2.

angonekton Short lived organisms inhabiting temporary pools on rocks, in tree stumps and similar sites.

ångstrom (Å) A derived unit of length equal to 0.1 millimicrons, or 10^{-10}m; see Appendix 13.

anheliophilous Thriving in diffuse sunlight; **anheliophile, anheliophily**.

anholocyclic parthenogenesis Reproduction by parthenogenesis alone, as when the sexual reproduction phase has been lost from an alternation of generations cycle; acyclic parthenogenesis; *cf.* parthenogenesis.

anhydrobiosis Dormancy induced by low humidity or by desiccation; **anhydrobiose**.

animal 1: Any member of the kingdom Animalia. 2: Any multicellular eukaryotic organism exhibiting holozoic nutrition and having capacity for spontaneous movement and rapid motor response to stimulation.

animal exclusion theory In biological oceanography, that patchiness *q.v.* results from the deliberate avoidance of dense phytoplankton concentrations by zooplankton.

animalcule An archaic term for any microscopic organism.

animalculism Encasement theory *q.v.*; the belief that the embryo was contained within the sperm cell.

anion A negatively charged ion; *cf.* cation.

anirotype Cheirotype *q.v.*

aniso- Prefix meaning unequal.

anisogametic Producing gametes (anisogametes) of dissimilar size, shape or behaviour; heterogametic; **anisogametism, anisogamety**; *cf.* isogametic.

anisogamous Having gametes of dissimilar size, shape or behaviour; **anisogamete**.

anisogamy The fusion of gametes of dissimilar size, shape or behaviour; anisomerogamy; heterogamy; *cf.* isogamy.

anisogonous Used of hybrids that do not show equal expression of both parental characters; *cf.* isogonous.

anisohologamy The fusion of hologametic *q.v.* gametes that are not quite identical in some respect.

anisohydric Hydrolabile *q.v.*

anisomerism 1: Serial repetition of parts that are more or less different. 2: Reduction in both the number and degree of differentiation of serially homologous structures.

anisomerogamy Anisogamy *q.v.*

anisomorphic Differing in shape or size; **anisomorphy**.

anisoploidy The condition of having an odd number of chromosome sets in somatic cells; **anisoploid**.

anisosymbiosis Symbiosis *q.v.* in which one partner (the macrosymbiont) is larger than the other (the microsymbiont); *cf.* isosymbiosis.

anlage A rudiment or primordium from which a given organ or structure develops during ontogeny.

annidation The survival of a competitively inferior mutant which occupies a niche not available to the parent type.

anniversary wind A local wind or larger wind system that recurs annually.

annotinous Used of growth added during the previous year; one year old.

annual Having a yearly periodicity; living for one year; *cf.* biennial, perennial.

annual heat budget The total heat entering a lake between the time of lowest and the time of highest heat content.

annual succession The cyclical pattern of changes in biota occurring during the year.

annual turnover The total biomass produced in one year.

annuation Fluctuations in abundance or behaviour resulting from annual changes in environmental factors.

annulled name An originally available name that has been suppressed by the ICZN and consequently becomes unavailable for purposes of priority *q.v.* and homonymy *q.v.*

annulled work A publication that the ICZN has ruled must not be used for purposes of nomenclature.

annulment The suppression by the ICZN of an available name as unavailable for the purposes of priority and homonymy, and the ruling of a work as unavailable.

anoestrus A non-breeding period; **anoestrous, anoestrum**; *cf.* dioestrus, monoestrous, polyoestrous.

anomalistic tide cycle A tidal cycle of about 27.5 days duration measured from perigee to perigee, during which the moon has completed one orbit around the Earth.

anomalous Abnormal; departing from type; **anomaly**.

anombrophyte A plant that avoids direct rain.

anonymous Pertaining to an author whose name is not known, or to a publication whose author is not known, and to a proposal contained in an anonymous work.

anorgoxydant An organism deriving energy from the oxidation of inorganic substrates.

anorthogenesis Changes in direction of evolution based on preadaptation.

anosmic Having no sense of smell; **anosmatic**.

anoxic Used of a habitat devoid of molecular oxygen; **anoxia, anoxicity**; *cf.* normoxic, oligoxic.

anoxybiotic Anaerobic *q.v.*; **anoxybiontic; anoxybiosis**.

antagonism 1: Opposition of force or principle; antagonist. 2: The inhibition of one species by the action of another; amensalism.

antarctic Pertaining to the area within the Antarctic circle.

Antarctic bridge The proposed southern land-bridge connecting Patagonia, Antarctica and Australia (and possibly New Zealand) from the Cretaceous to the Oligocene; Archinotis, Palaeantarctis.

Antarctic Circumpolar current A major cold surface current that flows eastwards through the Southern Ocean producing a circumglobal antarctic circulation; West Wind Drift; see Appendix 6.

Antarctic convergence That region of the oceans of the southern hemisphere (about at 50° to 60° S) where cold antarctic surface water moving north-east meets and sinks below warmer water moving south-east.

Antarctic divergence The region of extensive upwelling close to Antarctica where water from intermediate depths replaces surface water drifting north-east in the Antarctic Circumpolar current.

Antarctic kingdom One of the six major phytogeographical areas characterized by floristic composition; comprising New Zealand region, Patagonian region and South Temperate Oceanic island region; see Appendix 4.

Antarctogaea The Australian zoogeographical region, excluding New Zealand and Polynesia (Ornithogaea).

ante- Prefix meaning before, in front of.

antecedent moisture The moisture in the soil prior to a period of significant precipitation resulting in runoff *q.v.*

antecedent platform theory A theory of coral atoll and barrier reef formation that postulates the presence of a submarine platform from which reefs and atolls grow upward to the surface without changes in sea level.

antenatal Before birth; during gestation.

antennation The act of touching or probing with antennae.

antepisematic Pertaining to a character or trait aiding recognition and which implies a threat; *cf.* episematic.

anthecology The study of pollination and the relationships between insects and flowers.

anthelietum An alpino-arctic community.

anthesis The period of flowering; the time at which flower buds open; the period of maximum physiological activity in plants.

anthesmotaxis The organization and arrangement of the parts of a flower.

-antho- Combining form meaning flower; **-anthic-**.

anthogenesis The production of male and female offspring by asexual forms.

anthophilous Attracted to, or feeding on, flowers; **anthophile, anthophily**.

anthoptosis The fall of flowers.

anthotaxis The arrangement of the parts of a flower; **anthotaxy**.

anthotropism An orientation response of a flower to a stimulus; **anthotropic**.

anthracobiontic Living or growing on a burned or scorched substratum.

anthracriny The decomposition of organic matter into humus by soil organisms.

anthrageny The decomposition of plant material to form peat.

anthropic Pertaining to the influence of man; **anthropeic, anthropical**.

anthropocentric Interpreting the activities of other organisms in relation to human values; **anthropocentrism**.

anthropochorous Having propagules dispersed by the agency of man; **anthropochore, anthropochoric, anthropochory**.

Anthropogene Quaternary *q.v.*

anthropogenesis The descent of man; the origin (phylogeny) and the development (ontogeny) of man.

anthropogenic Caused or produced through the agency of man; **anthropogenous**.

anthropogenic climax A vegetational climax community *q.v.* resulting from human intervention.

anthropology The study of man; **anthropological**.

anthropometry The study of the morphometrics of the human body.

anthropomorphic Attributing human form or personality to other organisms or objects; humanization; **anthropomorphism**.

anthropophilous 1: Thriving in close proximity to man; androphilous; **anthropophile, anthropophily**. 2: Used of biting or blood-sucking insects that feed on man; **anthropophilic**.

anthropophyte A plant introduced unintentionally into an area by man during cultivation.

Anthropozoic Geological time since the appearance of man; Psychozoic.

anti- Prefix meaning against.

antibiosis The antagonistic association between two organisms in which one species adversely affects the other, often by production of a toxin; **antibiotic**.

antibody An immunoglobulin serum protein that binds with a specific antigen.

antiboreal Pertaining to cool or cold temperate regions of the southern hemisphere; austral; notalian; *cf.* boreal.

antiboreal convergence The subtropical convergence; the second major region of convergence in the oceans of the southern hemisphere about 10° north of the Antarctic convergence *q.v.*

anticodon A triplet of nucleotides in transfer RNA which pairs with a specific triplet (codon) in messenger RNA during translation in the ribosome.

anticryptic Used of protective coloration facilitating attack or capture of prey; *cf.* procryptic.

anticyclonic Used of a region of high atmospheric pressure at sea level; also used of the wind system around such a high pressure centre that has a clockwise rotation in the northern hemisphere or an anticlockwise motion in the southern hemisphere; **anticyclone**; *cf.* cyclonic.

antifeedant An allelochemic *q.v.* produced by one organism that prevents or deters another organism from feeding upon it.

antigen A substance which induces formation of antibodies *q.v.*

antigenic 1: Having the properties of an antigen; **antigenicity**. 2: Exhibiting sexual dimorphism.

antigenic drift A gradual change in the antigenic composition of some viruses.

antigeny Sexual dimorphism *q.v.*; **antigenic**.

anti-hybridization mechanism An alternative term for the mechanism of reproductive isolation *q.v.*

Antillean subregion A subdivision of the Neotropical region; West Indian Islands subregion; see Appendix 3.

Antilles current A warm surface ocean current that flows north in the western North Atlantic forming the westerly limb of the North Atlantic Gyre; see Appendix 6.

antimorph A gene producing the opposite effect to the wild type.

antimutagen An agent that retards or interferes with the production of mutations.

antipathetic symbiosis A symbiosis that is either advantageous or obligatory to one of the symbionts.

antiphyte That generation of a plant showing alternation of generations in which the gametes are produced.

antipodean Pertaining to opposite sides of the globe; **antipodes**.

antiserum Blood serum containing specific antibodies.

antithetic Used of a life cycle exhibiting either an alternation of haploid and diploid generations or an alternation of morphologically dissimilar generations.

antitropic wind A small scale local wind, such as land and sea breezes, or fall winds *q.v.*

antitropism The orientation of growth in a plant so that it twines in the opposite direction from that followed by the sun across the sky; **antitropic**; *cf.* eutropism.

antitype Paratype *q.v.*

antonym A word having the opposite meaning to another word.

ap. Abbreviation of Latin ***apud***, meaning in the work of; used in citing the work of an author contained within the publication of another author.

apandrous Lacking, or having non-functional, male reproductive organs; **apandry**.

apatetic coloration A misleading coloration pattern, such as coloration resembling physical features of the habitat; camouflage.

aperiodicity The irregular occurrence of a phenomenon; **aperiodic**.

aphanic species Sibling species *q.v.*

aphanism The loss or reduction of an organ or structure in the adult of a descendant stock which was present in both young and adult of the ancestral stock; the organ develops and becomes functional until a particular stage when it is rapidly destroyed.

aphantobiont An ultramicroscopic organism.

aphaptotropic Not influenced by or responsive to, a touch or contact stimulus; **aphaptotropism**.

aphasic lethal A lethal mutation in which death may occur at random during development.

aphelion The point in the Earth's orbit furthest from the sun; *cf.* perhelion.

apheliotropism An orientation response away from sunlight; negative heliotropism; aheliotropism; apohliotropism; apoheliotropism; **apheliotropic**.

aphercotropism An orientation response resulting in a turning away from an obstruction; **aphercotropic**.

aphidicolous Living amongst aggregations of aphids; **aphidicole**.

aphidivorous Feeding on aphids; **aphidivore**, **aphidivory**.

aphotic Used of an environment or habitat having no sunlight of biologically significant intensity.

aphotometric Pertaining to organisms or structures not affected by light.

aphotaxis 1: The absence of a directed response to a light stimulus in a motile organism; **aphototactic**. 2: Negative phototaxis *q.v.*

aphototropism 1: An orientation response away

from light; negative phototropism; **aphototropic**.
2: The absence of an orientation response to
light.

aphydrotaxis 1: The directed reaction of a motile
organism away from moisture; **aphydrotactic**. 2:
The absence of a directed response to moisture
in a motile organism.

aphyletic trogloxene An organism only
occasionally found in caves and which does not
reproduce within caves; *cf.* phyletic trogloxene.

aphytal zone The area of a lake floor devoid of
plant growth.

Aphytic The period of geological time before the
appearance of plant life; *cf.* Archaeophytic,
Caenophytic, Eophytic, Mesophytic,
Palaeophytic.

apivorous Feeding on bees; **apivore, apivory**.

aplano- Combining form meaning non-motile,
without power of locomotion.

aplanogamete A non-motile gamete;
cf. planogamete.

aplanetic Having non-motile spores; **aplanetism**.

aplasia The failure of development, or arrested
development, of a structure.

apo- Prefix meaning away from.

apocratic species An opportunistic species.

apoendemic A polyploid of restricted geographical
distribution derived from a more widespread
diploid; a type of neoendemic *q.v.*

apogamete A gamete produced by apomixis *q.v.*

apogamy 1: Apomixis *q.v.* 2: Direct production of
a plant from the gametophyte by budding,
without the formation of gametes;
apogamous.

apogean tides The tides of decreasing amplitude
occurring at the time the moon is furthest from
the Earth.

apogee The point in the orbit of the moon or
other satellite furthest from the Earth;
cf. perigee.

apogenic Sterile.

apogenotype A type specimen fixed through
substitution; as when a genus is renamed through
homonymy, the type species automatically
becomes type of the new genus.

apogeny Sterility.

apogeotropism An orientation movement against
gravity; negative geotropism; ageotropism;
apogeotropic.

apogynous Lacking, or having non-
functional, female reproductive organs;
apogyny.

apoheliotropism Apheliotropism *q.v.*

apolegamy Selected breeding; **apolegamic**.

apomeiotic spory The formation of spores from
spore mother cells without chromosome

reduction during meiosis; generative apospory;
gonial spory.

apomictic Pertaining to apomixis *q.v.*

apomictic parthenogenesis Ameiotic
parthenogenesis *q.v.*

apomictic speciation The development of a new
clonal or agamospermous microspecies from an
interspecific hybrid either by vegetative
propagation or by agamospermous seed
formation respectively.

apomixis Reproduction without fertilization, in
which meiosis and fusion of gametes are partially
or totally suppressed; including apogamy *q.v.*,
apospory *q.v.* and parthenogenesis *q.v.*; **apomict,
apomictic**.

apomorphic Derived from and differing from an
ancestral condition; apomorphic character;
apomorph, apomorphy; *cf.* plesiomorphic,
stasimorphic.

apophyte A native weed which appears after
cultivation.

aposematic coloration Coloration having a
protective function; sometimes used in a
restricted sense for warning coloration only;
aposematism; aposeme; *cf.* proaposematic
coloration, pseudoaposematic coloration,
synaposematic coloration.

aposporogony The absence of spore production
(sporogony).

apospory The production of a diploid
gametophyte from a sporophyte without spore
formation; **aposporous**.

apostatic Differing markedly from the norm.

apostatic selection Frequency dependent selection
in which a predator selects the most abundant
morph in a polymorphic population resulting in a
balanced polymorphism.

aposymbiotic Symbiont-free;
used of an organism normally found in a
symbiosis that has been deprived experimentally
of its partner.

apothermotaxis The absence of a directed
response to a temperature change;
apothermotactic.

apotype A specimen used to supplement the
description of a type; hypotype.

apotypic Diverging from the basic or typical form.

apotypic state The derived character state in a
transformation series *q.v.*; *cf.* plesiotypic state.

apparent noon The time at which the sun crosses
the meridian; sun noon.

appeasement display A display that serves to
prevent attack by reducing the opponent's attack
drive and releasing other specific non-aggressive
behaviour.

appetitive behaviour 1: A variable introductory

phase of an instinctive behaviour pattern or sequence; *cf.* consummatory act. 2: Behaviour aiding in the satisfaction of a need in response to an internal condition.

application The use of a name to denote a taxon.

approved Promulgated by the ICZN *q.v.* for purposes of nomenclature.

approximate Close to the desired result; **approximation**.

aptosochromatosis Change of colour in birds, without moulting.

apud (*ap*.) Latin, meaning in the work of; used in citing an author who introduced a new name in the work of another author.

aqua- Combining form meaning water.

aquaculture The cultivation of aquatic organisms; **aquiculture**.

aquatic Living in or near water; used of plants adapted for a partially or completely submerged life.

aquaticolous Living in water or aquatic vegetation; **aquaticole**.

aquatosere An ecological succession commencing in a wet habitat, leading to an aquatic climax.

aqueous desert An area of the sea floor or the bed of a lake more or less devoid of macroscopic organisms, typically with unstable sediment.

aquifer Permeable underground rock strata which hold water.

aquiherbosa The herbaceous communities of ponds and swamps; subdivided into emersiherbosa, sphagniherbosa and submersiherbosa *q.v.*

Aquilonian region A biogeographical region comprising Europe, Asia north of the Himalayas, Africa north of the Tropic of Cancer and America north of latitude 45° N.

aquiprata The community of herbs, grasses and bryophytes influenced by the presence of ground water in damp meadows.

arable Used of land that is cultivated or fit for cultivation.

Arber's law That any structure disappearing from a phylogenetic lineage during the course of evolution is never regained by descendants of that line; *cf.* Dollo's law.

arbitrary In taxonomy, used of a scientific name lacking formal derivation with regard to etymology (a name comprising an arbitrary combination of letters); used of an etymologically incorrect gender assigned to a name.

arboreal Living in trees; adapted for life in trees; arboricolous.

arboreous desert An area of sparsely scattered

trees with little or no vegetation between; desert forest.

arboretum A botanic garden or parkland dominated by trees.

arboricide A chemical that kills trees.

arboricolous Living predominantly in trees or large woody shrubs; dendricolous; arboreal; **arboricole**.

arboriculture The cultivation of trees.

arborvirus An arthropod-borne virus; **arbovirus**.

arbusticolous Living predominantly on scattered shrubs and perennial herbs with shrub-like habit; **arbusticole**.

Archaean A geological period (*ca.* 4500–2400 million years B.P.); a subdivision of the Precambrian; **Archean**; see Appendix 1.

archaebacteria One of the three primary kingdoms (urkingdoms *q.v.*) of living organisms, comprising all methanogenic bacteria; *cf.* eubacteria, urkaryote.

archaeophyte A plant taken into cultivation in prehistoric times.

Archaeophytic The period of geological time during the evolution of the earliest plants, the algae; Algophytic; Proterophytic; *cf.* Aphytic, Caenophytic, Eophytic, Mesophytic, Palaeophytic.

Archaeozoic A geological period preceding the Cryptozoic, within the early Precambrian era; sometimes used for the whole of the early Precambrian or Archaean period *q.v.*; Palaeolaurentian, **Archeozoic**; see Appendix 1.

archaeozoology The study of animal remains from archaeological sites.

archallaxis The addition of a new feature during the early period of embryonic morphogenesis *q.v.*, typically resulting in a major alteration of subsequent development.

Archatlantis Meso-Atlantic bridge *q.v.*

arche- Prefix meaning first, beginning, earliest, primitive.

Archean Archaean *q.v.*

archegenesis Abiogenesis; the spontaneous evolution of life from non-living precursors; archebiosis.

archetype The hypothetical ancestral type; the earliest common ancestor; architype; Bauplan; arquetype; praeform; **archetypal, archetypic**.

archi- Prefix meaning first: arche-.

archibenthal zone The continental slope; the sea floor from the edge of the continental shelf to the continental rise; **archibenthic zone**; see Appendix 7.

archicleistogamic Having reduced reproductive organs within permanently closed flowers; *cf.* archocleistogamic.

Archigalenis The proposed land-bridge *q.v.* across the north Pacific.

archigenesis Archegenesis *q.v.*

archigony The origin of life; archegenesis.

Archinotis Antarctic bridge *q.v.*

archipelago A group of islands.

architomy Reproduction in certain annelids involving body fission with subsequent regeneration; *cf.* paratomy.

archocleistogamic Having mature reproductive organs within permanently closed flowers; *cf.* archicleistogamic.

arctic 1: Pertaining to the area within the Arctic Circle. 2: Pertaining to a climate with a mean temperature below 0° C in the coldest month and below 10° C for the warmest month.

Arctic and Subarctic region A subdivision of the Boreal Kingdom; see Appendix 4.

Arctic brown soil A soil having a shallow profile, acidic crumb-structured yellow-brown A-horizon, B-horizon absent, C-horizon more or less neutral.

Arctic period The earliest period of the Blytt-Sernander classification (*ca.* 14 000–10 000 years B.P.) characterized by primarily tundra vegetation and cold late-glacial conditions.

arctic zone Latitudinal zone between 66.5° and 72° in both northern and southern hemispheres.

Arctogaea The biogeographical area comprising the Palaearctic, Nearctic, Ethiopean and Oriental regions; Megagaea; **Arctogea**; *cf.* Neogaea, Notogaea, Palaeogaea.

ardium An ecological succession following irrigation.

are (**a**) A metric unit of area equal to 100 square metres.

area 1: The overall geographical distribution of a taxon. 2: The range occupied by a community or other group.

area-species curve Species-area curve *q.v.*

arena An area used for communal courtship display; lek.

arenaceous Derived from or containing sand; having the properties of sand; growing in sand; sandy.

arenicolous Living in sand; psammobiontic; sabulicolous; **arenicole**.

areogeography Study of the distributions of plants and animals.

arescent Becoming dry or arid.

argillaceous Comprising clay-sized particles; having the properties of clay.

argillicolous Living on or in clay; **argillicole**.

argillophilous Thriving in clay or mud; **argillophile**, **argillophily**.

argodromilic Pertaining to slow flowing streams.

argotaxis Passive movement due to surface tension; argotactic.

arheic region An area in which no rivers arise; *cf.* endorheic region, exorheic region.

arhythmic Not displaying diurnal periodicity; non-rhythmic.

arid Used of a climate or habitat having a low annual rainfall of less than 250 mm, with evaporation exceeding precipitation and a sparse vegetation.

aridity index A relative measure of aridity, calculated from the ratio of moisture deficit (d) to potential evapotranspiration (PE) expressed as a percentage, $I = 100d/PE$; or from the formula $I = 12P/(T + 10)$ where P is the average monthly rainfall in millimetres and T the average monthly temperature in °C, in which case values less than 20 indicate increasing dryness, greater than 20 increasing wetness.

arido-active Used of plants employing desiccation avoidance mechanisms such as improved water uptake, efficient water conduction or reduction of transpiration.

arido-passive Used of plants sensitive to drought which can survive dry periods by the production of desiccation resistant seeds or structures.

arido-tolerant Used of plants able to tolerate drying out of protoplasm without damage.

aristogenesis Orthogenesis *q.v.*

aristogenic Eugenic *q.v.*

arithmetic growth Growth of an organism or population by linear increase in size or number of individuals.

arithmetic mean Average; of n values, calculated as the sum (Σn) divided by the number (n); mean.

arithmetic series A series of numbers in which the difference between any two successive numbers is the same; arithmetic progression.

arithmotype An isotype *q.v.* which belongs to a different taxon from the holotype.

arms race The sequence of mutual counter adaptations of two coevolving organisms, such as a predator and its prey.

aromorphosis Evolution characterized by an increase in the degree of integration and organization without marked specialization; *cf.* allomorphosis.

arquetype Archetype *q.v.*

array A list, usually of numbers, in some specified order.

arresting factor Any factor that maintains vegetation at a subclimax *q.v.* stage.

arrhenogenic Producing offspring that are entirely, or predominantly male; **arrhenogeny**; *cf.* allelogenic, amphogenic, monogenic, thelygenic.

arrhenoidism The manifestation of typically male characters in females.

arrhenotoky Parthenogenesis in which unfertilized eggs produce haploid males and fertilized eggs produce diploid females; haplodiploidy; **arrhenotokous**; *cf.* thelytoky.

artenkreis A species complex; a group of closely related species replacing one another geographically; a superspecies.

Artenpaare A species pair.

arthrogenous evolution The theory of evolution by preadaptation, that postulates a creative principle in living matter that manifests itself in response to environmental stimuli.

Article A section of the Code *q.v.*, comprising a mandatory rule or rules; *cf.* recommendation.

articulation The ratio of areas of inlets and bays to the total area of a lake; *cf.* insularity.

artifact Not natural; an artificial structure; **artefact**.

artificial character A character selected arbitrarily, without consideration of phylogenetic relationships.

artificial classification A classification structured for convenience, using easily observed phenotypic characters and not necessarily indicating phylogenetic relationships; key classification; *cf.* natural classification.

artificial key An identification key based on convenient phenotypic characters and not indicating phylogenetic relationships; *cf.* natural key.

artificial parthenogenesis The activation of an egg by artificial treatment.

artificial selection Selection by man; domestication; selective breeding.

arundineous Having an abundance of reeds; reedy.

arvum A grain field community.

asaccharolytic Used of microorganisms that are unable to break down carbohydrates, alcohols or polysaccharides.

ascaphus reflex Thanatosis *q.v.*

Ascension and St Helenan region A subdivision of the African subkingdom of the Palaeotropical kingdom; see Appendix 4.

ascriptum (*ascr.*) Latin, meaning ascribed to.

asexual 1: Lacking functional sex organs. 2: Used of reproduction not involving the fusion of gametes, including budding, fission, gemmation, parthenogenesis, spore formation and vegetative propagation; zelotypic.

asexual reproduction Reproduction without the formation of gametes and in the absence of any sexual process; monogenesis; monogony.

aspect 1: The degree of exposure of a site to environmental factors. 2: The seasonal changes in the appearance of vegetation.

aspect society A plant community dominated by a single species or group of species, at a particular season.

aspection The periodic changes in the appearance of vegetation within an area or community; the periods are usually recognized as seasons; **aspectation**.

asphyxia Death from lack of oxygen.

asporogenic Not producing, or developing from, spores; **asporogenous**.

asporous Without spores.

assemblage A group of fossils occurring together in the same stratigraphic level (an assemblage zone).

assembly 1: A gathering of social organisms for group activity. 2: The smallest functional community of plants or animals.

assimilation 1: The total energy flow of a heterotrophic individual, population or trophic level. 2: Gross production of autotrophs. 3: In ecological energetics, that part of the consumption *q.v.* of an individual, population or trophic level which is incorporated into the metabolic system for production *q.v.* and respiration *q.v.* per unit time, per unit area; metabolizable energy; expressed as assimilation (A) = production (P) + respiration (R), or A = consumption (C) − rejecta (FU).

assimilation efficiency, coefficient of In ecological energetics, the ratio of total food absorbed to total food ingested.

assimilation quotient The ratio of volume of carbon dioxide absorbed by a plant to the volume of oxygen liberated.

associate type Any of two or more type specimens listed in the original description of a taxon in the absence of a designated holotype; cotype; syntype.

association 1: A ranked category in the classification of vegetation; a climax plant community characterized by two or more dominant species which have the life-form *q.v.* typical of the formation to which the association belongs; the first word of a two-word association name ends in -*etum*. 2: Sometimes used in general to indicate a large assemblage of organisms in a particular area, with one or two dominant species, or to refer to any group of plants growing together and forming a small unit of natural vegetation.

association, coefficient of 1: An index of the frequency of co-occurrence of two species; calculated as the number of samples in which both occur divided by the number of samples in which they could be expected to occur by chance; index of association. 2: In statistics, a measure of the degree of dependence for data grouped in a contingency table *q.v.*, ranging from -1 for complete negative association through 0 for independence to $+1$ for complete positive association.

association coefficient A function that measures the degree of agreement between pairs of OTU's over an array of characters.

association element The constant species *q.v.* of an association.

association table A list of the species occurring in a number of stands of an association, together with additional bionomic data.

associes 1: An intermediate and unstable stage in the development of an association; a stage in an ecological succession. 2: A developmental stage of a consocies *q.v.*

associule A microcommunity of associes rank in a microsere *q.v.*

assortative mating Sexual reproduction involving the non-random pairing of individuals which are more closely alike than the average (positive) or less alike than the average (negative) in respect of one or more traits; assortive mating; *cf.* disassortative mating.

assortment The distribution of chromosomes to the germ cells during meiosis.

astatic Changing suddenly or drastically.

asterisk (*) A symbol used in taxonomy to denote categories of infraspecific rank; usually placed between the second and third words of the trinomen.

asthenobiosis A period of reduced metabolic activity.

astogeny 1: The development of a colony by budding. 2: The developmental history of a colonial organism.

astronomical tide The theoretical tide produced by the gravitational effects of the sun and moon, excluding meteorological factors; astronomic tide.

Astronomical unit (AU) The mean distance between the Earth and Sun; 149.6×10^6 km.

asymmetry Skewness *q.v.*

asymptotic distribution The limiting form of a probability or frequency distribution dependent on a parameter, as that parameter tends to a limit; **asymptote**.

asymptotic population The carrying capacity *q.v.* or maximum sustainable population size for a given habitat or area.

asynapsis The failure of homologous chromosomes to pair during meiosis; asyndesis.

asynethogametism Gametic incompatibility; the

inability of two apparently compatible gametes to fuse; aethogametism; *cf.* synethogametism.

asyngamy Reproductive isolation by virtue of different breeding or flowering seasons; **asyngamic**.

atavism A reversion to an ancestral character state not evident in recent generations, possibly due to a recessive gene or to complementary genes; the aberrant individual is sometimes referred to as a throwback; **atavistic**.

ateliosis Dwarfism; reduced size but with normal proportions.

athalassic Used of waters or water bodies that have not had any connection to the sea in geologically Recent times, all ions in solution are thus derived from the substratum or atmosphere.

athermobiosis Dormancy induced by relatively low temperatures; **athermobiose**.

Atlantic Equatorial countercurrent A warm surface ocean current that flows east in the tropical Atlantic, forming the northerly part of the equatorial circulation; see Appendix 6.

Atlantic Equatorial undercurrent A strong subsurface ocean current that flows east across the Atlantic at the equator.

Atlantic North American region A subdivision of the Boreal kingdom; see Appendix 4.

Atlantic North Equatorial current A warm surface ocean current that flows west in the equatorial Atlantic and forms the southerly limb of the North Atlantic Gyre; see Appendix 6.

Atlantic period A period of the Blytt-Sernander classification *q.v.* (*ca.* 7500–4500 years B.P.) characterized by oak, elm, linden and ivy vegetation and by the driest and warmest postglacial conditions.

Atlantic South Equatorial current A warm surface ocean current that flows west in the equatorial Atlantic and forms the northerly limb of the South Atlantic Gyre; see Appendix 6.

Atlantis Meso-Atlantic bridge *q.v.*

Atlanto-equatorial bridge Afro-Brasilian bridge *q.v.*

atlas The presentation of information by means of illustrations rather than by a description.

atmobios Organisms living in the air; aerial organisms.

atmophyte An epiphyte that obtains water by aerial assimilation over its entire surface.

atmosphere 1: The gaseous envelope surrounding the Earth, mainly consisting of oxygen and nitrogen. 2: A derived fps unit of the pressure of air at sea level; equal to a force of 14.695 916 pounds per square inch or 1.01325×10^5 N m^{-2}; see Appendix 13.

atokous Without offspring; non-reproductive; vegetative; *cf.* epitokous.

atrophic Used of organisms or life cycle stages that do not feed.

atrophy Marked reduction in size and functional significance.

attenuation 1: A reduction in strength or intensity; weakening. 2: The loss of virulence of a pathogenic microorganism.

Atterberg grade scale A decimal scale of particle sizes that uses 2 mm as a reference point and a fixed ratio of 10 to separate grades; subdivisions are calculated as the geometric means of the grade limits, for example 0.2, 0.6, 2.0, 6.3, 20.0.

atto- (a) Prefix used to denote unit $\times$ 10^{-18}.

attribute One of a set of descriptive terms; a character.

attrition Abrasion *q.v.*

atypical Unusual; abnormal.

atypicality A measure of the degree of difference of one OTU from a central or typical OTU.

AU Astronomical unit *q.v.*

auctorum (*auct.*) Latin, meaning of authors.

auctorum non (*auct. non*) Latin, meaning not of authors; used when citing a misapplied name.

aucuparious Attractive to birds.

Aufnahme A phytosociological record; a visual description of the vegetation together with appropriate habit and habitat data.

aufwuchs A community of aquatic organisms and associated detritus adhering to and forming a surface coating on submerged stones, plants and other objects; fouling organisms; periphyton.

augmented distance An estimate of the total number of fixed nucleotide mutations separating two nucleic acid sequences which is inferred by the maximum parsimony method *q.v.* and adjusted for variations in density throughout the phylogenetic tree; *cf.* phylogenetic density.

aulophyte A non-parasitic plant living within a hollow cavity in another plant.

aural Pertaining to hearing or to ears.

auroral Pertaining to the dawn.

austral Southerly; pertaining to cool or cold temperate regions of the southern hemisphere; antiboreal; *cf.* boreal.

Australian kingdom One of the six major phytogeographical areas characterized by floristic composition; comprising Central, North & East and South & West Australian regions; see Appendix 4.

Australian region A zoogeographical region comprising Australia, Maluku (the Moluccas), Sulawesi, Papua–New Guinea, Tasmania, New Zealand and the oceanic islands of the South Pacific; subdivided into Australian, Austro-Malayan, New Zealand and Polynesian subregions; Hawaii and Polynesia are sometimes excluded and regarded as distinct regions; see Appendix 3.

Australian subregion A subdivision of the Australian region, comprising Australia and Tasmania; see Appendix 3.

Austro-Malayan subregion A subdivision of the Australian region; see Appendix 3.

autallogamy Autoallogamy *q.v.*

autamphinereid An autotrophic amphibious plant.

autapomorphic character A derived (apomorphic) character unique to a given species or other monophyletic group.

autapomorphy The possession of a unique derived character by a species or monophyletic taxon; *cf.* synapomorphy, symplesiomorphy.

autarkean society A simple, independent nomadic, or sparsely distributed, human society; *cf.* allelarkean society.

autecic Autoecious *q.v.*; **autecious**.

autecology The ecology of individual organisms or species; idioecology; **autoecology**; *cf.* synecology.

autephaptomenon Autotrophic organisms which live attached to a surface; **autophaptomenon**.

authigenic Derived from sea water and precipitated directly onto the sediment rather than settling through the water column.

authigenic mineral A mineral formed during the process of sedimentation, and thus one from which the date of deposition can be determined.

author 1: The person who first published a name which satisfies the criteria of availability or valid publication. 2: The person to whom is attributed a work, a scientific name, a nomenclatural act or a nominal taxon.

authority The name of the author cited after a scientific name; author citation.

auto- Prefix meaning self, automatic, same one, self-operating.

autoallogamy The condition of a species in which some individuals are adapted to cross fertilization and others to self-fertilization; homodichogamy; allautogamy; **autallogamia, autallogamy**.

autoantibiosis Self-inhibition of growth; growth retardation due to nutrient depletion by the same organism.

autobiology Idiobiology *q.v.*

autobiosphere That part of the eubiosphere *q.v.* in which energy is fixed by photosynthesis in green plants; *cf.* allobiosphere.

autocatalysis The promotion of a reaction by one of its products; positive feedback.

autochorous Having motile spores or propagules disseminated by the action of the parent plant; **autochore, autochory**.

autochthonous Endogenous; produced within a given habitat, community or system; indigenous;

endemic; native; **autochthon, autochthone, autochthony**; *cf.* allochthonous.

autochtone 1: An indigenous organism; native; *cf.* allochtone. 2: Endogenous food material in a cave system derived from within the cave itself.

autocoprophagous Used of an organism that consumes its own faeces; refection; **autocoprophage, autocoprophagy**.

autodeme A deme *q.v.* composed predominantly of self-fertilizing individuals.

autoecious Used of a parasite that passes through the different stages of its life cycle within the same host individual; autoxenous; autecic; **autoecius**; *cf.* heteroecious.

autoecology Autecology *q.v.*

autogamy 1: The process of self-fertilization or self-pollination; orthogamy. 2: The fusion of two reproductive nuclei within a single cell, derived from a single parent. 3: Reproduction in which a single cell undergoes reduction division producing two autogametes which subsequently fuse. 4: In protistans, the division of the micronuclei to produce eight or more nuclei, of which two fuse and give rise to a new macronucleus.

autogenesis 1: Abiogenesis *q.v.* 2: Reproduction or origin within a given system or individual; autogeny; autogony; **autogenetic**.

autogenetic plankton Planktonic organisms that breed within the region within which they normally live; **autogenic plankton**.

autogenic Occurring within a given system; produced by the activities of living organisms within a system and acting upon that system.

autogenic succession The replacement of one community by another as a result of intrinsic changes in the environment brought about by the community itself; *cf.* allogenic succession.

autogenotype In taxonomy, a genotype *q.v.* by original designation; *cf.* apogenotype.

autogenous Used of a female insect that does not have to feed in order to facilitate maturation of her eggs; *cf.* anautogenous.

autogony Autogenesis *q.v.*

autograft Autoplastic grafting *q.v.*

autograph A document or text in the handwriting of the author, or a mechanical copy of such a document or text; *cf.* manuscript.

autoheteroploid An aneuploid *q.v.* derived by incomplete multiplication of the basic chromosome set of a single species; *cf.* alloheteroploid.

autoicous Having male and female inflorescences on the same plant; monoecious.

autoinfection The direct reinfection of a host

individual by larval offspring of an existing parasite.

autolycism A symbiosis *q.v.* between birds and other animals, including man, that is not strictly mutualistic or parasitic, such as the use of buildings as nest sites or of farming practices as feeding opportunities.

autolysis The breakdown of a cell by its own enzymes; **autolytic**.

automictic parthenogenesis Parthenogenesis *q.v.* in which meiosis is preserved and diploidy reinstated either by the fusion of haploid nuclei within a single gamete or by the formation of a restitution nucleus; parthenogamy.

automimicry Mimicry in which a palatable morph (the mimic) resembles a non-palatable morph (the model) of the same species; the polymorphism with respect to palatability can arise from a choice of different food plants by model and mimic; intraspecific mimicry.

automixis Obligatory self-fertilization, by autogamy *q.v.*, paedogamy *q.v.* or automictic parthenogenesis *q.v.*; **automictic**.

autonastic Pertaining to growth curvature promoted by endogenous factors; **autonastism, autonasty**.

autonereid An autotrophic water plant.

autonomic Spontaneous; used of movement or growth stimulated, enhanced or retarded by an internal stimulus; **autonomism**; *cf.* paratonic.

autonyctinasty Autonyctitropism *q.v.*

autonyctitropism The spontaneous assumption of the position usually adopted during the night; autonyctinasty; **autonyctitropic**.

autonym An automatically established name, applied to a nominate subordinate taxon.

autoorthotropism The tendency of a plant to grow in a straight line; **autoorthotropic**; *cf.* autoskoliotropism.

autoparasite A parasite living on another parasite; hyperparasite; superparasite.

autoparthenogenesis The development of an unfertilized egg that has been activated by a chemical or physical stimulus.

autopelagic Epipelagic; used of planktonic organisms living continually at the sea surface; *cf.* allopelagic.

autophagous Used of precocious offspring capable of locating and securing their own food; **autophage, autophagy**.

autophaptomenon Autephaptomenon *q.v.*

autophene An abnormal phenotype produced by genetic mutation which will not be modified on transplantation to a wild-type host; *cf.* allophene.

autophilous Self-pollinated; **autophily**.

autophyte A plant capable of synthesizing complex

organic substances from simple inorganic substrates, typically by photosynthesis; **autophytic**.

autoplastic grafting Transplanting tissue from one site to another within the body of the same individual; autograft; *cf.* heteroplastic grafting, homoplastic grafting.

autoploid Used of an organism with the characteristic chromosome set of its species.

autopolyploid A polyploid variety resulting from the multiplication of the chromosome set of a single species; panautopolyploid; *cf.* allopolyploid.

autopotamic 1: Pertaining to organisms adapted to stream environments and which complete their life cycles in streams; *cf.* eupotamic, tychopotamic. 2: Originating in fresh water; **autopotamous**.

autosexing The use of sex-linked morphological characters to identify the sex of an immature individual before the development of adult sexual dimorphism.

autoskoliotropism The tendency of a plant structure to grow in a curve; **autoscoliotropism**; **autoscoliotropous, autoskoliotropic**.

autosome Any chromosome other than sex chromosomes; euchromosome.

autospory Autotomy *q.v.*

autosyndesis The pairing of homologous chromosomes from the same parent as in an autopolyploid; *cf.* allosyndesis.

autotherm An organism that regulates its body temperature independent of ambient temperature changes; endotherm; homoiotherm; *cf.* allotherm.

autotilly Self-amputation of a part; as practised by some spiders.

autotomy Self-amputation of an appendage or other part of the body.

autotoxin Any substance produced by an organism that is toxic to itself.

autotrophic 1: Capable of synthesizing complex organic substances from simple inorganic substrates; including both chemoautotrophic and photoautotrophic organisms. 2: Used of any organism for which environmental carbon dioxide is the only or main source of carbon in the synthesis of organic compounds by photosynthesis; holophytic; holotrophic; phytotrophic; **autotroph, autotrophy**; *cf.* heterotrophic.

autotrophic lake A lake in which all or most of the organic matter present is derived from within the lake and not from drainage off the surrounding land; *cf.* allotrophic lake.

autotropism The tendency to grow in a straight line unaffected by external factors; **autotropic**; *cf.* tropism.

autotype 1: The type, by original designation, of a taxon. 2: A specimen designated by the author of a species subsequent to the original publication as being identical to the holotype; heautotype.

autoxenous Used of a parasite that passes through the different stages of its life cycle in the same host individual; autoecious; ametoecious; monoxenous; *cf.* heteroxenous.

autozygous Used of a zygote having two or more alleles at a given locus that are identical by descent; **autozygote**; *cf.* allozygous.

autumnal Pertaining to the autumn.

auxesis 1: Growth; sometimes used in a restricted sense to refer to the special case of isauxesis (isometric growth); *cf.* heterauxesis. 2: Growth in size by increasing cell volume rather than numbers of cells.

auxiliary character Any character used to confirm the identification of a taxon; confirmatory character; *cf.* diagnostic character.

auxin A plant hormone promoting or regulating growth.

auxoautotrophic Used of an autotrophic organism requiring certain specific additional nutrients for growth; **auxoautotroph**.

auxoheterotrophic Used of an organism that is unable to synthesize all substances necessary for growth and development.

auxotrophic Used of a microorganism having a biochemical deficiency and requiring supplementary growth factors not needed by the wild type (prototroph); **auxotroph, auxotrophy**.

available 1: In taxonomy, used of a work or name that must be considered for purposes of nomenclature; validly published. 2: Legitimate *q.v.* and with its type falling within the range of variation of the taxon.

available name A valid or invalid name that conforms to the relevant articles of the Code and which has not been annulled by the ICZN.

available nutrient That proportion of a given nutrient in a soil that can be readily taken up by plants.

available water That part of the soil water that can be readily taken up by plants; *cf.* holding capacity.

available work A work published after the starting point *q.v.*, that conforms to the provisions of the Code and that has not been annulled by the ICZN.

average The arithmetic mean.

average distance The mean distance between individual plants; determined by dividing the square root of an area by the number of individuals of a given species in that area.

avicennietum A mangrove swamp community.

aviculture The practice of keeping birds in captivity for scientific study.

avifauna The bird fauna of an area or period.

avowed substitute In taxonomy, a name explicitly proposed as a substitute for an existing name.

avulsion The rapid erosion of a shore by wave action during a storm; the sudden isolation of a piece of land by flood water or by a change in the course of a river.

axenic Used of a pure culture free from contaminant organisms; *cf*. dixenic, monaxenic.

axiom An established principle or self-evident truth.

axiom of extensionality An axiom from set theory, that classes having the same members are the same class.

azoic Without life.

Azoic The earliest geological period; the subdivision of the Precambrian era preceding the Archaeozoic; see Appendix 1.

azonal soil A soil lacking a well defined profile; a soil maintained in a permanently immature state by persistent deposition, truncation or erosion; abnormal soil; *cf*. intrazonal soil, zonal soil.

azotification Nitrogen fixation.

azygospore A spore produced from an unfertilized female gamete; aboospore; parthenospore.

azygote An organism produced by haploid parthenogenesis.

B

b.y. A billion years; an aeon.

back mutation A mutation of a gene so that it assumes a function lost by an earlier mutation; reverse mutation; *cf.* forward mutation.

backcross A cross between a hybrid offspring and one of its parents; the notation for backcross generations is B_1, B_2. . .

background genotype The residual genotype; that part of the genotype not primarily responsible for producing the phenotype.

background radiation Natural radiation, comprising cosmic radiation (elementary particles from outer space) and terrestrial radiation (derived from the decay of naturally occurring isotopes).

back-pull The logarithm of the tension (F), measured in centimetres of water, by which water is held in a soil; pF factor.

backshore The zone of a typical beach profile above mean high water; also used for the zone covered only in exceptionally severe storms.

bactericide An agent that destroys bacteria; **bactericidal**.

Bacteriological Code International Code of Nomenclature of Bacteria *q.v.*

bacteriology The study of bacteria; **bacteriological**.

bacterioneuston The bacterial flora at the sea surface; the bacterial component of the neuston *q.v.*

bacteriophage 1: A virus infesting and usually lysing bacteria; often abbreviated to phage. 2: An organism that feeds on bacteria.

bacteriophagous Feeding on bacteria; **bacteriophage**, **bacteriophagy**.

bacteriostatic An agent inhibiting the growth of bacteria but not destroying them.

badland An arid or semi-arid area with scanty vegetation and marked surface erosion.

Baker's rule That self-pollinating species are at an advantage in colonizing remote habitats because a population can be established from a single pioneer individual.

balance of nature The state of equilibrium in nature resulting from the constant interaction between organisms and their environment.

balanced hermaphrodite population A population composed exclusively of hermaphrodites; *cf.* unbalanced hermaphrodite population.

balanced lethality The maintenance of two non-allelic recessive lethal genes in different homologous chromosomes; when interbred half

the offspring are homozygous for one lethal or the other and die, the other half are heterozygous for the lethal genes and survive.

balanced load That component of genetic load *q.v.* in which a decrease in the overall fitness of the population occurs due to the segregation of inferior genotypes but whose alleles are maintained because they increase fitness in other combinations; segregational load.

balanced polymorphism Polymorphism in which the genetically distinct forms are more or less permanent components of the population, maintained by selection in favour of diversity as in the case of the selective superiority of the heterozygote over both homozygotes; *cf.* polymorphism.

balancing selection Selection that maintains a balanced polymorphism in a population.

bald An elevated treeless area within a region of forest vegetation, usually grassy.

Baldwin effect The genetic assimilation of acquired characters; the reinforcement or replacement of environmentally induced phenotypic adaptations by similar hereditary characters under the influence of selection.

ballistic Used of dehiscence in which the seeds are catapulted from the fruit.

Bancroft's law A generalization that organisms and communities move towards a state of dynamic equilibrium with their environment.

bar 1: A derived cgs unit of pressure equal to 1×10^6 dynes per square centimetre, or 10^5 newtons per square metre; see Appendix 13. 2: A submerged ridge of alluvial deposit in shallow water, produced by the action of water or wind currents.

bar chart A chart comprising rectangles (bars) of equal width; used to illustrate frequency distributions with the lengths of the bars proportional to the frequencies of occurrence.

barachorous Having propagules dispersed by their own weight; **barachore**, **barachory**.

baroclinic atmosphere An atmosphere in which surfaces of pressure and density intersect at some level(s); strong baroclinicity implies the presence of large horizontal temperature gradients and hence of strong thermal winds; *cf.* barotropic atmosphere.

barokinesis A change of linear or angular velocity in response to a change in pressure; **barokinetic**.

barophilic Thriving under conditions of high

hydrostatic or atmospheric pressure; **barophilous, barophile, barophily.**

barotaxis A directed reaction of a motile organism to a pressure stimulus; **barotactic.**

barotropic atmosphere The hypothetical atmosphere in which surfaces of pressure and density coincide at all levels; *cf.* baroclinic atmosphere.

barotropism An orientation response to a pressure stimulus; **barotropic.**

barren 1: Incapable of producing offspring; unproductive; infertile; incapable of producing seeds or fruit. 2: Devoid of vegetation, or of fossils.

barrial A mud flat.

barrier Any obstruction to the spread of an organism or to gene flow between populations; an unfavourable biological, climatic, geographical or other factor which prevents successful dispersal and establishment of a species.

barymorphosis Those structural changes in organisms that result from the effect of pressure or weight; **barymorphic.**

barysphere The mantle and core of the Earth beneath the lithosphere *q.v.*

basal area 1: A cross sectional area of a tree determined from the diameter of the trunk at breast height (DBH), 1.4 m above ground. 2: The total area of ground covered by trees measured at breast height. 3: The actual surface area of soil covered or occupied by a plant measured close to the ground; basal cover; ground cover; *cf.* cover.

base exchange capacity The total capacity (in milliequivalents) of a soil for holding replaceable cations.

basic Rich in alkaline minerals and typically rich in nutrients as well; alkaline; *cf.* acidic.

basic number The number of chromosomes in a basic chromosome set, denoted by x, and representing the lowest monoploid number in a polyploid series.

basic type 1: The primary type *q.v.* of a taxon. 2: The chromosome complement characteristic of a species.

basifixed Innate *q.v.*

basifuge A plant unable to tolerate basic soils; calcifuge.

basikaryotype Basic number *q.v.*

basionym The original name of a taxon subsequently replaced by another using the same stem, as a result of a change in rank or position of the taxon; **basinym.**

basking range The temperature range within which a reptile basking in direct sunlight remains inactive although alert.

basophilous Thriving in alkaline habitats; oxyphobous; **basophilic, basophile, basophily.**

Batesian mimicry The close resemblance of a palatable or harmless species (the mimic) to an unpalatable or venomous species (the model) in order to deceive a predator (the operator); *cf.* mimicry.

bathile Pertaining to the floor of a lake more than 25 m below the surface.

bathophilus Thriving in deep water habitats; **bathophilic, bathophile, bathophily.**

bathyal 1: Pertaining to the sea floor between 200 and 4000 m; bathybenthic; see Appendix 7. 2: Pertaining to the deep sea; bathybic, **bathyl.**

bathybic Pertaining to life on the deep sea floor; bathyal.

bathylimnetic Pertaining to the deep water regions of a lake; profundal; see Appendix 8.

bathylittoral That part of the marine sublittoral zone that is devoid of algae.

bathymetric contour A line on a chart connecting points on the ocean or lake floor of equal depth; depth contour; isobath.

bathymetry The measurement of ocean or lake depth and the study of floor topography; **bathymetric.**

bathypelagic Living in deep water below the level of light penetration but above the abyssal zone, between 1000 and 4000 m.

bathyphilous 1: Thriving in lowlands. 2: Thriving in the deep sea; bathophilus; **bathyphilic, bathyphile, bathyphily.**

bathyphytia A lowland plant community.

bathyplanktohyponeuston Planktonic organisms that migrate regularly between deep water and the surface and spend about equal amounts of time in the hyponeuston *q.v.* and bathyplankton *q.v.*; **bathyplanktohyponeustonic, bathyplanktohyponeustont.**

bathyplankton Planktonic organisms which undergo diurnal vertical migration, moving up towards the surface at dusk and down away from the surface at dawn; **bathyplanktonic;** *cf.* epiplankton.

bathyscaphe A free moving manned vehicle for deep ocean exploration; *cf.* bathysphere.

bathysmal Pertaining to great ocean depths.

bathysphere A spherical manned chamber suspended from a mother ship and lowered into the deep sea for underwater exploration; *cf.* bathyscaphe.

Bauplan The generalized, idealized archetypal bodyplan of a major taxon; typus; archetype.

Baventian glaciation An early glaciation of the Quaternary Ice Age in the British Isles; see Appendix 2.

Bayesian method A method for calculating the probability that a given unknown organism belongs to a specified taxon.

B-chromosome A supernumary or accessory chromosome differing from normal A-chromosomes *q.v.* in morphology, genetic effectiveness and pairing behaviour.

beach The gently sloping strip of land along the margin of a body of water that is washed by waves or tides sufficiently to inhibit all or most plant growth.

Beaufort wind scale A scale of wind speeds; see Appendix 21. The wind speed (V) is calculated as $V = 1.87\ B^{3/2}$ where B is the Beaufort number.

bed A unit layer in a sequence of rock strata; the smallest formal unit of lithostratigraphic classification.

bed load The quantity of rock and debris moved along a stream or river by the flow of water; *cf.* silt load.

bedding plane The division plane that separates individual beds or strata of a sedimentary or stratified rock.

Beestonian glaciation A glaciation of the Quaternary Ice Age in the British Isles, with an estimated duration of 100 thousand years; see Appendix 2.

behaviour Any observable action or response of an organism; the response of an organism, group or species to environmental factors.

behaviour hypothesis A hypothesis of population self-regulation, stating that mutual interactions involving spacing behaviour prevent unlimited increase in population density and that this spacing behaviour is not an inherited trait; *cf.* polymorphic behaviour hypothesis, stress hypothesis.

behavioural scaling 1: The range of adaptive behaviour patterns expressed by a given organism or society; **behavioural scale**. 2: The organization of a society into individual territories at low densities but into a dominance system at high densities.

bell-shaped distribution A frequency distribution having the shape of a cross section through a bell, as in a normal distribution.

belt A narrow strip or area of vegetation.

belt transect A narrow belt of predetermined width set out across a study area, and within which the occurrence or distribution of plants and animals is recorded.

Benguela current A cold surface ocean current that flows north off the east coast of Africa and forms the easterly limb of the South Atlantic Gyre; see Appendix 6.

benth- Prefix meaning sea or river bed, ocean depth.

benthic Pertaining to the sea bed, river bed or lake floor; benthal; benthonic; *cf.* pelagic.

benthogenic Derived from the benthos *q.v.*

benthohyponeuston Aquatic organisms that migrate regularly between the sea bed and the surface, spending about the same time in benthic and hyponeustonic *q.v.* habitats; **benthohyponeustonic, benthohyponeustont**.

benthon Benthos *q.v.*

benthonic Benthic *q.v.*

benthophyte A plant living at the bottom of a water body or on the bed of a river.

benthopleustophyte Any large plant resting freely on the floor of a lake but capable of drifting slowly with the currents.

benthopotamous Living on the bed of a river or stream.

benthos Those organisms attached to, living on, in or near the sea bed, river bed or lake floor; benthon; *cf.* endobenthos, haptobenthos, herpobenthos, psammon, rhizobenthos.

Berger-Parker dominance index A measure of the diversity of a habitat or community expressing the proportion of the total number of individuals (N_{tot}) due to the dominant species; calculated as $d = N_{max}/N_{tot}$, where N_{max} is the abundance of the dominant species.

Bergmann's rule The generalization that geographically variable species of warm-blooded vertebrates tend to be represented by larger races in the colder parts of their range than in the warmer parts; size rule.

berm The large deposits of dry loose sediment above the high tide line on a beach.

beta diversity A measure of the rate and extent of change in species along a gradient from one habitat to others; *cf.* alpha diversity, gamma diversity.

beta karyology In the study of chromosomes, the determination of chromosome numbers, the lengths of chromosome arms and the identification of sex chromosomes; *cf.* karyology.

beta taxonomy The arrangement of species into hierarchical systems of higher categories or taxa; macrotaxonomy; *cf.* alpha taxonomy, gamma taxonomy.

B-horizon The upper subsoil horizon with an accumulation of clay, humus, iron and sesquioxides as a result of leaching *q.v.* and translocation *q.v.* from upper layers; denoted by Bt, Bh, Bi and Bs respectively; illuvial layer; see Appendix 12.

bi- Prefix meaning two, twice, double.

biaiometamorphosis A disadvantageous change in form in response to a stimulus; biaiomorphosis.

bias 1: A systematic error in sampling inherent in the sampling technique. 2: Error introduced when the expected value of an estimator is not equal to its actual value or when the probability of rejecting the null hypothesis is not at a minimum when the null hypothesis is true.

bibliographic error An error in the citation of a paper; *cf.* misquotation.

bibliography An exhaustive list of references to any given subject; commonly used in publications as a synonym of references or literature cited; **bibliographic**.

bibulous Highly absorbent; capable of absorbing water.

bicentric Having two centres of distribution or evolution; used of a distributional range having two zones of concentration separated by an impoverished region.

biennial Lasting for two years; occurring every two years; requiring two years to complete the life cycle; *cf.* annual, perennial.

biferous Producing fruit twice a year, or producing two crops per season; **bifer**.

biflorous Flowering in both spring and autumn; **biflorus**.

big-bang reproduction Semelparous *q.v.* reproduction.

bigeneric Pertaining to a hybrid or cross involving two different genera; *cf.* multigeneric.

bight A large indentation in the coastline or continental shelf margin (seabight).

billion A thousand million, 10^9; this USA value has been adopted internationally; formerly a value of 10^{12} (a million million) was widely used.

bimodal Used of a population or frequency distribution having two modes or peaks; *cf.* modality.

binary character A character having only two states, recorded as present or absent ($+$ and $-$, or 1 and 0); all-or-none character; two-state character; *cf.* multistate character.

binary fission Division into two parts; the commonest form of asexual reproduction in protistans.

binary name Binomen *q.v.*

binary nomenclature Binomial nomenclature, principle of *q.v.*

binom 1: A taxon that is poorly defined in terms of biosystematic data but which is clearly characterized morphologically. 2: Agameon *q.v.*

binomen The two word scientific designation of a species, comprising a generic name and a specific epithet; binary name; binomial name; binominal name; binomina.

binomial distribution A mathematical distribution, each value of which gives the probability that a particular event will occur, that event having a constant probability of occurrence from test to test; used as a model when each individual in a sample may exhibit a character in either of two alternative states.

binomial name Binomen *q.v.*

binomial nomenclature, principle of That the species name consists of two words (a binomen), genus-group and family-group names of only one word and subspecies names of three words (a trinomen); binominal nomenclature; binary nomenclature.

binomialism The system of binomial nomenclature.

binominal name Binomen *q.v.*

-bio- Combining form meaning life, living, organism.

bioassay Biological assay; the use of an organism for assay purposes; any quantitative biological analysis.

biocenosis Biocoenosis *q.v.*

biochemical oxygen demand (BOD) The amount of oxygen required to degrade the organic material and to oxidize reduced substances in a water sample; used as a measure of the oxygen requirement of bacterial populations and serving as an index of water pollution; biological oxygen demand.

biochemical profile A description of a microorganism based on characters determined by biochemical techniques.

biochore A group of similar biotopes *q.v.*; a component of a biocycle and the biosphere.

biochron 1: The total duration of existence of a given taxon. 2: The total time span of a given biostratigraphic unit.

biochronologic unit A period of time based on biostratigraphic or objective palaeontological data.

biochronology The dating of biological events using biostratigraphic or palaeontological methods.

biocide A chemical toxic or lethal to living organisms.

bioclast A single fossil fragment.

bioclastic sediment A sediment composed of broken fragments of organic skeletal material.

bioclimate Microclimate *q.v.*

bioclimatic law 1: Any climatic rule *q.v.* 2: Hopkin's bioclimatic law *q.v.*

bioclimatology The study of climate in relation to fauna and flora; ecoclimatology.

biocoenology The qualitative and quantitative study of communities of organisms; biocenology;

biocenotics; biocoenotics; cenobiology; coenobiology.

biocoenosis A community or natural assemblage of organisms; often used as an alternative to ecosystem but strictly it is the fauna/flora associations *per se* excluding physical aspects of the environment; biocenosis; life assemblage; **biocenose, biocoenose**; *cf*. thanatocoenosis.

biocoenotic zoogeography The study of the distribution of animal communities (zoocoenoses); biocenotic zoogeography.

biocontrol Biological control *q.v.*

biocycles The major subdivisions of the biosphere, land, salt water and fresh water.

biodegradable Capable of being decomposed by natural processes; *cf*. non-biodegradable.

biodemography The integrated study of ecology and genetics of populations.

bioecology The study of living organisms in relation to their environment; ecology.

bioenergetics The study of energy flow through ecosystems; ecological energetics.

bioerosion Erosion resulting from the direct action of living organisms.

biofacies A subdivision of a sedimentary unit based on a distinctive assemblage of fossils.

biogenesis The principle that all living organisms have derived from previously existing living organisms; **biogenetic**; *cf*. abiogenesis.

biogenetic Produced by the activity of living organisms; **biogenic**.

biogenetic law The theory that an individual during its development passes through stages that resemble adult forms of its successive ancestors so that the ontogeny of an individual recapitulates the phylogeny of its group; Haeckel's law; recapitulation.

biogenic Produced by the action of living organisms; **biogenetic**.

biogenic meromixis The mixing of a lake caused by an accumulation in the monimolimnion *q.v.* of salts liberated from the sediment by biological activity; *cf*. crenogenic meromixis, ectogenic meromixis.

biogenic salts The soluble salts essential to life processes.

biogenous Living on or in other organisms; parasitic; symbiotic.

biogeny The science of evolution including both ontogeny and phylogeny.

biogeochemical cycle The cyclical system through which a given chemical element is transferred between biotic and abiotic parts of the biosphere, as for example in the carbon and nitrogen cycles; biogeochemistry.

biogeochemistry The study of mineral cycling and of organism–substrate relationships.

biogeocoenosis Ecosystem *q.v.*

biogeographical realms The major divisions of the terrestrial environment characterized by their distinctive biota; see Appendices 3 and 4.

biogeographical region Any geographical area characterized by distinctive flora and/or fauna; *cf*. biome, province.

biogeography The study of the geographical distributions of organisms, their habitats (ecological biogeography) and the historical and biological factors which produced them (historical biogeography); chorology; geonemy.

biogeosphere That part of the lithosphere *q.v.* within which living organisms can occur; eubiosphere.

bioglyph Trace fossil *q.v.*

bioherm 1: Any organism contributing to the formation of a coral reef. 2: A mound-like accumulation of fossil remains on the site where the organisms lived; **biohermal**.

biohydrology The study of the interactions between organisms and the water cycle *q.v.*; *cf*. hydrobiology.

bioid Any simple prebiotic chemical system capable of evolving from one steady state to a more stable one by mutation.

biolith A rock of organic origin.

biological Pertaining to living organisms or life processes.

biological amplification The concentration of a given persistent substance by the organisms in a food chain so that the level of the substance present in the body increases at each successive trophic level; biological magnification.

biological clock An inherent physiological mechanism for measuring time or maintaining endogenous rhythms; escapement clock.

biological control The control of a pest by the introduction, preservation or facilitation of natural predators, parasites or other enemies, by sterilization techniques, by the use of inhibitory hormones or by other biological means; biocontrol.

biological efficiency The ratio of the productivity of an organism or population to the energy consumed.

biological fitness Fitness *q.v.*

biological indicator A microorganism which is used as an indicator of chemical activity in, or the chemical composition of, a natural system.

biological interchange The interchange of elements between organic and inorganic states as a consequence of biological activity; involving the processes of biological mineralization *q.v.* and immobilization *q.v.*

biological magnification Biological amplification *q.v.*

biological mineralization The biological decomposition of organic compounds and liberation of inorganic minerals; *cf.* immobilization.

biological oxygen demand (BOD) The amount of oxygen required to degrade the organic material and oxidize reduced substances in a water sample; used as a measure of the oxygen requirement of bacterial populations and serving as an index of water pollution; biochemical oxygen demand.

biological races The sympatric populations of a species which differ biologically but not morphologically and in which interbreeding is inhibited by different food or host preferences or behaviour cycles.

biological rhythm A regular periodicity exhibited by biological processes.

biological role All actions or use of a structure by an organism in relation to its life history, behaviour and environment; excluding the functional properties of that structure; *cf.* function.

biological species Groups of actually or potentially interbreeding natural populations genetically isolated from other such groups by one or more reproductive isolating mechanisms; biospecies.

biological species concept The concept of a species as a population or series of populations of freely interbreeding organisms reproductively isolated from other such populations; *cf.* genetic species concept, morphological species concept.

biological spectrum A list of plant species in a given region or community together with percentage abundances of individual species.

biological time scale A time scale based on organic evolution giving the relative order of a sequence of evolutionary events in geological time.

biological weathering Changes in soil structure and composition resulting from biotic activity, as part of the soil forming process.

biologically accommodated communities Communities structured primarily by biological interactions and typical of environments in which physiological stresses are low; stability-time hypothesis *q.v.*

biology The science of life; the study of living organisms and systems; **biological**.

bioluminescence Light produced by living organisms and the emission of such biologically produced light; **bioluminescent**; *cf.* phosphorescence.

biolysis Death and subsequent tissue degradation of an organism.

biomass Any quantitative estimate of the total mass of organisms comprising all or part of a population or any other specified unit, or within a given area at a given time; measured as volume, mass (live, dead, dry or ash-free weight) or energy (calories); standing crop; standing stock.

biomass-days In ecological energetics, a quantity calculated as mean biomass multiplied by the number of days in a given study period.

biome A biogeographical region or formation; a major regional ecological community characterized by distinctive life forms and principal plant (terrestrial biomes) or animal (marine biomes) species; see Appendix 5.

biomechanics Study of the mechanics of the structure of living organisms.

biometeorology Study of the effects of atmospheric conditions on the flora and fauna.

biometry The application of statistical methods to biological problems and the mathematical analysis of biological data; **biometrics**.

biomineralization The process by which organisms produce intracellular or extracellular skeletal structures comprising crystalline or amorphous inorganic substances at least in part.

bion Biont *q.v.*

bionomic axes Used in the context of a niche as a multidimensional space, for those axes that relate to resources, such as food and space, for which there may be competition with neighbouring species; resource axes; *cf.* scenopoetic axes.

bionomics Ecology; the study of organisms in relation to environment; **bionomy**.

biont An individual organism; **bion**.

-bionta The ending of the name of a subkingdom in botanical nomenclature.

bionumeric code A numerical code in which separate categories of a hierarchy are assigned to different groups of digits.

biophagous Consuming or destroying other living organisms; **biophage, biophagy**; *cf.* saprophagous.

biophilous Thriving on other living organisms, often used specifically of plant parasites; **biophile, biophily**.

biophotogenesis Bioluminescence *q.v.*

biophysics The application of physics to the study of living organisms and systems.

biophyte A plant that feeds on other living organisms; a parasitic or predatory plant.

biopoiesis The origin of life, including the abiotic synthesis of macromolecular systems and the transformation (eobiogenesis) of these systems into the first living organisms (eobionts).

biorealm A group of similar biomes *q.v.*

bios Flora and fauna; biota.

biosequence A sequence of related soils that differ primarily as a consequence of differences in soil biota as soil forming factors; *cf.* catena.

bioseries The evolutionary sequence of changes in any heritable character.

bioserotype (bioser.) A biochemical variant of a serotype *q.v.*

bioseston Plankton, nekton and suspended organic particulate matter derived from living organisms, as the biological component of seston *q.v.*

biosome An ecologically controlled biostratigraphic unit; a sediment deposited under uniform biological conditions.

biospace Realized niche *q.v.*

biospecies 1: A biological species *q.v.* 2: A species defined primarily on biological characters; *cf.* morphospecies.

biospeleology The study of subterranean life; biospeology; **biospeleological**.

biospeology Biospeleology *q.v.*

biosphere The global ecosystem; that part of the Earth and atmosphere capable of supporting living organisms; ecosphere; ecumene; subdivided into biocycles, biochores and biotopes; also subdivided into the eubiosphere and parabiosphere.

biostasis The capacity of an organism to tolerate environmental fluctuations without exhibiting adaptive changes.

biostratigraphic unit A stratum or group of strata characterized by one or more distinctive fossils; biostratic unit.

biostratigraphic zone A biostratigraphic unit defined and characterized solely by the one or more fossil taxa from which it derives its name.

biostratigraphy The study and classification of rock strata based on their fossil content; stratigraphic palaeontology; **biostratigraphic**.

biostratinomy The study of the relationship between fossils and their environments.

biostrome An accumulation of fossils that are distinctly bedded but do not form a mound-like or reef-like structure (bioherm); a fossil bed having no pronounced topographical relief.

biosynthesis The production of organic compounds by living organisms.

biosystem Ecosystem *q.v.*

biosystematics The study of evolution and the classification of living organisms, based on biological information at population level such as genetic variability, hybridization, breeding strategies, competition and local adaptations; biosystematy; genonomy.

biota The total flora and fauna of a given area; bios.

biotaxis The directed reaction of a motile organism towards (positive) or away from (negative) a biological stimulus; **biotactic**.

biotelemetry Study of the behaviour and activity of organisms using remote detection and transmission equipment; radio tracking.

biothermal Pertaining to the interrelationship of temperature and living organisms.

biotic Pertaining to life or living organisms; caused or produced by, or comprising living organisms.

biotic adaptation Changes in form or physiology presumed to have arisen as a result of interaction with other organisms.

biotic area Any large area that can be delimited from adjacent areas on the basis of the flora and fauna.

biotic climax A plant community maintained at climax by some biotic factor such as grazing.

biotic community 1: A group of interacting species coexisting in a particular habitat; community. 2: In palaeontology, a group of species frequently found together.

biotic environment The biological environment comprising all organisms affecting the life of a particular individual, population or species; *cf.* abiotic environment.

biotic factors Environmental factors resulting from the activities of living organisms; *cf.* abiotic factors.

biotic potential The maximum potential rate of increase of an organism or population under ideal conditions; reproductive potential; *cf.* environmental resistance.

biotic province A continuous area containing one or more communities of living organisms which is distinct from adjacent areas.

biotic pyramid A diagrammatic representation of the successive trophic levels of a food chain in the form of a pyramid of numbers, with primary producers at the base and the sequence of consumers leading to the apex.

biotic succession An orderly sequence of communities replacing each other in time; ecological succession.

biotically sympatric Used of populations that occupy the same habitat within the same geographical area; biotic sympatry; *cf.* neighbouringly sympatric.

biotope 1: The smallest geographical unit of the biosphere or of a habitat that can be delimited by convenient boundaries and is characterized by its biota. 2: The location of a parasite within the host's body.

biotrophic Used of a parasite deriving nutrients

from the tissues of a living host; **biotroph**,
biotrophy.

biotrophic symbiosis A symbiosis in which one
symbiont uses its living partner as a food source;
cf. necrotrophic symbiosis.

bioturbation The mixing of a sediment by the
burrowing, feeding or other activity of living
organisms, forming a bioturbated sediment.

biotype 1: A group of genetically identical
individuals. 2: In microbiology, an infraspecific
category distinguishable by biochemical
characters; serotype; biovar.

biotype depletion The decrease in the number of
biotypes with increasing distance from the centre
of genetic diversity.

biovar Biotype *q.v.*

biozone A general term for any biostratigraphic
unit.

biparous 1: Producing two offspring in a single
brood; ditokous; **biparity**; *cf.* uniparous,
multiparous. 2: Having produced only one
previous brood.

bipolarity The occurrence of species or forms in
both northern and southern hemispheres but not
in the intervening equatorial belt; **bipolar**.

Birge's rule The generalization that the
thermocline is a transition zone in which the
temperature decreases at the rate of at least 1° C
per metre of depth.

birth rate The number of offspring produced per
head of population per unit time; natality.

bisect A profile of vegetation and soil showing the
natural distribution of growth above and below
ground; a cross section of vegetation and soil as
revealed by a trench extending down to the
deepest plant roots.

bisexual 1: Used of a population or generation
composed of functional males and females;
gonochoristic. 2: Used of an individual
possessing both male and female functional
reproductive organs; hermaphrodite; totipotent.
3: Used of a flower possessing both male and
female reproductive organs; perfect;
cf. unisexual.

bit The basic quantitative unit of information used
in digital computers, giving the solution to a
binary choice.

bitumenization Carbonification *q.v.*

bitypic Used of a taxon comprising only two
immediately subordinate principal taxa, as in a
family or genus comprising two genera or species
respectively; *cf.* monotypic, polytypic.

bivalent A pair of homologous chromosomes
synapsed during the first meiotic division; disome.

bivariate analysis The simultaneous analysis of
two variables.

biverbal Pertaining to a name comprising two
words that is not a binomen *q.v.* according to the
provisions of the Code.

bivoltine Having two generations or broods per
year; digoneutic; *cf.* voltine.

black box A conceptual unit whose function can
be evaluated without specifying the content.

black earth Chernozem *q.v.*

black mud A terrigenous marine sediment, black
in colour, rich in hydrogen sulphide and having a
high organic content; typical of poorly ventilated
anaerobic basins.

blackwater Water rich in humic acids and with
low nutrient concentrations.

blastaea A hypothetical adult ancestral animal
identical in form to the embryonic blastula (a
hollow sphere of cells occurring early in
ontogeny before invagination to form the
gastrula); *cf.* gastraea.

blastochorous Pertaining to a plant that is
dispersed by means of offshoots; **blastochore**.

blastogenesis Asexual reproduction by budding or
gemmation.

blending inheritance Inheritance in which parental
traits appear to blend in the offspring, with no
apparent segregation in subsequent generations;
cf. particulate inheritance.

bletting The change in consistency, without
putrefaction, of some fruits.

bloom An explosive increase in the density of
phytoplankton within an area; algal bloom.

blow off The loss of humus and loose top soil by
wind action.

blowout A small area of land from which all or
most of the soil has been removed by wind
erosion; an excavation in sandy ground produced
by wind action.

blue mud A terrigenous marine sediment, blue in
colour, formed under mildly reducing conditions,
similar to a green mud.

Blytt-Sernander climatic classification A
classification of late glacial and postglacial
climate based on stratigraphic evidence from
plant remains; comprising the Arctic, Preboreal,
Boreal, Atlantic, Subboreal and Subatlantic
periods.

Bodenvagen An omnicole; a species with no
preferred substratum.

bog An area of wet peaty substrate rich in organic
debris but low in mineral nutrients, with a
vegetation of ericaceous shrubs, sedges and
mosses.

bog lake A dystrophic *q.v.* lake.

Bog soil An intrazonal soil with a muck or peat
surface layer and an underlying peat horizon
over an impervious grey-blue clay, formed in

boggy or swampy conditions; typical of glaciated basins.

bolochorous Having propagules dispersed by propulsive mechanisms; **bolochore, bolochory**.

bonding The formation of a close relationship between two or more individuals; **bond**.

bone bed A stratum rich in fragments of fossil vertebrates.

bonitation A measure of the ecological success of a species, usually in terms of numerical abundance.

bora wind A cold wind descending from elevated snow-covered plateaux or mountains onto lowland or coastal areas.

bore A tidal flood with an abrupt front, characteristic of shallow estuaries with a high tidal range.

boreal Pertaining to cool or cold temperate regions of the northern hemisphere, the northern coniferous zone and taiga; *cf*. antiboreal.

Boreal kingdom One of the six major phytogeographical areas characterized by floristic composition; comprising Atlantic North American, Arctic and Subarctic, Euro-Siberian, Macaronesian, Mediterranean, Pacific North American, Sino–Japanese and Western and Central Asiatic regions; see Appendix 4.

Boreal period A period of the Blytt-Sernander classification (*ca*. 9000–7500 years B.P.), characterized by pine and hazel vegetation and by relatively warm and dry conditions.

botany The study of plants; phytology; **botanical**.

botryology The science of organizing items or concepts into groups or clusters.

bottleneck A sudden decrease in population density with a corresponding reduction of total genetic variability.

bottleneck effect A conceptual occurrence of genetic drift in populations reduced in size through fluctuations in abundance.

bottom water The dense, deep oceanic water mass formed by the cooling and sinking of surface waters at high latitudes.

boulder A large sediment particle greater than 256 mm in diameter and too big to be handled easily; see Appendix 11.

boundary layer The layer of fluid adjacent to a physical boundary in which the fluid motion is significantly affected by the boundary and has a mean velocity less than the free stream value.

B.P. Before present.

brachiation Progression in trees by swinging from branch to branch using the fore-limbs; **brachiating**.

brachy- Prefix meaning short.

brachychimous Pertaining to short winters; *cf*. brachytherous.

brachymeiosis A simplified form of meiosis completed in a single division by suppression of the second division.

brachytherous Pertaining to short summers; *cf*. brachychimous.

brachytheroxerochimous Pertaining to short summers with dry winters.

brachyxerochimous Pertaining to short dry winters.

bracketed key A dichotomous key in which contrasting parts of a couplet are numbered and presented together, without intervening couplets, although the brackets joining each couplet (from which the key derives its name) are now omitted; bracket key; parallel key; *cf*. dichotomous key, indented key.

brackets Square brackets, a pair of printed marks placed around a word or words; used to enclose a superspecies-group name, to indicate a misidentification in a synonymy, and to enclose the name of the author of any nomenclatural act that was originally published anonymously but for which the author later became known; *cf*. parentheses.

brackish Pertaining to water of salinity intermediate between fresh water and sea water; brackish water may be classified according to salinity as mesohaline, oligohaline or polyhaline.

brady- Prefix meaning slow.

bradyauxesis Heterauxesis *q.v.* in which *a* from the equation $y = bx^a$ is less than unity, so that a given structure is relatively smaller in large individuals than in small ones.

bradygenesis Retarded phylogenetic development; bradytely; **bradygenetic**; *cf*. tachygenesis.

bradytelic Evolving at a relatively slow rate; bradygenetic; **bradytely**; *cf*. horotelic, tachytelic.

branch 1: One of the four major animal groups in the early Cuvierian classification. 2: A lineage. 3: A line on a phylogenetic tree *q.v.* or dendrogram *q.v.*

branch swapping algorithm A step-wise procedure for removal and relocation of branches of a phylogenetic tree, with a view to generating alternative trees; *cf*. algorithm.

branchicolous Living on the gills of fishes or other aquatic organisms; **branchicole**.

Brandenburg glaciation A subdivision of the Weichselian glaciation; see Appendix 2.

Braun-Blanquet cover scale A scale for estimating cover of a plant species, comprising six categories; +(under 1%), 1 (1–5%), 2 (6–25%), 3 (26–50%), 4 (51–75%), and 5 (76–100%).

Brazil current A weak, warm surface current that

flows south off the coast of Brazil and forms the westerly limb of the South Atlantic Gyre; see Appendix 6.

Brazilian subregion A subdivision of the Neotropical region; Tropical South American subregion; see Appendix 3.

breakthrough A genotype, homozygous for a given recessive lethal gene, that survives the lethal crisis; escaper.

breed 1: To reproduce. 2: To propagate organisms under controlled conditions. 3: A group of organisms related by descent; an artificial mating group having a common ancestor; used especially in genetic studies of domesticated species; **breeding**.

breeding size The number of individuals in a population that are actually involved in reproduction during a given generation; breeding population size.

breeding system The mode, pattern and extent to which individuals interbreed with others from the same or different taxa.

breeding true Producing offspring phenotypically identical to the parents.

brephic stage A pre-adult stage; a larval stage; **brephnic**.

Brillouin index An index of the specific diversity of a system derived from information theory; the divesity index

$$D = 1/N \times \log_2 N!/(n_1!n_2! \ldots n_s!),$$

where N is the total number of individuals observed in a sample and $n_1, n_2, \ldots n_s$ is the number of individuals of the species 1, 2 . . . s.

British system fps system *q.v.*

British Thermal unit The fps unit of heat defined as the quantity of heat required to raise the temperature of 1 pound of water by 1° F; equal to 1.055×10^3 joules; see Appendix 13.

'broken-stick' distribution A distribution of numbers of individuals between species in a community, so that the numbers in any one of n species tend to correspond to the lengths of a line broken at random into n segments by $n - 1$ points designated at random as dividing points along the line; the distribution expected is that the rth rarest species S_r in a population of S_s species and N_s individuals has an abundance given by $N_r = N_s/S_s \, x_{i=1}^r \, 1/S_s - i + 1$; Macarthur's broken-stick model.

Bronze Age An archaeological period dating from about 5 thousand years B.P. in Europe; a period of human history characterized by the use of bronze tools.

brood 1: The offspring of a single birth or clutch of eggs; any group of young animals that are being cared for together by an adult. 2: To incubate eggs; **brooder, broodiness**.

brood parasitism The use of a host species to brood the young of another species (the parasite).

broticolous Living in close proximity to man, in dwelling houses or other buildings; **broticole**.

brotion An ecological succession resulting from the activities of man.

brotochorous Having propagules dispersed by the agency of man; **brotochore, brotochory**.

brown clay A pelagic sediment comprising an accumulation of volcanic and aeolian dust with less than 30% biogenic material; red clay.

Brown Earth Brown Forest soil *q.v.*

Brown Forest soil An intrazonal soil with a mull horizon but no pronounced illuvial layer, derived from calcium rich parent material under deciduous forest in temperate conditions; typically with negligible litter, dark brown friable A-horizon and lighter coloured B-horizon; Brown Earth.

Brown podzolic soil A zonal soil similar to Podzol soil *q.v.* but lacking the highly leached intermediate horizon; typically with moderate layer of litter and duff, dark greyish brown A-horizon and a granular light brown acidic B-horizon.

Brown soil A zonal soil with dark brown to greyish brown, slightly alkaline A-horizon and lighter coloured B-horizon over calcareous material, formed in temperate to cool arid climates.

brown water Dystrophic *q.v.* water.

Brownian movement The continuous random motion of microscopic particles immersed in a fluid, resulting from their bombardment by molecules of the fluid.

browse 1: To feed on parts of plants; **browser**. 2: The edible plant material within reach of browsing animals.

browse line A line marking the height to which browsing animals have been feeding.

Bruckner's cycle A climatic cycle of about 34.8 years' duration (ranging from 25 to 50 years), comprising an alternation between a warm dry period and a cold damp period.

brunification The appearance of a uniform brown coloration in a soil due to the formation of the mineral goethite from iron released by oxidative weathering.

Brunizem Prairie soil *q.v.*

bryochore That area of the Earth's surface covered by tundra.

bryocolous Living on or in moss; **bryocole**.

bryology The study of mosses and liverworts; hepaticology; muscology; **bryological**.

bryophilous Thriving in habitats rich in mosses and liverworts; **bryophile, bryophily**.

budding 1: A type of asexual reproduction in which a new individual develops as a direct growth from the body of the parent, and may subsequently become detached. 2: The division of insect colonies; colony fission.

buffer 1: A substance which stabilizes the pH of a solution against the addition of acidic or alkaline material. 2: To protect a system from change by external factors; **buffering**. 3: Anything that reduces an impact.

buffer species A plant or animal acting as an alternative food supply for another organism and thereby buffering the effect of the predator on its normal prey.

buffered populations Populations interacting in such a way as to maintain the density of each within certain limits.

buffering The amelioration of environmental conditions by vegetation or topographic features; **buffer**.

buffering gene Polygene *q.v.*

bulbifery Asexual reproduction by the formation of bulbils in the axils of leaves, or in the floral branches.

Bulletin of Zoological Nomenclature The official organ of the International Commission on Zoological Nomenclature.

bunt order A dominance hierarchy *q.v.* established by butting.

burden The total number or biomass of parasites infecting a given host individual.

bushel A unit of volume equal to 8 gallons; 36.369 dm^3.

C

c. Abbreviation of the Latin *cum*, meaning with.

C³ plant A plant employing the pentose phosphate pathway of carbon dioxide assimilation during photosynthesis; most green plants belong to this category.

C⁴ plant A plant employing the dicarboxylic acid pathway for carbon dioxide assimilation during photosynthesis and capable of utilizing lower carbon dioxide concentrations than C³ plants.

ca. Abbreviation of the Latin *circa*, meaning about.

caballing Downward displacement of surface ocean water in regions where surface water masses converge.

cable A measure of length equal to 1/10 of a nautical mile (608 ft) in UK, or 120 fathoms (720 ft) in USA; sometimes taken as 100 fathoms (600 ft).

cacogenesis The condition of being unable to hybridize; kakogenesis.

cacogenic Dysgenic *q.v.*

cadavericolous Living on dead bodies; also used of organisms feeding on dead bodies or carrion; **cadavericole.**

caducous Used of structures shed or falling off early; evanescent; fugacious.

caenodynamism The replacement of complex functions by simpler ones; **caenodynamic.**

Caenogaea A zoogeographical region incorporating the Nearctic, Oriental and Palaearctic; Cainogaea; Kainogaea; *cf.* Eogaea.

caenogenetic 1: Of recent origin. 2: Used of transitory adaptations developed in the early ontogenetic stages of an organism; **caenogenesis.**

caenomorphism A change from a complex to a simpler form in living organisms; **caenomorphic.**

Caenophytic The period of geological time since the development of the angiosperms in the mid to late Cretaceous; Cainophytic; Cenophytic; Neophytic; *cf.* Aphytic, Archaeophytic, Eophytic, Mesophytic, Palaeophytic.

caenophyticum The plant life of the Cenozoic; Cainophyticum; *cf.* mesophyticum, palaeophyticum.

Caenozoic Cenozoic *q.v.*

caespiticolous Living in grassy turf or pastures; cespiticolous; **caespiticole.**

Cainogaea Caenogaea *q.v.*

Cainozoic Cenozoic *q.v.*

calcareous Rich in calcium salts; pertaining to limestone or chalk; growing on or having an affinity for chalky soil.

calcicole A plant growing in soils rich in calcium salts; calcipete; calciphyte; gypsophyte; **calcicolous;** *cf.* calcifuge.

calcification 1: The deposition of calcium salts in living tissue. 2: The replacement of organic material by calcium salts during fossilization.

calcifuge A plant which is intolerant of soils rich in calcium salts; basifuge; calciphobe; **calcifugous;** *cf.* calcicole.

calcimorphic Used of an intrazonal soil having an accumulation of calcium salts.

calcipete Calcicole *q.v.*

calcipetrile Pertaining to the plant community occurring on a basic rock substratum.

calciphilic Thriving in environments rich in calcium salts; boring into calcareous shells; **calciphile, calciphily.**

calciphobe Calcifuge *q.v.*

calciphyte A plant of soils rich in calcium salts; calcicole; calciphite.

calcivorous Used of plants living on limestone or in soils rich in calcium salts; **calcivore, calcivory.**

calcosaxicolous Inhabiting rocky limestone areas; **calcosaxicole.**

California current A cold surface ocean current that flows south along the coast of California, forming the easterly limb of the North Pacific Gyre; see Appendix 6.

Californian subregion A subdivision of the Nearctic region; Western subregion; see Appendix 3.

caliology The study of burrows, nests, hives, tubes and other domiciles constructed by animals.

calobiosis Symbiosis in social insects, in which one species lives in the nest of, and at the expense of, another.

calorie (cal.) A derived cgs unit of heat defined as the amount of heat required to raise the temperature of 1 gram of water by 1°C at a pressure of 1 atmosphere; equivalent to 4.1855 joules at 15°C; see Appendix 13.

caloritropism Thermotropism *q.v.*; **caloritropic.**

CAM Crassulacean acid metabolism; a carbon dioxide fixation pathway in some plants, especially succulents, which fix large amounts of carbon dioxide in the dark and store malate within their cells.

CAM plant A plant employing crassulacean acid metabolism, typically a succulent.

Cambrian The earliest geological period of the Palaeozoic era (*ca.* 570–504 million years B.P.); Primordial; see Appendix 1.

Camin-Sokal hypothesis A special case of the maximum homology hypothesis *q.v.* in which back mutations to prior character states are not permitted.

camnium An ecological succession caused by cultivation.

campo A habitat comprising grassy plains with scattered bush and small trees, characteristic of parts of South America.

camptotropism The tendency to return to the natural position if forcibly displaced from it; **camptotropic**.

Canadian subregion A subdivision of the Nearctic region; Subarctic subregion; see Appendix 3.

canalization The buffering property of developmental pathways that tends to produce a standard phenotype despite environmental fluctuations and underlying genetic variability; developmental homeostasis.

canalizing selection Selection for phenotypic characters, the development of which is largely unaffected by environmental fluctuations and underlying genetic variability.

Canaries current A cold surface ocean current that flows south in the eastern North Atlantic, forming the easterly limb of the North Atlantic Gyre; see Appendix 6.

candela The standard SI base unit of luminous intensity, defined as the perpendicular luminous intensity In of a 1/600 000 square metre area of a full radiator at the temperature of freezing platinum at a pressure of 101 325 newtons per square metre; see Appendix 13.

cannibalism Intraspecific predation.

canonical analysis A type of ordination *q.v.* in which group means are represented by points in multidimensional space and interpoint distances are equal to the square root of the corresponding Mahalanobis D^2.

canopy The uppermost continuous stratum of foliage in forest vegetation formed by the crowns of the trees; subdivided into two or three canopy levels in energy rich rain forests.

cantharophilous Pollinated by beetles; **cantharophile, cantharophily**.

canyon In oceanography, a deep, steep-sided submarine valley.

Cape Horn current The Antarctic Circumpolar current in the region of Cape Horn; see Appendix 6.

Cape region South African region; see Appendix 4.

capillary water The water held in soil pore spaces and readily available to plants and soil organisms.

capitalist A plant having stored food reserves.

capneic Thriving under conditions of high carbon dioxide concentration.

caprification The pollination of fig flowers by fig insects (caprifers).

carbamate A biodegradable organic herbicide.

carbon-14 A radioactive isotope of carbon containing 8 neutrons compared with 6 in carbon-12 and having a half-life of 5700 years; used in radiometric carbon dating.

carbon assimilation The incorporation of inorganic carbon from carbon dioxide into organic compounds by photosynthesis.

carbon cycle The biogeochemical cycle of carbon, involving fixation of inorganic carbon dioxide by photosynthesis to form organic complexes through to its ultimate return to the atmosphere by respiration and decomposition.

carbon dating A method of estimating the age of archaeological material and subrecent fossils by measuring levels of carbon-14.

carbon dioxide compensation point That concentration of carbon dioxide below which, in a nonlimiting light intensity, photosynthesis just compensates for respiration and the value of nett photosynthesis is zero.

carbonate compensation depth (CCD) The oceanic depth at which the rate of solution of carbonate equals the rate of input at that level, typically a depth of about 5000 m; *cf.* lysocline.

carbonate dissolution depth The oceanic depth below which the solubility of calcium carbonate increases to the point that dissolution of calcareous shells begins; typically a depth of about 4000 m; *cf.* lysocline.

carbonicolous Living on burnt or scorched substrates; **carbonicole**.

Carboniferous A geological period within the Palaeozoic (*ca.* 365–290 million years B.P.), divided into Pennsylvanian and Mississippian; see Appendix 1.

carbonification The formation of coal from plant material; bitumenization; coalification; incarbonization; incoalation.

carbonization An unusual type of fossilization in which organic material has been reduced to a carbon film residue.

carboxyphilic Thriving in carbon dioxide rich habitats; **carboxyphile, carboxyphily**.

carcinogenic Cancer producing.

carcinology The study of crustaceans; **carcinological**.

cardinal points The temperature minimum (cold limit), temperature optimum and temperature maximum (heat limit) for nett photosynthesis over a range of temperatures.

Caribbean current A warm surface ocean current that flows north-west in the Caribbean Sea, derived in part from the Atlantic South Equatorial current and giving rise to the Florida current through the Gulf of Mexico; see Appendix 6.

Caribbean region A subdivision of the Neotropical kingdom; see Appendix 4.

carnivorophyte A flesh eating plant; usually an insectivorous plant.

carnivorous Flesh eating; creophagous; sarcophagous; **carnivore, carnivory**.

carpelotaxis The arrangement of carpels in a flower or fruit.

carpogenous Growing on or in fruit.

carpohylile Pertaining to a dry forest community.

carpolochmium A dry thicket community.

carpophagous Feeding on fruit or seeds; **carpophage, carpophagy**.

carpostrate A plant that is dispersed by means of its fruit.

carpotropic Pertaining to an orientation movement of a fruit stalk after fertilization.

carr A mire containing scrub vegetation; fen woodland.

carrier 1: An organism carrying a disease or infectious agent without showing typical symptoms and which is capable of passing the infection to another individual; *cf.* vector. 2: An individual that passes on a character or trait to the next generation without necessarily exhibiting the trait itself.

carrion Dead putrefying flesh.

carrying capacity The maximum number of organisms that can be supported in a given area or habitat; usually denoted by K; the upper asymptote of the logistic equation *q.v.*; K value.

caryogamy Karyogamy *q.v.*

caryotype Karyotype *q.v.*

case A problem of zoological nomenclature referred to the ICZN *q.v.* for a decision; normally comprising an application published in the Bulletin of Zoological Nomenclature, followed by comments from other workers and finally voted upon by members of the Commission whose majority decision is published as an Opinion *q.v.*

cast A rock formed within a cavity, as in the replacement of a fossil by mineral infiltration; *cf.* mould.

caste Individuals within a colony belonging to a particular morphological type or age group, or that share a particular behaviour pattern and perform the same specialized function.

caste polyethism The division of labour roles within a society according to caste; *cf.* age polyethism.

casting An object that has been cast off or voided by an organism, such as a faecal pellet or worm casting.

castration The removal of, or interference with the function of, the testes (geld, emasculation) or ovaries (spey).

casual 1: A plant growing in a community with which it is not normally associated. 2: An alien species that survives for only a short period in the habitat to which it was introduced.

casual society A temporary group formed within a society, typically having a high rate of turnover of members; causal group; *cf.* demographic society.

cata- Prefix meaning down, against; kata-.

catabatic Used of winds blowing down a slope, may be warm (foehn wind) or cold (fall wind).

catabolism That part of metabolism involving the degradation of complex substances into less complex ones with the resultant liberation of energy; katabolism; **catabolic**; *cf.* anabolism.

catadromous Migrating from fresh water to sea water, as in the case of fishes moving into the sea to spawn; katadromous; **catadromy**; *cf.* anadromous, oceanodromous, potamodromous.

catagenesis Regressive evolution towards loss of independence from, and control over, the environment; katagenesis; *cf.* anagenesis, stasigenesis.

catalogue An index; a compilation of taxonomic literature within an alphabetical list of species.

catamenial Monthly; menstrual; **catamenia**.

cataplasia An evolutionary state characterized by decreasing vigour; **cataplasis**; *cf.* anaplasia, metaplasia.

catarobic Used of an aquatic habitat in which slow decomposition of organic matter is taking place but in which conditions have not yet become anaerobic.

catastrophism The doctrine that fossil faunas were the result of catastrophic changes which had periodically exterminated large numbers of species, so that past cataclysmic geological or climatic events have had a major impact on the course of evolution; convulsionism; *cf.* uniformitarianism.

catathermal Pertaining to a period of decreasing temperature; *cf.* anathermal.

catch crop A quick-growing crop which is cultivated between the rows of a main crop, between the main crops of an ordinary crop rotation or in place of a main crop which has failed.

catch-mark-recatch method Lincoln index *q.v.*

catchment A natural drainage basin which channels rainfall into a single outflow.

category A group or level within a hierarchy of classification.

catena 1: A connected series of related events or objects. 2: A series of soils of similar age and derived from similar parent material, found under similar climatic conditions but showing different properties due to variations in soil forming factors; *cf.* biosequence, chronosequence, climosequence, clinosequence, thermosequence, toposequence.

catenary arc The curve assumed in equilibrium by a flexible cord suspended at each end.

Cathaysia A continental landmass comprising most of China and South East Asia, formed from Eurasia *q.v.* after the break-up of Pangaea; *cf.* Angara, Euramerica.

cation A positively charged ion; *cf.* anion.

cation exchange Exchange of positively charged ions held by the soil absorbing complex with other cations.

cation-exchange capacity A measure of the total exchangeable cations that can be held by a soil; expressed as milliequivalents per 100 grams of soil at pH 7.

caulicolous Growing on plant stems; **caulicole**.

caulocarpous Used of a plant producing fruit in each of two or more successive years; polytokous.

causal agent The organism responsible for a given disease; aetiologic agent; causal organism; causative agent.

cavernarius Living in caves; cavernicolous.

cavernicolous Living in subterranean caves or passages; cavernarius; **cavernicole**.

CCD Carbonate compensation depth *q.v.*

cedis incertae Incertae cedis *q.v.*

ceiling leaf area index The maximum leaf area index *q.v.* at which the rate of death of leaves at the base of the canopy due to low light intensity is equal to the rate of appearance of new leaves.

cell division The cleavage of a cell into two parts, typically involving mitotic division of the nucleus.

cell theory That all living organisms are composed of cells that divide during growth and reproduction.

cell-mediated immunity Immunity in which the antigen is bound to receptor sites on the surface of sensitized T-lymphocytes produced in response to a prior immunizing experience with that antigen, and in which there is no antibody involvement.

cellular Composed of cells; pertaining to cells; *cf.* acellular.

Celsius scale (°C) A scale of temperature with the melting point of ice as zero and boiling point of water as 100 degrees at normal atmospheric pressure; originally proposed in reverse; equivalent to Centigrade scale.

cenobiology Biocoenology *q.v.*

cenogenous Coenogenous *q.v.*

Cenophytic Caenophytic *q.v.*

cenospecies Coenospecies *q.v.*

Cenozoic A geological era (*ca* 65–0 million years B.P.) comprising the Quaternary and Tertiary periods; the age of mammals; Caenozoic; Cainozoic; Kainozoic; see Appendix 1.

cenozone Coenozone *q.v.*

cenozoology The study of extant animals.

centi- (c) Prefix meaning hundredth; used to denote unit $\times 10^{-2}$

Centigrade scale (°C) A scale of temperature that takes the melting point of ice as zero and the boiling point of water as 100 degrees; now abandoned in favour of the Celsius scale.

centimetre A cgs unit of length equal to 10^{-2} metres.

Central American bridge The land-bridge *q.v.* between North and South America that currently exists and was probably also present during the Mesozoic although the continents were separated for much of the Tertiary.

Central Australian region A subdivision of the Australian kingdom; see Appendix 4.

Central and East African subregion East African subregion of the Ethiopian region; see Appendix 3.

Central subregion Rocky Mountain subregion of the Nearctic region; see Appendix 3.

centre of dispersal 1: A centre of origin *q.v.* 2: An area from which a population spreads outwards.

centre of diversification The region supporting the greatest number of cultivated varieties of a plant.

centre of genetic diversity An area having a marked concentration of congeneric species.

centre of origin The area in which a given taxon originated and from which it has subsequently spread; centre of dispersal.

centric Used of a chromosome having a centromere; *cf.* acentric.

centrifugal Turning or tending away from the centre or axis; *cf.* centripetal.

centrifugal selection Disruptive selection *q.v.*; selection for the phenotypic extremes.

centripetal Turning or tending towards the centre or axis; *cf.* centrifugal.

centripetal selection Stabilizing selection *q.v.*; selection for the mean or intermediate phenotype.

centroid The point in character space the coordinates of which are the mean values of each character over a given cluster of OTUs; the centre of gravity of a cluster of OTUs.

centrotype In numerical taxonomy, that OTU closest to the geometric centre of the cluster of OTUs to which it belongs.

cephalic index A relative measure of the human skull giving the width as a percentage of the length along the antero-posterior axis.

cephalodiate Used of a lichen containing two different species of symbiotic algae.

ceremony A complex pattern of display behaviour used to promote and maintain social bonds in some groups.

cerophagous Wax-eating; **cerophage, cerophagy.**

Cerozem Sierozem *q.v.*

certation Differential growth of pollen grains of differing genetic constitution (from different plants) when placed on the same stigma.

cespiticolous Caespiticolous *q.v.*; **cespiticole.**

cet. Abbreviation of the Latin *cetera*, meaning the rest, the remainder.

cetology The study of cetaceans; **cetological.**

Ceylonese and South Indian subregion Ceylonese subregion of the Oriental region; see Appendix 3.

Ceylonese subregion A subdivision of the Oriental region; Ceylonese and South Indian subregion; see Appendix 3.

cf. Abbreviation of the Latin *confer*, meaning compare (with).

cgs system A system of scientific units based on the centimetre-gram-second; see Appendix 13.

chaetoplankton Planktonic organisms having spiny outgrowths to reduce their rate of sinking.

chaetotaxy The pattern or distribution of bristles, chaetae, setae, spines etc., used as taxonomic characters.

chalicad A gravel-slide plant.

chalicium A plant community of gravel slide habitats; **chalicodium.**

chalicophilous Thriving on gravel slides; chalicodophilous; **chalicodophile, chalicodophily, chalicophile, chalicophily.**

chalicophyte A plant inhabiting gravel slides; **chalicad.**

challenge Exposure to an antigen after specific immunization against that antigen.

challenge display A high intensity aggressive display performed by a male to a conspecific male.

Challenger unit A measure of abundance of deep sea benthic animals equal to 6667 animals (the total number of benthic organisms from depths greater than 900 m taken during the three years of the Challenger expedition).

chamaephyte A perennial plant having renewal buds at or just above ground level (up to 250 mm); usually low growing woody or herbaceous plants common in dry or cold climates and having buds on aerial branches near the ground; sometimes divided into active chamaephyte, passive chamaephyte and suffructicose chamaephyte; *cf.* Raunkiaerian life forms.

change, mandatory Mandatory change *q.v.*

Chaparral biome A mild temperate region with a Mediterranean climate of cool moist winters and long dry summers with a vegetation of varied sclerophyllous evergreen shrubs and a diverse fauna; see Appendix 5.

character 1: Any feature or attribute of an organism which forms the basis for comparison; *cf.* characteristic. 2: Any feature or trait transmitted from parent to offspring.

character coding The practice of recording character states within an assemblage by labelling with an unambiguous set of symbols (a characterizing code).

character convergence A process whereby two recently evolved species interact in such a way that either or both converge towards the other with respect to one or more traits; *cf.* character displacement.

character cost A relative measure of the practical difficulty in assessing particular characters or character states.

character couplet A pair of contrasting statements describing one or more characters; as in the couplets of a dichotomous key.

character displacement The tendency for enhanced character divergence in the sympatric populations of two species that are partly sympatric and partly allopatric in their distributions, owing to the selective effects of competition; character divergence; *cf.* character convergence.

character gradient Cline *q.v.*

character hierarchy A system of sequentially dependent characters *q.v.*

character index A numerical value which quantifies the degree of difference between two taxa.

character nesting The arrangement of characters and their dependent characters *q.v.* into a hierarchy.

character progression The sequential transformation of a character or trait over the geographical distribution of a population or species.

character release An increase in the variation of

certain phenotypic characters associated with the ecological release *q.v.* of a species.

character space The hypothetical space occupied by a taxon, having as many dimensions as characters used to define the taxon.

character state Any of the range of values, conditions or expressions of a particular taxonomic character.

character state tree The linear or branching sequence of character states in a transformation series.

character weighting The practice of assigning different importance values to taxonomic characters, usually giving preference to unambiguous and easily observed characters.

characteristic A distinguishing feature; often used loosely as a synonym of character *q.v.* although, more precisely, it refers to the distinctive state or expression of that character.

characteristic fossil A fossil species or genus diagnostic of a stratigraphic unit or time unit; index fossil; diagnostic fossil.

characteristic species A plant species that is almost always found in a particular association *q.v.* and is used in the delimitation of that community; guide species; index species; indicator species.

characterizing code Character coding *q.v.*

character-set minimization Selection of the smallest subset of available characters that is adequate for a particular purpose; test-set reduction.

chart datum The permanently established surface from which depth soundings or tide levels are referenced; formerly referred to low water *q.v.* but more recently to lowest astronomical tide *q.v.*; datum; datum plane; tidal datum.

chasmochomophyte A plant growing on detritus in rock fissures and crevices; chasmophyte.

chasmocleistogamic Used of plants bearing some flowers that open and are pollinated after opening and others that remain unopened and within which self-pollination takes place; chasmodichogamic; **chasmocleistogamy**.

chasmodichogamic Used of plants bearing both cleistogamic and chasmogamic flowers in the same inflorescence; chasmocleistogamic; **chasmodichogamy**.

chasmogamy Pollination occurring after the opening of a flower; **chasmogamic, chasmogamous**; *cf.* cleistogamy.

chasmophilous Thriving in rock crevices and fissures; **chasmophile, chasmophily**.

chasmophyte A plant growing on rocks, rooted in detritus and debris in crevices and fissures; chasmochomophyte.

checklist A list of species arranged within a simple classification for convenience of reference.

cheironym An unpublished scientific name; manuscript name; chironym.

cheiropterophilous Chiropterophilous *q.v.*

cheirotype A type specimen of a species designated by a manuscript name (cheironym); chirotype; quirotype.

chelation The combination of a substance with a metal ion, having the effect of keeping the element in solution and rendering it non-toxic.

chemical weathering Chemical changes of rock minerals during the soil forming process.

chemicofossil Molecular fossil *q.v.*

chemiotaxis Chemotaxis *q.v.*

chemoautotrophic Obtaining metabolic energy by the oxidation of inorganic substrates, such as sulphur, nitrogen or iron; as exhibited by some microorganisms; chemotrophic; **chemoautotroph**; *cf.* photoautotrophic.

chemocline The boundary zone in a lake between the deep stagnant water (monimolimnion) and the overlying region of freely circulating water (mixolimnion).

chemoheterotrophic Chemoorganotrophic *q.v.*

chemokinesis A change of linear or angular velocity in response to a chemical stimulus; **chemokinetic**.

chemolithotrophic Used of organisms that obtain energy from oxidation/reduction reactions and use inorganic electron donors; **chemolithotroph**; *cf.* chemoorganotrophic.

chemonasty A response to a diffuse chemical stimulus; a change in the structure or position of an organ in response to a diffuse chemical stimulus; **chemonastic**.

chemoorganotrophic Used of organisms that obtain energy from oxidation/reduction reactions and use organic electron donors; chemoheterotrophic; **chemoorganotroph**; *cf.* chemolithotrophic.

chemostat A steady state laboratory culture apparatus.

chemosynthesis The synthesis of organic compounds using chemical energy derived from the oxidation of simple inorganic substrates; **chemosynthetic**; *cf.* photosynthesis.

chemotaxis The directed reaction of a motile organism towards (positive) or away from (negative) a chemical stimulus; chemiotaxis; **chemotactic**.

chemotaxonomy The application of biochemical techniques and the use of biochemical characters in taxonomy.

chemotrophic Chemoautotrophic *q.v.*

chemotropism An orientation response to a chemical stimulus; **chemotropic**.

chemozoophobous Used of plants which protect themselves from herbivorous animals by the production of noxious chemical substances (allelochemics); **chemozoophobe**.

cheradium A sandbar community.

cheradophilous Thriving on wet sandbars; **cheradophile, cheradophily**.

cheradophyte A plant living on a sandbar.

Chernozem A zonal soil with a thick black A-horizon and calcium carbonate rich surface layer, over a lighter intermediate B-horizon and a calcareous C-horizon, formed in cool subhumid climates under tall prairie grasses; black earth; chernozyom; czernozem; tchornozem; tschernosem; tschernosion.

chernozyom Chernozem *q.v.*

chersium A plant community of dry wasteland.

chersophilous Thriving in dry wasteland habitats; **chersophile, chersophily**.

chersophyte A plant growing on dry wasteland or in poor shallow soil.

Chestnut soil A zonal soil with a thick dark brown, slightly alkaline A-horizon over a lighter coloured B-horizon and a calcareous lower horizon; formed in temperate to cool, subhumid to semi-arid climates beneath mixed tall and short grasses.

chianophilous Thriving under prolonged snow cover; chionophilous; **chianophile, chianophily**; *cf.* chianophobous.

chianophobous Intolerant of prolonged snow cover; chionophobous; **chianophobe, chianophoby**; *cf.* chianophilous.

chiasma The X-shaped configuration indicating the point at which paired chromatids have exchanged segments by crossing over *q.v.*; **chiasmata**.

chiasma interference The more frequent (negative) or less frequent (positive) occurrence of chiasmata in chromosome pairs during meiosis than would be expected by chance alone.

Chilean subregion A subdivision of the Neotropical region; South Temperate American subregion; see Appendix 3.

chimaera An organism having tissues of two or more genetic types, formed by mutation, abnormal chromosome segregation, or by artificial grafting; usually used of a plant, rarely of an animal; chimera; mosaic.

chimnochlorous Pertaining to plants with thin herbaceous leaves that persist over winter.

chimonophilous Thriving during the winter; used of plants that exhibit maximum development during the winter; **chimnophilous, chimnophile, chimnophily, chimonophile, chimonophily**.

chimopelagic Pertaining to deep water marine organisms which occur in surface waters only during the winter; chimnopelagic.

chimous Pertaining to winter or to cold climatic conditions.

chionium A plant community associated with snow.

chionophilous Thriving in snow covered habitats; chianophilous; **chionophile, chionophily**; *cf.* chionophobous.

chionophobous Intolerant of snow covered habitats; chianophobous; **chionophobe, chionophoby**; *cf.* chionophilous.

chionophyte A plant living in or under snow.

chironym Cheironym *q.v.*

chiropterophilous Pollinated by bats; cheiropterophilous; **chiropterophile, chiropterophily**.

chirotype Cheirotype *q.v.*

chi-squared statistic (χ^2) A statistic calculated as the sum of a set of terms, each term being the quotient of the squared difference between an observed frequency and the corresponding expected frequency divided by the expected frequency.

chi-squared test A statistical method, using the chi-squared statistic, for testing the degree to which observed frequencies or values differ from the frequencies or values expected from the specific hypothesis being tested; used as a goodness-of-fit test when data are arranged in a classification having a single variable, or as a test of independence or association in a classification having two or more variables.

chitinolytic Capable of degrading chitin.

Chitty hypothesis Polymorphic behaviour hypothesis *q.v.*

chledad A plant living on rubbish heaps; chledophyte.

chledium A plant community on waste ground or rubbish heaps.

chledophilous Thriving in wasteland habitats and rubbish heaps; chomophilous; **chledophile, chledophily**.

chledophyte A plant growing on rubbish heaps; chledad.

chlorinated hydrocarbon A synthetic contact insecticide, such as DDT, which is relatively resistant to biodegradation and persists in the food web, often accumulating in non-target organisms.

chlorinity A measure of chloride and bromide ion concentration in sea water (in grams per kilogram, or parts per thousand); used in estimating salinity. (Salinity = 1.80655 Chlorinity.)

chlorosity The chlorine content of 1 litre of sea water.

chomophilous Thriving in wastelands and rubbish heaps; chledophilous; **chomophile, chomophily**.

chomophyte A plant growing in detritus and other litter.

-chore A neutral suffix or prefix meaning to spread; used to mean region or agent of dispersal; **-chorous, -choric**.

C-horizon The lowest layer of a soil profile above the bedrock (R-horizon), comprising mineral substrate with little or no structure; containing gleyed layers or layers of accumulation of calcium carbonate or calcium sulphate in some soils; parent material; see Appendix 12.

chorography The practice of mapping or describing geographical regions, or the physical configuration of such a region.

chorology The description and delimitation of the distributional ranges of taxa; biogeography; *cf.* faunistics.

choronomic Used of the influence of local geography or environment on an organism.

chorotobiontic Graminicolous *q.v.*

chorotype A fossil specimen from the same stratum as the type of the species, but from a neighbouring locality; local type.

-chorous Suffix meaning dispersed by the agency of; aerochorous, androchorous, anemochorous, anthropochorous, autochorous, barachorous, blastochorous, bolochorous, brotichorous, clitochorous, crystallochorous, endozoochorous, entomochoric, gynochoric, hydrochoric, myrmecochorous, saurochorous, synzoochorous, zoochorous.

chresard That portion of the total soil water that is available to plants; growth water; *cf.* echard, holard.

chrom- Prefix meaning colour.

chromatic adaptation The ability of blue-green algae to modify their pigmentation to complement the incident light intensity and so maximize absorption.

chromatid One of the two distinct longitudinal filaments of a reduplicated chromosome apparent during mitosis and meiosis.

chromatotropism An orientation response to stimulation by light of particular spectral composition; **chromatotropic**.

chromomorphosis A change in form due to the influence of light of a particular spectral composition.

chromosomal chimaera A plant mosaic comprising cells of differing chromosomal composition.

chromosomal incompatibility The failure of interbreeding between members of sympatric populations due to differences in chromosome composition of the two populations.

chromosomal inheritance Inheritance of characters through the chromosomes of the nucleus; Mendelian inheritance; *cf.* cytoplasmic inheritance.

chromosomal polymorphism Polymorphism of chromosome types or structure, as in sex chromosomes and chromosome inversions; *cf.* polymorphism.

chromosomal sterility Hybrid sterility resulting from the lack of homology between parental chromosomes.

chromosome A deeply staining nuclear body composed largely of DNA and protein, and comprising a linear sequence of genes; karyomite.

chromosome aberration Any mutation of a chromosome which produces a change in the gene sequence.

chromosome complement The actual number of chromosomes in a nucleus.

chromosome deletion A chromosome mutation involving loss of genes.

chromosome insertion A chromosome mutation involving the gain of duplicate genes.

chromosome inversion A chromosome mutation involving alteration or reversal of a sequence of genes.

chromosome map A plan of an individual chromosome showing the gene sequence as an arrangement of nucleotide bases.

chromosome mapping Determination of the gene sequences on a chromosome.

chromosome number The number of chromosomes characteristic of a given organism; chromosome set.

chromosome pair The pair of homologous chromosomes that become intimately associated during meiosis and mitosis; chromosome pairing.

chromosome set The minimum viable chromosome complement in eukaryotes; basic number.

chromosome substitution The artificial replacement of one or more chromosomes by homologous or homoeologous chromosomes from another strain or closely related species.

chromosome theory of heredity The theory that chromosomes are the carriers of the genes and that their meiotic behaviour is the basis for Mendel's laws *q.v.*

chromosome translocation A mutation involving the transfer of a segment of one chromosome from one member of a homologous pair to the other.

chronic Mild; of long duration or frequent occurrence; *cf.* acute.

chronistic Pertaining to, or in relation to, time or a time scale.

chronistics The study of time relationships of objects or events.

chrono- Combining form meaning time.

chronocline A gradual change in a character or group of characters over an extended period of geological time; *cf.* cline.

chronofauna A geographically restricted animal community that maintains its basic character over a geologically significant period of time.

chronogenesis The time sequence of occurrence of organisms in stratified rock.

Chronology of the Quaternary Ice Age See Appendix 2.

chronometry The science of precise time measurement.

chronosequence A sequence of related soils that differ as a result of time as a soil forming factor; *cf.* catena.

chronospecies 1: A species which is represented in more than one geological time horizon. 2: The successive species replacing each other in a phyletic lineage which are given ancestor and descendant status according to the geological time sequence; palaeospecies.

chronostratigraphy The study of the age relationships of rock strata; **chronostratigraphic**.

chronotropism An orientation response due to age; used particularly with reference to the movement of leaves in plants; **chronotropic**.

chrymosymphily An amicable relationship between ants and lepidopterous larvae, based on the scent produced by the larvae; **chrymosymphile, chrymosymphilous**.

chtonophyte A terrestrial plant obtaining water from the soil through roots.

chylophyte A terrestrial plant rooted on a physically dry and hard substratum.

circa (*ca.*) Latin, meaning about, approximately.

circadian rhythm A biological rhythm having a periodicity of about 1 day length (24 h); diurnal rhythm.

circalittoral The lower subdivision of the marine sublittoral zone below the infralittoral zone dominated by photophilic algae; sometimes used for the depth zone between 100 and 200 m; circalitoral; see Appendix 7.

circalunadian rhythm A biological rhythm having a persistent periodicity of about 1 lunar day (24.8 h), usually bimodal.

circamonthly rhythm A biological rhythm having a periodicity of about 29.5 days.

circannian Circannual *q.v.*

circannual rhythm A biological rhythm having a periodicity of about 1 year (365.25 days).

circasemiannual rhythm A biological rhythm having 2 cycles within a period of approximately 1 year.

circle of vegetation The highest category in the classification of plant communities, including all those communities within a natural vegetational region.

circular overlap The overlap of the distributions of the two populations representing the ends of a circumglobal cline of intergrading populations in which adjacent populations interbreed but the sympatric terminal populations (ring species) do not interbreed.

circum- Prefix meaning around, about.

circumantiboreal Circumaustral *q.v.*

circumaustral Distributed around the high latitudes of the southern hemisphere; circumantiboreal; *cf.* circumboreal.

circumboreal Distributed around the high latitudes of the northern hemisphere; *cf.* circumaustral.

circumlocal Used of a local character species *q.v.* that has a broad range of distribution extending beyond the limits of the vegetation unit of which it is a characteristic species.

circumneutral species A species thriving in habitats of about neutral pH.

circumneutrophilous Thriving in conditions of about neutral pH; **circumneutrophile, circumneutrophily**.

circumnutation Growth movements of a plant about an axis.

circumpolar Distributed around the north or south polar regions.

circumregional Used of a regional character species *q.v.* that has a broad geographical range extending beyond the limits of the vegetation unit of which it is a characteristic species.

circumscissile dehiscence Spontaneous opening of a ripe fruit along a plane at a right angle to the longitudinal axis of the fruit.

circumscription The defined limits of a taxon as determined by an author; also the sum of the individuals within those limits; **circumscribe**.

circumtropical Occurring throughout the tropics; cosmotropical.

cis- Prefix meaning on the same side as.

cisalpine Situated on the south side of the Alps; *cf.* transalpine.

cistern epiphyte An epiphyte lacking roots and gathering water between the bases of its leaves.

cis-trans Two non-allelic mutations in a given chromosome pair can be arranged in two possible configurations, either both on one chromosome (cis) or one on each chromosome (trans).

cistron A functional gene; that part of a DNA

sequence that codes for a single cellular function, such as the synthesis of an enzyme.

cit. Abbreviation of Latin *citatus*, meaning to cite, cited.

city block distance Manhattan distance *q.v.*

clade A branch of a cladogram; a monophyletic group of taxa sharing a closer common ancestry with one another than with members of any other clade.

cladism Cladistic method *q.v.*

cladistic Pertaining to a clade or holophyletic group.

cladistic affinity The degree of recency of common ancestry; *cf.* patristic affinity.

cladistic distance The number of branching points between any two points in a phylogenetic tree; cladistic difference; *cf.* patristic distance, phenetic distance.

cladistic method A method of classification employing phylogenetic hypotheses as the basis for classification and using recency of common ancestry alone as the criterion for grouping taxa, rather than data on phenetic similarity; cladism; cladistics; *cf.* evolutionary method, omnispective method, phenetic method.

cladistic species concept The concept of a species as an entity delimited in time by successive speciation events, originating by speciation from an ancestral species and existing until it splits into two new daughter species; represented by the distance between two successive branching points on a cladogram.

clado- Prefix meaning branch, offshoot; klado-.

cladogenesis A branching type of evolutionary progress involving the splitting and subsequent divergence of populations; evolutionary diversification; dendritic evolution; **cladogenetic**; *cf.* phyletic evolution.

cladogram 1: A branching diagram representing the relationships between characters or character states, from which phylogenetic inferences can be made. 2: A diagrammatic representation of the evolutionary branchings of a lineage during divergence from its ancestral stock; species are represented by line segments, extant species denoted by upper case letters, stem species by lower case, branching points correspond to speciation events and are individually denoted by numbers; *cf.* phenogram, phylogenetic tree.

cladoptosis The periodic shedding of twigs.

clan A small dense group of plants growing within a restricted area and produced vegetatively or by seed from a single parent plant.

clandestine 1: Used of evolutionary changes not evident in the adult stage of an organism. 2: Used of adult characters of a descendant

species derived from embryonic characters of the ancestral species.

class 1: A rank within the hierarchy of taxonomic classification; the principal category between phylum or division and order; in botanical nomenclature a class has the ending *-opsida* or *-phyceae*, in zoology typically *-a*; see Appendix 10. 2: A ranked category in the classification of vegetation, comprising one or more orders and having the ending *-eae*. 3: In statistics, any category or group of like observations.

class boundary In statistics, the dividing line between adjacent classes of a frequency distribution.

class frequency In statistics, the number of observations in a particular class of a frequency distribution.

class limit In statistics, the upper or lower values of a class.

class mark In statistics, the midpoint of a class; the mean of the limits of a class.

classical In taxonomy, pertaining to a name that is derived from Latin or ancient Greek.

classification 1: A process of establishing, defining and ranking taxa within hierarchical series of groups; *cf.* artificial classification, natural classification. 2: A hierarchical series of groups or taxa; see Appendix 10.

clastic 1: Comprising fragmented pre-existing material; used of sediments that are formed of rock fragments or of clay minerals. 2: Producing or undergoing fragmentation.

clastizoic Used of rock comprising fragmented animal remains.

clastotype A part or fragment of a type specimen.

claustral colony founding A behaviour pattern exhibited by some social insects, in which the queen or royal pair isolate themselves in a closed cell and produce the first worker generation using their own body tissue as the source of nutrients.

clay Sediment particles between 0.002 and 0.004 mm in diameter and having colloidal properties; sometimes used for all sediment particles less than 0.004 mm in diameter; fine clay; see Appendix 11.

cleisto- Prefix meaning closed; kleisto-.

cleistogamy The condition of having flowers, typically small and inconspicuous, which remain unopened and within which self-pollination takes place; kleistogamy; **cleistogamic, cleistogamous**; *cf.* chasmogamy.

cleptobiosis Kleptobiosis *q.v.*

cleptoparasitism Kleptoparasitism *q.v.*; **cleptoparasite**.

climactic Pertaining to a climax *q.v.*

climagraph A chart plotting one climatic factor against another, commonly a diagram of mean monthly temperature plotted against mean monthly rainfall or humidity; hytherograph; climograph.

climate The long term average condition of weather in a given area; *cf.* weather.

climateric A critical phase or period in the life cycle of an organism.

climatic climax A more or less stable community in which the major factors affecting the vegetation are climatic, typical of zonal soils; regional climax; *cf.* climax.

climatic factor Any aspect of the prevailing weather conditions that influences the biota of an area.

climatic province A region characterized by a particular set of climatic conditions.

climatic regions The major divisions of the Earth's surface, polar, temperate, subtropical and tropical.

climatic rule Any generalization describing a trend in geographical variation of animals which can be correlated with a climatic gradient; ecogeographical rule; *cf.* Allen's law, Bergmann's rule, Gloger's rule, heart-weight rule, Hopkin's bioclimatic law, Rensch's laws.

climatic zone A latitudinal region characterized by a relatively homogeneous climate.

climatoid Pertaining to epiphytes with a narrow climatic and wide edaphic amplitude; *cf.* substratoid.

climatology The study of climate; *cf.* meteorology.

climatypes Two or more related populations that have differentiated mainly in response to differing climatic factors; climatic ecotypes.

climax A more or less stable biotic community which is in equilibrium with existing environmental conditions and which represents the terminal stage of an ecological succession; sometimes used as a synonym of formation *q.v.*; *cf.* climatic climax, disclimax, edaphic climax, physiographic climax, plagioclimax, polyclimax, postclimax, preclimax, proclimax, serclimax, subclimax, zootic climax.

climax area The region occupied by a climax community.

climax dominant The dominant species of a climax community.

climax species Any plant characteristic of a climax community.

climax unit A component of a climatic climax.

climograph Climagraph *q.v.*

climosequence A sequence of related soils that differ in certain properties as a result of the effect of climate as a soil forming factor.

clinal Having graded character states.

cline A character gradient; continuous variation in the expression of a character through a series of contiguous populations; **clinal**; *cf.* chronocline, coenocline, ecocline, ethocline, genocline, morphocline, nothocline, ontocline, phenocline, ratio cline, sociocline, topocline.

clinodeme A deme *q.v.* that forms part of a graded sequence of demes distributed over a given geographical area.

clinograde In limnology, the type of distribution in which concentration increases with depth, as in the oxygen concentration of most small productive lakes during summer stratification; *cf.* heterograde, orthograde.

clinolimnion That part of the hypolimnion of a lake in which the rate of heating falls exponentially with depth.

clinology The study of the decline of organisms or groups after acme *q.v.*

clinosequence A group of related soils that differ primarily as a result of the effect of the degree of slope as a factor influencing soil formation; *cf.* catena.

clinotaxis A directed reaction or orientation response of a motile organism to a gradient of stimulation; **clinotactic**.

clinotropism An orientation response to a gradient of stimulation; klinotropism; **clinotropic**.

clisere An ecological succession resulting from a major change in climate; *cf.* sere.

clisto- Cleisto- *q.v.*

clitochorous Dispersed by gravity; **clitochore, clitochory**.

clod A compacted mass of soil between about 5 and 250 mm in diameter.

clone An assemblage of organisms derived by asexual or vegetative multiplication from a single sexually derived individual; **clonal**; *cf.* genet, ortet, ramet.

clonodeme A deme *q.v.* composed of predominantly vegetatively reproducing individuals.

clonotype A specimen from a vegetatively propagated part of a plant from which the type specimen was also obtained; a specimen obtained from the original type by asexual reproduction.

close pollination Pollination between flowers on the same plant or within the same clone.

closed community An assemblage of plants which cover the entire ground area of the habitat in which they live, effectively preventing the establishment of other species; closed association.

closed population A population with no input of

genetic variation except that introduced by mutation.

closed society A society in which alien organisms are rarely admitted.

closest individual method A method of plotless sampling of vegetation in which a measurement is taken of the distance from the sampling point to the nearest individual, and the procedure repeated.

clumped distribution Contagious distribution *q.v.*

clusium An ecological succession on flooded soil; clysium.

cluster analysis A method that groups those variables within a set of variables that are highly correlated and that excludes from clusters those that are negatively correlated or uncorrelated; used in numerical taxonomy as a procedure for arranging OTUs into homogeneous clusters based on their mutual similarities.

clutch The number of eggs laid at any one time; clutch size.

clysium Clusium *q.v.*

clysotremic Pertaining to tide pools.

CO₂ compensation concentration The minimum atmospheric carbon dioxide concentration allowing nett photosynthesis.

co- Prefix meaning together, sharing, jointly, with.

coacervate Massed or heaped together.

coacervation The aggregation of organic molecules to produce particulate matter; coacervate.

coaction Interaction; reciprocal action between members of a group or community; coactors.

coadaptation 1: The evolution of mutually advantageous adaptations in two or more interactive species. 2: Selection by which harmoniously interacting genes accumulate in the gene pool of a population; internal balance; coadaption.

coaetaneous Of the same age; existing or appearing simultaneously; coetaneous.

coal ball A calcareous nodule containing fossilized plant remains, typically found in coal measures.

coalification Carbonification *q.v.*

coancestry, coefficient of A measure of the probability that two homologous genes, one from each of two individuals, are identical by descent.

coarse-grained environment The environment experienced by an organism with poor homeostatic mechanisms, low behavioural plasticity and mobility; experienced as sets of alternative, disconnected conditions; *cf.* fine-grained environment.

coarse-grained resources Resources distributed in such a manner that a consumer encounters

them in a proportion different from that in which they actually occur; *cf.* fine-grained resources.

coarse-grained species In the context of feeding strategies, used of coexisting related species that are specialized and morphologically dissimilar and hence utilize different food sources in the habitat; *cf.* fine-grained species.

cobble A sediment particle between 64 and 256 mm in diameter; large gravel; see Appendix 11.

coccolith ooze A pelagic sediment comprising at least 30% calcareous material predominantly in the form of coccolith remains.

Code The system of rules and recommendations regulating nomenclature, published in the most recent edition of the International Code of Zoological Nomenclature, the International Code of Botanical Nomenclature and the Bacteriological Code.

Code of ethics A list of recommendations on the propriety of certain taxonomic actions, set out in an appendix to the Code.

codemic Belonging to the same deme *q.v.*

codominance 1: The condition of having two equally dominant species in a plant community; semidominance; **codominant**. 2: The condition of having alleles that are neither completely dominant nor completely recessive so that each is expressed to some extent in the heterozygote; incomplete dominance; semidominance.

codon The unit of genetic coding; a triplet of adjacent nucleotides in DNA or in messenger RNA, which specify a particular amino acid; coding triplet.

coefficient Commonly used in statistics to denote constants as distinguished from variables; also used to denote a dimensionless description of a distribution or set of data.

coefficient of alienation Alienation, coefficient of *q.v.*

coefficient of assimilation efficiency Assimilation efficiency, coefficient of *q.v.*

coefficient of association Association, coefficient of *q.v.*

coefficient of coancestry Coancestry, coefficient of *q.v.*

coefficient of coincidence Coincidence, coefficient of *q.v.*

coefficient of community Community, coefficient of *q.v.*

coefficient of consanguinity Consanguinity, coefficient of *q.v.*

coefficient of difference A measure of the difference between two populations calculated as the difference between two means $(\bar{x}_1 - \bar{x}_2)$

divided by the sum of their standard deviations $(s_1 + s_2)$.

coefficient of ecological efficiency Ecological efficiency, coefficient of $q.v.$

coefficient of kinship Kinship, coefficient of $q.v.$

coefficient of parentage Parentage, coefficient of $q.v.$

coefficient of photosynthetic energy utilization efficiency Photosynthetic energy utilization efficiency, coefficient of $q.v.$

coefficient of relationship Relationship, coefficient of $q.v.$

coefficient of selection Selection, coefficient of $q.v.$

coefficient of variation A measure of the amount of variation within a population, calculated as the standard deviation (s) expressed as a percentage of the mean ($\bar{x}$); C. of V. $= s \times 100/\bar{x}$.

coelozoic Living in the lumen of an organ of the host individual.

coeno- Prefix meaning sharing, in common; ceno-; koino-.

coenobiology Biocoenology $q.v.$; cenobiology.

coenocline A sequence of communities distributed along an environmental gradient.

coenogamete A syncytium $q.v.$ that subsequently fragments to form gametes.

coenogamodeme A biosystematic unit comprising all the hologamodemes which are capable of exchanging genes to some extent, but not with freedom, and which is capable of hybridizing with other coenogamodemes producing sterile hybrid offspring; *cf.* deme.

coenogenesis Descent from a common ancestor; syngenesis; **coenogenetic**.

coenogenous Producing offspring oviparously at one season of the year and ovoviviparously at another; cenogenous.

coenomonoecious Trimonoecious $q.v.$

coenosis An assemblage of organisms having similar ecological preferences.

coenosite An organism that habitually shares food with another organism.

coenosium A plant community.

coenospecies A group of ecospecies capable of limited genetic exchange by forming fertile hybrids; cenospecies.

coenotrope Behaviour common to all members of a group or species.

coenotype A plant community of constant composition but without faithful $q.v.$ or differential species $q.v.$

coenozone A zone characterized by a particular fossil faunal or floral assemblage; cenozone.

coetaneous Coaetaneous $q.v.$

coevolution The interdependent evolution of two or more species having an obvious ecological

relationship, usually restricted to cases in which the interactions are beneficial to both species; also used for the evolutionary interaction between species having some degree of interdependence, such as a parasite and its host.

coexistence The occurrence of two or more species in the same area or habitat, usually used of potential competitors.

cohabitants Two or more organisms living together.

cohort 1: A group of individuals of the same age recruited into a population at the same time; age class. 2: A rank within a hierarchy of taxonomic classification; a category comprising a group of families; cohors; see Appendix 10.

coincidence, coefficient of In genetics, the proportion of observed number of double crossovers to the expected number.

coition Copulation; sexual intercourse; coitus.

cold blooded Poikilothermic $q.v.$

cold hardiness The tolerance of an organism to low temperatures.

cold monomictic Used of a lake with a summer overturn and in which the temperature of the water never rises above 4° C; *cf.* warm monomictic.

cold resistance The ability to survive exposure to temperatures below 0° C; hardiness.

cold temperate zone A latitudinal zone extending between 45° and 58° in both northern and southern hemispheres.

collaplankton Planktonic organisms rendered buoyant by gelatinous or mucous envelopes.

collateral type Any specimen used in the description of a species other than the primary type.

collective Used in botany to refer to names or epithets denoting hybrids or groups of hybrids.

collective group 1: In zoology, an assemblage of nominal species of which the generic positions are uncertain and which is accorded a collective name which is treated as a generic name in the meaning of the Code, but with no type species; collective. 2: A species aggregate $q.v.$

collective species Superspecies $q.v.$

colloid 1: A dispersion of one substance within another, having the properties of both a solution and a suspension; the system is composed of two phases, a continuous phase (the dispersion phase) and a discontinuous phase (the dispersed phase) which consists of discrete particles; colloidal system. 2: Very fine sediment particles less than 0.002 mm in diameter; see Appendix 11.

colluvial deposit Material transported to a site by gravity; as in rock deposits at the base of a scree slope; **colluvium**; *cf.* alluvial deposit, aeolian deposit.

colonist 1: Used of a plant introduced unintentionally as a result of the activities of man, typically a weed of cultivation, and now found only in more or less artificial habitats. 2: An organism, typically *r*-selected *q.v.*, which invades and colonizes a new habitat or territory.

colonization 1: The successful invasion of a new habitat by a species. 2: The occupation of bare soil by seedlings or sporelings.

colonization curve A graphical representation of the change in numbers of species coexisting in a given habitat over time.

colony 1: Used loosely to describe any group of organisms living together or in close proximity to each other; more precisely it refers to an integrated society in which members may be specialized subunits, such as a continuous modular society. 2: A group of organisms that have recently become established in a new area, or that are occupying a particular site.

colony fission Multiplication of a colony by the departure of one or more member groups to establish new colonies whilst leaving the parent colony still viable; *cf.* hesmosis, sociotomy, swarming.

colony odour The particular odour found on the bodies of social insects, providing a basis for colony recognition; colony odor.

-colous Suffix meaning to inhabit, inhabiting; agaricolous, algicolous, alsocolous, amanthicolous, ammocolous, amnicolous, ancocolous, aphidicolous, aquaticolous, arboricolous, arbusticolous, arenicolous, argillicolous, branchicolous, broticolous, bryocolous, cadavericolous, caespiticolous, calcicolous, calcosaxicolous, carbonicolous, caulicolous, cavernicolous, corticolous, crenicolous, culmicolous, dendrocolous, deserticolous, domicolous, epicolous, fimicolous, folicaulicolous, folicolous, forbicolous, fructicolous, fruticolous, fungicolous, gallicolous, gelicolous, geocolous, graminicolous, halicolous, herbicolous, humicolous, hydrocolous, hygrocolous, hylacolous, lapidicolous, latebricolous, leimocolous, lichenicolous, lignicolous, limicolous, limnicolous, lochmocolous, lucicolous, luticolous, merdicolous, monticolous, muscicolous, myrmecolous, nemoricolous, nervicolous, nidicolous, nivicolous, nomocolous, omnicolous, orgadocolous, ovariicolous, paludicolous, patocolous, pergelicolous, perhalicolous, petricolous, petrimadicolous, phreatocolous, piscicolous, planticolous,

pontohalicolous, poocolous, potamicolous, pratinicolous, psicolous, psilicolous, radicicolous, ramicolous, ripicolous, rubicolous, rupicolous, sabulicolous, sanguicolous, sepicolous, siccocolous, silicolous, silvicolous, sphagnicolous, spongicolous, stagnicolous, subgeocolous, sylvicolous, tegulicolous, telmicolous, termiticolous, terricolous, thamnocolous, thinicolous, tiphicolous, torrenticolous, tubicolous, umbraticolous, vaginicolous, viticolous, xerocolous.

comb. nov. Abbreviation of Latin *combinatio nova*, meaning new combination; used to indicate a new name resulting from a change in rank or position of an epithet from an earlier name (the basionym), as in the transference of a species name to a different genus producing a combination that differs from all previously published combinations; the addition of the nominate subgenus does not constitute a new combination.

combination A scientific name comprising a genus-group name followed by one (species) or more (subspecies) names peculiar to the taxon.

combined description A description of a new species assigned to a monotypic genus that utilizes the same character states to diagnose both the species and the genus; *descriptio generico-specifica*.

comm. Abbreviation of the Latin *communicavit*, meaning he (she or it) communicated.

commensalism Symbiosis in which one species derives benefit from a common food supply whilst the other species is not adversely affected. **commensal**.

commercial synonym A name used for a cultivar in place of the internationally accepted name in countries where the latter is commercially unacceptable.

commiscuum A biosystematic unit comprising one or more coenospecies *q.v.* capable of interbreeding and producing hybrids with partial fertility; *cf. comparium, convivium*.

Commission The International Commission on Zoological Nomenclature; ICZN.

common name Colloquial or vernacular name.

commonality, principle of That the character state having the widest distribution among the taxa comprising a higher taxon is considered to be the most primitive.

communal Pertaining to the cooperation between members of the same generation in nest building but not in care of the brood.

commune A society or group of conspecific organisms which have a social structure and consist of repeated members or modular units with a high level of coordination, integration and genotypic relatedness.

communicable Used of a disease which is readily transferred from one organism to another, usually used of diseases of humans and domestic animals.

communication Any action of one organism that modifies the behaviour pattern of another organism; the passage of information.

communicavit (comm.) Latin, meaning he (she or it) communicated.

community Any group of organisms belonging to a number of different species that co-occur in the same habitat or area and interact through trophic and spatial relationships; typically characterized by reference to one or more dominant species.

community, coefficient of The ratio of the number of species common to two communities to the total number of species occurring in each of the communities; index of similarity.

community concept That vegetation is composed of well defined, discrete integrated units which can be combined to form abstract classes reflective of natural phytosociological entities.

community matrix Guild matrix *q.v.*

community mosaic The arrangement of microstands within a given area.

community production (PP) The quantity of dry matter formed by the vegetation covering a given area; primary production, expressed in tons of organic dry matter per hectare of ground covered (t ha^{-1}) or in grams per square metre (g m^{-2}).

community production rate (PPR) The rate of production of organic dry matter by a community or stand of plants, calculated from the growth rate of individual plants (nett assimilation rate, NAR) and from a parameter reflecting the density of the stand (leaf area index, LAI), by the formula PPR = NAR × LAI.

companion species Species not restricted to any particular vegetation unit; indifferent species.

comparability of characters In taxonomy, the requirement for uniform descriptions of taxa for purposes of meaningful comparison; consistency of characters.

comparium A biosystematic unit comprising one or more coenospecies *q.v.* capable of producing fully fertile hybrids; *cf. commiscuum, convivium.*

compartment Any convenient division of an ecosystem that can be quantified for purposes of study.

compartmentalization The extent to which social groups are subdivided into discrete operational units.

compatible 1: Capable of cross fertilization.

2: Used of plants having the capacity for self-fertilization. 3: Able to coexist.

compensation depth 1: The depth at which primary production and respiratory utilization are equal so there is no nett production; compensation level; compensation point. 2: The ocean depth at which the rate of solution of calcium carbonate increases to the point that dissolution of calcareous shells begins; typically a depth of about 4000 m, below which calcareous sediments are not found.

compensation light intensity The light intensity at compensation depth.

compensation period The time required for a plant exposed to light to replace the carbohydrates utilized in respiration during the preceding dark phase.

competition The simultaneous demand by two or more organisms or species for an essential common resource that is actually or potentially in limited supply (exploitation competition), or the detrimental interaction between two or more organisms or species seeking a common resource that is not limiting (interference competition); **competitor**.

competitive associations Coexisting groups of species, the distributional boundaries of which are determined by competition for space or resources.

competitive exclusion The exclusion of one species by another when they compete for a common resource that is in limited supply.

competitive exclusion principle The principle that two species having identical ecological requirements cannot coexist indefinitely; complete competitors cannot coexist; Gause's hypothesis; Grinnell's axiom.

competitive release Expansion of the habitat range and food preferences due to a reduction in intensity of interspecific competition.

compilospecies A species or species complex that hybridizes with various related species and takes on some of the characteristics of each of these species in different areas.

complemental male A male which lives permanently attached to a female, often small and degenerate except for well developed reproductive organs.

complementary genes Two or more non-allelic factors collectively required to produce a single phenotypic trait; complementary factors; reciprocal genes.

complementary interaction Production by two or more interacting genes of a phenotypic effect different from that produced by any one of the genes independently.

complementary patterns The distribution patterns of two faunal assemblages which appear complementary, such that one group is dominant and expanding and the other declining or relict, suggesting a direct interaction.

complementary society A community of two or more plant species growing in close proximity which minimize competition, typically by temporal segregation of their main reproductive and vegetative phases.

complementation The cooperative interaction of mutant genes in double mutants, giving a phenotype closer to the wild type *q.v.* than any one of the mutant genes could produce alone; *cf.* intracistronic complementation, intercistronic complementation.

complementation map A diagrammatic representation of the complementation pattern of a series of mutants occupying a short chromosomal segment.

complete dominance The situation in which the trait produced by one allele is fully expressed (dominant) over the trait of another (recessive); *cf.* incomplete dominance.

complete flower A flower that has all four parts, pistil, stamen, petal and sepal; *cf.* incomplete flower.

complex In taxonomy, a term for any group or assemblage of related taxa, often used when the true separation or status of the individual units is uncertain, as in species-complex.

complex character A phenotypic character which is determined by more than one gene and, consequently, is not transmitted to offspring as a simple unit.

component analysis A type of multivariate analysis which represents a *p*-dimensional variation as due to a number of components so that as few components account for as much of the variation as possible; ideally the components are linear functions of the original variates.

component bar chart A bar chart in which the bars are subdivided into lengths proportional to the size of the component they represent.

composite species 1: A species comprising two or more subspecies; polytypic species. 2: In palaeontology, a species represented by a group of specimens that were obtained from two or more localities and that are not all of the same geological age.

composition A list of the species that comprise a community, or any other ecological unit.

compound In taxonomy, a word or name formed by the union of two or more words.

compound epithet An epithet formed from two or more words.

compound nest A nest formed by two or more species of social insects in which individuals from the separate colonies may intermingle but in which the broods remain separate; *cf.* mixed nest.

compression 1: A process of fossilization in which the organic remains have been flattened by overlying strata. 2: A decrease of niche width due to interspecific competition.

conceptive Capable of being fertilized.

conchology The study of shells; **conchological**.

conchometry The study of the morphometrics of shells.

concolorous Having uniform coloration; **concolour**.

concomitant immunity Premunition *q.v.*

concomitant mortality Death during exposure to a harmful substance or stimulus; *cf.* consequential mortality, delayed mortality.

concordance Characters which are the same in the OTUs under comparison.

concordant Used of twins or members of a group that share a given trait; *cf.* discordant.

concrescent Coalesced; growing together.

condensation An increase in the rate of development during phylogeny either by deletion of ontogenetic stages or by accelerated development, so that descendants pass through the ancestral part of ontogeny faster than did their ancestors.

condensed formula Formula *q.v.*; the name of an intergeneric hybrid formed by combining parts of the parent generic names preceded by a × or + sign.

condition factor A numerical index representing the relationship between body weight and length of a fish.

conditional In taxonomy, used with reference to the proposal of a scientific name which is made with definite reservations about its status.

conditioning The process of modifying the behaviour of an organism such that it responds to a given stimulus with a behaviour pattern normally associated with some other stimulus when the two stimuli are applied concurrently for a number of times.

conduction 1: The transfer of genetic material in bacterial conjugation. 2: The transfer of internal energy directly from one elementary particle to another.

confamilial Belonging to the same family.

confer (cf.) Latin, meaning compare (with).

confidence limits The upper and lower extremes of a confidence interval set to define the proportion of cases in which, on average, the assertion made is true; so that 95% confidence limits indicate that there is a 95% probability that the parameter being estimated lies within these limits; degree of confidence.

confirmatory characters Auxiliary characters *q.v.*

conformational population stability The phenomenon of a population retaining a constant response to a particular environmental state, but fluctuating under natural conditions as the environmental states fluctuate.

confused name *Nomen confusum q.v.*

confusing coloration Protective coloration that confuses the operator (predator) by presenting a different image according to the state of motion or rest of the signal sender (prey).

congeneric Belonging to the same genus; **congener**.

congenetic Having the same origin.

congenital Present at time of birth; geneogenous; used of a condition that has resulted from an embryonic aberration.

congested Crowded together; packed close together.

conglobate To roll up, as in some woodlice and millipedes; **conglobation**.

congregate To collect together or assemble into a group.

congruence The degree of correspondence between different classifications.

congruent Matching or coincident; **congruence**; *cf.* incongruent.

conifruticeta A forest dominated by coniferous shrubs.

conilignosa A forest dominated by trees and shrubs with needle-like foliage.

coniophilous 1: Thriving on substrates enriched by dust containing excreta. 2: Dust-loving; used of lichens which thrive when covered with a coat of dust; **coniophile, coniophily**.

conisilvae Coniferous forests.

conjugation 1: Union of gametes, nuclei, cells or individuals. 2: A type of sexual reproduction during which two cells unite but with only limited genetic exchange between the cells.

conlocal Used of a local character species *q.v.* that has a geographical range coinciding with that of the vegetation unit.

connascent Produced at the same birth.

connate water Interstitial water.

connectedness 1: The extent of the communication network (number and direction of links) between and within societies. 2: The number and frequency of interactions between the member species of a community.

conodrymium A community of evergreen plants.

conophorium A coniferous forest community.

conophorophilous Thriving in coniferous forests; **conophorophile, conophorophily**.

conregional Used of a regional character species *q.v.* that has a geographical range exactly coinciding with that of the vegetation unit.

cons. Abbreviation of the Latin *conservandum*, meaning to be conserved.

consanguineous Used of individuals sharing a recent common ancestor.

consanguinity 1: Relationship by descent from a common ancestor. 2: Relationship between igneous rocks derived from a common parent magma.

consanguinity, coefficient of Kinship, coefficient of *q.v.*

consecutive hermaphrodite An organism having functional male and female reproductive organs which mature at different times, either the male organs mature first (protandrous hermaphrodite) or the female organs (protogynous hermaphrodite); sequential hermaphrodite; consecutive sexuality; *cf.* synchronous hermaphrodite.

consecutive sexuality Consecutive hermaphrodite *q.v.*

consequential mortality Death during the recovery period following exposure to a harmful substance or stimulus; *cf.* concomitant mortality, delayed mortality.

conservation 1: The planned management of natural resources; the retention of natural balance, diversity and evolutionary change in the environment; *cf.* preservation. 2: A taxonomic procedure to conserve a scientific name which would otherwise contravene the provisions of the Code and thus be unavailable; **conserved name**.

conservative Used of characters exhibiting a low rate of change during evolution, and of taxa retaining many ancestral characters or character states.

conserved name *Nomen conservandum q.v.*

consistence A measure of cohesion of soil particles.

consistency of characters Comparability of characters *q.v.*

consociation A ranked category in the classification of vegetation; a small climax community or vegetative unit dominated by one particular species (the physiognomic dominant) which has the life form characteristic of the association *q.v.*

consociation dominant The dominant species of a consociation having the life form characteristic of the association; physiognomic dominant.

consociation subdominant A species that is subdominant at the level of a consociation but is dominant within the society.

consocies A stage in an ecological succession; part

of an association *q.v.* lacking one or more of its dominant species; a group of mores *q.v.*

consociule A micro-consociation *q.v.*

consolidated Compacted; made firm.

consortism The symbiotic relationship between the two partners (consortes) of a consortium.

consortium A group of individuals of different species, typically of different phyla, living in close association.

conspecific Belonging to the same species; *cf.* heterospecific.

constancy A measure of the distributional pattern of a plant, given by the number of different sampled plots which contain that particular species, usually expressed on the scale: r (less than 1%), I (1–20%), II (21–40%), III (41–60%), IV (61–80%), and V (81–100%); *cf.* fidelity.

constant extinction, law of Law of constant extinction *q.v.*

constant plankton The perennial permanent planktonic organisms of an area.

constant species 1: A species invariably present in a given community. 2. A species occurring in at least 50% of samples taken from a given community.

constipated Crowded together.

Constitution The constitution of the International Commission on Zoological Nomenclature or of the International Commission on Botanical Nomenclature.

constitutional type Habitus *q.v.*

constructive species Plant species that contribute to the development or persistence of the community.

consumer An organism that feeds on another organism or on existing organic matter, including herbivores, carnivores, parasites and all other saprotrophic and heterotrophic organisms; *cf.* primary producer.

consumer ingestion efficiency Efficiency of consumer ingestion (ECI) *q.v.*

consummatory act A behavioural act which constitutes the terminal phase of an instinctive behaviour sequence; *cf.* appetitive behaviour.

consumption (C) In ecological energetics, the total intake of food or energy by a heterotrophic individual, population or trophic unit per unit time; expressed as Consumption (C) = production (P) + respiration (R) + rejecta (FU); energy income; energy intake; ingestion.

consumptive use Evapotranspiration *q.v.*

contagious disease Any disease transmitted through physical contact.

contagious distribution A distribution pattern in which values, observations or individuals are more aggregated or clustered than in a random distribution, indicating that the presence of one individual or value increases the probability of another occurring nearby; aggregated distribution; clumped distribution; overdispersed distribution; superdispersed distribution; *cf.* random distribution, uniform distribution.

contamination The introduction of an undesirable agent such as a pest or pathogen, into a previously uninfested situation; contaminate, contaminant.

contaminative antigen An antigen common to both host and parasite, borne by the parasite but genetically of host origin.

contest competition Competition in which the resource is unequally partitioned between the competitors so that a successful organism obtains all it requires and an unsuccessful one obtains insufficient for survival and reproduction; *cf.* scramble competition.

contiguous Having boundaries that make contact but areas that do not overlap.

continental Pertaining to a continent or to the climatic conditions prevailing in the interior regions of a continent.

continental climate Any climate in which the difference between summer and winter temperatures is greater than the average range for that latitude because of distance from an ocean or sea.

continental drift The theory that the continental land masses have drifted apart over the course of geological time; displacement theory; epeirophoresis theory; *cf.* plate tectonics.

continental island An island that is close to, and geologically related to, a continental land mass, and was formed by separation from the continent; *cf.* oceanic island.

continental rise The gently sloping seabed from the continental slope to the abyssal plain, with an average angle of slope of about 1°; see Appendix 7.

continental sea Epicontinental sea *q.v.*

continental shelf The shallow gradually sloping seabed around a continental margin, not usually deeper than 200 m and formed by submergence of part of the continent; see Appendix 7.

continental slope The steeply sloping seabed leading from the outer edge of the continental shelf to the continental rise, with an average angle of slope of about 4° and a maximum of about 20° near the upper margin; see Appendix 7.

Continental Southeast Asiatic region A subdivision of the Indo-Malaysian subkingdom of the Palaeotropical kingdom; see Appendix 4.

continentality The overall ratio of land area to sea area.

contingency table A table consisting of two or more rows and columns of data in which observations or individuals are classified according to two variables; tests of independence such as the χ^2 test *q.v.* can be used to measure the relationship between the variables.

contingent symbiosis A symbiosis in which one species takes refuge in or on another species for shelter only.

continuity In taxonomy, the principle that continuity of usage of a particular name should take precedence over the priority of publication in determining which of two or more competing scientific names should be adopted.

continuous modular society A society *q.v.* in which the modular units or members are physically united to each other; equivalent to a colony.

continuous variable A variable that can theoretically assume any value between two given limits; *cf.* discrete variable.

continuous variation The more or less continuous series of infinitely small steps exhibited by a character or trait having a continuous spectrum of values from one given extreme to another; *cf.* discontinuous variation.

continuum A gradual or imperceptible intergradation between two or more extreme values.

continuum concept That vegetation changes continuously and is not differentiated, except arbitrarily, into phytosociological units.

contra- Prefix meaning opposite, against.

contranatant Swimming, moving, or migrating against the current; *cf.* denatant.

control A parallel experiment or test carried out to provide a standard against which an experimental result can be evaluated.

controlling factor Density dependent factor *q.v.*

convection A mode of heat transfer within a fluid, involving the movement of substantial volumes of the fluid from one place to another as a result of changes in the density of heated and non-heated areas.

convective rain Rainfall produced by convection, resulting from solar radiation heating the ground; *cf.* frontal cyclonic rain, orographic rain.

conventional behaviour Any behaviour by which members of a population reveal their presence and allow others to assess the density of the population; epideictic display.

convergence 1: Convergent evolution. 2: An oceanic region in which surface waters of different origins come together and where the

denser water sinks beneath the lighter water, as in the Antarctic convergence; *cf.* divergence.

convergent evolution The independent evolution of structural or functional similarity in two or more unrelated or distantly related lineages or forms that is not based on genotypic similarity; homoplasy; convergence; convergent adaptation.

convivium A biosystematic unit comprising those members of a *commiscuum q.v.* prevented from interbreeding by a geographical barrier; *cf. commiscuum, comparium.*

convulsionism Catastrophism *q.v.*

cooperation Protocooperation *q.v.*

coordinates In statistics, the numbers which locate the position of a variable point in space; in three dimensions the three Cartesian coordinates (x, y, z) with axes at right angles are usually used to locate a point.

coordination 1: The division of effort within a group without the establishment of leadership or of a dominance hierarchy. 2: In taxonomy, the principle that a name established at one rank in a family-group, genus-group or species-group is simultaneously available for a taxon having the same type, at another rank within the same group; coordinate.

Cope's rule The generalization that there is an evolutionary trend towards increasing body size within a phyletic series; Cope's law.

cophenetic correlation coefficient The measure of the agreement between similarity values implied by a phenogram, and those of the original similarity matrix.

coprobiont Any animal (coprozoite) or plant (coprophyte) living or feeding on dung; **coprobiontic**.

coprolites Fossilized faecal material.

coprology The study of animal faeces; **coprological**.

coprophagous Feeding on dung or faecal material; koprophagous, merdivorous; scatophagous; **coprophage, coprophagy**.

coprophilous Thriving on or in dung or faecal material; coprophilic; **coprophile, coprophily**.

coprophyte A plant living on dung or faecal material; **coprophytic**.

coprozoite An animal living on or in dung or faecal material; **coprozoic**.

copulate 1: To engage in sexual intercourse; the act of introducing spermatozoa into the female; coition; **copulation, copulatory**. 2: Conjugate; to fuse together, as in the case of gametes.

copy error A mutation which results from an error in the DNA nucleotide sequence during replication.

coquina A deposit of drifted shells.

cordillera An entire mountain system, including all subordinate ranges, interior plateaus and basins.

core A vertical column of bottom sediment; core sample.

core area That area within the home range *q.v.* which is most commonly utilized or frequented.

Coriolis force The deflecting force of the Earth's rotation that causes a body of moving water to be deflected to the right in the northern hemisphere and to the left in the southern hemisphere.

cormophylogeny The development of family groups or races.

corr. Abbreviation of the Latin *correctus*, meaning corrected (by).

corrasion Erosion by the mechanical action of a moving agent such as running water, wind or glacial ice.

correct name In botanical nomenclature, the name by which a taxon should properly by known; equivalent to valid name in zoological nomenclature.

corrected original spelling In taxonomy, the corrected version of an incorrect original spelling.

correctus (*corr.*) Latin, meaning corrected (by); used in botanical nomenclature to indicate a corrected spelling of a name; equivalent to a justified emendation *q.v.* in zoological nomenclature.

correlate To establish a mutual relationship or dependence between characters, factors or variables.

correlated characters Those phenotypic characters that exhibit some degree of association, either because they relate to an integrated gene complex or because they share a functional correlation.

correlated response The change in one character occurring as the consequence of selection for an apparently independent character, due to gene linkage or to a gene affecting more than one character.

correlation In statistics, the degree of relationship between two or more variables.

correlation coefficient A measure of the linear relationship between two quantitative variables, indicating the degree to which they vary together; denoted by r, values range from -1 (perfect negative correlation), through 0 (no linear relationship) to $+1$ (perfect positive correlation).

corridor A more or less continuous connection between adjacent land masses which has existed for a considerable period of geological time.

corticicolous Corticolous *q.v.*; **corticicole**.

corticolous Inhabiting or growing on bark; corticicolous; **corticole**.

coryphilous Thriving in alpine meadows; coryphophilous; **coryphile, coryphily**.

coryphium An alpine meadow community.

coryphophilous Coryphilous *q.v.*; **coryphophile, coryphophily**.

coryphophyte A plant inhabiting alpine meadows.

cosmogenous Derived from outer space; cosmic.

cosmology A branch of metaphysics concerned with the natural laws governing the universe as an orderly system; **cosmological**.

cosmopolitan Having a worldwide distribution, effect or influence; ecumenical; pandemic; ubiquitous; **cosmopolite**.

cosmotropical Circumtropical; occurring throughout the tropics.

cost of selection The concept of an accumulating cost to a population (in terms of genetic deaths) of replacing one allele by another in the course of evolution; cumulative substitional load.

coterie A social group of animals which defends a common territory against members of other coteries.

coterminous Used of organisms having similar distributions.

cotidal line The line that delimits the area covered by the tide at any given time.

cotransduction The simultaneous transduction *q.v.* of two or more genes when the transduced element contains more than one locus.

cotype In taxonomy, formerly used for each specimen of a type series in the absence of a designated holotype; syntype.

counter- Prefix meaning moving against or in the opposite direction.

counteracting selection The operation of selection pressures at two or more levels of organization, such as individual, family or population, so that certain genes are favoured at one level but disfavoured at another; *cf.* reinforcing selection.

counterevolution The evolution of traits within a population in response to adverse interactions with another population.

countershading The condition of an animal having a darkly coloured dorsal surface and a lighter ventral surface, that serves to disrupt the silhouette in a gradient of downwelling light.

couplet Character couplet; a pair of contrasting descriptive statements.

coupling configuration The linkage relationship of two dominant genes found on one chromosome with the alternative recessive alleles on the homologous chromosome; *cf.* repulsion configuration.

court 1: The group of workers that surround a queen in an insect colony; retinue. 2: That part of a communal display area defended by a particular male.

courtship Any behavioural interaction between males and females that facilitates mating.

courtship ritual A genetically determined characteristic behaviour pattern involving the production and reception of a complex sequence of visual, auditory and chemical stimuli by the male and female prior to mating.

COV Crossover value q.v.

covariance The expected value of the product of the deviations of two random variables from their respective means.

covarions Those codons that at any one time could change the amino acid coded without a significantly deleterious effect.

cover 1: Plant material, living (vegetative cover) and dead (litter cover), on the soil surface. 2: The area of ground covered by vegetation of a particular plant species; expressed as a scale (Braun-Blanquet scale, Domin scale), or as a percentage.

cover-stratification diagram A graphic representation of the coverage in each of the horizontal strata of stratified vegetation, the number of species in each stratum is usually included to complete the diagram.

cover type The existing vegetation of an area.

coverage That part of a sampled area covered by a particular plant species or individual plant canopy; typically expressed as a percentage.

covert A shelter; hiding place.

cran A unit measure of herrings by volume equal to 37.5 gallons (170.5 dm^3) and representing about 750 fishes.

crash A precipitous decline in the size of a population; cf. flush.

creationism The belief that all forms of life were created de novo, and have undergone little subsequent change; fundamentalism; creationist.

crèche A group of young animals which have left their nests.

cremnium A plant community associated with cliffs; cremnion.

cremnophilous Thriving on cliffs; cremnophile, cremnophily.

crenic Pertaining to a spring and the adjacent brook water flowing from the spring.

crenicolous Living in springs, or in brook water fed from a spring; crenicole.

crenium A plant community associated with spring water.

crenogenic meromixis Mixing in a lake caused by a saline spring delivering dense water into the bottom of a lake displacing the mixolimnion q.v.; cf. biogenic meromixis, ectogenic meromixis.

crenon The spring-water biotope; cf. stygon, thalasson, troglon.

crenophilous Thriving in or near a spring; crenophilic; crenophile, crenophily.

creode A canalized developmental pathway.

creophagous Carnivorous; used particularly of carnivorous plants; creophage, creophagy.

crepitation A defence mechanism in insects in which fluid is explosively discharged.

crepuscular Active during twilight hours; of the dusk and dawn; vespertine; cf. diurnal, nocturnal.

Cretaceous A geological period of the Mesozoic era (ca. 140–65 million years B.P.); see Appendix 1.

cretaceous Pertaining to chalk or limestone.

criterion A predetermined standard on which a decision may be based.

critical depth The depth at which the average light intensity of the water column equals compensation light intensity q.v.

critical group A group of organisms having some variation but in which taxonomic subgroups are difficult to delineate.

critical region The region of sample space containing all outcomes for which the null hypothesis is rejected; cf. acceptance region.

critical velocity The speed of water sufficient to move particles of a given size along a channel bed.

Cromerian interglacial An interglacial period of the Quaternary Ice Age in the British Isles; see Appendix 2.

cron A unit of time equal to one million (10^6) years.

cronism Actual or attempted eating of dead or moribund young by parents; kronismus.

crop growth rate (CGR) The rate of production of a homogeneous stand, calculated from the growth rate of individual plants (nett assimilation rate, NAR) and a parameter reflecting the density of the stand (leaf area index, LAI), according to the formula CGR = NAR × LAI.

cropper An organism that feeds actively on other organisms from the surrounding environment.

cross fertilization The union of male and female gametes from different individuals of the same species; allogamy; allomixis; staurogamy; xenogamy; cf. self-fertilization.

cross infection The transfer of a disease organism from one individual to another of the same species.

cross pollination Transfer of pollen from one

flower to the stigma of a flower on another plant of the same species.

crossing The mating of individuals of different breeds, races or strains in order to promote genetic recombination; hybridize; cross; crossbreed.

crossing over An exchange of corresponding segments between chromatids of homologous chromosomes during meiosis, rarely also during mitosis.

crossover unit A 1% crossover value between a pair of linked genes; used in chromosome mapping.

crossover value (COV) The percentage of gametes that exhibit crossing over of a particular gene.

crotovina A former animal burrow in one soil horizon that has been filled with organic matter or sediment from another horizon; krotovina.

crowding effect The reduction in the size of individual worms in a parasitic infection, an effect proportional to the number of worms present.

crown The highest part or layer; typically used of the uppermost foliage of a tree.

crown class The trees occupying a particular stratum in a forest canopy, comprising dominant, codominant, intermediate and over-topped crown classes.

crown cover The canopy formed by forest trees; crown canopy.

crown density The proportion of a sampled woodland area having complete crown cover.

crymad A plant of polar regions; crymophyte.

crymium A plant community of polar regions; crymophium.

crymnion Cryoplankton; the planktonic organisms of perpetual ice and snow.

crymophium Crymium *q.v.*

crymophilous Thriving in polar habitats; crymophilic; cryophilic; **crymophile, crymophily**.

crymophyte A plant of polar regions; crymad.

cryo- Prefix meaning cold, pertaining to cold.

cryochore Those regions of the Earth's surface perpetually covered by snow.

cryoconite Organisms and wind blown detritus that induce surface melt pits in glaciers.

cryogenic lake A lake formed by local thawing in an area of permanently frozen ground.

cryophilic Thriving at low temperatures; crymophilous; frigophilic; psychrophilic; **cryophilous, cryophile, cryophily**.

cryophylactic Resistant to low temperatures.

cryophyte 1: A plant growing on ice or snow. 2: A plant thriving at low temperatures.

cryoplankton Planktonic organisms of persistent snow, ice and glacial waters.

cryotropism An orientation response to the stimulus of cold or frost; **cryotropic**.

cryoturbation The physical mixing of soil materials by the alternation of freezing and thawing.

crypsis Concealment.

cryptic Used of coloration and markings that resemble the substratum or surroundings and aid in concealment; *cf.* phaneric.

cryptic polymorphism Genetic polymorphism controlled by a recessive gene, such that the different forms are not phenotypically distinguishable.

cryptic polyploidy The condition of having little or no phenotypic manifestation of differences in ploidy levels; semicryptic polyploidy.

cryptic species Sibling species *q.v.*

cryptobiosis The condition in which all external signs of metabolic activity are absent from a dormant organism.

cryptobiotic 1: Used of organisms which are typically hidden or concealed in crevices or under stones; **cryptobios**. 2: Cryptobiosis *q.v.*

cryptodepression That part of a lake basin below mean sea level.

cryptoendomitosis Polyploidization occurring within the intact nuclear envelope of somatic cells in the absence of detectable mitotic stages.

cryptofauna The fauna of protected or concealed microhabitats.

cryptogenic Of indeterminate descent; used of a fossil species of unknown or obscure phylogenetic descent from species occurring in earlier geological formations; **cryptogene**; *cf.* phanerogenic.

cryptogram In taxonomy, a shorthand summary of the properties of a virus taxon, comprising four pairs of symbols: (1) type of nucleic acid/strandedness, (2) molecular weight of nucleic acid/percentage of nucleic acid, (3) outline of particle/outline of nucleic acid and most closely surrounding protein, (4) host/vector.

cryptohybrid A hybrid displaying unexpected phenotypic characters.

cryptomere A hidden recessive hereditary factor.

cryptophyte A perennial plant with renewal buds below ground or water level, including geophyte *q.v.*, helophyte *q.v.* and hydrophyte *q.v.*; *cf.* Raunkiaerian life forms.

cryptophytium A community of geophytes *q.v.* and hemicryptophytes *q.v.*

cryptosphere The microhabitat occupied by cryptozoic organisms.

Cryptozoic The period of geological time during which only the most primitive life forms are found; a subdivision of the Precambrian era,

preceding the Proterozoic period; see
Appendix 1.

cryptozoic Pertaining to small terrestrial animals
(cryptozoa) inhabiting crevices, living under
stones, in soil or litter.

cryptozoology The study of cryptozoic organisms.

crystallochorous Dispersed by the agency of
glaciers; **crystallochore, crystallochory**.

CSMP A special purpose computer modelling
language designed to solve dynamic deterministic
models.

C-S-R **triangle** A three-component system of
ecological strategies conceptualized as a triangle
with the three extremes representing competitive
species (*C*-strategists), stress-tolerant species
(*S*-strategists) and ruderal species (*R*-strategists).

C-**strategist** Within the *C-S-R* triangle *q.v.* of
ecological strategies, a species typically with
large body size, rapid growth, relatively long life
span, relatively efficient dispersal, devoting only
a small proportion of metabolic energy to the
production of offspring or propagules; a
competitive species.

cteinophyte A parasitic plant which destroys its
host; cteinotrophic plant.

cteinotrophic Used of parasitic organisms that
destroy their hosts.

ctetology The study of acquired characters;
ctetological.

cull To reduce the number of animals in a
breeding group by removing inferior stock or by
killing selected animals.

culmicolous Growing on the stems of grasses;
culmicole.

cultigen An organism known only in cultivation;
cultivar *q.v.*

cultivar (CV) A variety of a plant produced and
maintained by cultivation; also the corresponding
taxon; *cultivarietas*; cultigen.

cultivarietas (*cult.*) Latin, meaning cultivated.

cultivation The preparation and use of land for
crop production.

culto-type A live culture produced from the
holotype of a species; cultype.

cultural desert An area largely devoid of fauna
and flora as a result of urbanization.

cultural eutrophication Over enrichment of a body
of water with nutrients as a result of cultivation
or other activities of man.

cultural forest Plantations of alien tree species,
frequently supporting a limited natural
vegetation.

cultural prairie An area of cultivated grainfield or
grassland.

cultype Culto-type *q.v.*

cumaphyte A surf plant; cymaphyte.

cumatophytic Pertaining to structural
modifications of plants in response to wave
action.

cumulative distribution A distribution showing the
number of observations above or below a given
value.

cumulate To gather together; to combine into
one; **cumulative**.

cuniculine Living in burrows resembling those of
rabbits.

cuprophyte A plant adapted to, or tolerating high
copper levels in the soil; frequently used as an
indicator of this particular soil type.

Curie point The temperature at which the
magnetic polarity of basalt cooling from the
molten state is fixed parallel to the terrestrial
magnetic field.

current 1: A non-tidal horizontal movement of the
sea. 2: In limnology, a continuous flow of water.

cursorial Adapted for running; running.

cuticular transpiration The loss of water vapour
through the cuticle of leaves; *cf.* stomatal
transpiration.

cutting An artificially detached portion of a plant
used for vegetative propagation.

CV Cultivar *q.v.*; abbreviation of the Latin
cultivarietas, meaning cultivated.

cyanogenic Used of living organisms that produce
hydrocyanic acid; cyanogenous.

cybernetics The study of control and
communication in systems, including information
systems and feedback control systems;
kybernetics.

cycleology The study of cyclic phenomena in
geology and palaeontology.

cyclic evolution A pattern of evolution involving
an initial phase of rapid expansion, followed by a
long period of relative stability and a final brief
episode of degeneration and over specialization
leading to extinction.

cyclic parthenogenesis Reproduction by a series of
parthenogenetic generations alternating with a
single sexually reproducing generation; holocyclic
parthenogenesis; monocyclic parthenogenesis;
cf. parthenogenesis.

cyclic succession An ecological succession in which
the original community is ultimately restored.

cyclical selection Sequential reversal of the
direction of selection resulting from cyclical
environmental changes.

cyclogeny Cyclomorphosis *q.v.*

cyclomorphosis Cyclical changes in form, such as
seasonal changes in morphology; cyclogeny.

cyclonic Used of a region of low atmospheric sea-
level pressure; also of the wind system around
such a low pressure centre that has a clockwise

rotation in the northern hemisphere and an anticlockwise motion in the southern hemisphere; **cyclone**; *cf.* anticyclonic.

cyesis That period between fertilization and birth in animals that produce a single offspring per brood.

cymaphyte A surf plant; cumaphyte.

cyrioplesiotype A typical specimen among several plesiotypes *q.v.*

cytocatalytic evolution Evolution initiated by an abrupt mutation resulting in the formation of polyploids and aneuploids.

cytodeme A local interbreeding group which differs in cytological characters from other such groups within the same taxon.

cytodiaeresis Mitosis *q.v.*

cytogamy 1: The fusion of two cells. 2: The fusion of both male nuclei of one of two conjugating organisms with the female nucleus of the other; double autogamy.

cytogenesis Cell production and development.

cytogenetics The study of chromosomal mechanisms, their behaviour and their effect on inheritance and evolution.

cytogenic Reproducing by cell division.

cytokinesis Division of the cytoplasm, usually following meiosis or mitosis; *cf.* karyokinesis.

cytology The study of the structure and function of living cells.

cytophagous Feeding on cells; **cytophage**, **cytophagy**.

cytoplasmic androgamy The fertilization of a male gamete by the cytoplasm of a female gamete.

cytoplasmic factor Any genetic factor which is resident in the cytoplasm of a cell.

cytoplasmic gynogamy The apparent fertilization of a female gamete by the cytoplasm of the male gamete.

cytoplasmic inheritance Inheritance of characters through cytoplasmic factors rather than through the nuclear chromosomes; *cf.* chromosomal inheritance.

cytotaxis 1: The movement of cells in relation to each other; used to describe cells separating (negative cytotaxis) or aggregating (positive cytotaxis); **cytotactic**. 2: The arrangement of cells within an organ.

cytotaxonomy Classification based on a study of cell structure, with special reference to chromosome structure.

cytotype A subspecific variety that has a chromosome complement differing from the standard for the species, either in chromosome number or structure.

cytozoic Intracellular; used of an organism living within a host cell.

Czernozem Chernozem *q.v.*

D

2,4-D 2,4-dichlorophenoxyacetic acid; a translocated hormone weed killer, used to control broad leaved herbaceous plants, and as a defoliant.

dam In animal breeding, the female parent; *cf.* sire.

dance A highly stylized repetitive movement, often associated with courtship.

dark respiration Respiration in photosynthetic plants occurring during the night.

Darling effect Fraser Darling effect *q.v.*

Darlington's law That the fertility of an allopolyploid is inversely proportional to the fertility of the original hybrid; Darlington's rule.

darwin A unit measure of evolutionary rate of change, representing an increase or decrease in any given character by a factor of 2.7 per million years; Haldane's evolutionary unit.

Darwinian fitness The relative competitive ability of a given genotype measured as the proportional contribution of the genotype to the gene pool of the subsequent generation.

Darwinism The view of evolution expressed by Charles Darwin; evolution occurring by means of natural selection (the struggle for life) acting on spontaneous variation arising within populations and species, resulting in the survival of the fittest; Darwinian evolution.

data 1: Units of information; facts; observations; results of an experiment or study; **datum**.

data matrix A tabulation of data, such as taxonomic characters, to show differences between categories (taxa) often in a machine readable form; comparison chart; data chart; data table.

date of publication In taxonomy, the earliest date at which copies of a work were first issued but not necessarily the date printed on the work itself; the date on which a name was first published so as to satisfy the requirements of the Code.

datum Chart datum *q.v.*; datum plane.

dauermodification Modification of a phenotype as a result of environmental change; the novel phenotype may be propagated vegetatively in the absence of the original stimulus but the modified character gradually weakens and is eventually lost.

daughter Any of the offspring of a given generation, applicable to both sexes.

day (d) 1: Solar day; the mean time taken for the sun to reach any given position on consecutive days. 2: Sidereal day; the mean time taken for one revolution of the Earth.

day-neutral A plant in which the flowering response is not dependent on photoperiodism.

DBH Abbreviation of diameter at breast height (1.4 m) of a tree; measured either over the bark (DBHOB) or under the bark (DBHUB); *cf.* basal area.

D-coefficient The reciprocal of the affinity index *q.v.*; used as a measure of ecosociological distance (D) between samples and calculated as $D = \sqrt{(a \times b)}/c$, where a and b are the numbers of species occurring in only community A or B respectively and c is the number of species common to both.

DDT Dichlorodiphenyltrichloroethane; a persistent organochlorine insecticide.

de novo Latin, meaning arising anew.

De Vriesianism The belief that evolution in general and speciation in particular are the results of drastic and sudden mutational changes; mutationism.

dear enemy phenomenon The recognition of territorial neighbours with the result that aggressive interactions are minimized between them, with the more intense aggression directed towards strangers.

death The total and irreversible cessation of all life processes.

death feint Thanatosis *q.v.*

death point The upper or lower limit of tolerance for any internal or environmental factor at, or beyond, which death occurs.

death rate Mortality; the number of deaths in a population per unit time.

debris Accumulated plant and animal remains.

deca- Prefix meaning ten, tenfold; used to denote unit × 10; deka-.

decalcification Removal of calcium carbonate from the soil by leaching.

decay Organic decomposition during which almost total oxidation of organic substances occurs; *cf.* mouldering, putrefaction.

decay of variability The reduction of heterozygosity due to the loss and fixation of alleles that accompanies genetic drift.

decennial Pertaining to periods of 10 years; *cf.* millennial, secular.

deci- Prefix used to denote unit × 10^{-1}.

deciduilignosa Tree and shrub communities in

which leaves are shed during unfavourable periods or seasons.

deciduous Used of structures that are shed at regular intervals, or at a given stage in development.

deciduous plant A plant that sheds its leaves during the unfavourable season of the year.

decimal reduction time The time required to reduce a viable population to 10% of its original size.

decision tree A general term for an identification dendrogram which is based on a series of alternative choices.

declaration A provisional amendment or clarification to the Code published by the International Commission on Zoological Nomenclature.

déclinant A subdivision of a subassociation.

decomposer Any organism that feeds by degrading organic matter.

decomposition Metabolic degradation of organic matter into simple organic and inorganic compounds, with consequent liberation of energy; **decomposer**.

decoy A crop of low value cultivated to attract pests away from a more valuable crop.

dedifferentiation The process of regression to a less specialized structure or condition.

deductive method A scientific method involving the formulation of theories or hypotheses from which singular statements (predictions) are deduced that can be tested; *cf.* inductive method.

deem In taxonomy, to consider something to be what it strictly is not; as in the case of a work deemed to be published by the author rather than by the actual publisher.

deep scattering layer (DSL) A more or less well defined layer present in most oceanic waters which reflects sound from echosounding equipment; produced by stratified populations of organisms which scatter sound waves and are recorded on an echo-sounder as a horizontal layer; false bottom.

defaecate To discharge faeces or egesta *q.v.*; **defaecation**, **defecate**, **defecation**.

defaunated Depleted of animals or of symbiotic microorganisms; defaunation.

definition In taxonomy, a statement or diagnosis of the characters distinguishing a taxon.

definitive host The host in which a parasite attains sexual maturity; primary host; *cf.* intermediate host.

deflation Erosion of surface layers by wind action; aeolian erosion.

deflected succession An ecological succession that

is redirected to an alternative climax by the influence of environmental factors.

deflection display Distraction display *q.v.*

defoliant A chemical, such as 2,4-D and 2,4,5-T, that causes leaves to fall from plants.

deforestation The permanent removal of forest and undergrowth.

degeneration Loss or reduction of structure or function during the course of evolution or ontogeny.

degradation Breakdown into smaller or simpler parts; reduction of complexity.

Degraded Chernozem soil A zonal soil with very dark brown or black A-horizon over a grey leached horizon and a brown lower horizon; formed in cool climates under forest-prairie transitional vegetation.

degree days Units used in the measurement of the duration of a life cycle or a particular growth phase of an organism; calculated as the product of time and temperature average over a specified interval.

degree of accumulation Accumulation, degree of *q.v.*

degree of confidence Confidence limits *q.v.*

degree of succulence A measure of the water storage capacity of a plant; calculated as the water content at saturation divided by the surface area.

degrees of freedom In statistics, the number of independent deviations from the mean on which an estimate of population variance can be based.

degression Regression to a less specialized condition; dedifferentiation; **degressive**.

dehiscence Spontaneous opening of ripe plant structures to liberate seeds or spores; **dehiscent**.

deimatic behaviour The adoption of an intimidating posture by one animal in order to frighten another; dymantic behaviour.

deka- Prefix meaning ten, tenfold; used to denote unit ×10; deca-.

delayed mortality Death occurring after the recovery period, following exposure to a harmful substance or stimulus when the animal appeared to be healthy; *cf.* concomitant mortality, consequential mortality.

deleterious 1: Having an adverse effect. 2: Used of a trait which impairs survival, or of a mutation that reduces fitness.

deletion The loss of a segment from a chromosome.

deletion mapping The use of overlapping deletions to map the positions of particular genes on a chromosome.

delimitation 1: A statement of the character states which define the limits of a taxon; diagnosis.

2: In phytosociology, the determination of the extent of a given syntaxon.

deliquescent Becoming fluid by the uptake of water from air.

delitescence The incubating period of a pathogenic organism.

delta A deposit of fine alluvial sediments at the mouth of a river.

delta karyology In the study of chromosomes, the determination of satellite DNA locations, nucleolar organizers and 5-s rRNA loci; *cf.* karyology.

deltaic Pertaining to a delta; used of the succession or cycle of processes involved in the formation of a delta.

deltaic sequence A sediment profile in the vicinity of a delta typically exhibiting a gradation of particle sizes.

deme A local interbreeding group; panmictic unit; also used loosely to refer to any local group of individuals of a given species; used as a neutral term in combination with qualifying prefixes, including agamodeme, autodeme, clinodeme, clonodeme, coenogamodeme, cytodeme, ecodeme, endodeme, gamodeme, genodeme, genoecodeme, hologamodeme, merogamodeme, phenodeme, plastodeme, plastoecodeme, serodeme, topodeme, topogamodeme, and xenodeme.

demersal Living at or near the bottom of a sea or lake but having the capacity for active swimming; **demersed**.

demographic society A stable society in which natural processes of birth and death are largely responsible for the age structure of the society; typically a relatively closed society; *cf.* casual society.

demography The study of populations, especially of growth rates and age structure.

denatant Swimming, moving, or migrating with the current; *cf.* contranatant.

denaturation An alteration in the structural properties of a protein producing a change in enzymic, hormonal or other biochemical activity.

dendrad An orchard plant.

dendritic evolution Cladogenesis *q.v.*

dendrium An orchard community.

dendrochore That part of the Earth's surface covered by trees.

dendrochronology A method of dating using annual tree-rings; tree-ring chronology.

dendroclimatology The determination of past climatic conditions from the study of the annual growth rings of trees.

dendrocolous Living in, or growing on, trees; arboricolous; **dendrocole**.

Dendrogaea The Neotropical region *q.v.* excluding temperate South America.

dendrogram A branching diagram in the form of a tree used to depict degrees of relationship or resemblance.

dendrology The study of trees.

dendrophagous Feeding on wood; hylophagous; lignivorous; xylophagous; **dendrophage, dendrophagus, dendrophagy**.

dendrophilous Thriving in trees; living in orchards; **dendrophile, dendrophily**.

denitrification The release of gaseous nitrogen or the reduction of nitrates to nitrites and ammonia by the breakdown of nitrogenous compounds, typically by microorganisms when the oxygen concentration is low; on a global scale thought to occur primarily in oxygen deficient environments in the oceans.

denizen A plant species found growing in the wild but considered to have been originally introduced for cultivation; established alien.

density The number of individuals or observations within a given area or volume.

density dependence A change in the influence of an environmental factor (a density dependent factor) that affects population growth as population density changes, tending to retard population growth (by increasing mortality or decreasing fecundity) as density increases or to enhance population growth (by decreasing mortality or increasing fecundity) as density decreases; *cf.* inverse density dependence.

density dependent factor Density dependence *q.v.*

density independent factor Any factor affecting population density, the influence of which is independent of population density.

denudation 1: Erosion of surface material to expose underlying rock. 2: The removal of surface vegetation; denude.

denuded quadrat A quadrat from which all the original organisms have been removed.

deoxygenated Used of an environment depleted in free oxygen; anaerobic; **deoxygenation**.

deoxyribonucleic acid DNA *q.v.*

depauperate Impoverished; moribund.

dependence The condition of a variable that is influenced or controlled by another variable; **dependent**; *cf.* independence.

dependent character A character that is applicable only if another character is present in a particular state; for example hairiness of leg applies only to an organism in which legs are present.

depletion effect The decline in the rate of immigration of species to an island as the number of resident species increases.

deplumation Moulting in birds.

deposit feeder Any organism feeding on fragmented particulate organic matter in or on the substratum; a detritivorous organism.

depuration The cleansing activity of shellfish effected by immersion in clean water prior to harvesting resulting in the removal of pathogenic organisms and pollutants potentially harmful to man.

derived Used of a character or character state not present in the ancestral stock; apomorphic; *cf.* plesiomorphic.

derived similarity The common possession of derived homologous characters or character states; synapomorphy.

dermatophyte A fungal skin parasite; epidermophyte; dermophyte, dermatophyton.

dermatotrophic Living and feeding on skin; **dermatotroph, dermatotrophy**.

dermatozoon An animal parasite of skin; dermozoon.

desalinization The removal of salts; as in the leaching of saline soils.

description A full statement of the character states of a taxon; *cf.* diagnosis.

descendants 1: All the individuals derived from the sexual union of two individuals. 2: The derivatives of a prototype, ancestor or other source.

descr. Abbreviation of the Latin *descriptione*, meaning description.

descriptio generico-specifica Combined description *q.v.*

desert An arid area with insufficient available water for dense plant growth; more precisely, an area of very sparse vegetation in which the plants are typically solitary and separated from each other by more than their diameter on average.

desert biome Desert: a region of environmental extremes having less than 250 mm annual rainfall, high evaporation and low humidity; typically with a large proportion of barren ground, sparse vegetation of solitary plants or small aggregations, and nocturnal animals.

desert forest Arboreous desert *q.v.*

Desert soil A very shallow light coloured zonal soil overlying calcareous material, formed under arid conditions; typically with a pale grey or brown shallow A-horizon, a slightly darker neutral or alkaline B-horizon and a calcareous and often compacted lower horizon; characteristically with sparse vegetation.

desert wind A dry wind blowing from a desert region, hot in summer and cold in winter.

deserticolous Living mostly on open ground in an arid or desert region; eremobiontic; **deserticole**.

desertification The development of desert conditions as a result of human activity or climatic changes; exsiccation.

desiccation Removal of water; the process of drying; **desiccate**.

desiccation avoidance Delaying of desiccation by various mechanisms that enable an organism to maintain a favourable tissue water content despite dryness of air or the soil; **desiccation avoidant**.

desiccation tolerance The capacity to tolerate drying out of protoplasm without damage; **desiccation tolerant**.

designation 1: In zoology, the act of fixing the type of the name of a species-group or genus-group taxon. 2: In botany, the general name or formula of a taxon.

det. Abbreviation of the Latin *determinavit*, meaning he (she or it) identified, he (she or it) determined.

determinate growth Growth that is limited during the life span of an individual, so that the organism reaches a maximum size after which growth ceases; *cf.* indeterminate growth.

determination 1: Identification of an organism or fossil by reference to an existing classification. 2: The activation of a specific developmental pathway from the range of possible pathways available.

deterministic model A mathematical model in which all the relationships are fixed and the concept of probability is not involved, so that a given input produces one exact prediction as an output; *cf.* stochastic model.

detrimental Used of a mutation that decreases the viability of the mutant genotype.

detriophagous Feeding on detritus; detritivorous; **detriophage, detriophagy**.

detritivorous Feeding on fragmented particulate organic matter; **detritivore, detritivory**.

detritus Fragmented particulate organic matter derived from the decomposition of plant and animal remains; organic debris; **detrital**.

deuterogenesis The second phase of embryonic development post gastrulation; *cf.* protogenesis.

deuterotoky Parthenogenesis in which both male and female offspring are produced; amphitoky; **deuterotokous**.

deuterotype A replacement type specimen.

devalid name In taxonomy, a name that is not valid because it was published before the starting date of the group concerned; devalidate; *cf.* pre-starting point.

development 1: Ontogeny; regulated growth and differentiation of an individual, including cellular differentiation, histogenesis and organogenesis.

2: The orderly sequence of progressive changes that result in increasing complexity of a biological system.

developmental homeostasis Canalization *q.v.*

developmental plasticity The capacity for developmental variation as a result of environmental influences on the genotype during growth.

developmental sterility Sterility resulting from abnormal development of reproductive structures in a first generation hybrid.

Devensian glaciation The most recent glaciation of the Quaternary Ice Age in the British Isles with an estimated duration of about 70 thousand years; Newer glaciation; see Appendix 2.

deviant index A measure of the extent to which an individual departs from the population norm.

deviate Each individual deviation of a distribution, computed as the observation minus the mean.

deviation 1: Departure of an observation from a given or expected value. 2: The gradual divergence during embryology of the ontogenetic pathways of related animals as development proceeds from the generalized to the specialized.

Devonian A geological period within the Palaeozoic (*ca.* 413–365 million years B.P.); see Appendix 1.

dew point The temperature at which liquefaction of a vapour begins.

D-horizon Any layer underlying a soil profile that differs from the parent material of the soil; R-horizon; see Appendix 12.

di- Prefix meaning two, twice.

dia- Prefix meaning across, through.

diabatic Non-adiabatic *q.v.*

diachronous Belonging to different geological periods.

diacmic Exhibiting two abundance peaks per year; *cf.* monacmic, polyacmic.

diacritic marks Marks used to modify the phonetic value of a letter or combination of letters.

diadromous Migrating between fresh water and sea water; *cf.* anadromous, catadromous.

diagenesis The chemical and physical processes, in particular compaction and cementation, involved in rock formation after the initial deposition of a sediment.

diagenetic deposit A pelagic sediment largely comprising minerals crystallized in sea water, such as zeolites and manganese nodules.

diageotropism Orientation at right angles to the direction of gravity; typically resulting in growth parallel to the soil surface; geodiatropism; geoparallotropism; transgeotropism; zenogeotropism; **diageotropic**.

diagnosis. A formal statement of the character

states which distinguish one taxon from another; *cf.* description.

diagnostic character Any character or character state that unambiguously differentiates one taxon from others; key character; peculiar character; *cf.* auxiliary character.

diagnostic fossil Characteristic fossil *q.v.*

diaheliotropism Orientation response of part or all of a plant aligning itself at right angles to the direction of incident sunlight; diheliotropism; transheliotropism; **diaheliotropic**.

dialect A local geographical variant of behaviour involved in communication.

diallelic Used of a polyploid with two different alleles at a given locus.

diamesogamy Fertilization through an external agency; **diamesogamous**.

diapause A resting phase; a period of suspended growth or development, characterized by greatly reduced metabolic activity, usually during hibernation or aestivation.

diaphototaxis An orientation response of an organism or structure at right angles to the direction of incident light; **diaphototactic**.

diaphototropism Orientation at a right angle to the direction of incident light; paraphototropism; **diaphototropic**.

diaspore Any part of an organism produced either sexually or asexually that is capable of giving rise to a new individual; propagule.

diastem A gap in the fossil record, denoted by a bedding plane, representing a period of time during which no deposition took place.

diatomaceous ooze A pelagic sediment comprising at least 30% siliceous material mainly in the form of diatom tests, generally restricted to high latitudes or areas of upwelling; diatom ooze; *cf.* ooze.

diatropism Orientation at a right angle to the direction of stimulus; **diatropic**.

diauxic growth Growth in two phases separated by a period of relative inactivity.

dibasic tetraploid A hybrid tetraploid derived from two diploid species with different basic numbers *q.v.*

dichlorvos 2,2-Dichlorovinyl dimethyl phosphate (DDVP); an organophosphorous insecticide and acaricide of relatively short persistence.

dichodynamic Used of a hybrid that displays the phenotypic characters of both parents to an equal extent; homodynamic hybrid.

dichogamy The maturation of male and female reproductive organs at different times in a flower or hermaphroditic organism, thereby preventing self-fertilization; heteracme; **dichogamic**, **dichogamous**; *cf.* adichogamy.

dichopatric Pertaining to populations or species having geographical ranges separated to the extent that individuals from the two populations never meet and gene flow is not possible; macrodichopatric; **dichopatry**; *cf.* allopatric, parapatric, sympatric.

dichothermy The condition in which a minimum occurs in a temperature curve of a water column, as in a lake in which a cold intermediate layer persists in summer at a depth of 75–200 m; *cf.* mesothermy, poikilothermy.

dichotomous character A character that exists in only two states.

dichotomous data Binary data; data existing in only a present or absent state.

dichotomous key An identification key constructed as a sequence of alternative choices, each pair forming a character couplet; diagnostic key; sequential key; *cf.* bracketed key, indented key.

dichotomy Division into two parts or categories; **dichotomous**.

dichotypic Dimorphic *q.v.*; used particularly of plants having two types of leaves or flowers.

diclinous Having male and female organs in separate flowers on the same or different plants; **diclinism**, **dicliny**; *cf.* monoclinous.

diecdysis Continuous moulting, with one ecdysis grading rapidly into the next.

diecious Dioecious *q.v.*

diecodichogamic Used of a population of dichogamic plants in some individuals of which the male flowers mature first while in others the female flowers mature first; **diecodichogamy**.

diel Daily; pertaining to a 24 h period.

dientomophilous Used of a plant pollinated by two different insect species and having two kinds of flowers each adapted for one of the insect pollinators; **dientomophily**.

differentia Diagnostic characters; Linnaeus' polynomial species diagnosis.

differential species. A plant species that can be used as an indicator of a particular vegetational type because of its relatively high fidelity *q.v.*

differentiation 1: The process of specialization and the progressive diversification of structure or function; integrated cellular specialization during embryonic development. 2: The diagnosis *q.v.* of a taxon.

diffuse competition Simultaneous interspecific competition between numerous species each having a small degree of niche overlap with other species; *cf.* intense competition.

diffusion 1: Passive movement of molecules in solution from a region of high concentration to a region of low concentration. 2: The intermixing of particles as a result of movement caused by thermal agitation.

diffusion pressure deficit (DPD) The amount by which the diffusion pressure of the water in soil is less than that of pure water (under atmospheric pressure) at the same temperature.

digametic Having two kinds of gametes, one producing male offspring and the other female; heterogametic; **digamety**.

digenesis The regular alternation of sexual and asexual generations; metagenesis; heterogenesis.

digenetic Pertaining to a symbiont requiring two different hosts during its life cycle; *cf.* monogenetic, trigenetic.

digenic Used of characters or traits controlled by the integrated action of two genes; *cf.* monogenic, oligogenic, polygenic, trigenic.

digenomic Containing two independently derived genomes.

digenous 1: Digenetic *q.v.* 2: Bisexual *q.v.*

digeny Sexual reproduction.

digested energy In ecological energetics, the total intake of energy by an individual, population or trophic unit (consumption) less that lost as faeces or regurgitated food (egesta); expressed as digested energy (D) = consumption (C) $-$egesta (F) or D = assimilation (A) $+$excreta (U).

digitigrade Walking with only the digits in contact with the ground.

digoneutic Producing two broods per season or year; bivoltine; **digoneutism**; *cf.* monogoneutic, polygoneutic, trigoneutic.

digonic Producing male and female gametes in different gonads of the same individual; **digony**; *cf.* syngonic.

diheliotropism Diaheliotropism *q.v.*

dihybrid cross A cross between two individuals heterozygous for two pairs of alleles; *cf.* monohybrid cross.

dilatant Used of a sediment that becomes more solid as a result of agitation or pressure; **dilatancy**; *cf.* thixotropic.

diluvial 1: In the geological time scale, the present time; Anthropozoic. 2: Pertaining to a flood.

dimegalic Used of gametes exhibiting marked size dimorphism; **dimegaly**.

dimictic Used of a lake having two seasonal overturn periods of free circulation, with accompanying disruption of the thermocline; *cf.* mictic.

dimonoecious Used of an individual plant having hermaphrodite, male, female and neuter flowers.

dimorphic 1: Pertaining to a population or taxon having two genetically determined, discontinuous morphological types; ditypic; **dimorphism**,

dimorphous, **dimorphy**; *cf*. monomorphic, polymorphic. 2: Having two different types of flower on a single plant; dichotypic; *cf*. monomorphic.

dimorphous life cycle A life cycle involving two different types of individual bearing different types of reproductive structures, such as the alternation of sporophyte and gametophyte generations; *cf*. monomorphous life cycle.

dinomic Used of a species restricted to two different biogeographical regions.

dioeciopolygamy The presence of both unisexual and hermaphrodite individuals in the same species; *cf*. monoecious polygamy.

dioecious Used of plants or plant species having male and female reproductive organs on different individuals; unisexual; **diecious**, **dioecy**; *cf*. monoecious, trioecious.

dioestrus The quiescent period between breeding periods in polyoestrous animals; **dioestrous**, **dioestrum**; *cf*. anoestrus, monoestrus, polyoestrus.

dioxin Tetrachlorodibenzoparadioxin (TCDD); a highly toxic and environmentally persistent product of the manufacture of 2,4,5-T *q.v.*

dipheny The occurrence in a population of two discontinuous phenotypes (phenes) that lack genetic foundation; **diphenic**, **diphenism**; *cf*. monopheny, polypheny.

diphygenic Exhibiting two types of development.

diphyletic Derived from two distinct ancestral lineages; used of a group comprising descendants of two different ancestors; **diphyly**; *cf*. monophyletic, polyphyletic.

diplanetic Having two motile stages during a single life cycle; **diplanetism**; *cf*. monoplanetic, polyplanetic.

diplo- Prefix meaning double, twofold.

diplobiont 1: A plant flowering twice in a single season. 2: An organism exhibiting a regular alternation of haploid and diploid generations during the life cycle; a diplohaplontic organism; *cf*. haplobiont.

diplogenesis 1: An abnormal duplication of a structure. 2: The postulated changes in germ-plasm accompanying modifications attributed to use and disuse of body organs and structures.

diplohaplontic Used of an organism exhibiting a regular alternation of haploid and diploid generations during the life cycle.

diploid Having a double set of homologous chromosomes, typical of most organisms derived from fertilized egg cells; synkaryotic; *cf*. haploid, monoploid, polyploid.

diploid apogamy Euapogamy *q.v.*

diploidization The attainment or restoration of the diploid state, for example, by the fusion of postmeiotic haploid nuclei.

diplont 1: The diploid stage of a life cycle; diplophase, 2: An organism having a life cycle in which the direct products of meiosis act as gametes; only the gametes of diplonts are haploid; *cf*. haplont.

diplontic life cycle A life cycle characterized by a diploid adult stage producing haploid gametes by meiosis, the zygote forming by fusion of a pair of gametes.

diplophase The diploid stage of a life cycle; diplont; zygophase; *cf*. haplophase.

diplophyte A diploid plant; sporophyte; *cf*. haplophyte.

diploses The doubling of the chromosome number; establishment of the zygotic chromosome number.

diplotype Genoholotype *q.v.*

direct competition Exclusion of one individual or species from a resource by direct aggressive behaviour or by the use of toxins by other individuals or species; *cf*. indirect competition.

direct gradient analysis The analysis of continuously intergrading plant communities in which samples are taken in positions arranged along an environmental gradient.

direct life cycle Homogonic life cycle *q.v.*

direction A statement published by the International Commission on Zoological Nomenclature, completing an earlier decision.

directional selection Selection for an optimum phenotype resulting in a directional shift in gene frequencies of the character concerned and leading to a state of adaptation in a progressively changing environment; dynamic selection; progressive selection; *cf*. disruptive selection, stabilizing selection.

directive coloration Surface markings which divert the attention or attack of a predator to the non-vital parts of the prey body.

directive species A species that attracts a predator, of which it is not a normal prey item, to an area rich in prey species.

disassortative mating Sexual reproduction in which individuals differing in one or more traits mate more frequently than would be expected by chance alone; negative assortative mating; exogamy; heterogamy; disassortive mating; *cf*. assortative mating.

disclimax A disturbed climax; an ecological succession maintained below climax by rapid expansion of introduced species, climatic instability, fire, grazing or by the activities of man.

discolorous Having a non-uniform coloration.

discontinuity A marked interruption in an otherwise continuous sequence of variation, populations, objects or events; disjunction.

discontinuous distribution The occurrence of a species in two or more separate areas, but not in intervening regions.

discontinuous modular society A society *q.v.* in which the members or modular units are separated from each other; equivalent to aggregation or population.

discontinuous variation Variation in which individuals of a sample fall into two or more overlapping classes; *cf.* continuous variation.

discordant Used of twins or members of a group that do not share a particular trait; *cf.* concordant.

discordant elements Elements mistakenly thought to constitute a single organism, for example fragments of a plant, and on which a scientific name was based but which were subsequently found to belong to more than one taxon.

discrete character Qualitative character *q.v.*

discrete signal A communication signal which can be either on or off, but which does not show intermediate states; *cf.* graded signal.

discrete variable A variable that can only assume certain values; *cf.* continuous variable.

diserotization Interference with, or inhibition of, copulation by low temperatures.

disharmonic Used of a fauna or flora that is unusual in its particular combination of species.

disharmony law That the logarithm of a dimension of any part of an animal is proportional to the logarithm of the body length of the animal.

disjunct Distinctly separate; used of a discontinuous range in which one or more populations are separated from other potentially interbreeding populations by sufficient distance to preclude gene flow between them.

disjunction The separation of the chromosomes during nuclear division.

disome Bivalent *q.v.*

disoperation A symbiosis in which one or both participants is adversely affected.

dispermy The penetration of a single ovum by two spermatozoa at the time of fertilization; **dispermic**; *cf.* monospermy, polyspermy.

dispersal 1: Outward spreading of organisms or propagules from their point of origin or release. 2: The outward extension of a species' range, typically by a chance event; accidental migration.

dispersalist model A model explaining biogeographical distributions of whole biotas primarily in terms of dispersal of taxa across or around existing barriers.

disperse phase The discontinuous phase of a colloid *q.v.*

dispersion 1: The pattern of distribution of organisms or populations in space. 2: The non-accidental movement of individuals into or out of an area or population, typically a movement over a relatively short distance and of a more or less regular nature; *cf.* migration. 3: In statistics, the distribution or scatter of observations or values about the mean or central value.

dispersion medium The continuous phase of a colloid *q.v.*

disphotic zone Dysphotic zone *q.v.*

displacement activity 1: The performance of a behavioural act outside the particular functional context of behaviour to which it is normally related. 2: Behaviour resulting from the activation by the specific action potential *q.v.* of one instinct, of the action pattern belonging to another instinct, appearing when a charged instinct is denied the opportunity for adequate discharge through its own consummatory acts *q.v.*; ambivalence.

displacement theory Continental drift *q.v.*

displacement volume A measure of plankton biomass; the volume of a plankton sample which has been drained of water.

display A behaviour pattern or signal modified in the course of evolution to convey information.

disruptive coloration A distinct coloration which delays recognition of the whole animal by attracting the attention of the observer to certain elements of the colour pattern.

disruptive selection Selection for phenotypic extremes in a polymorphic population, which preserves and accentuates discontinuity; centrifugal selection; diversifying selection; *cf.* directional selection, stabilizing selection.

dissemination Scattering or spreading, as of infectious agents, seeds, spores or other propagules; distribution; **disseminate**.

disseminule A disseminated propagule; a seed, fruit, spore or other structure modified for dispersal.

dissociation A mutation which increases the chromosome number.

dissogony Sexual maturation at two separate stages of a life cycle, with an intervening period during which no gametes are produced; dissogeny.

dissophyte A plant with xerophytic leaves and stems, and a mesophytic root system.

distance Taxonomic distance; any measure of dissimilarity between taxa.

distillation A mode of fossilization involving removal of volatile organic matter, leaving a carbon residue.

distraction display An elaborate pattern intended to attract the attention of an aggressor away from other more vulnerable members of a group; deflection display; paratrepsis.

distress call Fright cry *q.v.*

distribution 1: Dissemination *q.v.* 2: The geographical range of a taxon or group. 3: The spatial pattern or arrangement of the members of a population or group.

distrophyte A plant living in firm moist soil.

ditokous Producing two offspring per brood; biparous; *cf.* monotokous, oligotokous, polytokous.

ditopogamy Heterogamy *q.v.*

ditypic Having two distinct morphs; exhibiting marked sexual dimorphism; used of characters possessing two alternative states; dimorphic; **ditypism**.

diurnae Pertaining to day-flying insects; *cf.* nocturnae.

diurnal 1: Active during daylight hours; *cf.* crepuscular, nocturnal. 2: Lasting for one day only.

diurnal rhythm A biological rhythm having a periodicity of about 1 day length (24 h); circadian rhythm.

diurnation The daily fluctuation in the composition of a plant community.

divergence 1: The acquisition of dissimilar characters or traits by related organisms; divergent evolution. 2: A zone of oceanic upwelling where deep water rises and spreads out over the surface, as in the Antarctic divergence; *cf.* convergence.

divergence index Patristic distance *q.v.*

diversifying selection Disruptive selection *q.v.*

diversionary display Behaviour intended to attract the attention of a predator away from the more vulnerable members of a group; less elaborate and stereotyped than a distraction display *q.v.*

diversity 1: The absolute number of species in an assemblage, community or sample; species richness. 2: A measure of the number of species and their relative abundance in a community; low diversity refers to few species or unequal abundances, high diversity to many species or equal abundances. 3: The condition of having differences with respect to a given character or trait.

diversity index A measure of the number of species in a community and their relative abundances; such measures include the ratio between number of species and number of individuals, the Shannon-Wiener index, richness index, Brillouin index and Simpson index.

division 1: Separation into parts; fission. 2: A rank

in the hierarchy of plant classification; the principle category between Kingdom and class, takes the ending -*phyta*; see Appendix 10.

divisive classification A statistical method in which the body of data is divided into groups according to a predetermined criterion, such as maximizing the ratio of intergroup to intragroup variance.

divisive strategy A method of constructing hierarchies by successive division of high ranking units into smaller subordinate units; *cf.* agglomerative strategy.

dixenic Used of a mixed culture of one organism together with two other species; *cf.* axenic, monaxenic.

dixenous Used of a parasite utilizing two host species during its life cycle; **dixeny**; *cf.* heteroxenous, monoxenous, oligoxenous, trixenous.

dizygotic Used of twins derived from two separate fertilized eggs; fraternal (twins); *cf.* monozygotic.

DL$_{50}$ A measure of drought lethality, the degree of dryness that causes injury to 50% of a population.

DM Dry matter.

DNA Deoxyribonucleic acid; the primary genetic material of a cell; a polymer of the nucleotides adenine, guanine, cytosine and thymine, typically containing two polynucleotide chains in the form of a double helix; the sequence of nucleotide pairings in the chains is the basis of the genetic code; DNA molecules are the largest biologically active molecules known.

DOC Dissolved organic matter (carbon), expressed as grams carbon per litre.

Dogger A geological epoch of the middle Jurassic period (*ca.* 175 to 160 million years B.P.).

Dollo's law The general principle that evolution is irreversible and that structures and functions once lost are not regained; irreversibility rule; Dollo's rule; *cf.* Arber's law.

domestication The adaptation of plants and animals for life in intimate association with man.

domicile A home, nest, burrow, tube, den or other refuge; *cf.* hibernaculum.

domicolous Living in a tube, nest or other domicile; **domicole**.

Domin scale A scale for estimating cover and abundance of a plant species, comprising 11 categories: + (single individual), 1 (very few individuals), 2 (sparsely distributed, less than 1% cover), 3 (frequent but less than 4% cover), 4 (4–10% cover), 5 (11–25% cover), 6 (26–33% cover), 7 (34–50% cover), 8 (51–75% cover), 9 (76–90% cover) and 10 (91–100% cover).

dominance 1: The extent to which a given species predominates in a community because of its size, abundance or coverage, and affects the fitness of associated species. 2: The tendency of the phenotypic traits controlled by one allele (dominant) to be expressed over traits controlled by other alleles (recessives) at the same locus. 3: Physical domination initiated and sustained by aggression or other behavioural patterns of an individual.

dominance hierarchy A social order of dominance sustained by aggressive or other behaviour patterns; bunt order; hook order; peck order; dominance order.

dominance modifier A gene which modifies the dominance of another gene.

dominance order Dominance hierarchy *q.v.*

dominant 1: The highest ranking individual in a dominance hierarchy; alpha. 2: An organism exerting considerable influence upon a community by its size, abundance or coverage. 3: An allele which determines the phenotype of a heterozygote; *cf.* recessive.

dominule A dominant organism in a microhabitat, or of a community in a serule *q.v.*

Donau glaciation An early glaciation of the Quaternary Ice Age in the Alpine area, with an approximate duration of 260 thousand years, tentatively subdivided into four separate glacial periods; see Appendix 2.

dormancy 1: A state of relative metabolic quiescence, such as aestivation, cryptobiosis, diapause, hibernation and hypobiosis. 2: A state in which viable seeds, spores or buds fail to germinate under conditions favourable for germination and vegetative growth; dormant; *cf.* quiescence.

dosage compensation The effect produced by modifying genes that compensate for the difference between the dosage of major sex-linked genes present in males and females.

dosage effect The quantitative consequences on the phenotype of gene frequency.

double autogamy Cytogamy *q.v.*

doubling time The average time taken for the number of individuals in a population to double.

DPD Diffusion pressure deficit *q.v.*

drift 1: Genetic drift *q.v.* 2: Lateral displacement as a result of wind or water currents.

drimad A plant of an alkaline substratum.

drimium A plant community of an alkaline plain or salt basin.

drimophilus Thriving in salt basins or alkaline plains; drimyphilous; **drimophile, drimophily.**

drimyphilous Drimophilus *q.v.*; **drimyphile, drimyphily.**

driodad A dry thicket plant.

driodium A dry thicket community.

drive The complex of internal and external states and stimuli directed towards a particular goal or satisfying a particular need; a motivating, impelling internal physiological condition.

dromotropism An orientation response in climbing plants that results in spiral growth; **dromotropic.**

drosphilus Pollinated by dew; **drosophilous, drosophile, drosophily, drosphile, drosphily.**

drought 1: A prolonged chronic shortage, usually of water. 2: A period without precipitation during which the soil water content is reduced to such an extent that plants suffer from lack of water.

drought-evading Used of plants sensitive to drought which can survive dry periods by the production of desiccation resistant seeds or structures.

drought resistance The capacity to withstand periods of dryness, including both desiccation avoidance and desiccation tolerance.

dryad A shade plant.

drymium A woody plant community.

drymophytium A community of bushes and small trees.

drymyphyte Halophyte *q.v.*

dubiofossil A structure or object having a marked resemblance to a known fossil but of which the organic or inorganic nature cannot be ascertained or confirmed; problematic fossil; problematicum.

dubious name *Nomen dubium q.v.*

duff A product of litter decomposition; incompletely decomposed organic matter in which the original structure is no longer discernible.

dulosis Symbiosis in which workers of an ant species capture the brood of another species and rear them as slaves; **dulotic.**

dupli- Prefix meaning double, twofold.

duplicate genes Genes having an identical but non-cumulative effect.

durability The probability that a species or population will be represented by living descendants after a given extended period of time.

duriherbosa Permanent tall grassland vegetation.

durilignosa Broadleaved evergreen forest and scrub vegetation.

dwarfism The condition of being stunted, much smaller than normal; having restricted growth; microsomia; nanism; **dwarf.**

Dyar's law That head width increases in a regular geometric progression through a series of larval instars, in lepidopterans.

dymantic behaviour Deimatic behaviour *q.v.*

dynamic ecotrophic coefficient The ratio of the

consumption of a particular prey species by a predator to the production of that prey species.

dynamic selection Directional selection *q.v.*

dyne (dyn) The cgs unit of force; the force required to accelerate a mass of 1 gram by 1 centimetre per second per second; 1 dyne = 10^{-5} N.

dys- Prefix meaning bad, abnormal, insufficient, malfunction, difficult.

dysanthous Pertaining to the condition of being fertilized by pollen from another plant; **dysanthic**.

dyschronous Pertaining to plant species that do not overlap in their flowering periods.

dysgenesis The condition of infertility between hybrids which are themselves crossfertile with the parental stocks; **dysgenetic**.

dysgenic Pertaining to, or having, the capacity for decreasing the fitness of a race or breed; cacogenic; kakogenic; *cf.* eugenic.

dysgeogenous Used of rock that does not weather readily to form soil; *cf.* eugeogenous.

dysphotic zone The zone of intermediate light intensity in a water body, with insufficient light for photosynthesis but sufficient light for behavioural responses; **disphotic zone**; *cf.* euphotic zone, photic zone.

dysploid Having a chromosome complement that bears no obvious relationship to the polyploid series; aneuploid; **dysploidy**.

dysteleology Purposelessness; frustration of function; as when an insect obtains nectar from a flower by puncturing a nectary (a means not conducive to pollination) rather than via the floral opening.

dystric Pertaining to unhealthy soils; *cf.* eutric.

dystrophic 1: Used of fresh water bodies rich in organic matter mainly in the form of suspended plant colloids and larger plant fragments, but having low nutrient content; dystrophic lake; *cf.* eutrophic, mesotrophic, oligotrophic. 2: Used of species characteristic of dystrophic habitats.

dystrophic lake A dystrophic *q.v.* water body or lake; bog lake; brown water; humic lake.

dyticon An ooze-inhabiting community.

Dzierzon's rule That in social Hymenoptera all fertilized eggs produce females and all unfertilized eggs produce males.

E

e 1: The base of natural (Napierian) logarithms; 2.7183. 2: Latin, meaning from; *ex*.

e.p. Abbreviation of the Latin *ex parte*, meaning in part, partly.

-eae The ending of a name of a tribe in botanical nomenclature.

earlier synonym Older synonym *q.v.*

East African steppe region A subdivision of the African subkingdom of the Palaeotropical kingdom; see Appendix 4.

East African subregion A subdivision of the Ethiopian region; Central and East African subregion; see Appendix 3.

East Australia current A warm surface ocean current that flows south off the east coast of Australia, forming the western limb of the South Pacific gyre; see Appendix 6.

Eastern subregion Alleghany subregion of the Nearctic region; see Appendix 3.

ebb current The tidal current associated with a receding tide; *cf*. flood current.

ebb tide A receding tide; *cf*. flood tide.

ecad A plant or animal form produced in response to particular habitat factors, the characteristic adaptations not being heritable; a habitat form; ecophene; ecophenotype; oecad; oecophene.

ecballium An ecological succession resulting from timber felling.

eccentric Subcircular; not assuming a circular path; located away from the true geometric centre.

ecdemic Foreign; non-native; *cf*. endemic.

ecdysis The act of shedding or moulting the outer exoskeleton or cuticle.

ece Habitat; oikos.

ecesis The pioneer stage of dispersal to a new habitat; successful invasion and establishment by colonizing plants; oecesis; oecisis; oikesis; **ecize**.

echard That part of soil water not available for plant use; *cf*. chresard, holard.

echolocation The perception of objects using high frequency sound waves, used by some animals for navigation and orientation within the environment.

echo-sounding The use of sound waves (acoustic signals) to measure depth of water, plot the bottom profile, or locate submerged objects and dense aggregations of relatively small organisms.

ecize To undergo ecesis *q.v.*; to colonize.

eclectic Composed of theories or doctrines drawn from a number of different sources.

eclipse plumage Dull inconspicuous plumage of birds which alternates with a much brighter breeding plumage.

eclosion The emergence of an adult insect from the pupal case; sometimes also used for the hatching of an egg; **eclose**.

ecochronology The dating of biological events using palaeoecological evidence; oekochronology.

ecocide Any toxic substance that penetrates and kills an entire biological system.

ecoclimate The immediate climate of an individual organism; microclimate.

ecoclimatic Used of adaptations to ecoclimate *q.v.*

ecoclimatology The study of plants and animals in relation to climate; bioclimatology.

ecocline 1: A more or less continuous character variation in a sequence of populations distributed along an ecological gradient, with each population exhibiting local adaptation to its particular segment of the gradient; a gradient of ecotypes. 2: The differences in community structure resulting from changes in slope aspect around a mountain or ridge.

ecodeme A local interbreeding group occurring in a particular habitat.

ecodichogamic Used of dioecious organisms having different maturation times.

ecogeographical rule Any generalization describing a trend of geographical variation correlated with environmental conditions; climatic rule; *cf*. Allen's law, Bergmann's rule, Gloger's rule, heart-weight rule, Hopkin's bioclimatic law, Rensch's laws.

ecography Descriptive ecology.

ecogroup In palaeontology, the successive communities in a similar habitat.

ecological age The reproductive status (prereproductive, reproductive, postreproductive) of an organism.

ecological amplitude The range of a given environmental factor over which an organism or process can function; range of tolerance.

ecological bonitation A measure of the numerical abundance of an organism at a given place or time.

ecological efficiency The efficiency of transfer of energy from one trophic level to the next.

ecological efficiency, coefficient of The ratio of energy assimilated at a given trophic level to that assimilated at the previous trophic level; gross ecological efficiency.

ecological equivalence The situation in which two or more species can replace each other in the same ecological niche because of their similar ecological amplitudes; *cf.* ecological equivalent.

ecological equivalents Unrelated or distantly related species fulfilling similar ecological roles in different communities or geographical areas.

ecological isolation The absence of interbreeding between sympatric populations because of ecological barriers; a premating isolation *q.v.* mechanism; ecological segregation.

ecological longevity The average life span of an individual in a given population and under stated conditions.

ecological niche The concept of the space occupied by a species, which includes both the physical space as well as the functional role of the species; conceptualized as a multidimensional hypervolume which defines the biological space occupied by a species, and which is unique to the species; the dimensions of this space are the parameters of the niche; sometimes erroneously used as the equivalent of microhabitat; niche; *cf.* fundamental niche, realized niche.

ecological pressure The totality of all environmental factors that act as agents of natural selection; *cf.* prime mover.

ecological pyramid A model of the trophic structure of a community which takes the form of a pyramid of numbers, biomass or energy, in which producers form the base of the pyramid and successive levels represent consumers of higher trophic levels; food pyramid.

ecological race Ecotype; a local race of a species having conspicuous adaptive characters correlated with a given habitat type.

ecological release The enlargement of the niche *q.v.* of a species due to the removal of a competitor or other restricting species; *cf.* character release.

ecological segregation Ecological isolation *q.v.*

ecological sociology Synecology *q.v.*

ecological succession The gradual and predictable process of progressive community change and replacement, leading towards a stable climax community; the process of continuous colonization and extinction of species populations at a particular site; sere.

ecological valence Ecological tolerance or amplitude *q.v.*; ecological valency.

ecological valency, law of relativity of That the range of tolerance of a species is not constant throughout its entire range but varies from locality to locality according to variations in environmental factors.

ecological zoogeography The study of the distributions of animal species in relation to their life conditions and dispersal potential.

ecology The study of the interrelationships between living organisms and their environment; bioecology; oecology; oikology; **ecological**.

econ A local vegetational unit; **eca**.

ecoparasite A parasite restricted to a specific host or to a small group of related host species; oecoparasite.

ecophene All the naturally occurring phenotypes produced within a given habitat by a single genotype; ecad; oecophene.

ecophenotype A phenotype exhibiting non-genetic adaptations associated with a given habitat, or to a given environmental factor; ecad.

ecophysiology The study of the physiological adaptations of organisms to habitat or environment.

ecopotential Those aspects of the average genotype of a species that determine the ecological characteristics that develop in individuals of the species.

ecoproterandry The maturation of staminate flowers before pistillate flowers.

ecoproterogyny The maturation of pistillate flowers before staminate flowers.

ecospace Fundamental niche *q.v.*

ecospecies A group of populations or ecotypes having the capacity for free exchange of genetic material without loss of fertility or vigour, but having a lesser capacity for such exchange with members of other ecospecies groups; closely approximates to a biological species *q.v.*; oekospecies.

ecosphere Biosphere *q.v.*

ecostratigraphic unit A stratigraphic unit based on the nature of the environment at the time of deposition of the sediment; ecozone.

ecostratigraphy The study and classification of stratified rocks according to their mode of origin or to the environment at the time of deposition; oecostratigraphy; oekostratigraphy; **ecostratigraphic**.

ecosystem A community of organisms and their physical environment interacting as an ecological unit; the entire biological and physical content of a biotope; biosystem; holocoen.

ecosystem respiration (R_e) The total energy utilized during respiration *q.v.* by autotrophs (R_a) and heterotrophs (R_h); calculated as $R_e = R_a + R_h$.

ecosystematics The integrated study of the ecology, evolution and systematics of ecosystems.

ecotone The boundary or transitional zone between adjacent communities or biomes; tension zone.

ecotope 1: A particular habitat type within a larger geographical area. 2: The full range of adaptations of a species to external factors operating both within (niche components) and between (habitat components) communities.

ecotype 1: A locally adapted population; a race or infraspecific group having distinctive characters which result from the selective pressures of the local environment; ecological race. 2: A subunit within an ecospecies *q.v.* comprising individuals capable of interbreeding with members of that and other ecotypes within the ecospecies but remaining distinct through selection and isolation.

ecotypification The formation of ecotypes in a particular habitat.

ecozone An assemblage zone characterizing particular ecological conditions; ecostratigraphic unit.

ecronic Estuarine.

ectad Outwards; towards the exterior; **ectal**; *cf.* entad.

ectendotrophic Used of a parasite that feeds from both the exterior and interior of its host; **ectendotroph**.

ecto- Prefix meaning outside, outer.

ectocommensal A commensal symbiont that lives on the external surface of its host; *cf.* endocommensal.

ectocrine Exocrine *q.v.*

ectodynamorphic Pertaining to a soil type that is determined predominantly by the climate and vegetation, rather than by the properties of the parent material; *cf.* endodynamorphic.

ectogenesis 1: The generation of variation by extrinsic factors. 2: The development of an embryo *in vitro*.

ectogenic meromixis The mixing of a lake as a result of an external event or agency; as in the influx of saline water into a fresh water lake producing a density stratification; *cf.* biogenic meromixis, crenogenic meromixis.

ectogenous Arising or originating outside the organism or system; allochthonous; exogenous; **ectogenesis, ectogenic**; *cf.* endogenous.

ectohormone Pheromone *q.v.*

ectoophagous Used of an insect larva that hatches from an egg deposited on or near a supply of host eggs and feeds upon them; **ectoophage, ectoophagy**; *cf.* endoophagous.

ectoparasite A parasite that lives on the outer surface of its host; ectosite; epiparasite; episite; exoparasite; *cf.* endoparasite.

ectophagous Feeding on the outside of a food source; **ectophage, ectophagy**; *cf.* endophagous.

ectophloedic Growing on the outside of bark; epiphloedic; **ectophloedal, ectophloeodal, ectophloeodic**; *cf.* endophloedic.

ectophyte A plant parasite living on the outer surface of another organism; **ectophytic**; *cf.* endophyte.

ectopic Occurring in an abnormal place or manner.

ectosite Ectoparasite *q.v.*

ectosymbiosis A symbiosis in which both symbionts are external, neither lives within the body of the other; exosymbiosis; **ectosymbiont**; *cf.* endosymbiosis.

ectothermic 1: Poikilothermic; having a body temperature determined primarily by the temperature of the environment; cold-blooded; **ectotherm, ectothermal**. 2: Pertaining to a chemical reaction releasing heat energy; exothermic; *cf.* endothermic.

ectotrophic Obtaining nourishment externally without marked penetration into the food source; **ectotroph**; *cf.* endotrophic.

ectotropism Orientation movement away from the central axis; **ectotropic**; *cf.* endotropism.

ectozoon An animal ectoparasite.

ecumene The biosphere; any inhabited area.

ecumenical Cosmopolitan; worldwide in extent or influence; pandemic.

edaphic Pertaining to, or influenced by, the nature of the soil.

edaphic climax A more or less stable community, the structure and composition of which are determined largely by the properties of the soil or substratum; soil climax; *cf.* climax.

edaphic factors The physical, chemical and biological properties of the soil or substratum, which influence associated biota.

edaphic race A locally adapted population exhibiting adaptation to particular properties of the soil or substratum; edaphic ecotype.

edaphology The study of soils, particularly with reference to the biota and man's use of land for plant cultivation; pedology.

edaphon The soil flora and fauna; those organisms living in the interstitial water and pore spaces of soil; edaphonekton.

edaphonekton Edaphon *q.v.*

edaphophyte A plant living on or immediately beneath the surface of the soil.

edaphotropism An orientation response to a soil water stimulus; **edaphotropic**.

eddy A current of air or water flowing in a direction contrary to the main stream.

edge effect The effect exerted by adjoining

communities on the population structure within the marginal zone (ecotone), which often contains a greater number of species and higher population densities of some species than either adjoining community.

edge species A species found predominantly or commonly in the marginal zone of a community.

Ediacarian period The period of geological time immediately preceding the Cambrian (*ca.* 700–570 million years B.P.); the last of the Precambrian periods.

edition The total number of copies produced from one typesetting or from one set of plates; subsequent editions usually involve some revision of the text and resetting of the type, or production of new plates; *cf.* impression.

edobolous Having seeds dispersed by propulsion by means of turgor pressure; **edobole.**

edominant A secondary or accessory species that has little or no dominance in a community.

EDP Electronic data processing.

Eemian interglacial The interglacial period preceding the Weichselian glaciation in northern Germany and Poland; see Appendix 2.

EEZ Exclusive economic zone *q.v.*

effective breeding population The actual number of breeding individuals in a population.

effective lethal phase The stage in development at which a lethal gene is effective and causes death of the organism.

effective population number The number of individuals in an ideal population that would have the same rate of heterozygosity decrease as the real population under consideration.

effective population size The average number of individuals in a population which are assumed to contribute genes equally to the succeeding generation.

effective precipitation That part of total precipitation that becomes available for plant growth.

effective publication Publication in accordance with the requirements of the Code.

effete Functionless; no longer fertile or no longer functioning as a result of age.

efficiency of consumer ingestion (ECI) A measure of the food value of a given resource, calculated from body growth or production (P) and ingested food or consumption (C) by the formula $ECI = P/C$.

effluent The discharge of industrial or urban waste material into the environment; the outflow from a lake or river.

effodient Having the habit of digging.

egesta (F) In ecological energetics, that part of consumption expelled as faecal material or

regurgitated, and thus not absorbed; a component of rejecta *q.v.*

egestion Defaecation; the voiding of faecal or regurgitated matter.

egocentric Pertaining to characters or traits that favour the survival of the individual.

E-horizon A pale coloured layer below the A-horizon of a soil profile, having a lower organic content than the layers above and below; a zone of high eluviation; see Appendix 12.

eigenvalue In principle component analysis, the proportion of the total variance accounted for by the corresponding principle component.

Eimer's principle Epistacy *q.v.*

einstein A unit of amount of light quanta, equal to 1 mole photons.

Ekman spiral The phenomenon that causes the water of an oceanic gyre to pile up at its centre.

elaboration The process of producing or evolving complex structures from simpler ones; *cf.* reduction.

elaioplankton Planktonic organisms utilizing oil droplets for buoyancy.

elastic strain A reversible physical or chemical change in a living organism produced by stress.

Elbe glaciation A glaciation of the Quaternary Ice Age in northern Germany and Poland, with an estimated duration of 100 thousand years; see Appendix 2.

electrophoresis A technique for separating mixtures of organic molecules, based on their different rates of travel in an electric field.

electropism Electrotropism *q.v.*; **electropic.**

electrotaxis A directed reaction of a motile organism in response to an electric field; galvanotaxis; **electrotactic.**

electrotropism An orientation in reponse to an electric field; galvanotropism; electropism; **electrotropic.**

elements In taxonomy, the constituent parts that fall within the limits of a taxon, such as the species within a genus.

eleutherozoic Free-living.

elevation The vertical distance between a given point and a datum surface; altitude.

elfin forest A forest of higher elevations in warm moist regions, characterized by stunted trees with abundant epiphytes; *cf.* krummholz.

elimination In ecological energetics, any loss of biomass or energy by a population or trophic unit per unit time, per unit area or volume; including losses due to mortality, predation, emigration and moulting; the energy of eliminated biomass is available to, or utilized by, other trophic levels in the ecosystem.

elittoral The zone of the sea bed below the sublittoral, extending to the limit of light penetration; sometimes used for the sea bed below 40 m.

Elsterian glaciation A glaciation of the Quaternary Ice Age in northern Germany and Poland, with an estimated duration of 90 thousand years; see Appendix 2.

elutriation A method of removing interstitial organisms from a sediment sample by continuous flushing with water; **elution**.

eluvial layer The leached upper layer of a soil profile.

eluviation The translocation of suspended or dissolved soil material by the action of water; usually the removal of substances in solution is termed leaching *q.v.*

eluvium A sand dune community.

emasculation The removal of the male reproductive organs or the inhibition of male reproductive capacity; geld.

embryogenesis The development of an embryo; **embryogenetic**.

embryotrophy The process of nutrition of a mammalian egg before implantation at which point it becomes haemotrophic.

emend. **(em.)** Abbreviation of Latin *emendatus*, meaning amended; used in author citations when the author has changed the original spelling of a taxon without excluding the type of the name; the abbreviation precedes the name of the author effecting the change.

emendation In nomenclature, any demonstrably intentional change in spelling of a published scientific name, other than a mandatory change; any spelling that has been intentionally changed.

emendation, justified The correction of an incorrect original spelling.

emendation, unjustified Any emendation other than a justified emendation; the emended name is treated as a new name with its own authority and date.

emergent 1: An aquatic plant having most of the vegetative parts above water. 2: A tree which reaches above the level of the surrounding canopy; *cf.* submergent.

emergent evolution The appearance of entirely novel and unpredictable characters or traits by the rearrangement of existing potentialities.

emersed Pertaining to a plant or plant structure that projects above the water surface; amphibian; *cf.* submersed.

emersiherbosa Emergent herbaceous vegetation of ponds and swamps; *cf.* aquiherbosa.

emersion zone 1: The uppermost part of the eulittoral zone of a lake which is above water level for most of the year. 2: That part of the seashore covered only by extreme high tides.

Emery's rule The generalization that species of social parasites are very similar to their host species and are therefore presumably closely related to them phylogenetically.

emigration The movement of an individual or group out of an area or population; **emigrant**; *cf.* immigration.

emiocytosis The release of cellular constituents from cells by reverse pinocytosis.

emmenophyte An aquatic plant devoid of any floating parts; **emmophyte**; **emophyte**.

emophytic Pertaining to submerged vegetation; **emophyte**.

empathic learning Observational learning *q.v.*

empirical Based upon direct observation and experience rather than theory or preconception.

empirical taxonomy Classification of organisms based on observed phenotypic similarity.

empiricism A methodology based upon observation and experience rather than established theory.

enantiobiosis Inhibition of, or interference with, one species or population by the action of another; amensalism; antagonism.

enaulad A sand dune plant; enaulophyte.

enaulium A sand dune community.

enaulophilus Thriving in sand dunes; **enaulophile**, **enaulophily**.

enaulophyte A sand dune plant; enaulad.

encasement theory The theory that the embryo is contained within the sperm (spermism) or egg cell (ovism) and that it unfolds during development; incasement theory; preformation theory.

enculturation Passing on of cultural practices to younger members of a social group.

endangered species A species threatened with extinction.

endemic Native to, and restricted to, a particular geographical region; **endemicity**, **endemism**; *cf.* ecdemic.

endemic centre An area having a highly distinctive local biota, with 5–10% endemic species.

ending In taxonomy, the termination of a scientific name.

endo- Prefix meaning within, inside, inwards.

endobenthic Living within the sediment; boring into a solid substratum; *cf.* epibenthic, hyperbenthic.

endobenthos Organisms living within the sediment on the sea bed or lake floor; infauna.

endobiontic Used of organisms that live within the substratum; **endobiotic**, **endobiont**.

endocommensal A commensal symbiont that lives inside its host; *cf.* ectocommensal.

endodeme A local interbreeding group composed

of predominantly inbreeding but dioecious individuals; *cf.* deme.

endodynamorphic Pertaining to a soil type that is determined predominantly by the properties of the parent material; *cf.* ectodynamorphic.

endodyogeny A rare type of sexual reproduction exhibited by some protistans in which two daughter cells develop within a single parent cell which is destroyed in the process.

endoectothrix Growing in or on hair.

endogamy 1: Inbreeding; sexual reproduction between closely related individuals; endokaryogamy. 2: Pollination of a flower by pollen from another flower on the same plant; self-pollination. 3: Sexual reproduction in which there is a greater frequency of mating between related (or relatively closely related) individuals than would occur by chance alone (random mating); homogamy; positive assortative mating; *cf.* exogamy.

endogean An interstitial soil organism.

endogenous Arising or originating from within the organism or system; growing on the inside; autochthonous; **endogenetic**; *cf.* ectogenous, exogenous.

endogenous clock Any physiological system which shows an inherent sustained periodicity.

endokaryogamy. Endogamy *q.v.*

endolithic Growing within a rock or other hard inorganic substratum; petricolous; saxicavous; *cf.* epilithic.

endolithophagic Lithophagic *q.v.*

endolithophytic Pertaining to plants that penetrate rock or other hard inorganic substrata; **endolithophyte**, *cf.* epilithophytic, exolithophytic.

endomitosis Chromosomal replication within the nucleus of a cell that does not subsequently divide, resulting in polyploidy.

endomixis Self-fertilization in which male and female nuclei from one individual fuse; **endomictic**.

endomorphology Internal morphology; anatomy.

endoophagous Used of an insect larva that hatches from an egg deposited within a host egg and feeds upon the contents of that single egg; **endoophage, endoophagy**; *cf.* ectoophagous.

endoparasite An internal parasite which lives within the organs or tissues of its host; endosite; entoparasite; *cf.* ectoparasite.

endopelic Used of aquatic organisms that live within the sediment; *cf.* epipelic.

endopelos The community of aquatic organisms within the bottom sediment.

endopetrion Organisms living in the interstices of rock.

endophagous Feeding from within the food source;

entophagous; **endophage, endophagy**; *cf.* ectophagous.

endophenotypic Used of characters or components of the phenotype which are not directly adaptive and may not affect an organism's competitive abilities; *cf.* exophenotypic.

endophloedic Living or occurring within the bark; endophloic; hypophloeodal; **endophloedal, endophloeodal, endophloeodic**; *cf.* ectophloedic.

endophyllous Living or growing in leaves; *cf.* epiphyllous.

endophyte A plant living within another plant; entophyte; **endophytic**; *cf.* ectophyte.

endopolygeny The formation of many daughter cells, each surrounded by its own membrane, within a mother cell; **endopolygony; endopolygenic, endopolygonic**.

endopolyploidy 1: Endomitosis; multiplication division of the chromosomes without division of the nucleus. 2: The condition of having certain cells or tissues in a polyploid state within an otherwise diploid individual.

endopsammon The microscopic biota inhabiting sand and mud.

endorheic region An area in which rivers arise but do not reach the sea as they are lost in closed basins or dry courses; *cf.* arheic region, exorheic region.

endosite Endoparasite *q.v.*

endosymbiosis Symbiosis in which one symbiont (the endosymbiont, or inhabiting symbiont) lives within the body of the other (the inhabited symbiont); *cf.* ectosymbiosis.

endosymbiosis theory That the three classes of organelles (mitochondria, basal bodies/flagella/cilia, photosynthetic plastids) of eukaryotic cells evolved from free-living prokaryotic ancestors through a series of endosymbiotic relationships.

endothermic 1: Warm-blooded; maintaining a body temperature largely independent of the temperature of the environment; homoiothermic; **endotherm, endothermal**; *cf.* ectothermic. 2: Pertaining to a chemical reaction utilizing heat energy; *cf.* exothermic.

endotrophic Obtaining nourishment internally, as of a mycorrhiza *q.v.* in which the fungus penetrates the host root system; **endotrophism**, *cf.* ectotrophic.

endotropism Orientation movement of lateral organs towards the central axis; **endotropic**; *cf.* ectotropism.

endoxylic Living or growing within wood; *cf.* epixylic.

endozoic 1: Living within or passing through the body of an animal; entozoic; *cf.* epizoic. 2: Used of a method of seed dispersal in which seeds are ingested by an animal and later voided in the faeces.

endozoochorous Dispersed by the agency of animals, typically after passage through the gut; **endozoochore, endozoochory**; *cf.* epizoochorous.

energesis Catabolic processes involving the liberation of energy.

energetics The study of energy transformations within a community or system; ecological energetics.

energy The capacity to do work; involving thermal energy (heat), radiant energy (light), kinetic energy (motion) or chemical energy; measured in joules.

energy budget The balance of energy input and energy utilization within a community or system; expressed as consumption (C) =production (P) +respiration (R) +rejecta (FU).

energy flow The passage of energy into and out of an organism, population or system; the passage of energy through the different trophic levels of a food chain.

energy income Consumption *q.v.*; energy intake.

energy subsidy External energy introduced into a system that reduces the amount of energy required for maintenance of that system.

enhalid Growing in saltings or in loose soil under salt water.

enphytotic Used of plant diseases restricted to particular localities or geographical areas.

entad Inwards; towards the centre or interior; internally; **ental**; *cf.* ectad.

entelechy Orthogenesis *q.v.*

enterozoon A parasitic animal living within the gut of its host.

ento- Prefix meaning within, inner.

entomochoric Dispersed by the agency of insects; **entomochore, entomochory**.

entomogamous Pollinated by insects; entomophilous; **entomogamy**.

entomogenous Living in or on insects.

entomography Description of an insect or of its life history.

entomology The study of insects; **entomological**.

entomopathogenic Causing disease in insects.

entomophagous Feeding on insects; insectivorous; **entomophage, entomophagy**.

entomophilous Pollinated by, or dispersed by the agency of, insects; entomogamous; **entomophile, entomophily**.

entomophyte Any fungus growing on or in an insect.

entomosis A disease caused by an insect parasite.

entomotaxy The preservation and preparation of insects for study.

entomotomy The study of insect anatomy.

entoparasite Endoparasite *q.v.*

entophagous Feeding from within a food source; endophagous; **entophage, entophagy**.

entophytic Used of a plant (an entophyte) growing within another plant.

entozoic Living or passing through the body of an animal; endozoic; **entozoa, entozoon**.

entrainment Coupling or synchronization of a biological rhythm to an external time source (zeitgeber) causing the rhythm to display the frequency of the zeitgeber, as in a biological rhythm having a lunar frequency.

entropy A measure of randomness or disorder in a system; as an example of the application of this concept to a biological system in relation to diversity, areas occupied by a single species have low entropy because there is a high probability of finding that particular species there, whereas transitional zones between adjacent areas have high entropy because there is uncertainty as to which species any given individual will belong.

environment The complex of biotic, climatic, edaphic and other conditions which comprise the immediate habitat of an organism; the physical, chemical and biological surroundings of an organism at any given time.

environment form Ecad *q.v.*

environmental conditioning Modification of the environment through the activity of the biota.

environmental hormone An external chemical substance which exerts an excitatory or inhibitory effect on biological systems or processes; exocrine.

environmental resistance Limitation of the reproductive potential of a group or population by unfavourable environmental conditions, as in the case of overcrowding; *cf.* biotic potential.

environmental sex determination Phenotypic sex determination *q.v.*

environmental variance That part of total phenotypic variance resulting from exposure of individuals within a population to different environmental conditions; *cf.* genetic variance, phenotypic variance.

environmentalism Doctrine emphasizing the role of environmental factors rather than heredity in the development of biological traits, especially behavioural characteristics.

enzootic A disease occurring in a given animal species within a limited geographical area.

eobiogenesis The transformation of prebiotic macromolecular systems into the first living organisms (eobionts); biopoesis.

eobionts The earliest living organisms developed from prebiotic macromolecular precursors.

Eocene A geological epoch within the Tertiary period (*ca.* 54–38 million years B.P.); see Appendix 1.

Eogaea A zoogeographical region incorporating Africa, South America and Australasia; *cf.* Caenogaea.

Eogene Palaeogene *q.v.*

eolation Aeolation *q.v.*

eolian Aeolian *q.v.*

eon 1: An indefinitely long period of geological time; aeon. 2: A unit of time equal to 10^9 years.

eophyte A fossil plant from the earliest fossiliferous rocks.

Eophytic The period of geological time during which the Algae were abundant; *cf.* Aphytic, Archaeophytic, Caenophytic, Mesophytic, Palaeophytic.

eosere A major ecological succession within the climatic climax of a geological period; the ecological succession of vegetation of an era or aeon.

Eozoic The Precambrian era *q.v.*; sometimes used to refer to the earliest part of the Precambrian only.

epacme The period in phylogenetic or ontogenetic development of a group or organism just prior to the point of maximum vigour or adulthood; *cf.* acme, paracme.

epedaphic Pertaining to climatic factors or conditions.

epeiric sea A shallow sea covering part of a continental landmass, typically less than 200 m in depth; sometimes also used to include pericontinental seas *q.v.*; epicontinental sea.

epeirogenic Used of broad, mainly vertical, deformations of the Earth's crust which produce the major features of relief; **epeirogenesis**, **epeirogeny**.

epeirophoresis theory Continental drift *q.v.*

epembryonic Pertaining to the developmental stages following birth and the completion of the embryonic phase.

ephaptomenon Organisms which are adnate or attached to a surface; *cf.* planomenon, rhizomenon.

epharmonic convergence The appearance of morphological or structural similarity in groups that are not closely related.

epharmony The process of gradual adaptation to a changing environment; **epharmose, epharmosis**.

ephebic 1: Pertaining to the adult stage, between the juvenile and old age. 2: Pertaining to an evolutionary peak or acme *q.v.* 3: Used of features which first appeared in evolution in the adult stage; *cf.* neanic.

ephemer 1: An organism that survives only a short

period of time when introduced into a new area. 2: An organism that is sexually mature for a single day only.

ephemeral Lasting for only a day; short-lived or transient, as in an organism that grows, reproduces and dies within a few hours or days, or a flower that lasts for a day or less; **ephemerous**.

ephydate An aquatic plant having its vegetative parts floating at the surface but not normally projecting above the surface; *cf.* hyperhydate, hyphydate.

ephydrogamous Having water borne pollen grains transported at the water surface; or being pollinated by such pollen; *cf.* hyphydrogamous.

epi- Prefix meaning upon, on the surface of.

epibenthic Living at the surface of the substratum, on the sea bed or on the lake floor; **epibenthile**, **epibenthous**; *cf.* endobenthic, hyperbenthic.

epibenthos The community of organisms living at the surface of the sea bed or lake floor.

epibiontic Living attached to another organism but without benefit or detriment to the host; epicolous; **epibiotic, epibiont**.

epibios The epibenthic *q.v.* community.

epibiosis A symbiosis in which one organism lives upon the outer surface of another.

epibiotic 1: Used of an endemic species that is a relic of a former fauna or flora. 2: Epibiontic *q.v.*

epiclysile Pertaining to the tide pools of the upper shore.

epicolous Living attached to the surface of another organism but without benefit or detriment to the host; epibiontic; **epicole**.

epicontinental sea A shallow sea covering part of a continental landmass, typically less than 200 m in depth; sometimes also used to include pericontinental seas *q.v.*; continental sea; epeiric sea; inland sea.

epideictic display A synchronized communal display by which members of a population reveal their presence and allow others to assess population density; conventional behaviour.

epidemic Used of a disease affecting a high proportion of the population over a wide area; *cf.* epiphytotic, epizootic.

epidemiology The study of factors affecting the spread of diseases in populations or communities.

epidendric Pertaining to epiphytes growing on trees and shrubs.

epidermophyte Dermatophyte *q.v.*

epifauna The total animal life inhabiting a sediment surface or water surface; epibenthos; *cf.* infauna.

epigamic Used of any character or trait that serves to attract or stimulate individuals of the opposite sex during courtship, other than essential structures and behaviour of copulation; **epigamous**.

epigamic selection Sexual selection *q.v.*

epigean Living or growing at or above the soil surface; **epigaeic, epigeal, epigeous**.

epigeic Terrestrial; of the soil surface.

epigenesis 1: The theory that the embryo forms by successive gradual changes in the amorphous zygote; *cf.* preformation theory. 2: A change in the mineral character of a rock as a result of external influences.

epigenetic Pertaining to the interaction of genetic factors and the developmental processes through which the genotype is expressed in the phenotype.

epigenetic sequence The switching on and off of different genes in a precise order during development.

epigenetics The study of the causal mechanisms of development.

epigenotype The entire developmental complex of gene interactions which produces the phenotype.

epigenous Developing or growing on a surface.

epigeotropism An orientation response of a plant producing growth across the surface of the soil; **epigeotropic**.

epigeous Epigean *q.v.*

epilimnion The warm upper layer of circulating water above the thermocline in a lake; **epilimnetic, epilimnile**; *cf.* hypolimnion.

epilithic Growing on rocks or other hard inorganic substrata; petrophilous; *cf.* endolithic.

epilithophagic Lithophagic *q.v.*

epilithophytic Pertaining to a plant growing on the surface of stones, rocks and other hard inorganic substrata; exolithophytic; **epilithophyte**; *cf.* endolithophytic.

epilittoral 1: The zone on the seashore above the intertidal zone, which is influenced by the effects of salt spray; spray zone. 2: The paralimnetic zone of a lake entirely above water level and uninfluenced by spray.

epimeletic Used of social behaviour patterns in animals relating to the care of other individuals; *cf.* etepimeletic.

epimorphosis 1: A form of development in arthropods in which all larval forms are suppressed or passed within the egg prior to hatching and the juvenile hatches with the adult morphology. 2: Regeneration of a structure by an organism.

epinasty Downward curvature of a plant structure due to differential growth of upper and lower surfaces; **epinastic**.

epinekton Organisms attached to actively swimming (nektonic) forms but which are incapable of independent movement against water currents; **epinecton, epinectonic, epinektonic**.

epineuston Organisms living in the air on the surface film of a water body; a component of the neuston *q.v.*; supraneuston; **epineustic, epineustonic**; *cf.* hyponeuston.

epiontology The study of the developmental history of plant distributions.

epiorganism Any colony or society of organisms acting as a functional unit; superorganism.

epiparasite Ectoparasite *q.v.*

epipelagic zone The upper oceanic zone extending from the surface to about 200 m; see Appendix 7.

epipelic Used of those aquatic organisms moving over the sediment surface or living at the sediment/water interface; *cf.* endopelic.

epipetalous Living on petals.

epiphenomenon An event which occurs together with another but which is not causally linked and has no effect upon it.

epiphloedic Growing on the surface of bark; ectophloedic; **epiphloedal, epiphloeodal, epiphloeodic**.

epiphloeophyte A plant living on the surface of bark.

epiphyllous Used of an epiphyte (an epiphyll) growing on a leaf.

epiphyte 1: A plant growing on another plant (the phorophyte) for support or anchorage rather than for water supply or nutrients; aerophyte. 2: Any organism living on the surface of a plant.

epiphytology The study of the nature and ecology of plant diseases.

epiphyton Sparsely distributed periphyton *q.v.*

epiphytotic Pertaining to an epidemic disease amongst plants.

epiplankton 1: Planktonic organisms living within the surface 200 m (the epipelagic zone); **epiplanktonic**; *cf.* bathyplankton. 2: Organisms living attached to larger planktonic organisms or to floating objects.

epipleuston Organisms which move over the surface film of a water body with most or all of their bodies above the water; **epipleustonic**.

epipsammic Attached to sand particles.

epipsammon Organisms living on the surface of a sandy substratum or on the surface of the sand particles.

epirhizous Growing on the surface of roots.

epirrheology The study of the effects of exogenous factors on plants.

episematic Pertaining to a character or trait (an episeme) that aids in recognition, as in some colour markings; **episematism**; *cf.* antepisematic, proepisematic, pseudepisematic.

episite 1: An ectoparasite *q.v.* 2: A predator dependent upon utilizing a succession of prey organisms in order to complete its life cycle.

epistacy The manifestation of a greater degree of evolutionary modification in one of two related taxa; Eimer's principle; **epistasis, epistasy**.

epistasis 1: The interaction of non-allelic genes in which one gene (epistatic gene) masks the expression of another at a different locus; *cf.* hypostasis. 2: Epistacy *q.v.*

epiterranean Living or growing at the soil/air interface.

epithet In taxonomy, the second word of a binomial name of a species and the second and third words of a trinomial name of a subspecies; specific name; trivial name.

epitheta hybrida In taxonomy, specific epithets made up of parts of words from two or more different languages.

epitheta specifica rejicienda Rejected specific epithets.

epitokous Reproductive; having or producing offspring; *cf.* atokous.

epitrophy Increased growth on the upper side; **epitrophic**.

epitropism Geotropism q.v.; **epitropic**.

epixylic Living or growing on wood; **epixylous**; *cf.* endoxylic.

epizoic 1: Living attached to the body of an animal; used of a non-parasitic animal that lives attached to the outer surface of another animal; **epizoan, epizoism, epizoite, epizoon**; *cf.* endozoic. 2: Dispersed by attachment to the surface of an animal; epizoochorous.

epizoochorous Dispersed by attachment to the surface of animals; epizoic; **epizoochore, epizoochory**; *cf.* endozoochorous.

epizoon An organism living attached to the body of an animal; epizoite; epizoan.

epizootic Pertaining to an epidemic disease in animals.

epoch 1: A major interval of geological time; a subdivision of a period. 2: An event or time which marks the beginning of a new phase of development.

epoikophytic Used of incompletely naturalized plants which are almost entirely confined to roadside or pathway sites; **epoikophyte**.

epontic Used of an organism that lives attached to the substratum.

epsilon karyology In the study of chromosomes, the determination of the main distinctive loops on the basis of lampbrush chromosome analysis; *cf.* karyology.

Equatorial current A warm surface ocean current that flows west in the tropical Pacific Ocean; see Appendix 6.

equatorial tide The tides that occur at intervals of about two weeks when the moon is over the equator; at this time successive high and low tides show minimum inequality.

Equatorial zone A latitudinal zone extending about 15° on either side of the equator.

equilibrium community A community or biota in which the rate of extinction of species is equal to the rate of immigration of new species.

equilibrium model A model of the structure and evolution of island biotas in which species equilibrium is reached when the rate of extinction balances the rate of immigration or colonization by new species.

equilibrium population 1: A population in which gene frequencies are at equilibrium. 2: A population having a stable size in which death and emigration rates are balanced by birth and immigration rates.

equinoctial 1: Pertaining to the equinox. 2: Used of a plant which has flowers which open and close at particular times during the day; horological.

equinoctial tide A tide of high amplitude occurring when the sun is at or near the equinox.

equinox Either of the two occasions each year when the sun crosses the equator, producing day and night of equal duration.

equipotent Having an equal effect or potential.

equitability The property of a community that relates to the evenness of distribution of species or their relative abundances; maximum equitability indicates that all species are represented by a similar number of individuals and minimum equitability that one species only is dominant and all others are sparsely represented; evenness.

era Any of the major intervals of geological time; the sequence of eras being Precambrian, Palaeozoic, Mesozoic and Cenozoic; see Appendix 1.

eradicate To destroy totally; to exterminate; **eradication**.

eremacausis The process of humus formation by the oxidation of plant matter.

eremad A desert plant; eremophyte.

eremic Pertaining to deserts or sandy regions; **eremean**.

eremium A desert community.

eremobic Living in isolation; having a solitary existence.

eremobiontic Living in desert regions; deserticolous.

eremology The study of deserts.

eremophilous Thriving in desert regions; **eremophile, eremophily**.

eremophyte A desert plant; eremad.

erg 1: A sandy desert. 2: The cgs unit of work; the work done by a force of 1 dyne acting over a distance of 1 centimetre; 1 erg $= 10^{-7}$ J.

ergasialipophyte A relic plant from earlier cultivation.

ergasiaphyte A plant introduced by man for cultivation; **ergasiophyte**.

ergasiapophytic Used of plants (ergasiapophytes) that colonize cultivated fields.

ergatandromorphic Used of social insects in which worker and male characters are blended; **ergatandomorph**; *cf.* ergatogynomorphic.

ergatandrous Having worker-like males, as in some social insects.

ergatogynomorphic Used of social insects in which worker and female characters are blended; **ergatogynomorph**; *cf.* ergatandromorphic.

ergatogynous Having worker-like females, as in some social insects; **ergatogyne**; *cf.* ergatandrous.

ergatomorphic Resembling a worker; ergatoid.

ergatotype In taxonomy, a type of a worker caste of a social insect species.

ergodic Pertaining to a process in which any sequence or sample is representative of the whole.

ergodicity The property of populations for which averages over time can be substituted for averages over space.

ergonomics The quantitative study of work, performance and efficiency.

ericaceous Pertaining to a heath.

ericetal Growing on heathland or moorland; ericeticolous.

ericilignosa A plant community dominated by heaths; erifruticeta.

ericophyte A plant growing on heathland.

erosion Wearing away; weathering; the removal of the land surface by water, ice, wind or other agencies; **erode, erosive**; *cf.* accelerated erosion. geological erosion, gully erosion, natural erosion, normal erosion, rill erosion, sheet erosion and splash erosion.

erpoglyph A fossil worm cast.

err. typogr. Abbreviation of the Latin *errore typographico*, meaning by typographical error.

erratic 1: Pertaining to unattached organisms that are moved around by physical agencies. 2: A rock fragment transported into an area from outside, found either incorporated into the sediment or lying free; glacial erratic.

error 1: In statistics, the deviation of an obtained value from an expected value; *cf.* Type I and Type II error. 2: In nomenclature, an unintentional incorrect spelling, such as a typographical error.

ersatz A substitute; typically an artificial and inferior alternative.

erucivorous Feeding on caterpillars; **erucivore, erucivory**.

erythrism Excessive and abnormal redness in the coloration of an individual or population.

escape A cultivated plant which has become established in the wild.

escapement clock Biological clock *q.v.*

escaper Breakthrough *q.v.*

escatophyte A plant of a climax community.

esculent Edible.

esoteric Arising from within the organism or system; restricted to a small group.

éspeces jumelles Sibling species *q.v.*

essential element A chemical element which is essential to the life of an organism.

essentialism 1: Typology *q.v.* 2: Belief in the reality of underlying universal principles.

establish In taxonomy, to erect or make available; to publish a scientific name or nomenclatural act within the provisions of the Code.

established Growing and reproducing successfully in a given area; **establishment**.

estival Aestival *q.v.*

estivation Aestivation *q.v.*; **estivate**.

estrus Oestrus *q.v.*

estuary Any semi-enclosed coastal water, open to the sea, having a high fresh water drainage and with marked cyclical fluctuations in salinity; usually the mouth of a river.

et Latin, meaning and; used in nomenclature to connect the names of co-authors, often substituted by an ampersand.

et al. Abbreviation of Latin *et alii*, meaning and others; used in an author citation to indicate that a publication or scientific name was written or published jointly by more than two authors.

etepimeletic Used of social behaviour patterns in young animals serving to elicit care from adults; *cf.* epimeletic.

etheogenesis Parthenogenesis in which only males are produced from unfertilized male gametes.

Ethiopian region A zoogeographical region comprising that part of the African continent south of the Sahara, tropical Arabia, Madagascar and neighbouring islands; subdivided into East African, Malagasy, South African and West African subregions; Madagascar and its neighbouring islands are sometimes excluded and regarded as a separate region; see Appendix 3.

ethnobotany Study of the use of plants by the races of man.

ethnology Study of the character, history and culture of the races of man; ethnography.

ethnozoology Study of the use of animals and animal products by the races of man.

ethocline A graded series in the expression of a particular behavioural trait within a group of related species.

ethogram A list or description of the behavioural patterns of a given species, typically of individual behaviour or pair interactions.

ethological Pertaining to behaviour, particularly the species' specific aspects of behaviour.

ethological isolation The absence of interbreeding between members of different populations because of behavioural differences that preclude effective mating; a premating isolation *q.v.* mechanism.

ethology The study of animal behaviour; **ethological**.

ethospecies Species distinguished primarily by behavioural traits.

etiolation Abnormal growth of a green plant when grown in darkness; such plants are pale yellow due to the absence of chlorophyll and the stems are typically elongate bearing small leaves.

etiology Aetiology *q.v.*

-etum Suffix used in phytosociology to denote an association *q.v.* by combination with the root of the generic name of the dominant species of the association (the specific name being placed in the genitive case), for example the name of the red fir association, Abietum magnificae, is derived from the dominant species *Abies magnifica*.

etymology 1: The derivation and meaning of a word, as of a scientific name or epithet in taxonomy. 2: That branch of linguistics concerned with the ancestry of words and language.

euapogamy Development of a diploid sporophyte from one or more cells of the gametophyte without fusion of gametes; diploid apogamy.

euapospory The total failure to form spores.

euautochthony The accumulation of organic remains in the site, and more or less in the same relative positions, in which the organisms from which they were derived lived; *cf.* hypoautochthony.

eubacteria One of the three primary kingdoms (urkingdoms *q.v.*) of living organisms; comprising all typical bacteria; *cf.* archaebacteria, urkaryote.

eubiosphere The true biosphere; that part of the biosphere in which the physiological processes of living organisms can occur; comprising the allobiosphere and autobiosphere; biogeosphere; *cf.* parabiosphere.

eucaryotic Eukaryotic *q.v.*; **eucaryote**.

eucaval A troglophile *q.v.*

euchromosome An autosome *q.v.*

Euclidean distance In numerical taxonomy, the distance between OTUs calculated by extension of Pythagoras' theorem.

Euclidean interstand distance In stand ordination, a measure of the difference between two stands of plants; derived from the species scores for the two stands and from the number of species common to both stands.

Euclidean space Three-dimensional space.

euclonal Used of a society *q.v.* comprising genotypically identical modular growth units that may follow an independent existence if separated from the parent organism (ramets); *cf.* paraclonal, pseudoclonal.

eucoen Those members of a community that are unable to exist outside that community; *cf.* tychocoen.

eudominant A species more or less restricted to a particular climax community.

euephemerous Used of flowers that have a single opening period; **euphemeral**.

euepineuston Organisms that remain in the epineuston throughout life; **euepineustonic, euepineustont**.

eugenic Pertaining to, or having the capacity for, increasing the fitness of a race or breed; aristogenic; *cf.* dysgenic.

eugenics The science of breeding; the application of genetic principles to the improvement of the hereditary qualities of a race or breed.

eugeogenous Used of rock that weathers readily to form soil; *cf.* dysgeogenous.

eugeophyte A plant having a resting period during its life cycle during which it retains only its hypogean parts.

eugonic Prolific; growing profusely.

euhalobous Living in seawater with a salinity of 30–40 parts per thousand; used especially of phytoplankton.

euhaline Living only in saline inland water bodies.

euheterosis True hybrid vigour *q.v.*

euhydatophyte Any fully submerged plant species that never produces aerial structures.

euhydrophilous Thriving submerged in fresh water; **euhydrophile, euhydrophily**.

euhyponeuston Organisms that remain in the hyponeuston throughout life; **euhyponeustonic, euhyponeustont**.

eukaryotic Used of organisms the cells of which

have a discrete nucleus separated from the cytoplasm by a membrane, with DNA as the genetic material and with defined cytoplasmic organelles; **eucaryotic, eukaryote**; *cf.* prokaryotic.

eulittoral 1: The shore zone between the highest and lowest seasonal water levels in a lake; often a zone of disturbance by wave action; see Appendix 8. 2: The intertidal zone of the sea shore; see Appendix 7.

euneuston Organisms that remain in the neuston throughout life; **euneustonic, euneustont**.

eupelagic Used of planktonic organisms found only in open oceanic water away from the influence of the sea bed.

euperiphyton Sessile organisms attached to rooted aquatic plants; *cf.* pseudoperiphyton.

euphemeral Used of flowers that open and close within a single day; **euephemerous, euphemera**.

euphenics That branch of biology concerned with the improvement of genetically defective individuals after birth.

euphilous Used of a plant or flower that has morphological adaptations for attracting and guiding a specialized pollinator; **euphile, euphily**; *cf.* allophilous.

euphotic zone The surface zone in the sea or large lake with sufficient light penetration for nett photosynthesis to occur; *cf.* dysphotic zone, photic zone.

euphotometric Pertaining to leaves orientated to receive maximum light; *cf.* panphotometric.

euphototropism Orientation response of plant structures to maximize incident illumination; **euphototropic**.

euplankton 1: Holoplankton *q.v.* 2: Plankton of open waters.

euploid A polyploid with an exact multiple of the basic chromosome number; **euploidy**; *cf.* aneuploid.

eupotamic Pertaining to an aquatic organism thriving in both flowing and standing fresh water; *cf.* autopotamic, tychopotamic.

Euramerica A continental landmass comprising Europe and North America formed from Eurasia *q.v.* after the break up of Pangaea; *cf.* Angara, Cathaysia.

Eurasia The northern supercontinent formed by the break up of Pangaea in the Mesozoic (*ca.* 150 million years B.P.), comprising North America, Greenland, Europe and Asia, excluding India; Laurasia; *cf.* Gondwanaland.

euroky The ability to tolerate a wide range of environmental conditions.

European subregion A subdivision of the Palaearctic region; see Appendix 3.

europhyte A plant inhabiting leaf mould.

Euro-Siberian region A subdivision of the Boreal kingdom; see Appendix 4.

eurotophilous Thriving in or on leaf mould; **eurotophile, eurotophily**.

eury- Prefix meaning wide; *cf.* steno-.

euryadaptive Used of parasites tolerating a wide range of host species.

eurybaric Tolerant of a wide range of atmospheric pressure or hydrostatic pressure; *cf.* stenobaric.

eurybathic Tolerant of a wide range of depth; *cf.* stenobathic.

eurybenthic Living on the sea or lake bed over a wide range of depth; *cf.* stenobenthic.

eurybiontic Used of an organism tolerating a wide range of a particular environmental factor; *cf.* stenobiontic.

eurychoric Widely distributed; **eurychore, eurychorous**; *cf.* stenochoric.

eurycladous Euryhaline *q.v.*

eurycoenose Widespread; having a wide distribution; common; *cf.* stenocoenose.

euryhaline Used of organisms that are tolerant of a wide range of salinity; eurycladous; polyhalophilic; *cf.* holeuryhaline, oligohaline, polystenohaline, stenohaline.

euryhydric Tolerant of a wide range of moisture levels or humidity; *cf.* stenohydric.

euryhygric Tolerant of a wide range of atmospheric humidity; *cf.* stenohygric.

euryionic Having or tolerating a wide range of pH; *cf.* stenionic.

eurylume Tolerant of a wide range of light intensity; *cf.* stenolume.

euryoecious Tolerant of a wide range of habitats and environmental conditions; euryoekous; euryokous; euroky; *cf.* amphioecious, stenoecious.

euryphagous Utilizing or tolerant of a wide variety of foods or food species; euryphagic; pleophagous; plurivorous; polyphagous; **euryphage, euryphagy**; *cf.* stenophagous.

euryphotic Tolerant of a wide range of light intensity; *cf.* stenophotic.

euryplastic Showing a broad developmental response (phenotypic variation) to different environmental conditions; *cf.* stenoplastic.

eurysubstratic Tolerant of a wide range of substratum types; *cf.* stenosubstratic.

eurysynusic Used of plants having a wide distribution; *cf.* stenosynusic.

eurythermic Tolerant of a wide range of ambient temperatures; **eurytherm, eurythermal, eurythermous**; *cf.* amphithermic, stenothermic.

eurythermophilic Tolerant of a wide range of relatively high temperatures; **eurythermophile**; *cf.* stenothermophilic.

eurytopic 1: Tolerant of a wide range of habitats; physiologically tolerant; **eurytopy**. 2: Having a wide geographical distribution; *cf.* amphitopic, stenotopic.

eurytropic Used of organisms exhibiting a marked response or adaptation to changing environmental conditions; **eurytropism**; *cf.* stenotropic.

euryxenous Tolerating a wide range of host species; *cf.* stenoxenous.

eusexual Used of organisms that display a regular alternation of karyogamy and meiosis; *cf.* parasexual, protosexual.

eusocial Used of a social group in which members are fully integrated and cooperate in caring for young, with non-reproductive individuals assisting those involved in producing offspring, and in which different generations contributing to colony labour overlap; **eusociality**; *cf.* presocial.

euspory The formation of spores by normal meiotic divisions; **eusporic**.

eustatic 1: Pertaining to worldwide changes in sea level, such as would be produced by melting of continental glaciers, but excluding relative changes in level resulting from local coastal subsidence or elevation. 2: Changing only very slowly and gradually.

eusynanthropic Living on or in human habitations; *cf.* exanthropic, synanthropic.

eutelegenesis The improvement of breeding stock by artificial insemination.

eutely The condition of nuclear or cellular constancy in which the adult stage of an organism has the same number of nuclei or cells as the juvenile; **eutelic**.

euterrestrial Fully terrestrial; used of organisms adapted to living on land.

euthenics The branch of science dealing with the control of physical, biological and social environments for the well-being of humanity.

eutherophyte An annual seed plant.

eutraphent Used of an aquatic plant typical of water bodies with high nutrient concentrations; *cf.* mesotraphent, oligotraphent.

eutric Pertaining to healthy or fertile soils; *cf.* dystric.

eutripton Non-living suspended particulate matter (tripton) that is indigenous or native to the area in which it occurs; autochthonous tripton.

eutroglobiont Troglophile *q.v.*

eutroglophile A troglophile *q.v.* which reproduces in the cave habitat; *cf.* subtroglophile.

eutrogloxene An organism found only occasionally in underground caves and passages but not fully tolerant of cave conditions and not reproducing within the cave habitat; *cf.* subtrogloxene.

eutrophapsis In social insects, the feeding of young by a female in a previously prepared nest.

eutrophic 1: Having high primary productivity; pertaining to waters rich in the mineral nutrients required by green plants. 2: Used of a lake in which the hypolimnion becomes depleted of oxygen during the summer by the decay of organic matter sinking from the epilimnion; *cf.* dystrophic, mesotrophic, oligotrophic.

eutrophication Over enrichment of a water body with nutrients, resulting in excessive growth of organisms and depletion of oxygen concentration.

eutrophy The visitation of particular kinds of flowers only by certain specialized insect pollinators.

eutropism The orientation of growth in a plant so that it twines in a direction that follows the passage of the sun across the sky; **eutropic**; *cf.* antitropism.

eutropous Used of specialized insect species adapted to feed on particular kinds of flowers; *cf.* allotropous.

euxerophyte A true xerophyte with a well developed root system.

evanescent Ephemeral; short-lived; transient.

evaporation Loss of moisture in the form of water vapour; conversion to vapour; **evaporate**.

evapotranspiration The actual total loss of water by evaporation from soil and from water bodies, and transpiration from vegetation, over a given area with time; usually expressed in terms of equivalent rainfall; consumptive use; *cf.* potential evapotranspiration.

evenness Equitability; the component of species diversity that measures the relative abundance of species; calculated as $E = H/\log_2 S$, where S is the number of species and H is the Shannon–Wiener index of diversity *q.v.*

evergreen A plant, typically a tree or shrub, that has leaves all year round, and sheds them more or less regularly through all seasons.

eversporting Used of a race of plants that does not breed true.

evolution 1: Any gradual directional change; unfolding. 2: Any cumulative change in the characteristics of organisms or populations from generation to generation; descent or development with modification; **evolve**.

evolutionary biology The integrated science of evolution, ecology, behaviour and systematics.

evolutionary conservatism The preservation of ancestral similarity in species on diverging evolutionary pathways due to the retention of a high proportion of common ancestral alleles.

evolutionary method A method of classification

employing hypothetical reconstructions of evolutionary history incorporating both cladistic data (on the sequence of branching events) and morphological divergence data; synthetic method; *cf.* cladistic method, omnispective method, phenetic method.

evolutionary rate The speed of evolutionary change.

evolutionary species A lineage evolving separately from others and possessing its own unitary evolutionary role and tendencies; a distinct evolutionary sequence of ancestral and descendant populations; equivalent to a biological species viewed through the perspective of evolutionary time.

evolutionary tree Phylogenetic tree *q.v.*

evolve To change, produce or emit; to undergo gradual directional change; **evolution**.

evorsion Erosion caused by eddies and strong currents.

ex Latin, meaning from, according to; used in nomenclature to connect the names of two authors when the second author validly published a name proposed by, but not validly published by, the first author.

ex parte (*e.p.*) Latin, meaning in part.

exa (E) Prefix, used to denote unit $\times 10^{18}$.

exanthropic Living remote from human habitations; *cf.* eusynanthropic, synanthropic.

exaration Erosion by the action of ice.

excitability Sensitivity; the capacity of a living organism to respond to a stimulus; **excitable**.

excl. Abbreviation of Latin *exclusus*, meaning excluded; used in taxonomy to indicate those components included in a given taxon by one author but not thought to belong to that taxon by the present author and thus excluded.

excl. gen. Abbreviation of Latin *excluso genere*, meaning with the genus excluded.

excl. spec. Abbreviation of Latin *exclusa speciei*, meaning with the species excluded.

excl. specim. Abbreviation of Latin *exclusis specimenibus*, meaning with the specimens excluded; *excl. spec.*

excl. var. Abbreviation of Latin *exclusis varietatibus*, meaning with the varieties excluded.

exclamation mark (!) In taxonomic publication, sometimes cited after a herbarium specimen to indicate that it has been examined.

excluded name A scientific name that is not subject to the provisions of the Code.

exclusion principle Competitive exclusion principle *q.v.*

exclusive economic zone (EEZ) The 200 mile wide marine zone under the jurisdiction of the coastal state.

exclusive species A species totally or almost totally confined to one particular community; a species of high fidelity *q.v.*

exclusiveness Fidelity *q.v.*; the extent to which a species is restricted to a particular community and is excluded from others.

excreta (U) In ecological energetics, that part of the assimilated energy that is removed from the body as secretion, excretion or exudation; a component of rejecta *q.v.*

excretion The process of eliminating waste material from the body; **excreta**, **excrete**, **excretory**; *cf.* secretion.

exegete Any natural phenomenon or taxon used as an indicator of an ecological process; **exegetic**.

exemplar A random sample of a taxon.

exendotrophic Used of a flower pollinated by pollen from another flower of the same or different plant.

exhabited symbiont The organism that can be regarded as the host in a symbiosis in which neither symbiont is contained within the other; *cf.* exhabiting symbiont.

exhabiting symbiont The organism that lives upon another organism in a symbiosis in which neither symbiont is contained within the other; *cf.* exhabited symbiont.

exhaustive sampling In statistics, the examination of the complete population.

exobiology The study of extraterrestrial life.

exobiotic Living on the outer surface of another organism or on the surface of the substrate; **exobiont**.

exocrine Secreted externally; used of an external chemical substance which exerts an excitatory or inhibitory effect on biological systems or processes; ectocrine; environmental hormone; *cf.* endocrine.

exocytosis Extrusion of material from a cell by reverse pinocytosis.

exogamy 1: Outbreeding; sexual reproduction between individuals that are not closely related. 2: Pollination of a flower by pollen from a flower on a different plant; cross pollination. 3: Sexual reproduction in which there is a greater frequency of mating between unrelated (or relatively distantly related) individuals than would occur in random mating; disassortative mating; *cf.* endogamy.

exogenous Originating from outside the organism or system; due to, or triggered by external environmental factors; allochthonous; ectogenous; xenogenous; *cf.* endogenous.

exogenous plankton Planktonic organisms originating in another locality from that in which they have been found; allochthonous plankton.

exoisogamy The condition in which an isogamete undergoes fusion only with an isogamete from a different brood; **exoisogamous**.

exolithophytic Pertaining to plants that grow on the surface of rock or other hard inorganic substrata; epilithophytic; **exolithophyte**; *cf.* endolithophytic.

exomixis Fusion of gametes from different sources; xenomixis; **exomictic**.

exomorphology External morphology.

exon In eukaryotic chromosomes, that portion of a DNA strand within a gene that codes for a protein; *cf.* intron.

exoparasite Ectoparasite *q.v.*

exophenotypic Used of those characters or components of the phenotype which are adaptive and affect the organism's competitive abilities; *cf.* endophenotypic.

exophytic Pertaining to the outside of a plant.

exorheic region An area in which rivers arise and from which they flow to the sea; *cf.* arheic region, endorheic region.

exosymbiosis Ectosymbiosis *q.v.*; **exosymbiont**.

exotic Not native; alien; foreign; an organism or species that has been introduced into an area.

exotropism An orientation movement of lateral organs away from the main axis; **exotropic**.

exovation The act or process of hatching.

expanding Earth hypothesis That the Earth has increased its volume and hence crustal surface area as a result of global expansion from about 80% of the present dimensions during the last 180–200 million years.

expatriate An individual carried out of its normal range to another area in which it can survive but is unable to reproduce; terminal immigrant.

expected frequency In statistics, the frequency calculated on the basis of the appropriate theory.

experimental error Inaccuracy presumed to be due to chance variation or extraneous variables; variation not accounted for by the hypothesis.

experimental unit The subject or grouping to which a treatment is supplied.

exploitation 1: The utilization of a resource. 2: An interaction between organisms in which one benefits at the expense of another.

explosive evolution The splitting of a group or population into numerous lines of descent within a relatively short period of geological time; explosive radiation.

exponential growth Growth that is a simple function of the size of the growing entity, so that the larger the population the faster it grows.

exponential growth phase The period of maximum population growth; log. growth phase; *cf.* lag growth phase.

exponential trend A trend in a time series described by an equation of the form $y = ab^x$; which appears as a straight line when plotted on semi-logarithmic graph paper.

expression The manifestation of a character or trait; phenotypic expression; **expressivity**.

expressivity The degree of expression of a gene or gene combination in the phenotype; phenotypic expression.

exsic. (exs.) Abbreviation of Latin *exsiccatus*, meaning dried; used for dried preserved herbarium specimens.

exsiccate To dry up; to remove moisture; desiccate; **exsiccation**.

exsiccation Desertification; the development of desert conditions as a result of human activity or of climatic changes.

extant Existing or living at the present time; *cf.* extinct.

extinct No longer in existence; no longer living; *cf.* extant.

extinction 1: The process of elimination, as of less fit genotypes. 2: The disappearance of a species or taxon from a given habitat or biota, not precluding later recolonization from elsewhere.

extinction coefficient A measure of the degree of attenuation of light within a water column, or within the plant canopy for a given leaf area index *q.v.*

extinction curve A graph of the total number of species in a given habitat or area against either time or the extinction rate.

extinction, *K* Extinction occurring when the population has reached or exceeded carrying capacity *q.v.*; *cf.* extinction, *r*.

extinction, *r* Extinction occurring when colonists fail to establish themselves in a new site; *cf.* extinction, *K*.

extinction rate The number of species in a given habitat or area that become extinct per unit time.

extirpate To remove surgically; to destroy totally; to pull up by the roots; **extirpation**.

extraclinal Outside, or not forming part of, a cline *q.v.*

extraneous 1: Existing or originating outside the habitat, community or system. 2: Used of an organism that exists in the periphery of its species' range; *cf.* intraneous.

extrapolation The process of estimating a value beyond the range of actual values, on the basis of an equation obtained from the actual values; *cf.* interpolation.

extratropical Outside the tropics.

extrinsic Existing or having its origins outside an individual, group or system; extraneous.

extrinsic isolating mechanism Any environmental barrier that isolates potentially interbreeding populations; *cf.* intrinsic isolating mechanism.

extrorse dehiscence Spontaneous opening of ripe fruit from the inside outwards; *cf.* introrse dehiscence.

exudativorous Feeding on gum and other exudates from trees; **exudativore, exudativory.**

exude To ooze out, or diffuse out; **exudate, exudation.**

F

F₁ First filial generation; the offspring resulting from a parent cross.

F₂ Second filial generation; offspring resulting from a cross within the F₁ generation.

f.sp. Abbreviation of Latin *forma specialis*, meaning special form; used of a variant of a symbiotic species characterized by its host specificity.

fabric The physical properties of a soil determined by the spatial arrangement of the soil particles; soil fabric.

faciation A ranked category in the classification of vegetation; a subdivision of an association *q.v.* having more than one, but less than the total number, of dominant species because of local differences in climatic conditions.

facies 1: The general appearance or aspect of an individual, population or community; also the general aspect or appearance of sedimentary rocks or fossils. 2: A ranked category in the classification of vegetation; a local modification of an ecological community. 3: A variant of a biotope *q.v.*

facilitation Enhancement of the behaviour or performance of an organism or system by the presence or actions of others; social facilitation.

factor 1: Any causal agent; anything that is responsible for the independent inheritance of a character. 2: In statistics, any variable thought to influence the variable under investigation; or in a factor analysis, a linear combination of variables to which is attributed relevance.

factor analysis Any statistical method for resolving complex relationships by isolating and identifying causal factors and ranking them according to the extent to which each factor accounts for the variance observed in the data.

factor compensation The modification of the tolerance, or ecological amplitude, of a species to one factor by the interplay of one or more additional factors.

factorial The product of multiplying the number preceding the exclamation mark by all smaller integers in turn; i.e. $4! = 4 \times 3 \times 2 \times 1 = 24$.

facultative Contingent; assuming a particular role or mode of life but not restricted to that condition; used of organisms having the facility to live, or living, under atypical conditions; *cf.* obligate.

facultative anaerobe An aerotolerant organism that normally grows anaerobically but which is able to grow under aerobic conditions.

facultative gamete A motile spore (zoospore) that can function as a gamete.

facultative parthenogenesis The process by which some eggs develop parthenogenetically if not fertilized.

facultative thermophile An organism requiring a temperature between 50 and 65° C for optimum growth but capable of growth at lower temperatures; *cf.* obligate thermophile.

faecal-seston That part of the seston *q.v.* comprising the particulate faecal material; also used of the associated community of organisms; **fecal-seston**.

faeces Excrement; waste material voided through the anus; **feces, faecal**.

Fahrenheit scale (°F) A scale of temperature that has the freezing point of water at 32° and the boiling point of water at 212°.

Fahrenholz rule The generalization that there is usually a marked degree of parallelism in the phylogenetic development and speciation of a parasite group and its host group.

faithful species A species occurring only in a particular community; exclusive species.

Falkland current A cold surface ocean current that flows north of the east coast of Argentina, originating in part as a northerly deflection from the Antarctic Circumpolar current; see Appendix 6.

fall wind A strong cold wind descending from an elevated plateau or glacier.

false bottom Deep scattering layer *q.v.*

family 1: A group comprising parents, offspring and others closely related or associated with them. 2: A category comprising one or more genera or tribes of common phylogenetic origin, more or less separated from other such groups by a marked gap; a rank within the hierarchy of taxonomic classification, the principal category between order and tribe (or genus). 3: An ecological community comprising a single organism, usually occupying a small area and representing an early stage in a succession.

family-group In taxonomy, those categories within the hierarchy superfamily, family, subfamily, tribe and subtribe, and any additional intermediate categories.

family name The scientific name of a taxon of

family rank; usually with the ending -*idae* in zoology and -*aceae* in botany.

family tree A diagrammatic representation of the lineage of a family or group.

fancy name A term given to a cultivar that is neither the proper scientific name nor its common vernacular name; fancy epithet.

far red light Solar radiation in the spectral range 700–800 nm.

farinaceous Containing or comprising flour; starchy; **farinose**.

fascicle A bound part of a volume or serial that has been issued separately; commonly referred to as a part.

fathom A secondary fps unit of length ; used especially for measuring water depth; equal to 6 feet or 1.8288 m; see Appendix 13.

fauna 1: The entire animal life of a given region, habitat or geological stratum; **faunal**, **faunistic**. 2: A faunal work; *cf.* flora.

faunal assemblages, law of That similar assemblages of fossils indicate similar geological ages for the formations in which they occur.

faunal break An abrupt change from one fossil assemblage to another at a definite horizon in a stratigraphic sequence.

faunal province A zoogeographical subregion containing a distinct fauna which is more or less isolated from other regions by barriers to migration.

faunal region A region supporting a characteristic fauna; see Appendix 3.

faunal stage A time-stratigraphic unit *q.v.* based on a faunal zone or faunal assemblage.

faunal succession, law of That fossils succeed one another in a definite and recognizable sequence, each geological formation having a different total fossil composition from that above and below.

faunal work Fauna; a list of the animal species in a specified area, habitat or geological time horizon.

faunation The assemblage of animal species in a particular area; *cf.* vegetation.

faunichron The geological time unit corresponding to a faunizone *q.v.*

faunistics The study of all or part of the fauna of a particular locality or region; *cf.* chorology.

faunizone A biostratigraphic unit characterized by a particular fossil fauna; faunal zone; *cf.* florizone.

faunula The animal population of a small geographical area or microhabitat.

feces Faeces *q.v.*; **fecal**.

fecund Producing offspring; fruitful; proliferating.

fecundate To fertilize or pollinate.

fecundity The potential reproductive capacity of an organism or population, measured by the number of gametes or asexual propagules; *cf.* fertility.

fecundity ratio The ratio of pregnant or ovigerous females in a population, or other specified group, to the total number of females present.

federation A ranked category in the classification of plant communities, comprising one or more unions; equivalent in rank to alliance *q.v.*; the name of a syntaxon of federation rank usually has the ending -*ion*.

federion A community or assemblage comprising several species that are not mutually interdependent.

feedback Those elements of a control system or homeostatic mechanism linked by reciprocal influences, such that the source or input is continuously modified by the product of the process in order to maintain the necessary degree of constancy; a feedback loop may be negative and have a stabilizing or inhibitory effect or positive and have a disruptive or facilitative effect.

fell A bare rocky hillside or mountain slope.

fell-field A type of tundra ecosystem having sparse dwarfed vegetation and flat, very stony soil.

female The egg producing form of a bisexual or dioecious organism; symbolized by ♀; *cf.* male.

femto- (f) Prefix used to denote unit $\times 10^{-15}$.

femtoplankton Planktonic organisms less than 0.2 μm in diameter; see Appendix 9.

fen A eutrophic mire, with a winter water table at ground level or above, usually dominated by herbaceous grasses.

feral 1: Used of a plant or animal that has reverted to the wild from a state of cultivation or domestication. 2: Wild; not cultivated or domesticated.

fermentation The enzymatically controlled anaerobic breakdown of organic substrates; **fermentative**.

fertile Producing or having the capacity to produce offspring, fruit, pollen or abundant growth; fecund; prolific.

fertility The actual reproductive performance of an organism or population, measured as the actual number of viable offspring produced per unit time; birth rate; natality; *cf.* fecundity.

fertility ratio The number of offspring in a population in relation to the number of adult females.

fertilization The union of a male and a female gamete to form a zygote, but sometimes used more generally for the act of insemination, impregnation or pollination.

fetal Foetal *q.v.*

fetalization Foetalization *q.v.*

fide Latin, meaning according to; on the authority of; *teste*.

fidelity The degree of restriction of a plant species to a particular situation, community or association; assessed according to a five point scale: 5, exclusive; 4, selective; 3, preferential; 2, indifferent; 1, strange; *cf.* constancy.

field capacity The amount of water retained by a previously saturated soil when free drainage has ceased; expressed as percentage dry weight of the soil; *cf.* chresard, echard, holard.

field layer The horizontal vegetation stratum in woodland comprising all herbaceous plants, mosses and lichens, as well as woody plants not more than 500 mm in height.

fig. Abbreviation of Latin *figura*, meaning figure, illustration.

figure-type In taxonomy, an original figure or illustration of a specimen.

fil. Abbreviation of Latin *filius*, meaning son.

filial generation An offspring generation; *cf.* F_1, F_2.

filial regression The tendency of offspring of superior genotypes to revert to the average of the species; Galton's law of filial regression.

filter bridge A narrow temporary land-bridge or corridor having relatively uniform climatic conditions that acts as a selective filter of those organisms attempting to migrate along or across it.

filter feeder Any animal that feeds by filtering suspended particulate organic matter from water.

fimbricolous Growing in or on dung; **fimbricole**.

fimetarious Pertaining to dung; **fimetarius**.

fimicolous Growing in or on dung; **fimicole**.

fine-grained environment The environment experienced by an organism with good homeostatic mechanisms, high behavioural plasticity and mobility; experienced as a succession of interconnected conditions; *cf.* coarse-grained environment.

fine-grained resource Any resource distributed in such a manner that the consumer encounters it in the proportion in which it actually occurs; *cf.* coarse-grained resource.

fine-grained species In the context of feeding strategies, used of co-existing related species that are morphologically similar and utilize the same food resource but not in equal quantities or proportions; *cf.* coarse-grained species.

fire climax A more or less stable plant community, the structure and composition of which is dependent on regular burning.

firnification The process by which newly fallen snow becomes granular and compacted; **firn**.

first reviser In taxonomy, the first author to cite simultaneously published synonyms, including

different forms of the same name, or nomenclatural acts and to select one to take precedence over the other in the interest of nomenclatural stability.

fission Asexual reproduction or division by splitting into two (binary fission) or more (multiple fission) parts.

fissiparous Used of organisms that reproduce by division of the body into two or more equal parts; **fissiparism, fissiparity**.

Fitch algorithm A procedure for establishing the minimum number of times that a character has changed during its evolution; *cf.* algorithm.

fitness The relative competitive ability of a given genotype conferred by adaptive morphological, physiological or behavioural characters, expressed and usually quantified as the average number of surviving progeny of one genotype compared with the average number of surviving progeny of competing genotypes; a measure of the contribution of a given genotype to the subsequent generation relative to that of other genotypes.

fixation 1: In taxonomy, the determination of a type species or type specimen, whether by designation, indication or subsequent designation; *cf.* original fixation, subsequent fixation. 2: The trend towards homozygosity of genes having favourable pleiotropic effects; the attainment of homozygosity in a population which thereby becomes monomorphic with respect to a given allele.

fixed action pattern An inherited, relatively complex movement pattern within instinctive behaviour exhibited in response to a stimulus.

fixed continent theory That the continents have always existed and occupied the same relative positions as they do today; theory of the permanence of the continents.

fixity of species The theory that individuals of one species could give rise only to others like themselves, and thus, that all species were independently created and have not undergone subsequent change.

fjord A deep narrow inlet of the sea between steep slopes or mountains; fiord.

flark A local wet area of sparse, weakly peat-forming fen vegetation interspersed with drier areas or features.

flash colour The bright contrasting colour located on concealed surfaces of some animals; thought to attract attention when presented to a predator, thus promoting concealment of the animal when the colour is suddenly withdrawn.

F-layer The upper or humus layer of a soil profile, exhibiting initial stages of plant decomposition; fermentation layer; Of-layer; see Appendix 12.

floaters Those subordinate individuals forced to disperse to less favourable habitat areas because of their inability to establish a territory.

floating Used of plants lying flat on the water surface or occasionally immediately below it.

flocculation The aggregation of fine particles in the disperse phase of a colloid.

floccule An aggregation of microorganisms or particles in a liquid; flocculate.

flood current The tidal current associated with an incoming or flood tide; cf. ebb current.

flood plain The area of lowland along a water course that is subject to periodic flooding and sediment deposition.

flood tide Incoming tide; cf. ebb tide.

flora 1: A floral work; a published work describing the plant life of an area. 2: The plant life of a given region, habitat or geological stratum; floral, floristic; cf. fauna.

floral stage A time-stratigraphic unit q.v. based on a florizone or floral assemblage.

Florida current A warm surface ocean current derived from the Caribbean current system, that sweeps past Florida from the Gulf of Mexico and flows north-east into the Atlantic; see Appendix 6.

floristic composition A list of plant species of a given area, habitat or association q.v.; flora.

floristic element In phytogeography, a convenient term for any group of plants sharing a common feature of importance, such as migratory history.

floristic region A region supporting a characteristic flora; see Appendix 4.

floristics The study of species composition of vegetation, or of the plant species of a particular locality or region.

florizone A biostratigraphic unit characterized by a particular fossil flora; cf. faunizone.

florology The study of the production and development of vegetative formations.

florula 1: The plant species of a small area; a local flora. 2: An assemblage of plant fossils from a single stratum or group of adjacent strata; the floral assemblage of a small zone; florule.

flow Fluid movement; the quantity of fluid moving.

flowering The maturation of the floral organs and the expansion of their envelopes; florification, flossification.

fluctuate To move back and forth; to depart from the normal state in a more or less irregular manner.

flumineous Pertaining to running water.

flush An explosive increase in the size of a population; cf. crash.

flushing The deposition of dissolved substances in the upper layers of a soil profile by the action of water; cf. leaching.

flushing time A measure of the turnover time for fresh water in an estuary.

fluvial Pertaining to rivers and river action.

fluviatile Inhabiting rivers and streams; fluvial; lotic.

fluvioglacial Pertaining to streams flowing from a glacier, or to deposits laid down from glacial streams; glaciofluvial.

fluviology The study of rivers.

fluviomarine Inhabiting rivers and the sea.

fluvioterrestrial Inhabiting streams and the surrounding land.

fluviraption Erosion by running water or wave action.

flux 1: The amount of organic matter transferred from one compartment of a food web to another per unit time. 2: The quantity of substance that crosses a given surface in a unit time; flux is a scalar quantity.

flux density Flux per unit area at a point in space; calculated by dividing the flux by the surface area and allowing the area to tend towards zero; flux density is a vector quantity.

flyway A migration route of birds.

foehn wind A warm dry wind on the leeward side of a range of mountains; föhn wind.

foetalization Paedomorphosis resulting from the preservation of ancestral foetal characters in the adults of descendants; fetalization.

föhn wind Foehn wind q.v.

foliar Pertaining to leaves.

foliar theory That the various whorls of a flower are serially homologous with leaves on the stem.

folicaulicolous Living attached to leaves and stems; folicaulicole.

folicolous Living or growing on leaves; folicole, foliicolous, foliicole.

folivorous Leaf-eating; folivore, folivory.

fontaneous Pertaining to a fresh water spring; fontanus.

food chain A sequence of organisms on successive trophic levels within a community, through which energy is transferred by feeding; energy enters the food chain during fixation by primary producers (mainly green plants) and passes to the herbivores (primary consumers) and then to the carnivores (secondary and tertiary consumers).

food chain efficiency The efficiency of transfer of energy from one trophic level to the next; ecological efficiency.

food cycle Food web q.v.

food nexe Food web q.v.

food niche The position of an organism in a food web.

food pyramid A graphical representation of the food relationships of a community, expressed quantitatively as numbers, mass or total energy at each trophic level, with the producers forming the base of the pyramid and successive levels representing consumers of higher trophic levels; ecological pyramid.

food web The network of interconnected food chains of a community; food cycle; food nexe.

foot The fps unit of length, defined as one-third of the Imperial Standard yard (equal to 0.9144 m); equal to 0.3048 m; see Appendix 13.

forage 1: To search for food; **foraging**. 2: The plant material actually consumed by a grazing animal; *cf.* herbage.

foraging strategy The methods employed by an organism in its search for food.

foraminiferal ooze Pelagic calcareous sediment comprising more than 30% calcium carbonate predominantly in the form of foraminiferal tests; *cf.* ooze.

forb A broad-leaved herbaceous plant.

forbicolous Living on broad-leaved plants; herbicolous; **forbicole**.

forbivorous Feeding on broad-leaved plants; **forbivore, forbivory**.

foredune A dune ridge, more or less stabilized by the initial stages of the primary sere.

foreshore 1: Within a typical beach profile, the zone between mean high water and mean low water; also used of the upper intertidal zone that is covered only by exceptional spring tides. 2: The strip of land forming the margin of a lake; **fore-shore**.

forest A relatively large area of closely canopied trees; a small stand of trees may be termed a grove.

forgotten name *Nomen oblitum q.v.*

form 1: The essential shape and structure of an organism or group; configuration. 2: Any minor variant or recognizable subset of a population or species; morph. 3: The lowest category in the hierarchy of botanical classification; *forma*; see Appendix 10.

form genus A taxon of generic rank comprising one or more form species *q.v.*

form species One of a group of species having similar morphological characters but which are not confirmed as having common ancestry; morphospecies.

forma The lowest category in the hierarchy of botanical classification, applied to minor variations within a population; **form**.

forma specialis (f.sp.) Latin, meaning special form;

a variant of a symbiotic species characterized by its host specificity; *formae speciales*.

formation 1: The highest category in the classification of vegetation; a major regional climax vegetation type within a biome, comprising a number of associations. 2: A group of plant communities in a given continental or other geographical area exhibiting similar physiognomic, climatic and environmental characteristics. 3: Any sedimentary, igneous or metamorphic rock which is present as a defined bed or unit.

formation-type A group of geographically widespread plant communities of similar physiognomy and life form, related to major climatic and other environmental conditions; equivalent to a biome.

formenkreis 1: An aggregate of allopatric subspecies or species; superspecies. 2: In palaeontology, a group of related species or variants.

formula A hybrid plant name formed by connecting the generic names of the parents (A+B, or A×B), or by combining the names or parts of the names preceded by a similar sign (+AB or ×AB); condensed formula.

formulation Used in nomenclature to refer to the way in which a name is formed.

forward mutation The mutation of a gene from the normal or wild-type to a mutant state; *cf.* back mutation.

fossil An organism, fragment, impression or trace of an organism preserved in a rock; may be either a body fossil (such as a bone or shell) or a trace fossil (such as a burrow, track or imprint); *cf.* recent.

fossil fuel A fuel, such as coal, petroleum and natural gas, derived from the fossilized remains of organisms.

fossiliferous Used of a rock, sediment or horizon containing fossils.

fossorial Adapted for digging or burrowing into the substratum.

fototype Phototype *q.v.*

fouling An assemblage of organisms growing on the surface of floating or submerged man-made objects, that increases resistance to water flow or otherwise interferes with the desired operation of the structure.

founder effect That only a small fraction of the genetic variation of a parent population or species is present in the small number of founder members of a new colony or population; founder principle.

fps system A system of scientific units based on

the foot-pound-second, also known as the British System; see Appendix 13.

fracture zone In geology, a zone along which displacement has occurred; major fractures are found at right angles to the mid-oceanic ridges.

Frankfurt glaciation A subdivision of the Weichselian glaciation *q.v.*; see Appendix 2.

Fraser-Darling effect The stimulation of reproductive activity of a mating pair by the presence and activity of other members of the species; Darling effect.

frass Faecal matter and other fine animal debris found on the soil surface.

fraternal twins Dizygotic *q.v.* twins.

free floating Used of aquatic plants that are floating and not anchored to the substratum; pleustonic.

free-living Living independently of any host organism; non-symbiotic; eleutherozoic.

freemartin A sterile female twin, partially converted towards a hermaphrodite condition by some influence of its male twin.

free-running period Natural period *q.v.*

freeze drying A method of preserving biological material by dehydration from the frozen state under high vacuum; lyophilization.

frequency The number of items belonging to a category or class; the number of occasions that a given species occurs in a series of samples.

frequency curve A graphical representation of a continuous frequency distribution; the variate being the abscissa and the frequency being the ordinate.

frequency distribution An arrangement of data grouped into classes, each with its corresponding frequency of occurrence.

frequency polygon A graphical representation of a frequency distribution obtained by drawing a straight line between the successive class marks representing the class frequencies.

frequency-dependent selection Selection occurring in the situation in which the relative fitness of alternative genotypes is related to their frequency of occurrence within a population.

fresh water Water having a salinity of less than 0.5 parts per thousand, or alternatively, less than 2 parts per thousand.

Friedman's test A nonparametric method comprising a two-way analysis of variance; used to test for differences in mean level between several samples with each sample containing the same number of counts.

fright cry The sudden loud call made by a startled animal; distress call; mercy call; pain cry.

frigid climate A climate in which a permanently

frozen soil surface is covered more or less continuously with ice and snow; polar climate.

frigid zone That part of the Earth's surface within the polar circles.

frigideserta Tundra; the open communities of cold arctic or alpine regions; frigorideserta.

frigofugous Intolerant of cold conditions; **frigofuge**.

frigophilic Thriving in cold environments; cryophilic; **frigophile, frigophily**.

frigorideserta Frigideserta *q.v.*

fringing herbs Semi-emergent plants commonly fringing the edge of streams.

front The boundary zone between two air or water masses differing in properties, such as density, pressure, temperature or salinity.

frontal Pertaining to a boundary layer (front) separating two adjacent air or water masses; frontal surface.

frontal-cyclonic rain Rainfall produced by major cyclonic eddies and discontinuities; of particular significance in the equatorial rain belt and in temperate maritime regions; *cf.* convective rain, orographic rain.

fructicolous Living on or in fruits; **fructicole**.

frugivorous Feeding on fruit; **frugivore, frugivory**.

frutescence The time of maturity of a fruit.

fruticeta Shrub forest.

fruticolous Living or growing on shrubs; **fruticole**.

F-test A statistical test used to evaluate the probability that two sample variances are the same (i.e. the samples are drawn from the same population); variance-ratio test.

FU Used in ecological energetics to symbolize rejecta *q.v.*

fucivorous Feeding on seaweed; **fucivore, fucivory**.

fugacious Evanescent; lasting for a short time; falling early from the parent plant; caducous.

fugichnion A trace fossil, escape-structure or equilibrium structure, representing the activity of an organism directed towards the maintenance of a given depth in, or specific orientation towards, the substratum surface.

fugitive species An inferior competitor which is always excluded locally under interspecific competition, but which persists in newly disturbed habitats by virtue of its high dispersal ability; a species characteristic of temporary habitats.

function 1: All physical and chemical properties of a structure that relate to its form and organization, excluding the action or use of the structure which is more critically termed its role or biological role *q.v.* 2: In phytosociology, the special adaptive features of the components of vegetation, such as periodicities, dispersal mechanisms and physiological tolerances.

functional morphology Interpretation of the function of an organism or organ system by reference to its shape, form and structure.

functional mosaic A hybrid expressing one allele in some tissues or cells and the other allele in others.

functional response A change in the rate of predation by an individual predator in response to a change in density of the prey; *cf.* numerical response.

fundamental niche The entire multidimensional space that represents the total range of conditions within which an organism can function and which it could occupy in the absence of competitors or other interacting species; ecospace; preinteractive niche; prospective ecospace; *cf.* realized niche.

fundamentalism Creationism *q.v.*

fungicide A chemical used to kill or control fungi.

fungicolous Living in or on fungi; **fungicole**.

fungistasis Inhibition of fungal growth.

fungivorous Feeding on fungi; mycetophagous; mycophagous; **fungivore**, **fungivory**.

furiotile lake Any partially disjunct body of water, that connects with the mainstream only during high water.

furlong A unit of length equal to 220 yards (10 chains; 2.01168×10^2 m).

G

gale A very strong wind having a speed of 39–63 m.p.h. (17.4–28.2 m s⁻¹); see Appendix 21.

gallery forest A narrow strip of forest along the margins of a river in an otherwise unwooded landscape.

gallicolous Living in galls; gallicole.

galliphagous Feeding on galls; galliphage, galliphagy.

gallivorous Feeding on galls; gallivore, gallivory.

gallon 1: Imperial gallon q.v. 2: US gallon q.v.

Galton's law An early concept in genetics that half the inherited characters of an individual of bisexual parentage are derived equally from each parent (one quarter from each), one-quarter from the grandparents, one-eighth from the great grandparents, and so on.

Galton's law of filial regression That offspring of superior genotypes tend to revert to the average for the species; filial regression.

galvanotaxis A directed reaction of a motile organism to an electric current; electrotaxis; galvanotactic.

galvanotropism An orientation response to an electric current; electrotropism; galvanotropic.

-gam- Combining affix meaning sexual, united; used with reference to reproduction involving fertilization.

game theory A mathematical theory dealing with the determination of optimum strategies in competitive situations involving two or more competitors.

gamete A mature reproductive cell (usually haploid) which fuses with another gamete of the opposite sex, to form a zygote (usually diploid); the male gametes are known as sperm (spermatozoa) and the female gametes as eggs (ova); cf. agamete.

gametic incompatibility The absence of interbreeding because of physiological differences between the gametes which prevent fertilization or subsequent development.

gametic number The number of chromosomes in the nucleus of a gamete, usually half the somatic chromosome number.

gametic pool The total potential gamete production of any given generation of a sexually reproducing population.

gametogamy The fusion of a male and female gamete to form a zygote; gametogamic.

gametogenesis The process of gamete production; gametogeny; gametogenetic.

gametogenic Used of variations arising from spontaneous changes in the chromosomes of gametes.

gametophyte The haploid sexual phase of a plant which exhibits an alternation of generations, from which gametes are produced, usually by mitotic division; the haploid gametophyte is typically formed by meiotic division of a diploid sporophyte; gamophyte; haplophyte.

gametotoky Amphitoky q.v.

gametropism Orientation movements of plant structures immediately before or after fertilization; gametropic.

gamic Fertilized.

gamma diversity The richness in species of a range of habitats in a geographical area, dependent upon both the alpha diversity q.v. of the habitats it contains and the extent of the beta diversity q.v. between them.

gamma karyology In the study of chromosomes, the use of Giemsa and fluorescent techniques to locate the main C, G and Q bands; cf. karyology.

gamma link A dichotomously branching link in a food chain, involving two predators feeding on one prey species; cf. iota link, lambda link.

gamma taxonomy That aspect of taxonomy concerned with intraspecific populations and with phylogenetic trends; cf. alpha taxonomy, beta taxonomy.

gamobium The sexual phase in an alternation of generations; cf. agamobium.

gamodeme An assemblage of individuals forming a relatively isolated naturally interbreeding population.

gamogenetic Pertaining to the formation of gametes; produced by the union of gametes; used of reproduction by union of gametes; sexual; gamogenic; gamogonic; gamogenesis.

gamogony Formation of gametes by sporogony.

gamontogamy Hologamy q.v.

gamophase The haploid phase of a life cycle; haplophase; cf. zygophase.

gamophyte Gametophyte q.v.

gamosematic Pertaining to coloration, markings or behaviour patterns that assist members of a pair to locate each other.

gamotropic Used of flowers that alternate between open and fully closed; cf. agamotropic, hemigamotropic.

gamotropism An orientation response of gametes

to one another; the mutual attraction of gametes; **gamotropic**.

-gamy Suffix meaning marriage, fertilization, sexual union.

gasoplankton Planktonic organisms which make use of gas-filled vesicles or sacs for buoyancy.

gastraea The hypothetical adult ancestor of all higher animals, inferred from the existence of the gastrula (formed by invagination of a blastula) as a stage common to the early ontogeny of higher animals; *cf.* blastaea.

Gause's hypothesis That two species with the same ecological requirements cannot coexist in the same place indefinitely; that is they cannot form steady state populations if they occupy the same niche; Gause's exclusion principle; Gause's rule; Gause's law: Grinnell's axiom.

Gaussian curve A normal curve; the symmetrical statistical curve of the frequency distribution of a normally distributed population.

Gaussian distribution Normal distribution *q.v.*

geitonogamy Fertilization between different flowers on the same plant; **geitonogamic**; *cf.* xenogamy.

geld The removal of, in interference with the activity of, the testes of a male; emasculation; *cf.* spay.

gelicolous Living in geloid soils having a crystalloid content of between 0.2 and 0.5 parts per thousand; **gelicole**; *cf.* halicolous, pergelicolous, perhalicolous.

geloid soil Soil type characterized by low salt content, weak solutions, strong colloidal properties and a crystalloid content of less than 0.5 parts per thousand; *cf.* haloid soil.

gemma A bud or outgrowth capable of developing into an independent organism.

gemmation Asexual reproduction in plants by budding off a group of cells (a gemma) which later separates partially or completely from the parent to form a new individual; usually referred to in animals as budding.

gemmule theory The theory of heredity, that somatic cells contain particles (gemmules) that are responsible for carrying hereditary traits to the gonads from where they are transmitted to the offspring; *cf.* pangenesis.

gen. Abbreviation of the Latin *genus*, meaning genus.

-gen- Combining affix meaning race, genus, birth, gene, producer.

gen. et sp. nov. Abbreviation of the Latin *genus and species nova*, meaning new genus and species.

gene The basic unit of inheritance, comprising a specific sequence of nucleotides on a DNA chain, that has a specific function and occupies a specific locus on a chromosome; alternative forms of a gene are known as alleles; genetic factor.

gene complex The entire system of interacting genetic factors of an organism.

gene dosage The number of alleles present in a given genotype, dependent upon the state of ploidy.

gene flow The exchange of genetic factors within and between populations by interbreeding or migration.

gene frequency The proportion of one allele to the total of all alleles at the same locus in the gene pool.

gene locus The position of a gene on a chromosome; locus.

gene map A map of the relative positions of genes on a chromosome or fragment of a chromosome.

gene mutation A point mutation; any heritable change in a single gene.

gene pool 1: The total genetic material of a freely interbreeding population at a given time. 2: All the genes at a given locus in a population in a given generation.

gene substitution The replacement in a population of one allele by a mutant allele.

geneagenesis Parthenogenesis *q.v.*

genealogy 1: The study of ancestral relationships and lineages; geneology. 2: An ancestor–descendant lineage.

genecology The study of intraspecific variation and genetic composition in relation to environment.

geneogenous Congenital *q.v.*

genepistasis The phenomenon in which one of two related forms develops further than the other.

generalist A species having a broad habitat range or food preference; *cf.* specialist.

generation 1: Formation; production. 2: All the individuals produced within a single life cycle.

generation time 1: The average duration of a life cycle between birth and reproduction. 2: The mean period of time between reproduction of the parent generation and reproduction of the first filial generation.

generative apogamy The production of a sporophyte plant from a haploid gametophyte by asexual means.

generative apospory Apomeiotic spory *q.v.*

generative parthenogenesis The development of a haploid organism from a female gamete that has undergone a meiotic reduction division but has not been fertilized; haploid parthenogenesis.

generic Pertaining to a genus.

generic name A scientific name of a taxon of

genus rank; the first word of a binomen or trinomen.

generitype Genotype; the primary type of the type species of a genus.

genesiology The study of reproduction; **genesiological**.

-genesis Suffix meaning development, descent, origin, formation.

genet A unit or group derived by asexual reproduction from a single original zygote, such as a seedling or a clone; *cf*. ortet, ramet.

genetic Pertaining to the genes.

genetic assimilation The reinforcement, by alteration in genetic material, of phenotypic modifications which themselves had no genetic basis.

genetic background That part of the genotype other than the gene or genes under consideration.

genetic behavioural polymorphism hypothesis Polymorphic behaviour hypothesis *q.v.*

genetic code The biochemical basis of heredity; the sequence of nucleotide base pairs on the DNA polynucleotide chain which encodes the genetic information; within the code three successive nucleotide base pairs (a codon) code for a single amino acid.

genetic death 1: Selective elimination of a genotype carrying mutant alleles that reduce fitness *q.v.*; genetic extinction. 2: The state that ensues when an individual effectively becomes unable to produce offspring.

genetic disharmony Incompatibility between parental genes manifested in the progeny.

genetic distance A measure of the genetic disparity of two individuals or populations, in terms of the probability of the common possession of a given gene or character.

genetic drift The occurence of random changes in the gene frequencies of small isolated populations, not due to selection, mutation or immigration; drift; Sewall Wright effect.

genetic engineering Experimental alteration of the genetic constitution of an individual.

genetic equilibrium The state exhibited by an equilibrium population in which neither allele frequencies nor genotype distributions change from one generation to the next; the maintenance of more or less constant allele ratios in a gene pool through successive generations; Hardy-Weinberg equilibrium.

genetic extinction Genetic death *q.v.*

genetic factor Gene *q.v.*

genetic homeostasis The tendency of a population to move to genetic equilibrium and to resist sudden changes in genetic composition.

genetic load The average number of lethal equivalents (potential genetic deaths) per individual in the population, viewed as the accumulated depression of fitness from a theoretical optimum, caused by deleterious genes; comprising mutational load *q.v.*, segregational load *q.v.* and substitutional load *q.v.*

genetic polymorphism The co-occurrence of two or more alleles at the same locus in a population at frequencies that cannot be accounted for by recurrent mutation alone.

genetic recombination Recombination *q.v.*

genetic segregation Segregation *q.v.*

genetic species concept The concept of species as gene pools in which the genes reproduce asexually during DNA replication and generate phenotypes (organisms) that can reproduce sexually to produce new combinations of genes; *cf*. biological species concept, morphological species concept.

genetic system The organization of the genetic material and the reproductive strategy of a species.

genetic variance That part of phenotypic variance of individuals in a population produced by differences or changes in genetic constitution such as mutation or recombination; *cf*. environmental variance, phenotypic variance.

genetics The science of heredity and variation.

genetype Genotype *q.v.*

genholotype Genoholotype *q.v.*

-genic Suffix meaning producing, produced by.

genic balance The sum of the combined effects of all the genes that interact to produce a given phenotypic trait.

genocline 1: A graded series of genotype frequencies within the geographical range of a population. 2: A cline resulting from hybridization between adjacent but genetically distinct populations.

genocopy The phenotypic expression of one gene that resembles that of another, non-allelic gene.

genodeme A local interbreeding population characterized by genotypic features; *cf*. deme.

genoecodeme A local interbreeding population occurring in a particular habitat (an ecodeme) that is characterized by genotypic features.

genoholotype In taxonomy, the primary type of the type species of the genus, designated by the author in the original description of the genus; diplotype; genholotype.

genolectotype In taxonomy, the primary type of the type species of a genus selected from a genosyntype *q.v.* series subsequent to the original description; logotype.

genome The minimum set of non-homologous chromosomes required for the proper functioning of a cell; the basic (monoploid) set of chromosomes of a particular species; the gametic chromosome number; **genom**.

genomorph 1: A polyphyletic taxon of generic or subgeneric rank containing a group of superficially similar but not closely related species. 2: A species that differs from the genotype as a result of evolutionary changes over a prolonged period of time.

genonomy Biosystematics *q.v.*

genopathic disease A disease resulting from a genetic condition; congenital disease.

genophenes The different phenotypes of the same genotype.

genophyletic Pertaining to a group of taxa that are inferred to have greater genotypic similarity to one another than to members of any other groups; **genophyly**.

genophylistic Pertaining to phylogenetic relationships between taxa based on inferred overall genotypic similarity; **genophylist, genophylistics**.

genospecies An aggregate of interbreeding populations of mutually interfertile forms contributing to a common gene pool.

genosyntype Any individual from the syntype series of the type species of a genus.

genotype 1: The hereditary or genetic constitution of an individual; all the genetic material of a cell, usually referring only to the nuclear material: *cf.* phenotype. 2: All the individuals sharing the same genetic constitution; biotype. 3: The specimen on which a genus–group taxon is based; the primary type of the type species; generitype; genetype.

genotypic distance A measure of the disparity between two genotypes in terms of the probability that two individuals differ with respect to the gene at a particular locus.

genotypic sex determination The condition in which the sex of an individual is determined primarily by the balance of genetic factors in the zygote or spore.

genotypic sterility Hybrid sterility resulting from an imbalance in the genotype of the zygote.

genovariation A point mutation *q.v.*

gens A species group; a distinct evolutionary lineage.

genus A category in biological classification comprising one or more phylogenetically related, and morphologically similar species; a rank in the hierarchy of taxonomic classification forming the principal category between family and species; **genera**; see Appendix 10.

geo- Prefix meaning earth, terrestrial.

geoaesthesia The capacity of a plant to perceive and respond to gravity; graviperception; geoperception.

geobenthos 1: The sum total of all terrestrial life. 2: That part of the bottom of a stream or lake not covered by vegetation; *cf.* phytobenthos.

geobiology The study of the biosphere.

geobiont An organism spending its whole life in the soil; a member of the permanent soil fauna; terricole; **geobiontic**.

geobios The total life of the land; that part of the Earth's surface occupied by terrestrial organisms; **geobiontic**; *cf.* halobios, hydrobios, limnobios.

geobiotic Terrestrial.

geobotany Plant biogeography; the study of plants in relation to geography and ecology; phytogeography.

geocarpy Ripening of fruits underground.

geochron An interval of geological time; a rock-stratigraphic unit.

geochronologic unit A unit of geological time.

geochronology The science of dating and the study of time in relation to the Earth's history as revealed by geological data; geological chronology; **geochrony, geochronological**.

geochronometry Quantitative geochronology, usually with time measured in years.

geocline A graded sequence of morphological variation through a series of populations, resulting from spatial or topographical separation.

geocolous Living in the soil for part of the life cycle; **geocole**.

geocosmology The study of the origin and geological history of the Earth.

geocryptophyte A plant having perennating organs or renewal buds below the soil surface; geophyte.

geodiatropism Orientation at right angles to gravity; diageotropism; **geodiatropic**.

geodyte A ground living organism; geocole; terricole.

geoecology Environmental geology.

Geoffroyism The theory that adaptive genetic changes can be induced by environmental factors.

geographical barrier Any geographical feature that prevents gene flow between populations.

geographical equivalents Two or more taxa that occupy the same geographical area and have patterns of distribution that are nearly congruent; syngeographs.

geographical isolation The separation of potentially interbreeding populations by geographical barriers; a premating isolation mechanism *q.v.*

geographical parallel communities Geographically separated communities that occupy similar substrata or habitats and have similar taxonomic composition.

geographical race A race which is geographically separated from other populations of the same species; frequently regarded as a subspecies.

geographical speciation Allopatric speciation; speciation occurring during a period of geographical isolation.

geographical variation Any differences between spatially separated populations of a species.

geological chronology Geochronology *q.v.*

geological erosion Normal or natural erosion due to long term geological processes undisturbed by the activities of man; natural erosion.

geological period See Appendix 1.

geology Study of the structure, processes and chronology of the Earth.

geomalism The response of organisms to gravitational influences.

geometric mean A measure of location, lying between the arithmetic mean and the harmonic mean, calculated as the nth root of the product of n values.

geometric series A series of values in which the ratio between successive values is the same; as in 10, 100, 1000; geometric progression.

geometric similarity Isometric growth *q.v.*; increase or decrease in size without accompanying change of shape or form.

geonemy Biogeography *q.v.*

geonastic Growth curvature towards the ground; **geonasty**.

geonyctitropism Orientation movements in plants during darkness in response to gravity; **geonyctitropic**.

geoparallotropism An orientation movement of an organ or structure to bring it parallel to the soil surface; diageotropism; **geoparallotropic**.

geoperception The capacity of an organism to perceive and respond to gravity; geoaesthesia, graviperception.

geophagous Feeding on soil; deriving nutrients from soil or from the sediment; **geophage**, **geophagy**.

geophilic 1: Thriving or growing in soil; terricolous; **geophile, geophilous, geophily**. 2: Used of plants that fruit below the soil surface.

geophyte A perennial plant having perennating organs or renewal buds, such as corms or rhizomes, buried well below the soil surface; geocryptophyte; *cf.* Raunkiaerian life forms.

geoplagiotropism Orientation at an oblique angle to the soil surface; plagiogeotropism; **geoplagiotropic**.

geosere 1: The series of climax formations in an area throughout geological time. 2: An ecological succession commencing on a clay substratum.

geosol An assemblage of soil profiles having similar stratigraphic relationships and similar physical and chemical pedogenic features.

geostrophic Pertaining to the deflective force caused by the rotation of the Earth.

geostrophic flow The large scale circulation of oceanic water resulting in the elevation and depression of the central portions of a gyre, maintained by the opposing effects of gravity and the Coriolis force.

geosyncline A trench in the Earth's crust, usually submarine, into which sediments settle over long periods.

geotaxis A directed response of a motile organism towards (positive) or away from (negative) the direction of gravity; **geotactic**.

geothermal Pertaining to heat derived from the Earth's interior.

geotropism An orientation response to gravity; epitropism; helcotropism; **geotropic**.

geoxene An organism that becomes a temporary or accidental member of the soil fauna.

geratology The study of decline and senescence of populations.

germ cell A gamete *q.v.* or agamete *q.v.*

germ line The lineage of generative cells and their descendants that give rise to the gametes.

germ line theory Germ plasm theory *q.v.*

germ plasm The hereditary material transmitted to the offspring via the gametes.

germ plasm theory That there are two types of cells, germ plasm and somatic cells, and that germ plasm is potentially immortal through transmission from generation to generation; germ line theory.

germinal Pertaining to, or influencing, the germ cells.

germination The commencement of growth of a propagule or bud.

germule A unit of colonization or migration.

gerontic Pertaining to the later stage of phylogeny or ontogeny; gerontal.

gerontogeic Pertaining to the Old World; **gerontogaean, gerontogaeous**; *cf.* amphigeic, neontogeic.

gerontology The study of senescence, the processes and effects of ageing.

gerontomorphosis Evolutionary change resulting from adaptations and modifications of adult structures; **gerontomorphic**.

gestalt A pattern of biological phenomena in which the properties of the functional whole differ from those predicted from the sum of the component parts; holism; *cf.* reductionism.

gestalt morphology A modern school of plant morphology which aims to determine the principal structural types of the main organs or of whole plants irrespective of their taxonomic relationships and of a wide range of individual variants.

gestation The period of development of an embryo within the uterus of a viviparous animal, from conception to birth; **gestate**.

G-horizon A gley horizon.

giga- (G) Prefix used to denote unit $\times 10^9$.

gigantism The condition of being much larger than normal, or of exhibiting excessive growth; often associated with polyploidy; **gigantic**; *cf.* nanism.

Gipping glaciation Wolstonian glaciation *q.v.*

glacial Pertaining to those geological intervals characterized by cold climatic conditions and advancing ice sheets and caps; **glaciation**.

glacial marine deposit A marine sediment having a significant allochthonous terrigenous component derived from transportation by icebergs.

glacial refugium Pleistocene refuge *q.v.*

glacial relict A species that has survived from Pleistocene faunas and floras, typically in a restricted location or habitat (a Pleistocene refuge *q.v.*).

glaciofluvial Used of sediments transported by ice and deposited from the flowing meltwaters of a glacier; fluvioglacial.

glareal Growing on dry exposed and typically gravelly ground; **glarial, glareous**.

gley A soil type which is subject to periodic waterlogging because of poorly permeable C-horizon and hence is subject to gleying; **glei**.

gley horizon A soil horizon characterized by the deposition of iron and manganese compounds; G-horizon.

gleying In waterlogged soils, the removal of iron and manganese compounds from anaerobic surface layers of a soil to deeper layers where they are precipitated under oxidizing conditions; upper layers tend to be dull grey, deeper layers intensely mottled; denoted by a suffix *g* in soil profile diagrams; **gleyzation, gleyed**.

globigerina ooze A pelagic sediment found over extensive regions of the ocean floor, comprising more than 30% calcium carbonate in the form of foraminiferan tests of which *Globigerina* is the dominant genus; *cf.* ooze.

Gloger's rule The generalization that among warm-blooded animals those races living in warm and humid areas are more heavily pigmented than those in cool dry areas; pigments are typically black in warm humid environments, red

and yellow in dry areas, and generally reduced in cool areas; pigmentation rule; Gloger's law.

glycophyte A plant thriving in a soil of low salt concentration, typically less than 0.5% sodium chloride; *cf.* halophyte.

gnesiogamy Cross fertilization between two individuals of the same species; intraspecific zygosis.

gnotobiotic system An experimental biological culture system to which only preselected components (organisms and nutrients) are introduced; used for studying the dynamics of feeding in single or multi-species systems; **gnotobiota, gnotobiosis**.

goal orientation Telotaxis *q.v.*

-gon- Combining affix meaning sexual, gamete, reproduction.

gonad output In ecological energetics, that part of assimilation released as reproductive propagules.

Gondwanaland The southern supercontinent formed by the break up of Pangaea in the Mesozoic (*ca.* 150 million years B.P.); comprising the present South America, Africa, Arabia, Australia, Antarctica, India and New Zealand; Gondwana; *cf.* Eurasia.

gone Any germ cell formed by meiotic division.

gonial spory Apomeiotic spory *q.v.*

gonochorism 1: The fusion of male and female gametes produced from separate unisexual individuals; sexual reproduction. 2: The history or development of sex differentiation.

gonochoristic 1: Used of individuals having separate sexes; unisexual. 2: Used of populations having male and female individuals in the same population; bisexual; **gonochoric, gonochorism, gonochrism**.

gonogenesis The formation of germ cells by meiotic division.

gonosomatic index (GSI) The ratio of total gonad weight to total body weight of an individual or population, usually expressed as a percentage.

-gony Suffix meaning offspring.

goodness of fit The closeness of agreement between a set of observed frequencies and a set of expected frequencies; usually measured by a chi-squared test.

Gothlandian Silurian *q.v.*

GPP Gross primary production *q.v.*

gradation The sum total of all the processes involved in the modification of the Earth's surface by the transportation of material.

grade A level of organization and adaptation; also a delimitable unit of anagenetic advance or biological improvement.

graded signal A signal used in animal communication which varies in frequency and/or intensity, and as a result can convey quantitative information to the signal receiver; *cf.* discrete signal.

gradient 1: The rate of change of a variable with distance. 2: A regularly increasing or decreasing change in a factor, such as ambient temperature. 3: A character gradient (cline *q.v.*).

gradient analysis Ordination *q.v.*

gradualism Phyletic gradualism *q.v.*

graft To induce the union of a tissue (a graft) from one organism with that of another by artificial means or to transplant tissue from one site to another on the same organism; in zoology the two elements are referred to as donor and recipient or host, and in botany the part being introduced is the scion and the recipient is the stock; *cf.* autoplastic grafting, heteroplastic grafting, homoplastic grafting.

graft hybrid A hybrid produced by grafting two dissimilar plants.

grallatorial Adapted for wading.

gram The cgs unit of mass, equal to the mass of one-thousandth part of a cubic decimetre of water, or of 1 kilogram *q.v.*

gram calorie Calorie *q.v.*; gcal.

graminicolous 1: Growing on grasses. 2: Used of an animal spending most of its life in a grassy habitat; chorotobiontic; **graminicole**.

graminivorous Feeding on grass; **graminivore**, **graminivory**.

graminology The study of grasses; agrostology.

grand mean In statistics, the true mean of a population; denoted by μ, and estimated by the mean of all the observations in a sample or experiment.

granivorous Feeding on seeds; **granivore**, **granivory**.

granule A sediment particle between 2–4 mm in diameter; small gravel; see Appendix 11.

grassland An area of vegetation dominated by herbaceous grasses, sometimes used for any herb dominated vegetation.

gravel Sediment particles between 2–256 mm in diameter; sometimes used for a class of smaller particles between 2–4 mm in diameter; see Appendix 11.

graveolent Possessing a strong or offensive odour.

gravid Carrying eggs or young; ovigerous; pregnant.

graviperception The perception of gravity; geoaesthesia; geoperception.

gravitational water Water which drains by gravity through the soil and which is readily available to soil organisms and plants.

gray Grey *q.v.*

grazing Feeding on herbage, algae or phytoplankton, by consuming the whole food plant or by cropping the entire surface growth in the case of herbage.

Great Ice Age Pleistocene *q.v.*

Great Interglacial period The major interglacial period in the middle of the Quaternary Ice Age; see Appendix 2.

green mud A terrigenous marine sediment, green in colour, formed under mildly reducing conditions and containing iron in the ferrous state, and some organic matter.

greenhouse effect The tendency towards increasing temperature of the lower layers of the atmosphere caused by an increase in atmospheric carbon dioxide which, together with water vapour, absorbs radiated heat more efficiently than it absorbs the incident solar radiation of short wavelengths.

gregarious Tending to aggregate actively into groups or clusters.

Gregg's paradox The apparent contradiction of the basic premise of the Linnaean system of classification, that one taxon cannot belong to two categories, by the existence of monotypic taxa, such as a family in which the single taxon it contains (a genus) belongs simultaneously to two different categories (genus and family).

gressorial Adapted for walking.

Grey Brown podzolic soil A zonal soil with a thin dark litter layer over a grey-brown, moderately acidic A-horizon and an illuviated lower horizon; formed in humid temperate climates on young land surfaces, typically glacial deposits under deciduous forest.

Grey Desert soil A soil type similar to a Sierozem *q.v.*, but having a calcareous surface layer; formed in warm temperate arid climates.

Grey Wooded soil A woodland soil having a deep litter and duff layer, a deep weakly acidic brown to grey A-horizon, and a deep neutral to moderately acidic B-horizon.

Grinnell's axiom That two species with the same resource requirements are unlikely to form steady state populations since one will compete more successfully for the resource and displace the other; Gause's hypothesis.

grooming The act of cleaning and tidying the body surface to remove foreign matter and parasitic organisms, by licking, nibbling or other directed behaviour; *cf.* allogrooming, self-grooming.

gross ecological efficiency Ecological efficiency, coefficient of *q.v.*

gross primary production (GPP) The total assimilation of organic matter by an autotrophic individual, population or trophic unit, per unit time per unit area or volume; **gross primary productivity**; *cf.* nett primary production.

gross production 1: Total assimilation of organic matter by an individual, population or trophic unit per unit time per unit volume or area; **gross productivity**; *cf.* nett production. 2: Assimilation *q.v.*

gross secondary production The total assimilation of organic matter or energy by a primary consumer individual, population or trophic unit per unit time per unit area or volume; **gross secondary productivity**; *cf.* nett secondary production.

ground water All the water that has percolated through the surface soil into the bedrock; phreatic water.

ground-storey The lowest layer of vegetation in a stratified woodland or forest community, comprising small trees and shrubs, herbs and plant debris.

Ground-Water Laterite soil An intrazonal soil with hardpans rich in iron and aluminium, formed immediately above the water table.

Ground-Water Podzol soil An intrazonal soil with a mat of organic surface material over a thin acid humus layer, underlain by a pale grey leached horizon and a dark brown hardpan lower horizon, formed in humid cool to tropical climates under conditions of poor drainage with forest vegetation.

group 1: A general term for an assemblage of organisms which remain together and interact over a period of time. 2: In zoology, an assemblage of related or coordinate taxa, as in species-group or genus-group; in botany, an assemblage of cultivars *q.v.* in a species or species hybrid. 3: Those rocks formed during each of the major divisions of geological time.

group predation Cooperation between a number of predators in hunting and handling prey.

group selection Selection acting upon a group of two or more individuals by which characters can be selected for that benefit the group rather than the individual; *cf.* kin selection.

grove A small stand of trees, or fragment of former extensive woodland.

growth Increase in size, number or complexity; progressive development.

growth efficiency coefficient In energetics, the ratio of total energy production to total energy consumption.

growth form The characteristic appearance of a plant under a particular set of environmental conditions; life form.

growth water That portion of the total soil water available to plants; chresard.

grumusol A soil rich in clay that expands and contracts under conditions of high or low water content respectively.

guano An accumulation of sea bird droppings rich in phosphates and nitrates.

guest An animal living and/or breeding within the nest, domicile or colony of another species.

Guiana current A warm surface ocean current that flows north-west to link the South Equatorial current with the Caribbean current; see Appendix 6.

guide species Indicator species *q.v.*

guild A group of species having similar ecological resource requirements and foraging strategies, and therefore having similar roles in the community.

guild matrix A rectangular array of the competitive coefficients of the member species of a guild; often referred to as the community matrix.

Guinea current A warm surface ocean current that flows south off the west coast of North Africa, derived in part from the Canary current; see Appendix 6.

Gulf Stream The warm surface ocean current derived from the Antilles current and the Florida current that flows north-east into the Atlantic and forms the northern limb of the North Atlantic Gyre; see Appendix 6.

gully erosion The formation of gullies and narrow channels due to localized erosion by surface water; *cf.* erosion.

gumivorous Feeding on gum and other exudates from trees; **gumivore, gumivory**.

Günz glaciation A glaciation of the Quaternary Ice Age in the Alpine area, with an estimated duration of 100 thousand years; see Appendix 2.

Günz-Mindel interglacial An interglacial period of the Quaternary Ice Age in the Alpine area; see Appendix 2.

guttation The natural exudation of water from an uninjured plant surface.

guyot A submarine seamount rising from the abyssal plain.

gymno- Prefix meaning naked.

gymnogamy Fertilization.

-gyn- Combining affix meaning female, ovary, pistil.

gynandromorph An individual of mixed sex; a sexual mosaic having some parts genotypically and phenotypically male and others female; the most frequent condition is bilateral gynandromorphy in which the left and right halves of the body are of different sex; gynander; **gynandrism, gynandromorphism**.

gynecogeny Parthenogenesis *q.v.*; **gynecogenic**.

gynetype A female type specimen *q.v.*;
cf. androtype.

gynic Female; *cf.* andric.

gynochoric Pertaining to organisms dispersed by
motile females; **gynochore**, **gynochorous**.

gynodioecious Used of plants or plant species
having female (pistillate) and hermaphrodite
(perfect) flowers on separate plants in a
population or species; **gynodioecy**;
cf. androdioecious.

gynoecious Used of a plant having female flowers
only; *cf.* androecious.

gynogamete A female gamete; egg; ovum.

gynogenesis The process in which an egg develops
parthenogenetically after the egg has been
activated by sperm or pollen; pseudogamy;
cf. androgenesis.

gynogenous Producing female offspring only;
cf. androgenous.

gynomerogony The development of an egg
fragment prior to fusion with a male
nucleus.

gynomonoecious Having female (pistillate) and
hermaphrodite (perfect) flowers on the same
plant; **gynomonoecy**; *cf.* andromonoecious.

gynomorphic Having a morphological resemblance
to females; **gynomorphy**; *cf.* andromorphic.

gynopaedium A family group in which the female
parent remains with the offspring for some time;
cf. patrogynopaedium, patropaedium.

gynophyte A female plant.

gynopleogamy The condition of a plant species
having three sexual forms, with either perfect
flowers, pistillate flowers or staminate flowers.

gypsophilous Thriving on chalk or gypsum-rich
soils; **gypsophile**, **gypsophily**.

gypsophyte A plant inhabiting chalk or gypsum-
rich soils; gypsum plant.

gypsophytium A plant community of limestone
substrata.

gyre A circular or spiral system of movement,
characteristic of oceanic currents, and other
systems of water movement.

gyttja Sedimentary peat comprising predominantly
plant and animal residues precipitated from
standing water.

H

habit The external appearance, aspect or growth form of an organism; constitutional type; **habitus**.

habitat The locality, site and particular type of local environment occupied by an organism; ece; local environment; oike; oikos.

habitat form The characteristic growth form of an organism under a given set of environmental conditions; ecad.

habitat group A group of unrelated plant species living under broadly similar habitat conditions.

habitat-type A group of plant communities having similar habitat relationships.

habituation The simplest form of learning in which the reduction or loss of a response to a stimulus occurs as a result of repeated stimulation which is not followed by any kind of reinforcement *q.v.*

habitudinal segregation The separation and consequent reproductive isolation of populations occupying different habitats.

habitus The characteristic form and appearance of an organism; habit; constitutional type.

hadal zone Ocean depths below 6000 m; the trenches and canyons of the abyssal region; ultra-abyssal zone; see Appendix 7.

Hadean Early Precambrian; the geological period prior to the Archaeozoic (*ca.* 3400 million years B.P.); see Appendix 1.

hadopelagic Pelagic in ocean depths greater than 6000 m (the hadal zone); ultra-abyssopelagic.

Haeckel's law Biogenetic law *q.v.*

haematobium An organism living in blood; **haematobic**.

haematocryal Poikilothermic *q.v.*; cold-blooded.

haematocytozoon A parasite living within a blood cell.

haematogenesis Haemopoiesis *q.v.*

haematophagous Feeding on blood; sanguivorous; **haematophage, haematophagy, haemophagous**.

haematophyte Any member of the blood flora.

haematothermal Homoiothermic *q.v.*; warm-blooded; **haematothermic**.

haematozoon An animal parasite living free in the blood; *cf.* haematocytozoon.

haemocoelous viviparity A mode of reproduction in which eggs are retained and commence development, in the haemocoel of the female parent.

haemoparasite Any parasite inhabiting the blood of its host.

haemophagous Feeding on blood; haematophagous; **haemophage, haemophagy**.

haemopoiesis The process of blood formation; haematogenesis.

haemotrophic Obtaining nutrients from blood; used particularly with reference to the nourishment of an embryo by the maternal blood supply.

halarch succession Halosere *q.v.*

Haldane's evolutionary unit A unit measure (in darwins) of increasing body size with time on an evolutionary time scale.

Haldane's rule That in most cases of interspecific hybridization one sex is frequently absent, rare, intersexual or sterile in the offspring and that sex is the heterogametic sex; Haldane's law.

half tide level Mean tide level *q.v.*

Half-Bog soil An intrazonal soil with a dark brown black peaty surface layer over a grey mottled mineral layer, formed in humid, cool to tropical climates under conditions of poor drainage, with grasses, sedges or forest vegetation.

half-life 1: The survival time of half the individual components of an unstable system. 2: A measure of radioactive decay; the time taken for the level of radioactivity to reduce by one-half. 3: The time taken for an individual or biological system to eliminate one-half of a given substance introduced into it.

half-sibs Organisms with only one parent in common.

halibios Halobios *q.v.*

halic Pertaining to saline conditions.

halicolous 1: Living in haloid soils *q.v.* having a crystalloid content between 0.5 and 2 parts per thousand; *cf.* gelicolous, pergelicolous, perhalicolous. 2: Used of a plant living in a habitat with a high salt content; **halicole**.

haliplankton Marine or inland saltwater planktonic organisms.

halmyrolysis Chemical rearrangement, replacement and weathering of rocks or sediment on the sea floor.

halo- Prefix meaning salty; **hali-**.

halobenthos Marine benthos.

halobiont A marine organism or an organism living in a saline habitat; **halobion, halobiontic**.

halobios The total life of the sea; that part of the Earth's surface occupied by marine organisms; halibios; **halobiontic**; *cf.* geobios, hydrobios, limnobios.

halocline A salinity discontinuity; a zone of marked salinity gradient.

halodrymium A mangrove community.

haloid soil A soil type characterized by high salt content, concentrated solutions and with a crystalloid content of more than 0.5 parts per thousand; *cf.* geloid soil.

halolimnetic 1: Pertaining to salt lakes. 2: Halolimnic *q.v.*

halolimnic Used of marine organisms adapted to live in fresh water; **halolimnetic**.

halomorphic Used of an intrazonal soil having an accumulation of salts.

halonereid A marine plant.

haloneuston Marine or salt water neuston.

halophilous Thriving in saline habitats; **halophile**, **halophily**; *cf.* halophobic.

halophobic Intolerant of saline habitats; **halophobe**; *cf.* halophilous.

halophreatophyte A plant utilizing saline ground water.

halophyte A plant living in saline conditions; a plant tolerating or thriving in an alkaline soil rich in sodium and calcium salts; a seashore plant; drymyphyte; *cf.* glycophyte.

haloplankton Marine or inland saltwater plankton; haliplankton.

halosere An ecological succession commencing in a saline habitat; halarch succession.

halosphere The marine component of the biosphere.

haloxenic Tolerant of saline conditions although not normally associated with saline habitats; **haloxene**.

hamabiosis A symbiosis without obvious advantage to either symbiont.

hammada A rocky desert.

handbook A field guide or identification key not comprising taxonomic conclusions or nomenclatural data.

Hanson abundance scale A scale for estimating abundance of a plant species, comprising six categories: absent (0), scarce (1–4 plants per square metre), infrequent (5–14 m^{-2}), frequent (15–29 m^{-2}), abundant (30–99 m^{-2}) and very abundant (more than 100 m^{-2}).

hap- Prefix meaning once; hapax-.

hapanthous Hapaxanthic *q.v.*

hapantotype A set of preserved preparations of directly related individuals, or of individuals representing different stages in a life cycle.

hapaxanthic Having a single flowering phase during the life cycle; hapanthous, **hapaxanthous**; *cf.* pollakanthic.

haplo- Prefix meaning simple, single.

haplobiont 1: A plant flowering once per season. 2: An organism not exhibiting a regular

alternation of haploid and diploid generations during the life cycle; *cf.* diplobiont.

haplodiploidy The genetic system found in some animals in which males develop from unfertilized eggs and are haploid, and females develop from fertilized eggs and are diploid; **haplodiploid**, **haplodiplont**.

haplogenesis The evolution of new forms.

haploid Having only a single set of chromosomes; having the gametic chromosome number, denoted by *n*; hemiploid; **haploidy**; *cf.* diploid, monoploid, polyploid.

haploid apogamy Meiotic apogamy *q.v.*

haploid parthenogenesis The development of a haploid individual from a female gamete that has undergone meiotic reduction division but has not been fertilized; generative parthenogenesis.

haplometrosis The founding of a colony of social insects by a single fertile female; monometrosis; **haplometrotic**; *cf.* pleometrosis.

haplont 1: The haploid phase of a life cycle; haplophase. 2: An organism having a life cycle in which meiosis occurs in the zygote to produce the haploid phase; only the zygote of haplonts is diploid; *cf.* diplont.

haplophase The haploid stage of a life cycle; gamophase; haplont; *cf.* diplophase.

haplophyte A haploid plant; gametophyte; *cf.* diplophyte.

haplosis The establishment of the gametic (haploid) chromosome number, usually by meiosis.

haplotype The single species included in a genus at the time of its designation, thus becoming the type species of the genus.

haptobenthos Those aquatic organisms that live closely applied to, or growing on, submerged surfaces; *cf.* benthos.

haptonastic Used of growth movement of a plant in response to a touch or contact stimulus; **haptonasty**.

haptotropism An orientation response to a touch or contact stimulus; stereotropism; thigmotropism; **haptotropic**.

haptotype An icotype *q.v.* collected at the same time as the holotype of a species but possibly taken from a different plant.

hard pan A compacted layer in the B-horizon of a soil, typically rich in deposited salts, restricting drainage and root penetration.

hardening The process of increasing cold resistance in plants.

hardiness The ability to survive exposure to subzero temperatures; cold resistance.

hardware The actual electronic (and other)

components from which a computer is constructed; *cf.* software.

Hardy–Weinberg equilibrium The maintenance of more or less constant allele frequencies in a population through successive generations; genetic equilibrium.

Hardy–Weinberg law That allele frequencies will tend to remain constant from generation to generation and that genotypes will reach an equilibrium frequency in one generation of random mating and will remain at that frequency thereafter; demonstrating that meiosis and recombination do not alter gene frequencies.

harem A group of females associated with a single male that seeks to prevent other males from mating with them.

harmonic mean The reciprocal of the arithmetic mean of the reciprocals of a set of observations.

harmosis The total response of an organism to a stimulus, comprising both reaction and adaptation.

harpactophagous Feeding by preying on other animals; rapacious; **harpactophage, harpactophagy**.

Hawaiian region 1: A zoogeographical region comprising the Hawaiian Archipelago; regarded either as a subdivision of the Polynesian subregion or as a separate region; see Appendix 3. 2: A subdivision of the Polynesian subkingdom of the Palaeotropical kingdom; see Appendix 4.

hawking Feeding in flight.

hb. Herb. q.v.

heart–weight rule That the ratio of body weight to heart weight increases in races of animals found in cold regions compared with races from warmer regions, because of the necessity of maintaining a greater temperature differential between body and environment; Hesse's rule.

heat A period of sexual activity or receptiveness.

heath Vegetation characteristic of low fertility, acidic, poorly drained soils, dominated by small leaved shrubs of Ericaceae (heathers and heaths) and Myrtaceae (myrtles).

heautotype A specimen subsequently selected by the author of a species but not the type specimen on which the species was based; autotype.

heavy metal A metallic element of high specific gravity; including antimony, bismuth, cadmium, copper, gold, lead, mercury, nickel, silver, tin and zinc.

hebetic Pertaining to adolescence; juvenile.

hecistotherm Hekistotherm *q.v.*; **hecistothermic**.

hectare (ha) A metric unit of area; equal to 10 000 m² or 100 are; see Appendix 13.

hecto- (h) Prefix used to denote unit $\times\ 10^2$.

hedonic Pertaining to factors acting to stimulate sexual activity.

hekistoplankton Flagellated nanoplanktonic organisms.

hekistotherm A plant with a requirement for a temperature of less than 10° C in the warmest month, typically occurring in regions with a mean annual temperature below 0° C; sometimes used to refer to those organisms living above the tree line in areas of heavy snow; hecistotherm; **hekistothermic**; *cf.* megatherm, mesotherm, microtherm.

helad A marsh plant; helodad; helophyte.

helcotropism Geotropism *q.v.*; **helcotropic**.

heleoplankton The planktonic organisms of small ponds and marshy habitats; **heleoplanktonic**.

heli- Prefix meaning sun.

heliad A plant thriving in full sunlight; heliophyte; oread.

helic Pertaining to marshes or marsh communities.

heliophilous Thriving under conditions of high light intensity; **heliophil, heliophile, heliophily**; *cf.* heliophobous.

heliophobous Intolerant of high light intensity; shade-loving; heliophobic; skiophilic; umbrophilic; **heliophobe** *cf.* heliophilous.

heliophyllous Used of plants having leaves that can tolerate full sunlight; *cf.* sciophyllous.

heliophyte A plant that shows optimum growth under conditions of full sunlight; a species requiring full sunlight is sometimet called an obligate heliophyte and one that can tolerate full sunlight but grows best in shade is a facultative heliophyte; heliad; oread; *cf.* skiophyte.

heliophytium A plant community thriving in conditions of full sunlight.

heliosis Solarization *q.v.*

helioskiophyte A plant which thrives in both sunlight and shade, but which grows best in sunny conditions.

heliotaxis A directed response of a motile organism towards (positive) or away from (negative) sunlight; phototaxis; **heliotactic**.

heliothermic Used of organisms that maintain a comparatively high body temperature by basking in sunlight; **heliotherm**.

heliotropism An orientation response to sunlight; **heliotropic**.

helioxerophilous Used of desert organisms thriving in both strong sunlight and drought conditions; **helioxerophile, helioxerophily**.

helioxerophyllous Used of plants having leaves able to tolerate conditions of full sunlight and drought.

helium A marsh community.

helminthology The study of parasitic flatworms and round worms.

helobius Living in marshes; paludal; palustrine; uliginous; **helobious**.

helodad A marsh plant; helad; helophyte.

helodium An open swampy woodland community; helorgadium.

helodric Pertaining to a swamp thicket.

helodrium A swamp thicket community.

helohylium A swamp forest community.

helohylophilous Thriving in wet or swampy forests; **helohylophile, helohylophily**.

helokrene A marsh spring community.

helolochmium A meadow thicket community.

helolochmophilous Thriving in meadow thicket habitats; **helolochmophile, helolochmophily**.

helolochmophyte A meadow thicket plant.

helophilous Thriving in marshes; **helophile, helophily**.

helophyte 1: A perennial plant with renewal buds, commonly on rhizomes, buried in soil or mud below water level; limnocryptophyte; *cf.* Raunkiaerian life forms. 2: Any marsh or bog plant; helad; helodad; limnophyte.

heloplankton The floating vegetation of a marsh.

helorgadium An open swampy woodland habitat; helodium.

helorgadophilous Thriving in swampy woodlands; **helorgadophile, helorgadophily**.

helotism Symbiosis in which one symbiont enslaves the other.

helper In an avian communal breeding system, a bird that assists in nest building, care of young or gives other aid, but is not related to either parent or offspring.

hemera The period of deposition of a fossiliferous stratum.

hemeranthous Flowering only during the day; **hemeranthic, hemeranthy**; *cf.* nyctanthous.

hemerocology The study of the ecology of cultivated areas and culture communities.

hemerophilous Thriving in habitats influenced by the activities of man or under cultivation; **hemerophile, hemerophily**.

hemerophyte A cultivated plant.

hemi- Prefix meaning half.

hemialloploid A polyploid hybrid (allopolyploid) having chromosome sets derived from two species that are not fully intersterile.

hemiautophytic Used of a parasitic plant which is also capable of photosynthesis; **hemiautophyte**.

hemiautoploid A polyploid derived from intraspecific hybrids or by the differentiation of the chromosome sets of successful panautoploids *q.v.*

hemibathybial Used of planktonic organisms inhabiting the epipelagic zone.

hemiboreal Pertaining to the southern taiga *q.v.*

hemichimonophilous Used of plants that thrive under cold conditions and begin their growth even during frost; **hemichimonophile, hemichimonophily**.

hemicleistogamic Used of a plant having flowers that only open partially before pollination.

hemicryptophyte A perennial plant with renewal buds at ground level or within the surface layer of soil; typically exhibiting degeneration of vegetative shoots to ground level at the onset of the unfavourable season; *cf.* Raunkiaerian life forms.

hemiendobiotic Used of an organism that is typically found within its host, sometimes outside it.

hemiendophytic Used of a fungus that occurs both external and internal to its host at different times.

hemiepiphyte A plant that spends only part of its life cycle as an epiphyte, producing both aerial and subterranean roots at different times.

hemigamotropic Used of flowers that alternate between open and partially closed; *cf.* agamotropic, gamotropic.

hemigamy The activation of an ovum by a male nucleus without fusion; semigamy; **hemigamous**.

hemikaryotic Used of a cell with the haploid number of chromosomes; **hemikaryon**.

hemimetabolous Used of the pattern of development characterized by gradual changes, without metamorphosis or distinct separation into larval, pupal and adult stages; hemimetamorphic; hemimetabolic; **hemimetaboly**; *cf.* holometabolous.

hemimetamorphic Hemimetabolous *q.v.*

hemiparasite 1: A partial or facultative parasite that can survive in the absence of a host; meroparasite. 2: A parasitic plant that develops in the soil from free seeds; *cf.* holoparasite.

hemipelagic sediment An oceanic sediment intermediate between a pelagic and a non-pelagic sediment in terrigenous content.

hemiplankton Organisms that spend only part of their life cycle in the plankton; meroplankton; *cf.* holoplankton.

hemiploid Having half the somatic chromosome number; haploid.

hemisaprophyte A plant that may be autotrophic or saprotrophic at different times.

hemitropic Used of flowers adapted for pollination by particular insect species; also used of the insects that pollinate specifically adapted flowers.

hemizygoid parthenogenesis The development of an individual from an unfertilized haploid egg.

hemizygous Pertaining to unpaired genes or chromosomal segments in a diploid cell.

Hennigian systematics A method of classification based on the study of phylogenetic relationships between taxa in which morphological resemblance between species is not considered as a simple measure of phylogenetic relationship but is divided into the concepts of convergence, symplesiomorphy *q.v.* and synapomorphy *q.v.* of which only the latter category of resemblance is used to establish relationships; phylogenetic systematics.

Hennig's deviation rule That when an ancestral species splits, one of the daughter species diverges more strongly from the ancestral condition than the other.

hepaticology The study of liverworts.

hepodoche A secondary succession.

hepta- Prefix meaning seven, sevenfold.

herb A plant having stems that are not secondarily thickened and lignified (non-woody) and which die down annually; herbaceous.

herb. (**hb.**) Abbreviation of the Latin *herbarium*, meaning a herbarium.

herbaceous stratum The non-woody growth of the forest floor.

herbage Total vegetation available to a grazing animal; *cf.* forage.

herbarium 1: A collection of preserved (usually dried) plant specimens; *hortus siccus*. 2: The building in which such a collection is kept.

herbicide A chemical used to kill weeds or herbage.

herbicolous Living predominantly in herbaceous habitats; **herbicole**.

herbivorous Feeding on plants; phytophagous; **herbivore, herbivory**.

herbosa Communities of grasses and herbs; herbaceous vegetation.

hercogamous Used of a flower having the stamens and stigma positioned in such a way as to prevent self-pollination; **hercogamy, herkogamous, herkogamy**.

hereditary Having a genetic basis; transmitted from one generation to the next.

heredity The mechanism of transmission of specific characters or traits from parent to offspring.

heremetabolic Used of a pattern of development characterized by incomplete metamorphosis, with a resting phase at the end of the nymphal stage.

heritability 1: The capacity of being inherited. 2: That part of the phenotypic variability that is genetically based, usually expressed as the ratio of genetic variance to phenotypic variance.

heritage Those characters or traits of an organism that indicate adaptations to a different ancestral mode of life.

herkogamous Hercogamous *q.v.*

hermaphrodite Having both male and female reproductive organs in the same individual (animal) or the same flower (plant); androgyne; bisexual; the maturation of the male organs before the female is protandrous hermaphroditism, the female before the male is protogynous hermaphroditism, and simultaneously is synchronous hermaphroditism.

hermatypic Used of reef-forming corals that contain symbiotic algae within the polyps; **hermatype**.

herpesian Pertaining to amphibians and reptiles.

herpetogeny The history of colonization and evolution in the establishment of the modern amphibian and reptile faunas.

herpetology The study of amphibians and reptiles; **herpetological**.

herpism Creeping locomotion; used of protistans employing pseudopodia.

herpobenthos Those organisms growing or moving through muddy sediments; *cf.* benthos.

herpon Crawling organisms.

hertz (**Hz**) A derived SI unit of frequency, defined as the frequency of a periodic phenomenon of which the periodic time is 1 second; one cycle per second (1 cps); see Appendix 13.

hesmosis Colony fission in ants.

Hesse's rule Heart–weight rule *q.v.*

heteracme Dichogamy *q.v.*

heterauxesis Allometric growth; the differential growth of body parts (x and y), expressed by the equation $y = bx^a$, where a and b are fitted constants; change of shape or proportion with increase in size; allometry; heterogony; *cf.* bradyauxesis, isauxesis, tachyauxesis.

heterecious Heteroecious *q.v.*

hetero- Prefix meaning different, other, other than usual.

heteroallelic Used of genes having mutations at different mutational sites; *cf.* homoallelic.

heterobathmy of characters The mosaic distribution of relatively primitive and relatively derived characters in related species and species groups.

heteroblastic Having indirect development with distinct larval and adult stages.

heterochoric Used of a plant species occurring in two or more closely related associations *q.v.*

heterochromatism Change of colour.

heterochromosome Allosome *q.v.*

heterochrony 1: An evolutionary change in the onset or timing of development of a feature relative to the appearance or rate of development of the same feature in the ontogeny of an ancestor. 2: The appearance of a particular fauna or flora in two different regions at different times; **heterochronic, heterochronous**.

heteroduplex A double stranded nucleic acid molecule in which the strands are derived from separate sources.

heterodynamic 1: Having a life cycle comprising two or more distinct forms or stages exhibiting dissimilar biological activity. 2: Used of genes that simultaneously effect different developmental processes; *cf.* homodynamic.

heterodynamic hybrid A hybrid which displays the characters of one parent more than the other; segregate; *cf.* homodynamic hybrid.

heteroecious 1: Used of a parasite occupying two or more different hosts at different stages of the life cycle; heteroxenous; metoxenous; **heterecious, heteroecius, heteroecism**; *cf.* homoecious. 2: Used of a non host-specific parasite; metoecious. 3: Used of a unisexual organism in which male and female gametes are produced by different individuals.

heterofacial Showing regional character differentiation.

heterogameon A species comprising distinct races which produce morphologically stable populations when selfed but produce various types of viable and fertile offspring when crossed.

heterogamete 1: A gamete produced by the heterogametic sex *q.v.* 2: A gamete belonging to one of two distinguishable types; anisogamete.

heterogametic Having two kinds of gametes, one producing males and the other females; digametic; **heterogamy**; *cf.* homogametic.

heterogametic sex The sex which is determined by a pair of dissimilar sex chromosomes (X and Y, or Z and W) or by a single unpaired sex chromosome (X or Z); the sex which produces two types of sex-determining gametes; heterogamety; *cf.* homogametic sex.

heterogamy 1: The union of gametes (heterogametes) of different shape or size; anisogamy; ditopogamy. 2: Alternation of two sexual generations, one syngamic and the other parthenogenetic; heterogony. 3: The condition of a plant producing both male and female gametes from one kind of flower; **heterogamic, heterogamous**; *cf.* homogamy. 4: Disassortative mating *q.v.*

heterogeneity coefficient A measure of vegetation heterogeneity, calculated as the total number of

species in a table of records divided by the average number of species per record.

heterogeneity index A measure of genetic variance based on genotypic distance.

heterogeneous Having a non-uniform structure or composition; **heterogeneity**; *cf.* homogeneous.

heterogenesis 1: Spontaneous generation; abiogenesis; xenogenesis. 2: Alternation of generations; diagenesis; metagenesis. 3: The appearance of a mutant in a population; **heterogenetic**.

heterogenetic Derived from different ancestral stocks; heterogen.

heterogenetic association The pairing of chromosomes from different ancestors in an allotetraploid *q.v.*

heterogenic Pertaining to a population or gamete containing more than one allele at a given locus.

heterogonic life cycle 1: A cycle of alternating parthenogenetic and sexually reproducing phases. 2: A cycle involving an alternation of a parasitic and a free-living generation; indirect life cycle; *cf.* homogonic life cycle.

heterogony 1: Allometric growth *q.v.*; **heterogonic**. 2: Cyclic parthenogenesis with an alternation of one or more parthenogenetic generations with a bisexual (amphigonic) generation; heterogamy.

heterograde 1: A non-uniform gradient of a factor in the water column. 2: In limnology, a type of oxygen distribution found in some lakes in which a very striking maximum (positive) or minimum (negative) in oxygen concentration develops in the metalimnion during stratification; *cf.* clinograde, orthograde.

heterograft Tissue transplanted from one organism to another of a different species; xenograft.

heterogynistic Used of characters that are more strongly marked in females than in males; **heterogynism**.

heterogynous Having two kinds of females.

heteroicous Heteroecious *q.v.*

heterokaryon A multinucleate cell possessing genetically different nuclei; **heterocaryon, heterokaryote, heterokaryosis, heterokaryotic**; *cf.* homokaryon.

heteromesogamic Used of a species in which different individuals employ different modes of fertilization.

heterometabolic Having a pattern of development exhibiting an incomplete metamorphosis.

heteromixis Sexual reproduction in fungi involving the union of nuclei with different genetic constitutions, each from a different thallus; amphithallism; *cf.* homomixis.

heteromorphic Having different forms at different times or at different stages of the life cycle; used of a plant having an alternation of vegetatively dissimilar generations; **heteromorphous, heteromorphy**; *cf.* homomorphic.

heteromorphosis Regeneration of a part following injury, with a structurally different replacement part; xenomorphosis.

heteronomous Having a different growth pattern or specialization.

heteroparthenogenesis Parthenogenesis producing either offspring that reproduce parthenogenetically, or offspring that reproduce sexually.

heterophagous Feeding on a wide variety of food items; used of a parasite utilizing a wide variety of hosts; **heterophage, heterophagy**.

heterophasic Used of a life cycle comprising two or more distinct phases, as in an alternation of sexual and asexual generations.

heterophile reaction An antigen-antibody reaction in which the antibody was not specifically produced in response to the antigen.

heterophyadic Pertaining to a genetic polymorphism.

heterophyte 1: A plant occurring in a wide range of habitats. 2: A dioecious *q.v.* plant. 3: A parasitic plant devoid of chlorophyll.

heteroplanobios Organisms passively transported by flood water.

heteroplastic grafting Transplanting tissue (the heterograft) from one organism to another of a different species; *cf.* autoplastic grafting, homoplastic grafting.

heteroploid 1: Aneuploid *q.v.* 2: Used of a population comprising aneuploid, diploid and euploid individuals.

heteroselection Selection favouring heterozygotes; *cf.* homoselection.

heterosis Hybrid vigour; the selective superiority of heterozygotes; the increased vigour resulting from hybridization when measured against either parental stock; **heterotic**.

heterospecific Pertaining to a different species; *cf.* conspecific.

heterosporous Having two kinds of spores, such as microspores and megaspores, that give rise to distinct male and female gametophyte generations respectively; **heterosporic, heterospory**; *cf.* homosporous.

heterostyly A polymorphism of flowers in which the length of the style and stamens varies between individuals of a species, helping to prevent self-pollination; *cf.* homostyly.

heterosymbiosis Symbiosis between two organisms belonging to different species; interspecific symbiosis; *cf.* homosymbiosis.

heterothallism A mode of reproduction in fungi

and algae which involves the interaction of two thalli, each haploid thallus being self-incompatible; the thalli may exhibit distinct sexual differentiation into male and female (morphological heterothallism) or may be structurally similar (physiological heterothallism); **heterothallic, heterothallous**; *cf.* homothallism.

heterothermic Poikilothermic *q.v.*; cold-blooded; heterothermal; *cf.* homoiothermic.

heterothermic soil A soil that gains or loses heat readily; *cf.* homothermic soil.

heterotopic Occurring in a wide variety of habitats.

heterotopy A phyletic change in the site within the germinal layers, from which an organ differentiates during ontogeny.

heterotrophic 1: Obtaining nourishment from exogenous organic material; used of organisms unable to synthesize organic compounds from inorganic substrates; organotrophic; zootrophic; **heterotroph**; *cf.* autotrophic. 2: Used of plants occurring in a wide range of habitats on a wide variety of soil types, and of protistans utilizing a wide variety of food materials; **heterotrophy**.

heterotype In taxonomy, a type which has been derived by combining the characters of two different species.

heterotypic Differing from type, or from the normal condition; abnormal; *cf.* homotypic.

heterotypic synonyms In taxonomy, synonyms based on different nomenclatural types; subjective synonyms; taxonomic synonyms; *cf.* homotypic synonyms.

heteroxenous Used of a parasite occupying more than one host during its life cycle; heteroecious; **heteroxeny**; *cf.* dixenous, monoxenous, oligoxenous, trixenous.

heterozone An organism not restricted to a single habitat during development; *cf.* homozone.

heterozygote A diploid individual formed by the fusion of gametes carrying different alleles at a given locus and which produces two kinds of gametes differing with respect to the allele at that locus; *cf.* homozygote.

heterozygous Having two different alleles at a given locus of a chromosome pair; **heterozygosis, heterozygosity**; *cf.* homozygous.

heuristic Providing direction in the solution of a problem but otherwise unjustified or unjustifiable.

hexa- Prefix meaning six, sixfold.

hibernaculum The domicile in which an animal hibernates or overwinters; winter-quarters.

hibernal Pertaining to the winter; hiemal; *cf.* aestival, vernal.

hibernation The act or condition of passing the

winter in a torpid or resting state, typically involving the abandonment of homoiothermy in mammals; *cf.* aestivation.

hibernator An animal that hibernates; may be an obligate or permissive hibernator.

hidroplankton Planktonic organisms that achieve buoyancy by means of surface secretions.

hiemal Pertaining to the winter; hibernal; **hyemal**; *cf.* aestival, vernal.

hiemilignosa Tropical deciduous monsoon forest and bush vegetation.

hierarchy 1: A general integrated system comprising two or more levels, the higher controlling to some extent the activities of the lower levels. 2: A series of consecutively subordinate categories forming a system of classification; nested hierarchy. 3: Dominance hierarchy *q.v.* 4: Taxonomic hierarchy *q.v.*

high tide High water; the maximum height of the incoming tide.

high water (HW) High tide; the highest surface water level reached by the rising tide.

higher high water (HHW) The higher of two high waters during any tidal day in areas where there are marked inequalities of tidal height.

higher low water (HLW) The higher of two low waters *q.v.* during any tidal day in areas where there are marked inequalities of tidal height.

Hill–Robertson effect Selection of alleles at one locus producing gene frequency changes at other loci.

Himalayan subregion Indo-Chinese subregion of the Oriental region; see Appendix 3.

Hindostan subregion Indian subregion of the Oriental region; see Appendix 3.

hipotype Hypotype *q.v.*

histic Rich in peat.

histogenesis Tissue formation and development; the reorganization of body tissue during metamorphosis.

histogram A graph of a frequency distribution obtained by constructing rectangles whose bases coincide with the class intervals and whose areas are proportional to the class frequencies; when class intervals are equal the rectangle height is also proportional to class frequency.

histology The branch of anatomy dealing with the structure of tissues; **histological**.

histometabasis The process of fossilization in which the cellular structure of a plant or animal is preserved.

histophilous Inhabiting living host tissue; parasitic.

histophyte A parasitic plant.

historical zoogeography The study of the distributions of animal species and higher taxa in terms of their past history and the palaeogeographical evolution of the region in which they occur.

histozoic Living within the tissues of another organism; parasitic.

H-layer The heavily transformed humus layer of a soil with no macroscopic plant remnants and having a strong mineral particle intermix; humified layer; Oh-layer; see Appendix 12.

Hofacker and Sadler's law That an older male parent crossed with a younger female will produce more male progeny than a young male crossed with an older female.

holaedeotype The original specimen of a species to have its genitalia examined (aedeotype) and which is also the holotype of the species.

holandric Occurring only in males; used of a character carried on the Y chromosome in the heterogametic male sex; **holandry**; *cf.* hologynic.

Holarctic A zoogeographical region comprising the Palaearctic and Nearctic regions; see Appendix 3.

holarctic Distributed throughout North America, Asia and Europe; circumboreal.

holard The total water content of the soil; *cf.* chresard, echard.

holding capacity The difference between field capacity *q.v.* and permanent wilting percentage *q.v.*

holendophyte A parasitic plant which passes its entire life within its host.

holeuryhaline Used of organisms that freely inhabit fresh water, sea water and brackish water, or which establish populations in all three environments; *cf.* euryhaline, oligohaline, polystenohaline, stenohaline.

holism The concept that all physical and biological entities form a single unified, interacting system and that any complete system has a totality that is greater than the sum of the constituent parts; gestalt; **holistic**; *cf.* reductionism.

holo- Prefix meaning whole, complete, entire, total.

holobenthic Used of an organism confined to a benthic existence throughout the life cycle.

Holocene A geological epoch within the Quaternary period (*ca.* 10 000 years B.P. to the present time); Recent; see Appendix 1.

holocoen Ecosystem *q.v.*

holocoenotic Pertaining to the effect of combined or interacting environmental factors on an organism or biological system.

holocyclic parthenogenesis Reproduction by a series of parthenogenetic generations alternating with a single sexually reproducing generation; cyclic parthenogenesis; monocyclic parthenogenesis; *cf.* parthenogensis.

holoendemic A species which is not of recent origin but has retained a narrow geographical distribution.

hologamodeme A local interbreeding population comprising all those individuals which are able to interbreed with a high level of freedom under a given set of conditions; a unit capable of hybridizing with other hologamodemes to give hybrids showing some fertility; *cf.* deme.

hologamy A mode of reproduction in protistans involving gametes similar in size to the vegetative cell; gamontogamy; macrogamy; **hologamete**; *cf.* merogamy.

hologenesis 1: The theory that all species arose simultaneously throughout their modern distribution, without subsequent dispersal. 2: The origin of species by multiple mutation from a single extinct ancestor.

hological method A study of the totality of a system that concentrates on the operation of the whole rather than the specific contributions of the component parts; hological; *cf.* merological method.

hologynic Occurring only in females; used of sex linked characters in species with a heterogametic female sex; **hologyny**; *cf.* holandric.

holohomoiotype A homoeotype *q.v.* of the same sex as the holotype or lectotype.

hololectotype Lectotype *q.v.*

holometabolous Used of the pattern of development which is characterized by a distinct metamorphosis with well defined larval and adult stages; holometabolic; holometamorphic; **holometabolism, holometaboly**; *cf.* hemimetabolous.

holometamorphic Holometabolous *q.v.*

holomictic Used of a lake having complete free circulation throughout the water column at the time of winter cooling; *cf.* mictic.

holomorphosis Regeneration in which the entire part is replaced.

holomorphospecies In palaeontology, a species based on the collective morphological characters of assemblages of fossils from throughout the geographical range of a given stratigraphic level.

holoparalectotype A paralectotype *q.v.* belonging to the same sex as the lectotype; *cf.* alloparalectotype.

holoparasite An obligate parasite; a parasite that cannot survive away from its host; *cf.* meroparasite.

holoparatype A paratype of the same sex as the holotype; *cf.* alloparatype.

holopelagic Used of aquatic organisms that remain pelagic throughout the life cycle; *cf.* meropelagic.

holophyletic Used of a group of taxa comprising a single ancestral species or taxon and all the descendant species or taxa; **holophyly**.

holophytic Synthesizing organic compounds from inorganic substrates in the manner of a green plant; photosynthetic; autotrophic; holotrophic; **holophyte**; *cf.* holozoic.

holoplankton Those organisms that are permanent members of the plankton; euplankton; **holoplanktonic, holoplanktont**; *cf.* meroplankton.

holoplastotype In taxonomy, a cast of a holotype.

holoplesiotype In taxonomy, a plesiotype *q.v.* of the same sex as the holotype; *cf.* alloplesiotype.

holosaprophyte A plant deriving nourishment entirely from decomposing organic matter; obligate saprophyte.

holoschisis Amitosis *q.v.*

holotrophic 1: Used of a plant which synthesizes all organic compounds from inorganic substrates, typically by photosynthesis; autotrophic; holophytic. 2: Used of a predator that utilizes only one species of prey organism.

holotype In taxonomy, the single specimen designated or indicated as the type specimen of a nominal species by the original author at the time of publication or the single specimen when no type was specified but only one specimen was present.

holozoic Feeding entirely in the manner of an animal by ingesting complex organic matter; *cf.* holophytic.

Holsteinian interglacial An interglacial period in the middle of the Quaternary Ice Age in northern Germany and Poland; see Appendix 2.

homalochoric Used of species restricted to a single community.

home That part of the habitat of an organism that is utilized for resting and breeding; *cf.* hotel.

home range The area, usually around the domicile, over which an animal normally travels in search of food; home realm; *cf.* territory.

home site The location of the domicile or resting place regularly used by a particular animal.

homeochronous Homochronous *q.v.*

homeomorphy The condition of having the same or very similar morphology as a result of convergent evolution rather than common ancestry; the state of being structurally similar but phylogenetically distinct; homoeomorphy; **homeomorph, homeomorphic**.

homeostasis The maintenance of a relatively steady state or equilibrium in a biological system by intrinsic regulatory mechanisms; **homeostatic**.

homeotely Evolutionary changes of homologous structures; **homeotelic, homeotic**.

homeotic mutation Mutation from one state to another of a structure that exists in a series of states, such as the mutation of an insect wing into a haltere; homoeotic mutation.

homeotype Homoeotype *q.v.*

homeozoic Pertaining to separate geographical areas having similar faunal composition; homoeozoic.

homing The behavioural act of returning to an original location.

homiogamy The union of two gametes of the same sex; homoiogamy; **homiogamic**.

homiothermic Homoiothermic *q.v.*; **homiothermal**, **homiothermy**.

homo- Prefix meaning similar, alike, the same.

homoallelic Used of genes having mutations at the same mutational sites; *cf.* heteroallelic.

homobium The symbiosis between a fungus and an alga in a lichen.

homochronous Occurring at the same time, or age in successive generations; simultaneous; homeochronous.

homodichogamy The condition of a species in which some individuals exhibit homogamy *q.v.* and others dichogamy *q.v.*; autoallogamy; **homodichogamic**.

homodynamic 1: Having a life cycle that progresses from generation to generation without a resting stage that exhibits a distinct change of biological activity. 2: Used of genes that simultaneously effect the same developmental process; *cf.* heterodynamic.

homodynamic hybrid A hybrid that displays the characters of both parents to an equal extent; dichodynamic hybrid; *cf.* heterodynamic hybrid.

homodynamous Metamerically homologous.

homoecious 1: Used of a parasite occupying the same host throughout its life cycle; monoxenous; *cf.* heteroecious. 2: Host-specific.

homoeogamy The condition in which an antipodal cell and not the oospore is fertilized.

homoeologous chromosomes Chromosomes that are only partially homologous, such as those apparently derived from ancestral homologous chromosomes *q.v.* which exhibit reduced synaptic attraction.

homoeomorphy Homeomorphy *q.v.*

homoeosis The modification of one segmental appendage or structure to resemble a different appendage or structure of the same homologous series.

homoeothermic Homoiothermic *q.v.*; **homoeothermal**, **homoeothermy**.

homoeotic mutation Homeotic mutation *q.v.*

homoeotype A specimen compared with the type by an author other than the author of the species and determined as conspecific; **homeotype**; **homotype**.

homogamete The gamete produced by the homogametic sex *q.v.*

homogametic Having only one kind of gamete; **homogamy**; *cf.* heterogametic.

homogametic sex The sex which is determined by a pair of similar sex chromosomes (X); that sex which produces only one type of gamete; homogamety; *cf.* heterogametic sex.

homogamy 1: The simultaneous maturation of male and female reproductive organs in a flower or hermaphrodite organism; synacme; *cf.* dichogamy. 2: The condition of a plant producing only one kind of gamete on one kind of flower; **homogamic**, **homogamous**; *cf.* heterogamy. 3: Positive assortative mating *q.v.*

homogeneon A species that is genetically and morphologically homogeneous.

homogeneous Similar throughout; of uniform structure or composition; **homogeneity**; *cf.* heterogeneous.

homogenesis Non alternation of generations; the succession of morphologically similar generations; **homogenetic**, **homogeny**; *cf.* heterogenesis.

homogenetic Used of a group having a common origin or ancestral stock; **homogen**.

homogenous More or less alike owing to descent from common stock; *cf.* homologous.

homogonic life cycle A life cycle in which all generations are either parasitic or free-living; direct life cycle; *cf.* heterogonic life cycle.

homograft Homoplastic grafting *q.v.*

homoiogamy Homiogamy *q.v.*; **homoiogamic**.

homoiohydric Used of protistans, fungi and plants that can compensate for short-term fluctuations in water supply and rate of evaporation so that their water content can be maintained at a given level more or less independently of ambient humidity; *cf.* poikilohydric.

homoiologous Homologous *q.v.*

homoiosmotic Used of organisms that maintain a relatively constant internal osmotic pressure; *cf.* poikilosmotic.

homoiothermic Used of animals that regulate their body temperature independent of ambient temperature fluctuations; warm-blooded; endothermic; idiothermous; homoeothermic; homiothermic; homothermic; **homoiotherm**, **homoiothermal**, **homoiothermy**; *cf.* poikilothermic.

homokaryon A multinucleate cell in which the nuclei have identical constitutions; **homocaryon**; *cf.* heterokaryon.

homologous 1: Used of structures, traits or properties having common ancestry but not necessarily retaining similarity of structure, function or behaviour; homoiologous; **homologue, homology**; *cf.* analogous. 2: Used of characters that share an evolutionary transformation from the same ancestral character state; *cf.* transformation series. 3: Used of polypeptides or proteins that share the same or similar amino acid sequences.

homologous chromosomes Structurally similar chromosomes having identical genetic loci in the same sequence and which pair during nuclear division; homologues; *cf.* homoeologous chromosomes.

homolotropism Growth orientation in a horizontal direction; **homolotropic**.

homomixis Sexual reproduction in fungi involving the union of genetically similar nuclei from one thallus; *cf.* heteromixis.

homomorphic Used of a plant with a dimorphous life cycle in which the two different types of individuals in the life cycle are morphologically similar; isomorphic; **homomorphous, homomorphy**; *cf.* heteromorphous.

homonym In taxonomy, each of two or more identical names denoting different taxa of the family, genus or species-group, or of species or subspecies within the same genus; *cf.* primary homonym, secondary homonym.

homonymy 1: In taxonomy, the existence of two or more identical names; **homonymous**. 2: The principle that a junior homonym of an available name is invalid within the terms of the Code and should not be used.

homophylic Showing resemblance due to common ancestry.

homoplastic grafting Transplanting tissue (the homograft) from one individual to another of the same species; *cf.* autoplastic grafting, heteroplastic grafting.

homoplasy Structural resemblance due to parallelism or convergent evolution rather than to common ancestry; homoplastic similarity; nonhomologous similarity; homoplasia; homoplasty; **homoplast**.

homorophyletic Pertaining to a group of taxa sharing similar characters or traits that have been inherited from a common ancestral species; **homorophyly**.

homorophylistic Pertaining to a patristic relationship between taxa based on both reasonable genotypic similarity and practical or intuitive considerations; **homorophylist, homorophylistics**.

homoselection Selection favouring homozygotes; *cf.* heteroselection.

homosequential Used of closely related species in

which chromosome banding patterns and hence the gene sequences are considered to be identical.

homosporous Having only one kind of spore; isosporous; **homosporic, homospory**; *cf.* heterosporous.

homostyly The condition in which all the flowers of a species have styles of similar length; *cf.* heterostyly.

homosymbiosis Symbiosis between two organisms belonging to the same species; intraspecific symbiosis; *cf.* heterosymbiosis.

homotaxis The presence of similar communities, assemblages or species in different regions or strata; **homotaxial, homotaxy**.

homotenous Pertaining to the primitive form or condition; **homotene**.

homothallism A mode of reproduction in fungi and algae in which each thallus is self-fertile, not showing haploid incompatibility, and each produces both male and female sex cells; **homothallic, homothallous**; *cf.* heterothallism.

homothermic Homoiothermic *q.v.*; **homothermal, homothermous, homothermy**.

homothermic soil A soil that gains or loses heat slowly; *cf.* heterothermic soil.

homotopotype A homoeotype *q.v.* from the type locality.

homotropic Used of a flower fertilized by its own pollen.

homotropism Orientation of an organ or structure in the same direction as the body to which it belongs; **homotropic**.

homotype Homoeotype *q.v.*

homotypic Conforming to type, or the normal condition; *cf.* heterotypic.

homotypic association An assemblage of organisms of the same species occurring together as they are descendants of the same parent or parents.

homotypic synonyms In taxonomy, synonyms based on the same nomenclatural type; nomenclatural synonyms; obligate synonyms; objective synonyms; *cf.* heterotypic synonyms.

homoxenic Used of two or more different species of parasites occurring on the same host species; **homoxenia**; *cf.* alloxenic.

homozone An organism restricted to a single habitat during development; *cf.* heterozone.

homozygote An individual with identical alleles at the two homologous loci of a chromosome pair; derived from the union of gametes bearing identical alleles; *cf.* heterozygote.

homozygous Having identical alleles at a given locus of a chromosome pair; isogenic; *cf.* heterozygous.

homunculus A microscopic human form contained within the sperm, according to animalculism *q.v.*

honing The process of tooth sharpening.

hook order A social dominance hierarchy found in horned mammals established by the aggressive use of horns.

Hopkin's bioclimatic law The generalization that in temperate North America life history events tend to occur with an average delay of 4 days for each degree of latitude northwards, 5° of longitude eastwards and 400 feet (122 m) of altitude in spring and early summer; with corresponding earliness in late summer and autumn.

horary Lasting an hour or two; **horarius**.

horiodimorphic Used of an organism exhibiting a seasonal or periodic alternation of form; horodimorphic; **horiodimorphism**.

horizon Any horizontal stratum within a sediment, soil profile, water column or geological series.

horizontal classification A classification which groups species from a similar time interval of evolution (horizontal with respect to a phylogenetic tree) rather than those sharing a common lineage; *cf.* vertical classification.

hormesis The stimulus afforded by exposure to non-toxic concentrations of a potentially toxic substance.

hormone A chemical substance secreted by an endocrine gland, that produces a specific physiological response on some remote target tissue or system.

hormone weedkillers Synthetic organic herbicides that are similar in effect to natural plant hormones (auxins), promoting, retarding or modifying the growth and development of plants.

hornotinus Hortinus *q.v.*

hornus Hortinus *q.v.*

horodimorphic Horiodimorphic *q.v.*; **horodimorphism**.

horological 1: Pertaining to time, or the measurement of time. 2: Used of a flower that opens and closes at particular times during the day; equinoctial.

horotelic Evolving at the normal or average rate; **horotely**; *cf.* bradytelic, tachytelic.

hort. Abbreviation of Latin **hortulanorum**, meaning, of gardeners and of **hortorum**, meaning of gardens; used to indicate the name of the garden of origin (**ex horto**).

hortal Used of an ornamental plant that has escaped from cultivation and can be found growing wild.

hortinus Pertaining to the plant growth of the current year; **hornotinus, hornus**.

hortus siccus Herbarium *q.v.*

hospitating Offering refuge or acting as a host (hospitator) to another organism (the hospite).

hospitator An organism offering refuge or acting as a host to another organism.

hospite An organism taking refuge in, or inhabiting, another organism.

host 1: Any organism that provides food or shelter for another organism; the inhabited symbiont in an endosymbiosis *q.v.* or the exhabited symbiont in an exosymbiosis *q.v.*; may be a definitive host (infected by the mature adult stage) or an intermediate host (infected by developmental stages). 2: An animal that is the recipient of a tissue graft.

host race A genetic race of a parasite species occurring on a particular host species; **hostic race**.

host range All the host species that may be infected by a given parasite; host list.

host specificity The extent to which an adult parasite is restricted in the variety of host species utilized.

hotel That part of the habitat of an organism that is used for resting, breeding and feeding; *cf.* home.

hour A unit of time equal to 3600 seconds.

Hoxnian interglacial An interglacial period in the middle of the Quaternary Ice Age in the British Isles; see Appendix 2.

Hult-Sernander cover scale A scale for quantifying cover of a plant species, comprising 5 categories; 1 (0–6.25%), 2 (6.5–12.5%), 3 (13–25%), 4 (26–50%) and 5 (51–100%).

Humboldt current A cold surface ocean current that flows north off the Pacific coast of South America, arising from the Antarctic Circumpolar current and forming part of the South Pacific Gyre; see Appendix 6.

Humic Gley soil An intrazonal soil with a thick peaty surface layer over an underlying gley horizon, formed in forest swamps and wet meadows.

humic lake A lake rich in organic matter mainly in the form of suspended plant colloids and larger plant fragments but having low nutrient content; dystrophic lake.

humicolous Living on or in the soil; **humicole**.

humicular Saprophytic *q.v.*

humidity The amount of water vapour in the atmosphere.

humidity compensation point The minimum atmospheric relative humidity allowing nett photosynthesis.

humification The transformation of dead plant material into humus.

humus A dark brown amorphous material formed

by the partial decomposition of plant and animal remains; representing the organic constituent of soil although also found in colloidal suspension in water; denoted by the suffix *h* in a soil profile.

hundredweight (cwt) A unit of mass equal to 112 pounds (50.80 kg); long hundredweight. A short hundredweight of 100 pounds (45.36 kg) is used in the USA.

husbandry The cultivation, production and management of plants and animals; used especially with reference to domesticated species; farming.

hybrid 1: Offspring of a cross between genetically dissimilar individuals; in taxonomy, often restricted to the offspring of interspecific crosses. 2: A community comprising taxa derived from two or more separate and distinct communities.

hybrid breakdown Reproductive failure of F_2 hybrids.

hybrid inviability The failure of F_1 hybrids to reach sexual maturity.

hybrid name *Nomen hybridum q.v.*

hybrid speciation The origin of a new species as a direct result of interspecific hybridization.

hybrid sterility Sterility of F_1 hybrids due to failure of the reproductive organs (developmental sterility) or to chromosomal incompatibility (segregational sterility).

hybrid swarm A series of highly variable forms produced by the crossing and backcrossing of hybrids.

hybrid vigour Heterosis *q.v.*

hybrid zone The zone of overlap between adjacent populations, subspecies or species in which interbreeding occurs; hybrid belt.

hybridization Any crossing of individuals of different genetic composition, typically belonging to separate species, resulting in hybrid offspring; **hybridisation, hybridize.**

hydatoaerophyte A plant growing in permanent water but with part of the assimilative organs floating with a dry upper surface, or emerging above the water.

hydatophytium A submerged plant community.

hydrad Hydrophyte *q.v.*

hydrarch succession An ecological succession commencing in a habitat with abundant water; hydrosere; *cf.* mesarch succession, xerarch succession.

hydric Pertaining to water; wet.

hydro- Prefix meaning water.

hydroanemophilous Having air borne spores that are discharged after wetting of the spore producing structure; **hydroanemophily.**

hydrobiology The study of life in aquatic habitats.

hydrobios The sum total of all aquatic life; that

part of the Earth's surface occupied by aquatic organisms; *cf.* geobios, halobios, limnobios.

hydrocarpic Used of aquatic plants that are pollinated above water but develop below the surface.

hydrochamaephyte An aquatic plant producing renewal buds up to 250 mm from the lake floor; an aquatic chamaephyte.

hydrocharid Used of plants floating on or in water.

hydrochimous Pertaining to wet winters.

hydrochoric Dispersed by the agency of water; **hydrochore, hydrochorous, hydrochory.**

hydrochthophyte A plant growing normally in water and producing emergent leaves or shoots.

hydrocleistogamy Self-pollination in a flower that remains closed because it is submerged; **hydrocleistogamic.**

hydrocolous Living in an aquatic habitat; **hydrocole.**

hydrodynamic Pertaining to the action of water, waves or tides.

hydrogenic soil A soil formed under waterlogged conditions; hydromorphic soil.

hydrogeophyte An aquatic plant producing renewal buds on a buried rhizome; an aquatic geophyte.

hydroharmose The total reaction and adaptation of an organism (harmosis) to water.

hydrohemicryptophyte An aquatic plant producing renewal buds at the water/sediment interface; an aquatic hemicryptophyte.

hydroid Wet or watery.

hydrolabile Used of plant species that are unable to maintain a favourable water content throughout the day and are able to tolerate large water losses and corresponding increase in cell sap concentration; anisohydric; *cf.* hydrostable.

hydrologic cycle The global cycle of water movement, incorporating the oceans, atmosphere and land; water cycle.

hydromegathermic Used of organisms inhabiting warm wet environments, such as tropical rain forests; **hydromegatherm.**

hydromorphic 1: Adapted for aquatic life. 2: Pertaining to an intrazonal soil that has formed under waterlogged or poorly drained conditions; hydrogenic (soil).

hydromorphosis Structural modification of an organism in response to water or wet habitat conditions; **hydromorph, hydromorphic.**

hydronasty Orientation movement in plants in response to changes in atmospheric humidity; **hydronastic.**

hydroperiodism The control of vegetative

processes in plants by periodic dryness; seasonal hydroperiodism.

hydrophilous 1: Thriving in wet or aquatic habitats; *cf.* hydrophobic. 2: Pollinated by water borne pollen; **hydrophile, hydrophily**.

hydrophobic Intolerant of water or wet conditions; water repellent; **hydrophobe, hydrophoby**; *cf.* hydrophilous.

hydrophyte 1: A perennial plant with renewal buds below water and with submerged or floating leaves; *cf.* Raunkiaerian life forms. 2: Generally, any plant adapted to live in water or very wet habitats; water plant; hydrad; comprising ephydate, hyperhydate and hyphydate categories; **hydrophytic**; *cf.* hygrophyte, mesophyte, xerophyte. 3: A plant requiring a large amount of water for optimum growth.

hydrophytium A bog or swamp plant community.

hydroponics A system of plant culture in which the growing plants have their roots immersed in a nutrient rich solution or in an inert substrate which is irrigated with nutrients.

hydrosere An ecological succession commencing in a habitat with abundant water, typically on the submerged sediments of a standing water body; hydrarch succession.

hydrosphere The global water mass, including atmospheric, surface and subsurface waters.

hydrostable Used of plant species that are able to maintain a favourable water content throughout the day, with the water balance *q.v.* remaining near zero; isohydric; *cf.* hydrolabile.

hydrostatic pressure The pressure exerted by a column of water; pressure increases by about 1 atmosphere per 10 m of depth down a water column.

hydrotaxis The directed response of a motile organism towards (positive) or away from (negative) a water or moisture stimulus; **hydrotactic**.

hydrothermal Pertaining to hot water; used especially of hot water springs or ocean floor vents.

hydrotherophyte An aquatic annual plant; an aquatic plant that completes its life cycle within a single limited vegetative period, overwintering as seeds.

hydrotribium A badland community.

hydrotribophilus Thriving in badlands; **hydrotribophile, hydrotribophily**.

hydrotribophyte A badlands plant.

hydrotropic Exhibiting a trend towards greater water content in an ecological succession.

hydrotropism Orientation in response to water; **hydrotropic**.

hyemal Hiemal *q.v.*

hyetal Pertaining to rain or precipitation.

hygric Pertaining to moisture, or to moist or humid conditions.

hygro- Prefix meaning wet, humid, damp, moist.

hygrochastic Used of a fruit in which dehiscence is induced by moisture; **hygrochasic, hygrochasy**; *cf.* xerochastic.

hygrocolous Living in moist or damp habitats; **hygrocole**.

hygrodrymium A tropical rain forest community.

hygrokinesis A change in the rate of movement of an organism in response to a change in humidity; **hygrokinetic**.

hygroklinokinesis A change in the rate of random movement of an organism expressed as a change in the frequency of turning movements (rate of change of direction) in response to a humidity stimulus; **hygroklinokinetic**.

hygrometric Used of plant movements dependent on changes in moisture levels.

hygromorphic Structurally adapted for life in moist habitats.

hygroorthokinesis A change in the rate of random movement of an organism expressed as a change in the linear velocity in response to a humidity stimulus; **hygroorthokinetic**.

hygropetric Used of an organism living in the surface film of water on rocks; petrimadicolous; hygropetrobiont, hygropetrobios.

hygropetrobios The hygropetric fauna and flora; hygropetrical fauna.

hygrophilous Thriving in moist habitats; **hygrophile, hygrophilic, hygrophily**; *cf.* hygrophobic.

hygrophobic Intolerant of moist conditions; **hygrophobe, hygrophoby**.

hygrophorbium A moist pasture, fenland or low moorland community.

hygrophyte A plant living in a wet or moist habitat, typically lacking xeromorphic features; **hygrophytic**; *cf.* hydrophyte, mesophyte, xerophyte.

hygropoium A meadow community.

hygroscopic Readily absorbing and retaining moisture from the atmosphere; **hygroscopicity**.

hygrosphagnium A high moor community.

hygrotaxis The directed response of a motile organism to moisture; **hygrotactic**.

hygrotropism Orientation in response to humidity or moisture; **hygrotropic**.

hyl- Prefix meaning wood.

hylacolous Living among trees, or in woodland or forest; hylocolous, **hylacole**.

hylad A forest or woodland plant; hylophyte.
hylile Pertaining to forest or woodland.
hylium A forest community; hylion.
hylocolous Hylacolous *q.v.*; **hylocole.**
hylodad A plant of open woodland.
hylodium A dry open woodland community.
hylodophilus Thriving in dry open woodland; **hylodophile, hylodophilous, hylodophily.**
hylodophyte A dry open woodland plant.
hylogamy The fusion of a haploid nucleus with a diploid nucleus.
hylophagous Feeding on wood; dendrophagous; lignivorous; xylophagous; **hylophage, hylophagy.**
hylophilous Living or thriving in forests; **hylophile, hylophily;** *cf.* helohylophilous, hylodophilus.
hylophyte A wood or forest plant; hylad.
hylotomous Used of wood-cutting insects.
hypactile Pertaining to that part of the littoral zone on the seashore exposed by the tide for less than one-quarter of the tidal cycle.
hypautochthony The accumulation of plant remains within the same general area in which they lived but no longer at the exact site; *cf.* euautochthony.
hyper- Prefix meaning more than (effect or quantity), above, higher (position).
hyperallobiosphere The upper allobiosphere *q.v.* comprising the highest parts of the major mountain ranges, typically at altitudes greater than 5500 m; *cf.* hypoallobiosphere.
hyperanisogamic Used of organisms having large and active female gametes.
hyperbenthic Living above but close to the substratum; suprabenthic; **hyperbenthos;** *cf.* endobenthic, epibenthic.
hyperborean Pertaining to the frigid region of the far north.
hyperchimaera An organism having tissues of two or more genetic types in which the genetically different components are intimately mixed.
hypercyesis Superfoetation *q.v.*
hyperdispersed Overdispersed *q.v.*
hyperepiphyte An epiphyte attached directly to another epiphyte.
hypergamesis The utilization of excess spermatozoa by the female or the female gamete, for nourishment.
hyperhydate An aquatic plant which grows up through the water and has vegetative structures projecting above the water surface; *cf.* ephydate, hyphydate.
hypermetamorphic Used of an organism having two or more distinct metamorphoses in the life

cycle or having a protracted and extensive metamorphosis; **hypermetamorphosis.**
hypermorphic allele A mutant gene with an effect greater than that of the wild type *q.v.*; hypermorph; *cf.* amorphic allele, hypomorphic allele, neomorphic allele.
hypermorphosis The phyletic extension of ontogeny beyond its ancestral termination, so that adult ancestral stages become preadult stages of descendants.
hypermultivariate Used of the final representation of an analysis in which no worthwhile reduction of dimensionality is achieved, usually because each character contributes information about an independent dimension of variation.
hyperosmotic Pertaining to a solution that exerts a greater osmotic pressure than the solution being compared; *cf.* hypoosmotic, isosmotic.
hyperparasite An organism parasitic upon another parasite; superparasite; **hyperparasitic, hyperparasitism.**
hyperplasia Excessive growth due to increase in cell number; **hyperplastic;** *cf.* hypertrophy.
hyperploidy Having one or more chromosomes or chromosome segments in addition to the normal complement; **hyperploid;** *cf.* hypoploidy.
hypersaline Having a high salinity, well in excess of normal sea water, typical of enclosed bodies of water with high evaporation rates.
hyperspace A space of more than three dimensions; employed to conceptualize the relationship between a number of taxa or observations over a number of dimensions (characters or parameters).
hypertely Evolutionary overspecialization; progressive attainment of disproportionate size by an organism or structure; **hypertelic.**
hypertonic Used of a cell or tissue having a greater osmotic pressure than the solution in which it is immersed; *cf.* tonicity.
hypertrophiation Hypertrophication *q.v.*; **hypertrophiate.**
hypertrophication Over-enrichment with nutrients; hypertrophiation, **hypertrophicated.**
hypertrophy Excessive growth due to increase in cell size; *cf.* hyperplasia.
hyphalomyraplankton Planktonic organisms of brackish waters; the floating organisms of saline waters; **hyphalmyraplankton.**
hyphydate An aquatic plant living permanently below the water surface; *cf.* ephydate, hyperhydate.
hyphydrogamous Having water borne pollen transported below the water surface; **hyphydrogamic; hypohydrogamic;** *cf.* ephydrogamous.

hypnody A prolonged resting period (diapause) during development; **hypnodic**.

hypnoplasy Arrested development resulting in failure to reach normal size.

hypnosis A state of dormancy in seeds that retain the capacity for normal development; **hypnotic**.

hypnote A dormant organism.

hypnotic reflex Thanatosis *q.v.*

hypo- Prefix meaning lower than (effect or quantity), below, under (position).

hypoallobiosphere The lower allobiosphere, including ocean depths below the euphotic zone and hypogean habitats; *cf.* hyperallobiosphere.

hypobenthile Abyssal *q.v.*

hypobenthos The fauna of the sea floor below 1000 m; hypobenthic.

hypobiosis The condition in which only minimal outward signs of metabolic activity are present in a dormant organism.

hypobiotic Living in sheltered microhabitats.

hypocaloric Used of habitats or conditions of sparse or reduced food supply.

hypocarpogean Producing subterranean fruit; **hypocarpogenous**; *cf.* amphicarpogean.

hypoclysile Pertaining to low tide pools in which water temperature does not rise more than 3° C above that of the sea.

hypodigm The entire known material of a species available to a taxonomist.

hypodispersed Underdispersed *q.v.*

hypogean Living or germinating underground; hypogeal; **hypogeous, hypogaeous**.

hypogenesis Development without alternation of generations.

hypogenous Living or growing underneath an object or on its lower surface.

hypohydrogamic Hyphydrogamous *q.v.*

hypolimnion The cold bottom water zone below the thermocline in a lake; **hypolimnetic, hypolimnile**; *cf.* epilimnion.

hypolithic Living beneath rocks or stones.

hypomorphic allele A mutant allele with a lesser effect than that of the wild type *q.v.*; leaky gene; hypomorph; *cf.* amorphic allele, hypermorphic allele, neomorphic allele.

hyponasty Upward growth; **hyponastic**.

hyponeuston Organisms living immediately below the surface film of a water body; a component of the neuston; infraneuston; **hyponeustic, hyponeustonic**; *cf.* epineuston.

hyponym A name rejected in the absence of a type.

hypoosmotic Pertaining to a solution that exerts a lower osmotic pressure than the solution being compared; *cf.* hyperosmotic, isosmotic.

hypophloeodal Living or occurring immediately below the surface of bark; **hypophloeodic**.

hypophyllous Growing on the lower surface of leaves.

hypoplankton Planktonic organisms living close to the sea bed or lake floor.

hypoplasia Underdevelopment; lack of normal growth.

hypoplastotype A mould or model of a supplementary type *q.v.*

hypoploidy Having one or more chromosomes or chromosome fragments less than the normal complement; **hypoploid**; *cf.* hyperploidy.

hyporheic Pertaining to saturated sediments beneath or beside streams and rivers.

hypostasis The interaction of non-allelic genes in which one gene (the hypostatic gene) is masked by the expression of another at a different locus; *cf.* epistasis.

hypotelminorheic Pertaining to seeps or springs where ground water oozes to the surface, forming bogs or small streams.

Hypothermal period Little Ice Age *q.v.*

hypothermia The condition of a homoiothermic animal which has a lower than normal temperature; **hypothermic**.

hypothesis An assertion or working explanation that leads to testable predictions; an assumption providing an explanation of observed facts, proposed in order to test its consequences; **hypothetical**.

hypothesis of non-specificity That there are no distinct major groups of genes exclusively affecting one class of characters, such as morphological, physiological or ethological characters.

hypothetical concept In taxonomy, an imaginary taxon formulated by an author as a missing link but with no nomenclatural status.

hypotonic Used of a cell or tissue having a lower osmotic pressure than the solution in which it is immersed; *cf.* tonicity.

hypotrophic Used of an organism, such as a virus, that utilizes the substances of the host cell as a food source.

hypotype A specimen described and or figured to extend knowledge of a previously described species; hipotype.

Hypsithermal period A postglacial interval (*ca.* 9000–2500 years B.P.) during which mean temperatures were above those existing at the present time; equivalent to Boreal, Atlantic and Subboreal periods combined; Megathermal period; *cf.* Xerothermal period.

hysteresis Delay in adjustment of a process in response to a change in an associated process.

hysterogenic Used of a structure or trait that appears late in development.

hysterophyte A plant obtaining nourishment from dead or decaying organic matter; saprophyte; **hysterophytic**.

hysterotely The retention of larval characters in the pupal or adult stage, or of pupal characters in the adult; metathetely *q.v.*; **hysterotelic**.

hytherograph Climagraph; a chart of temperature and rainfall.

I

I₁ First inbreeding generation; subsequent inbreeding generations denoted $I_2, I_3 \ldots I_n$.

IAMS International Association of Microbiological Societies.

Iapetus ocean A lower Palaeozoic ocean which extended, in terms of present day geography, from Norway to Connecticut, USA.

IAPT The International Association of Plant Taxonomy.

ibid (**ib.**) Abbreviation of Latin **ibidem**, meaning in the same place; as in the same book or journal.

ICBN International Code of Botanical Nomenclature.

ichnite A fossil track or footprint; ichnolite.

ichnocoenosis An assemblage of trace fossils; **ichnocoenose**.

ichnofauna The animal traces of an area.

ichnoflora The plant traces of an area.

ichnofossil A trace fossil.

ichnogenus A higher taxon in the classification of traces.

ichnolite A fossil track or footprint; ichnite.

ichnology The study of traces made by organisms, both fossil and recent.

ichnospecies A subordinate taxon in the classification of traces.

ichthyofauna The fish fauna of a given region.

ichthyology The study of fishes; **ichthyological**.

ichthyoneuston The ichthyological component of the neuston, typically fish eggs or fry; **ichthyoneustont**.

ichthyophagous Feeding on fishes; piscivorous; **ichthyophage, ichthyophagy**.

ICNB International Committee on Nomenclature of Bacteria.

ICNV International Committee on Nomenclature of Viruses.

iconotype A drawing or photograph of a type specimen.

icoplastotype A cast of an icotype *q.v.*

icotype In taxonomy, a specimen used for identification, worked on by the original author, or collected from the type locality, but not one used for the published descriptions.

ICSB International Committee on Systematic Bacteriology.

ICTV International Committee on Taxonomy of Viruses.

ICZN The International Commission on Zoological Nomenclature.

-idae The ending of a name of a subclass in botanical, or of a family in zoological, nomenclature.

identical twins Monozygotic *q.v.* twins.

ideotype In taxonomy, a specimen examined by the author of a species but not collected from the type locality; idiotype.

idio-adaptation Allomorphosis *q.v.*

idiobiology That aspect of biology concerned with study of individual organisms; autobiology.

idiochromosome A sex chromosome *q.v.*

idioecology Autecology *q.v.*

idiogamy Self-fertilization; **idiogamous**.

idiogram A diagrammatic representation of the chromosome complement of a cell.

idiotaxonomy Taxonomic study of individuals, populations, species and higher taxa; traditional taxonomy; *cf.* syntaxonomy.

idiothermous Homoiothermic *q.v.*

idiotype 1: The total hereditary determinants of an individual, comprising both chromosomal and extrachromosomal factors. 2: In taxonomy, a specimen identified by the original author of a species but not collected by him from the type locality; ideotype.

idiotypic Sexual; pertaining to sex or sexual reproduction.

igneous rock Rock formed by solidification of molten magma; *cf.* metamorphic rock, sedimentary rock.

IJSB International Journal of Systematic Bacteriology.

illegitimate name *Nomen illegitimum q.v.*

Illinoian glaciation A glaciation of the Quaternary Ice Age in North America with an estimated duration of 100 thousand years; see Appendix 2.

illuminance The luminous flux per unit area falling perpendicularly on a surface.

illuvial layer A zone of illuviation; the B-horizon of a soil profile.

illuviation The process of deposition and precipitation of material in the B-horizon, that was leached or eluviated from the A-horizon of a soil.

imagination The final stage of development that produces the imago (adult).

imago The adult stage of an insect; **imagine**; *cf.* instar.

imbibition The non-active absorption or uptake of water; **imbibe**.

imbibitional water That part of the soil water held within the lattice structure of colloidal matter.

imitative behaviour The behaviour of one animal copying that of another; imitation.

immature 1: Used of the developmental stages of an organism preceding the attainment of sexual maturity. 2: Used of any incompletely differentiated system.

immature soil A recently formed soil in which profile development is incomplete for the prevailing climatic and biological conditions.

immersed Used of an aquatic plant growing entirely beneath the water surface; submersed.

immigration The movement of an individual or group into a new population or geographical region; **immigrant**; *cf.* emigration.

immigration curve A graph of the rate of immigration of species against either the number of resident species, or time.

immigration rate The number of new species arriving in a given area or habitat per unit time.

immiscible Used of liquids that cannot be mixed to a homogeneous state.

immission Any pollutant discharged into the environment at a particular place and time.

immobilization The conversion of an element from an inorganic to an organic state by an organism, *cf.* biological mineralization.

immunity The state in which a host is more or less resistant to an infective agent; used with reference to resistance arising from exposure to the agent and consequent stimulation of an immune response.

immunogenic Having an antigenic action; used of a substance producing an immunological response.

immunological distance A measure or estimate of the difference between two antigens determined by immunological techniques.

immunology The study of immunity and immunological reactions.

imperfect flower A unisexual flower; a flower lacking either pistils or stamens.

Imperial gallon (gal.) The fps unit of volume defined as the volume of 10 pounds of water of density $0.988\,859$ g ml^{-1}, weighed in air of density $0.001\,217$ g ml^{-1}; equal to 4.54609 dm^3; see Appendix 13.

impervious Not permeable; usually with reference to the passage of water or air; **impermeable**.

implantation 1: In mammals, the attachment of an embryo to the uterus wall; nidation. 2: The insertion of a graft into the host or stock.

importance value A measure of overall importance of a given species in a community, calculated as the sum of relative frequency *q.v.*, relative density *q.v.* and relative dominance *q.v.*

impoundment An artificially enclosed body of water; typically with fluctuating water levels and high turbidity.

impregnation Insemination *q.v.*; **impregnate**.

impression 1: The total number of copies of a work produced at one time; an edition *q.v.* may comprise two or more impressions. 2: A mode of fossil formation in which an imprint of an organism is produced featuring the size and general external morphology.

imprinting The imposition of a stable behaviour pattern in social species by exposure, during a particular period in early development, to one of a restricted set of stimuli whereby the broad supra-individual characteristics of the species come to be recognized as a 'species-pattern' and subsequently used as releasers *q.v.*

impriorable name *Nomen illegitimum q.v.*

in Used in nomenclature to link the names of two authors, the first being responsible for making available a name within the publication of which the second was the overall author or editor.

in adnot. Abbreviation of the Latin *in adnotatione*, meaning in an annotation.

in litt. Abbreviation of the Latin *in litteris*, meaning in correspondence.

in sched. Abbreviation of the Latin *in schedula*, meaning on a herbarium sheet or label.

in situ In the original location.

in syn. Abbreviation of the Latin *in synonymis*, meaning in synonymy.

Inabresis Afro-Brasilian bridge *q.v.*

inadmissable In taxonomy, used of a name or epithet which cannot be used within the provisions of the Code.

-inae The ending of a name of a subtribe in botanical nomenclature and of a subfamily in zoological nomenclature.

inapplicable character A dependent character *q.v.* which is meaningless because the character on which it depends is absent.

inbreeding Mating or crossing of individuals more closely related than average pairs in the population; endogamy; *cf.* outbreeding.

inbreeding coefficient 1: The probability that two alleles at the same locus in an individual are identical by descent. 2: The proportional loss of heterozygosity as a result of inbreeding.

inbreeding depression Reduction of fitness and vigour by increased homozygosity as a result of inbreeding in a normally outbreeding population.

inc. ced. Abbreviation of the Latin *incertae cedis q.v.*

incarbonization Carbonification *q.v.*

incasement theory Encasement theory *q.v.*

incertae cedis (inc. ced.) Latin, meaning of

uncertain seat (affinity); used of a category of uncertain taxonomic position; *incertae sedis*.

inch A derived fps unit of length equal to 25.4 mm.

incidence of infection The number of infected hosts as a proportion of the host population; *cf*. intensity of infection.

incidental mention In taxonomy, the use of a name by an author who demonstrates no intention of adopting or introducing it, but merely mentions it without comment on its suitability; such a mention does not fulfil the requirements of the Code.

incipient About to become or occur; nascent; used of the initial stage in the development of a structure or event.

incipient species Populations that are in the process of diverging to the point of speciation but which still have the potential to interbreed although they are prevented from doing so by a specific barrier.

incl. Abbreviation of the Latin *inclusus*, meaning included.

inclination Aspect; the direction of a slope face.

included niche A niche that occurs entirely within the larger niche of another more generalized species; the species occupying the included niche survives by being highly adapted or specialized, and therefore having competitive superiority, in some limited part of the niche.

inclusive fitness The sum of an individual's fitness, quantified as the reproductive success of the individual and its relatives, with the relatives devalued in proportion to their genetic distance.

incoalation Carbonification *q.v.*

incompatibility 1: The inability to unite, fuse or form any homogeneous or viable association; **incompatible**. 2: Failure of either cross fertilization or self-fertilization as a result of structural, physiological or genetic factors. 3: Failure of a graft to unite with the host or stock tissue.

incomplete dominance The partial expression of both alleles at a given locus so that the phenotype of the heterozygote is intermediate between both homozygotes; codominance; semidominance; *cf*. complete dominance.

incomplete flower A flower that lacks one or more of the four basic parts (pistil, stamen, petal and sepal); *cf*. complete flower.

incongruent Not matching or fitting; **incongruence**; *cf*. congruent.

incorrect original spelling In taxonomy, an original spelling that is incorrect under the provisions of the Code.

incorrect subsequent spelling In taxonomy, any

change in the spelling of an available name other than a mandatory change or an emendation.

incross Breeding or mating between homozygotes; *cf*. intercross.

incubation 1: The maintenance of eggs under conditions favourable for hatching; **incubate**. 2: The period between infection by a pathogen and the appearance of symptoms.

indeciduous Persistent; everlasting; evergreen; adeciduate; non-caducous; non-deciduate; **indeciduate**.

indehiscent Used of fruits that do not open spontaneously when ripe; **indehiscence**.

indented key A dichotomous key in which the first part of a contrasting couplet is followed by all subsequent couplets; each subordinate couplet being indented one step further to the right for clarity of presentation; yoked key; *cf*. bracketed key.

independence The condition of a variable that is not influenced or controlled by another variable; **independent**; *cf*. dependence.

independent assortment, law of Mendel's second law; that the random distribution of alleles to the gametes results from the random orientation of the chromosomes during meiosis.

independent samples In statistics, pairs or groups of samples in which the selection of one sample in no way affects selection of the other samples.

indeterminate Not definite, not identified, not classified.

indeterminate growth Growth that continues throughout the life span of an individual such that body size and age are correlated; *cf*. determinate growth.

index fossil A fossil apparently restricted to a narrow stratigraphic unit or time unit; characteristic fossil; diagnostic fossil; key fossil.

index of similarity The ratio of the number of species common to two communities to the total of species in both; coefficient of community.

Index, Official Official Index *q.v.*

index species A species indicative of a particular set of environmental conditions; characteristic species, guide species, indicator species.

index zone A reference stratum or strata identifiable by a particular palaeontological or lithological composition.

Indian Equatorial Countercurrent A warm surface ocean current that flows east in the equatorial Indian Ocean and forms the southern winter countercurrent to the North Equatorial current; see Appendix 6.

Indian North Equatorial current A warm surface ocean current that flows west in the equatorial Indian Ocean, generated largely by the north-easterly winds of the winter monsoon; see Appendix 6.

Indian region A subdivision of the Indo-Malaysian subkingdom of the Palaeotropical kingdom; see Appendix 4.

Indian South Equatorial current A warm surface ocean current that flows west in the central Indian Ocean and forms the northern limb of the South Indian Gyre and the southern limb of the North Indian Ocean circulation; see Appendix 6.

Indian subregion A subdivision of the Oriental region; sometimes also used to refer to the entire Oriental zoogeographical region; Hindostan subregion; see Appendix 3.

indication 1: In taxonomy, information published before 1931 that allows a name to be treated as available in the absence of a description or definition. 2: Published information that determines the type species of a generic name in the absence of an original designation of the type species.

indicator An organism, species or community characteristic of a particular habitat, or indicative of a particular set of environmental conditions.

indicator species A species, the presence or absence of which is indicative of a particular habitat, community or set of environmental conditions; characteristic species; guide species; index species.

indifferent species A species exhibiting no obvious affinity for any particular community; a species of low fidelity *q.v.*; companion species.

indigenous Native to a particular area; autochthonous; used of an organism or species occurring naturally in an environment or region; **indigen, indigenae, indigene.**

indirect competition Competition for a resource in which exploitation by one individual or species reduces the availability to other individuals or species; *cf.* direct competition.

indirect gradient analysis The analysis of continuously intergrading plant communities in which samples are compared in terms of species composition and arranged along axes, or in three-dimensional space, according to the similarity measurements.

indirect life cycle Heterogonic life cycle *q.v.*

indirect reference In taxonomy, a clear statement that the description or diagnosis of a taxon for which a new name is given has been effectively published in an earlier work.

individual A single organism or unit.

individual distance The minimal distance that an animal endeavours to maintain between itself and other individuals of the species.

individualism A symbiosis in which the two symbionts are intimately associated, appearing to form a single organism not resembling either of the partners; **individuation.**

individuation The process of development of a single functional unit, as of an individual or colony.

Indo-Chinese subregion A subdivision of the Oriental region; Himalayan subregion; see Appendix 3.

Indo-Malayan subregion A subdivision of the Oriental region; Malayan subregion; see Appendix 3.

Indo-Malaysian subkingdom A subdivision of the Palaeotropical kingdom; see Appendix 4.

induction 1: The determination of the developmental fate of one cell mass by another; **inductor.** 2: A hereditary transformation mediated by foreign DNA in a multicellular organism.

inductive method A scientific method involving the formulation of universal statements, such as hypotheses or theories, from singular statements, such as empirical observations or experimental results; *cf.* deductive method.

indurated soil Strongly compacted dense soil almost impermeable to water and roots; denoted by the suffix x in a soil profile.

induration The process of hardening, as of sediments or rocks rendered hard by compaction, heat, pressure or other agencies; **indurate.**

industrial melanism An increase in the frequency of melanistic (dark pigmented) morphs as a response to changes in selection, in habitats blackened by industrial pollution.

-ineae The ending of the name of a suborder in botanical nomenclature.

ineditus Latin, meaning unpublished; *ined.*

inert Inactive; physiologically quiescent.

infauna The total animal life within a sediment; endobenthos; *cf.* epifauna.

infection The process of invasion of a host by a parasite or microorganism; the pathological condition resulting from the growth of infectious organisms within a host; **infect.**

infectious disease Any disease transmitted without physical contact.

infertile Not fertile; non-reproductive; **infertility.**

infestation Invasion by parasites or pests.

infiltration The process by which water seeps into a soil, influenced by soil texture, aspect and vegetation cover.

influent Any organism influencing another organism or group of organisms.

information theory A branch of probability theory dealing with the likelihood of transmission of messages, accurate within specified limits, when the units of information comprising the message are subject to probabilities of transmission failure, random disturbance, etc., and with the measurement of information content of systems; it employs a quantitative measure of information and does not reflect the qualitative value (i.e. the meaning) of a message.

infra- Prefix meaning below, smaller; *cf.* supra-.

infradian rhythm A biological rhythm with a period less than 24 h; *cf.* ultradian rhythm.

infradispersed distribution Underdispersed distribution *q.v.*

infrageneric Below the rank of genus.

infrahaline Used of fresh water having a salinity of less than 0.5 parts per thousand.

infralittoral 1: The depth zone of a lake permanently covered with rooted or adnate macroscopic vegetation; often divided into upper (with emergent vegetation), middle (with floating vegetation) and lower zones (with submerged vegetation). 2: The upper subdivision of the marine sublittoral zone, typically dominated by photophilous algae, having a lower limit at the depth at which illumination is at about 1% of the surface level; sometimes used for the depth zone between low tide and 100 m; infralitoral; see Appendix 7.

infralittoral fringe The transitional marine zone immediately below the eulittoral, between the intertidal and sublittoral zones; sublittoral fringe; see Appendix 7.

infraneuston Planktonic organisms suspended from the undersurface of a water film; hyponeuston; **infraneustonic**.

infraorder A category in a hierarchy of classification below that of suborder; see Appendix 10.

infrared radiation That part of the solar radiation spectrum above the wavelength 780 nm.

infraspecific Within the species; pertaining to a taxonomic category below species level; intraspecific.

infrasubspecific Pertaining to a taxonomic category below subspecies level.

ingesta The total intake of substances into the body; **ingest**.

ingestion 1: The action of taking in food material; feeding. 2: Consumption *q.v.*

inhabited symbiont The host symbiont in an endosymbiosis *q.v.* which contains the other symbiont within its body; *cf.* inhabiting symbiont.

inhabiting symbiont The symbiont contained within the body of the host symbiont in an endosymbiosis *q.v.*; endosymbiont; *cf.* inhabited symbiont.

inheritance The transmission of genetic information from ancestors or parents to descendants or offspring.

inheritance of acquired characters Acquired characters, inheritance of *q.v.*

inhibition Any process that acts to restrain reactions or behaviour.

-ini The ending of a name of a tribe in zoological nomenclature.

inland sea Epicontinental sea *q.v.*

innate Inherited; present at birth; basifixed; used of behaviour that is instinctive and not learned; inborn behaviour.

innate immunity Inherited immunity; the state of having anatomical or physiological features that make an organism unsuitable as a host for a given parasite or pathogen.

innate releasing mechanism (**IRM**) An instinctive mechanism that releases a specific response to a sign stimulus.

inoculation The introduction or insertion of a pathogen into a host organism by a vector; or of an organism into an experimental culture.

inoculum The individual or group of individuals comprising the founders of a colony or newly established population; breeding stock.

inorganic Pertaining to, or derived from, non-biological material; used of non carbon-chain compounds; *cf.* organic.

input load A minor component of the genetic load *q.v.* representing the input of inferior alleles by mutation and immigration.

inq. Abbreviation of the Latin *inquilinus*, meaning naturalized.

inquilinism A symbiosis in which one organism (the inquiline) lives within another without causing harm to its host; also used for the relationship between a species living within the burrow, nest or other domicile of another species; **inquilinous**.

insecticide A chemical used to kill insects; used generally for any chemical agent used to destroy invertebrate pests.

insectivorous Feeding on insects; entomophagous; **insectivore, insectivory**.

insemination The introduction of sperm into the genital tract of a female; impregnation.

insessorial Adapted for perching.

insolation Exposure to solar radiation.

inspiration The process by which oxygen is absorbed by plants and animals.

instantaneous growth rate The growth rate of a population at any given point in time.

instantaneous speciation The isolation from the parent species of a single individual which is capable of establishing a new species population.

instar Any intermoult stage in the development of an arthropod; *cf.* imago.

instinct An inherited and adapted system of coordination within the nervous system as a whole, which when activated finds expression in behaviour culminating in a fixed action pattern *q.v.*; when 'charged' it shows evidence of specific action potential *q.v.* and readiness for release by an environmental releaser *q.v.*; instinctive.

instinctive behaviour Relatively complex, highly stereotyped behaviour exhibited in response to an environmental stimulus and directed towards a predictable end product.

insular Pertaining to islands; used of an organism that has a restricted or limited habitat or range.

insular endemic An organism that evolved in a restricted geographical area and has retained that limited range without subsequent spread to a wider area.

insulosity The percentage of the area of a lake, within the shore line, occupied by islands.

integration, law of Oligomerization *q.v.*

intense competition Interspecific competition between a small number of species having considerable niche overlap with each other; *cf.* diffuse competition.

intensity of infection The mean number of parasites per individual of the host population; *cf.* incidence of infection.

intention movement A preparatory action exhibited prior to a complete behavioural response.

inter- Prefix meaning between.

interbreeding Mating or hybridization between different individuals, populations, varieties, races or species.

intercalated Inserted between adjacent structures or strata.

interception That part of total precipitation retained on the surface of vegetation before reaching the ground and returned to the atmosphere by evaporation.

intercistronic complementation Complementation *q.v.* between pairs of heterozygous non-allelic mutations belonging to different functional units of a chromosome (cistrons), to give a phenotype closer to the wild type than any one mutant would produce alone; *cf.* intracistronic complementation.

intercompensation The correlation in the changes in relative importance of the different density dependent factors involved in population regulation.

intercross The breeding or mating of heterozygotes; *cf.* incross.

interdemic Between local interbreeding populations.

interdependent association Mutualism *q.v.*

interface The contact surface between two contiguous substances or zones.

interference Interaction that limits the access of one population to a resource it shares with another.

interferon A low molecular weight protein formed by cells in response to the action or stimulus of a virus.

interfertile Capable of interbreeding.

intergeneric Between genera.

intergenic Used of mutations involving more than one gene.

interglacial 1: A warm period between two Ice Ages or glaciations; see Appendix 2. 2: A period with a mean annual temperature equal to or greater than those of the Fladrian climatic optimum (*ca.* 2° C above present temperatures).

intergradation zone The boundary zone between two adjacent subspecies occupied by a population of intergrades.

intergrade An individual occurring on the boundary between adjacent subspecies and which possesses intermediate characters or traits.

interlittoral The shallow subtidal marine zone to a depth of about 20 m.

intermediate host The host occupied by juvenile stages of a parasite prior to the definitive host *q.v.* and in which asexual reproduction frequently occurs; secondary host.

intermicellar liquid The liquid continuous phase of a colloid *q.v.*

internal balance Coadaptation *q.v.*

internal selection Selection against mutant alleles that interfere with the integration of metabolic activities.

International Association of Microbiological Societies (IAMS) The body responsible for the organization of Microbiological Congresses and, through the ICSB, for the publication of the International Code of Nomenclature of Bacteria.

International Association of Plant Taxonomy (IAPT) The body responsible for the publication of the ICBN and for the organization of committees dealing with the nomenclature of plants, at Botanical Congresses.

International Bureau for Plant Taxonomy and Nomenclature The publisher of the International Code of Botanical Nomenclature, Taxon and Regnum Vegetabile.

International Code of Botanical Nomenclature (ICBN) The internationally adopted set of rules governing botanical nomenclature.

International Code of Nomenclature of Bacteria The internationally adopted set of rules governing the nomenclature of Bacteria; Bacteriological Code.

International Code of Zoological Nomenclature
The internationally adopted set of rules governing zoological nomenclature.

International Commission on Zoological Nomenclature (ICZN) The body that compiles Official Indexes and Official Lists, that rules on matters of priority and validity of taxonomic names in zoology and is responsible for the International Code of Zoological Nomenclature.

International Committee on Nomenclature of Bacteria (ICNB) The former name of the International Committee on Systematic Bacteriology.

International Committee on Nomenclature of Viruses (ICNV) The former name of the International Committee on Taxonomy of Viruses.

International Committee on Systematic Bacteriology (ICSB) The body responsible for the publication of the International Code of Nomenclature of Bacteria, and for the Judicial Commission *q.v.* and taxonomic subcommittees dealing with the nomenclature of particular groups of Bacteria.

International Committee on Taxonomy of Viruses (ICTV) The body responsible for the taxonomic subcommittees dealing with the nomenclature of viruses and for approval of newly proposed names of viruses.

International Journal of Systematic Bacteriology (IJSB) The official organ of the International Association of Microbiological Societies.

International Trust for Zoological Nomenclature (ITZN) The publishing organization of the International Commission on Zoological Nomenclature.

interpolation The process of determining the value of an intermediate function within the range of known values, without using the function equation; *cf.* extrapolation.

intersex An individual of a bisexual species possessing reproductive organs and secondary sexual characters partly of one sex and partly of the other, but which is of one sex only genetically.

interspecies A cross between two distinct species; interspecific hybrid.

interspecific Between two or more distinct species; used of a hybrid between two distinct species.

interspersed Scattered or spread irregularly amongst other objects or individuals; **intersperse, interspersion**.

interstade A climatic episode within a glacial stage during which a secondary retreat or standstill occurred; **interstadial**; *cf.* stade.

interstadial An interval of slightly warmer climate between two cold maxima (stadials) of a single glaciation.

intersterile Unable to interbreed; **intersterility**.

interstitial Pertaining to, or occurring within, the pore spaces (interstices) between sediment particles.

intertidal zone The shore zone between the highest and lowest tides; eulittoral zone; littoral; tidal zone; see Appendix 7.

interval 1: The time span between two events; chronologic interval. 2: The strata between stratigraphic markers; stratigraphic interval.

interval estimate An estimate of a population parameter comprising a range (defined by extreme values) in which the parameter may be considered to lie; *cf.* point estimate.

intervarietal hybrid A cross between two varieties of the same species.

interzonal Between two zones; used of pelagic species inhabiting two or more depth zones.

intra- Prefix meaning within.

intracistronic complementation Complementation *q.v.* of a pair of mutant alleles within the same functional unit of a chromosome (cistron) to give a phenotype closer to the wild type than either mutant would alone; *cf.* intercistronic complementation.

intrademic selection Selection occurring within a local interbreeding population.

intradiel Within the period of a single 24-h day.

intrageneric Within the same genus; between members of a genus.

intragenic Within the same gene.

intralocal Used of a local character species *q.v.* that has a limited geographical range within the wider range of the vegetation unit to which it belongs.

intraneous Used of individuals of a species that occur near the centre of distribution of the species; *cf.* extraneous.

intraregional Used of a regional character species *q.v.* that has a limited geographical range within the wider range of the vegetation unit to which it belongs.

intrasexual selection Sexual selection *q.v.*

intraspecific Within a species; between individuals or populations of the same species.

intraspecific mimicry Automimicry *q.v.*

intrazonal soil An immature soil showing more or less well developed characteristics that reflect the influence of some local topographical or physiographical factors, age or parent material (lithographic factors), over the basic effects of climate and vegetation as soil forming factors; transitional soil; *cf.* azonal soil, zonal soil.

intrinsic Originating or occurring within an individual, group or system; *cf.* extrinsic.

intrinsic isolating mechanism Any genetically determined factor that prevents interbreeding between individuals of sympatric species; *cf.* extrinsic isolating mechanism.

intrinsic rate of increase The potential rate of growth of a population in an infinite environment, designated by r; calculated as $r = b - d$, where b is the instantaneous birth rate and d is the instantaneous death rate; Malthusian parameter.

introgression The spread of genes of one species into the gene pool of another by hybridization and backcrossing; introgressive hybridization.

intron In eukaryote chromosomes, that portion of a DNA strand within a gene that does not code for a protein; *cf.* exon.

introrse dehiscence Spontaneous opening of a ripe fruit from outside inwards; *cf.* extrorse dehiscence.

intussusception The assimilation of new material and the incorporation of this material within the existing matter or fabric.

inundatal Liable to flooding; used of plants that occupy sites susceptible to inundation during wet weather.

invalid name *Nomen invalidum q.v.*

invasibility The relative freedom with which an immigrant species can invade an established community.

invasion The mass movement or encroachment of organisms from one area into another; **invader**.

inverse density dependence A change in the influence of an environmental factor (an inverse density dependent factor) that affects population density as population density changes, tending to enhance population growth (by decreasing mortality or increasing fecundity) as population density increases or to retard population growth (by increasing mortality or decreasing fecundity) as population density decreases; *cf.* density dependence.

inverse density dependent factor Inverse density dependence *q.v.*

inverse square law That the energy per second per unit area of any electromagnetic radiation received by a surface at right angles to the rays is inversely proportional to the square of the distance of the surface from the source.

inversion A structural change of chromosome resulting from the reversal of a segment such that the genes occur in a reversed sequence.

inversion race A naturally occurring population characterized by a particular chromosome inversion or inversions.

inviability A measure of the number of individuals failing to survive in one cohort or group, relative to another.

ionosphere The outer layer of the atmosphere more than 80 km above the Earth's surface, characterized by free electrically charged particles.

iota link A direct unbranching link in a food web, between a prey species and its unique predator or parasite; *cf.* gamma link, lambda link.

Iowan glaciation The penultimate glaciation of the Quaternary Ice Age in North America, with an estimated duration, together with the Wisconsin glaciation, of about 70 thousand years; see Appendix 2.

Ipswichian interglacial The interglacial period preceding the Devensian glaciation in the British Isles; see Appendix 2.

IRM Innate releasing mechanism *q.v.*

Irminger current A cold surface ocean current that flows west and north around the southern coast of Greenland, derived in part from a deflection of the North Atlantic Drift; see Appendix 6.

Iron Age An archaeological period dating from about 3000 years B.P. in Europe; a period of human history characterized by the smelting of iron for industrial purposes.

irreversibility rule Dollo's law *q.v.*

irritability Sensitivity; the ability to perceive and respond to stimuli; **irritable**.

irruption An irregular abrupt increase in population size or density typically associated with favourable changes in the environment and often resulting in the mass movement of the population.

isallobar A line on a chart or map connecting points of equal change of atmospheric pressure per unit time.

isallotherm A line on a chart or map connecting points of equal temperature variation within a given time interval.

isanomaly A line on a chart or map connecting points of equal anomaly, especially of equal magnetic anomalies; **isanomal, isonomaly**.

isauxesis The special case of heterauxesis *q.v.* in which a, from the equation $y = bx^a$, equals unity and in which no change in shape occurs with increasing size; auxesis; isometric growth.

isentropic Without change of entropy *q.v.*

island biogeography, theory of That the number of species inhabiting an island is a function of island area and distance from the mainland, and is determined by the relationship between immigration (greater on larger and nearer islands) and extinction (greater on smaller islands); MacArthur–Wilson theory of island biogeography.

island model migration Gene flow by periodic migration between groups of semi-isolated populations; *cf.* river model migration.

iso- Prefix meaning equal.

isoalleles Alleles which are so similar in their phenotypic effects that special techniques are required to distinguish them.

isobar A line on a chart or map connecting points of equal atmospheric or hydrostatic pressure.

isobase A line on a chart or map connecting sites of equal land surface depression, caused by the overburden of ice during periods of glaciation.

isobath A line on a chart or map connecting points of equal depth; bathymetric contour.

isobenth A line on a chart or map connecting points of equal biomass.

isoblabe A line on a chart or map connecting points of equal damage or infestation by a pest or pathogen.

isocheim The isotherm *q.v.* of the coldest month; isocheimenal; *cf.* isothere.

isochimous Pertaining to regions of similar low temperature.

isochronal line 1: A line on a chart or map connecting points (localities) at which a given event occurs simultaneously, as in the mean date of arrival of certain migratory species; isochron. 2: Isophene *q.v.*

isocoenoses Communities of plants and animals having similar life forms *q.v.*

isocommunity A community that shows a close structural and ecological resemblance to another natural community.

isocon A line on a chart or map connecting points of equal concentration of a given factor, such as salinity.

isocotype Adelfotype *q.v.*

isodeme A line on a map connecting points of equal population density; isodemic line.

isodynamic 1: Of equal strength; having the same energy content. 2: A line on a chart or map connecting points of equal magnitude or value of any factor or force.

isofag A line on a chart or map connecting points of equal degree of damage by pests or pathogens.

isofemale strain A strain derived from a single fertilized founder female obtained from a natural population.

isoflor A line on a chart or map connecting points supporting equal numbers of species.

isogametic Producing gametes of similar size, shape and behaviour; **isogametism**; *cf.* anisogametic.

isogamous Having gametes that are similar in size, shape and behaviour; having gametes

(isogametes) not differentiated into male and female; *cf.* anisogamous.

isogamy The fusion of gametes of similar size, shape and behaviour; isomerogamy; zygogamy; *cf.* anisogamy.

isogene A line on a map connecting points of equal gene frequencies.

isogeneic Having the same set of genes; homozygous; syngenic; isogenic; *cf.* allogenic.

isogenesis Evolution without essential changes in levels of biological integration, or in independence from the environment.

isogenetic Of the same origin; isogenous.

isogenic Homozygous *q.v.*; having the same set of genes; syngenic; isogeneic; *cf.* allogenic.

isogenotype A type species to which two generic names have been applied; **isogenotypic**.

isogenous Of the same origin; **isogenetic**.

isogonous Used of hybrids that show equal expression of parental characters; **isogonic**; *cf.* anisogonous.

isohaline A line on a chart or map connecting points of equal salinity.

isohel A line on a chart or map connecting points having an equal duration of sunshine.

isoholotype A plant specimen removed at a later date from the same plant from which the holotype was originally collected.

isohydric 1: A line on a chart or map connecting points of equal pH. 2: Hydrostable *q.v.*

isohyet A line on a chart or map connecting points of equal rainfall; isohyetal.

isohyp A line on a chart or map connecting points of similar elevation; contour line.

isokeraunic A line on a chart or map connecting places with equal frequency of occurrence of thunderstorms.

isolate A population or group of populations that is geographically, ecologically or otherwise separated from other populations, and in which interbreeding occurs.

isolating mechanisms Any intrinsic or extrinsic mechanism or barrier to the free exchange of genes between populations.

isolation The separation of two populations so that they are prevented from interbreeding whether by extrinsic (premating) or intrinsic (postmating) mechanisms.

isolume A line on a chart or map connecting points of equal light intensity.

isomegath A line on a chart or map connecting points having sediments with similar median particle sizes.

isomerism 1: Repetition of similar parts. 2: The possession of equal numbers of serially homologous structures.

isomerogamy Isogamy *q.v.*

isometric growth Isauxesis; growth in which relative proportions of the body parts remain constant with change in total body size; auxesis; geometric similarity; *cf.* allometric growth.

isomorphic Used of a plant having an alternation between diploid and haploid generations which are morphologically similar in appearance; homomorphic; **isomorphous**; *cf.* heteromorphic.

isomorphism Superficial similarity between individuals of different species or races; **isomorphic, isomorphous**.

isoneotype In taxonomy, a duplicate of a neotype *q.v.*

isonif A line on a chart or map connecting points of equal depth of snow.

isonome A line on a chart or map connecting points of equal abundance of a species.

isonym In taxonomy, one of two or more names based on the same nomenclatural type.

isoosmotic Having the same osmotic pressure; isosmotic; isotonic.

isopach A line on a chart or map connecting points of equal sediment thickness.

isopag A line on a chart or map connecting points having an equal number of days per annum during which ice is present.

isoparatype In taxonomy, a duplicate of a paratype *q.v.*

isoparthenogenesis Normal parthenogenesis *q.v.*

isophagous Used of predators or parasites which are selective with regard to food or hosts but which are not restricted to a single food type or host species.

isophane A single function of longitude, latitude and altitude used in biogeographical analysis of intraspecific variation.

isophene 1: A line on a chart or map connecting points having the same frequency of occurrence of a given phenotype or variant; phenocontour. 2: A line on a chart or map connecting points at which seasonal events occur at the same date; isophenal map; isochronal line.

isophenogamy Sexual reproduction occurring more frequently between individuals of the same phenotype than would be expected from random selection of mates from within the population.

isophenous Sharing or producing similar phenotypic effects; **isophenic**.

isophot A line on a chart or map connecting points of equal incident radiation.

isophotic Equally illuminated on all sides.

isophyte A line on a chart or map connecting points of similar vegetation height above ground level.

isophytochrone A line on a chart or map connecting points having similar lengths of growing season.

isopiptesis A line on a chart or map connecting points at which a given migratory species arrives on the same date.

isoplankt A line on a chart or map connecting points of equal plankton density, or of similar plankton species composition.

isopleth A line on a chart or map connecting points of equal concentration of a given factor.

isopract A line on a chart or map connecting points of equal expression of a given factor.

isopycnic 1: Having the same density. 2: A line on a chart or map connecting points of equal density.

isoseismal A line on a chart or map connecting points of equal earthquake intensity; isoseismic line.

isosmotic Pertaining to solutions that exert the same osmotic pressure; isoosmotic; *cf.* hyperosmotic, hypoosmotic.

isosporous Having asexually produced spores of only one kind; homosporous; **isosporic, isospory**; *cf.* heterosporous.

isosymbiosis Symbiosis between two organisms of equal size (isosymbionts); *cf.* anisosymbiosis.

isosyntype In taxonomy, a duplicate of a syntype.

isotelic Producing or tending to produce the same effect; homoplastic.

isothere The isotherm of the warmest month; *cf.* isocheim.

isotherm A line on a chart or map connecting points of equal temperature.

isotonic Used of a cell or tissue having the same osmotic pressure as the solution in which it is immersed; equimolecular; isoosmotic; *cf.* tonicity.

isotype 1: In taxonomy, a duplicate of a holotype from the single collection that contained the holotype. 2: A type described from two species of the same genus; isotypic. 3: A form occurring in a variety of localities.

isozoic Having a similar fauna.

iteration The procedure used in computation, in which the operations are repeated until a good fit is obtained; repetition; iterative method.

iterative adaptation The repeated independent production of an adaptive type at successive points in a lineage; iterative evolution.

iteroparous Used of organisms that have repeated reproductive cycles; **iteroparity**; *cf.* semelparous.

ITZN International Trust for Zoological Nomenclature.

J

Jaccard's coefficient A measure of the similarity in species composition between two communities (*A* and *B*) calculated as $S_j = c/(a + b - c)$, where *c* is the number of species common to both, and *a* and *b* are the numbers of species occurring only in communities *A* and *B* respectively.

jactitation A method of seed dispersal in which seeds are tossed from the fruit by jerking movements of the plant.

jarovization Vernalization *q.v.*

jet effect wind A wind that is intensified as a result of funnelling through a narrow canyon or mountain gap.

jordanon A microspecies, race or other infraspecific group.

Jordan's laws 1: That the closest relatives of a species are found immediately adjacent to it but isolated from it by a geographical barrier.
2: That individuals of a given fish species develop more vertebrae in a cold environment than in a warm one.

joule (J) A derived SI unit of energy and work, defined as the work done to displace a point, by the application of a force of 1 newton, through a distance of 1 metre in the direction of the force; also the work done when a current of 1 ampere flows through a resistance of 1 ohm for 1 second; see Appendix 13.

J-shaped distribution A frequency distribution with the general shape of a letter J, describing the rapid exponential phase of population growth.

Juan Fernandez region A subdivision of the Neotropical kingdom; see Appendix 4.

Judicial Commission The subcommittee of the International Committee on Systematic Bacteriology that is responsible for bacteriological nomenclature between Congresses.

jungle Dense seral vegetation especially characteristic of tropical regions with a high level of precipitation.

junior homonym The later published of two homonyms; preoccupied name; *cf.* senior homonym.

junior synonym In taxonomy, any of two or more synonyms other than the earliest published; later synonym; younger synonym; *cf.* senior synonym.

Jurassic A geological period of the Mesozoic era (*ca.* 210–140 million years B.P.); see Appendix 1.

justified emendation In zoological nomenclature, the corrected version of an incorrect original spelling.

K

K extinction Extinction, K *q.v.*

K selection Selection producing superior competitive ability in stable predictable environments, in which rapid population growth is unimportant as the population is maintained at or near the carrying capacity of the habitat; *cf.* r selection.

K strategist K-selected species *q.v.*

K value Carrying capacity *q.v.*

Kainogaea Caenogaea *q.v.*

Kainozoic Cenozoic *q.v.*

kairomone An interspecific chemical messenger, the adaptive benefit of which falls on the recipient rather than the emitter.

kakogenesis Cacogenesis *q.v.*

kakogenic Dysgenic *q.v.*

Kamchatka current Oyashio current *q.v.*; see Appendix 6.

Kanamori scale A scale for measuring the magnitude of seismic disturbances; *cf.* Richter scale.

Kansan glaciation A glaciation of the Quaternary Ice Age in North America with an estimated duration of 90 thousand years; see Appendix 2.

karstic Pertaining to irregular limestone strata permeated by streams, typically with sinks, caves and other subterranean passages.

karyallogamy Reproduction in which two morphologically identical gametes fuse; **karyallogamous**.

karyapsis Karyogamy *q.v.*

karyogamete The nucleus of a gamete.

karyogamy The fusion of gametic nuclei; caryogamy; karyapsis; **karyogamic**.

karyogenetic Heritable; not subjected to direct environmental influences.

karyogram A photographic representation of the chromosome complement (karyotype) of a cell.

karyokinesis Nuclear division; mitosis *q.v.*; *cf.* cytokinesis.

karyological Pertaining to the nucleus, especially the chromosomes.

karyology The branch of cytology dealing with the study of nuclei, especially the structure of chromosomes; karyology has been subdivided into alpha karyology *q.v.*, beta karyology *q.v.*, gamma karyology *q.v.*, delta karyology *q.v.*, epsilon karyology *q.v.*, and zeta karyology *q.v.*

karyomite Chromosome *q.v.*

karyomitosis Mitosis *q.v.*

karyotype 1: The chromosome complement of a cell, individual, or group; caryotype. 2: The structural characteristics of the chromosome set. 3: Those individuals having an identical complement of chromosomes. 4: Karyogram *q.v.*, idiogram *q.v.*

kata- Prefix meaning down, against; **cata-**.

katabatic wind A dense mass of cold air flowing down an incline in response to gravity.

katabolism Catabolism *q.v.*; **katabolic**.

katadromous Catadromous *q.v.*; **katadromy**.

katagenesis Catagenesis *q.v.*

kataklinotropism Negative clinotropism *q.v.*; **kataklinotropic**.

Katathermal period Little Ice Age *q.v.*

katatropism An orientation response downwards; **katatropic**.

katharobic Inhabiting pure water, or water of very low organic content; **katharobe, katharobia**.

kelvin (K) The standard SI base unit of temperature, defined as 1/273.16 of the thermodynamic temperature of the triple point of water; see Appendix 13.

kelvin scale (K) A temperature scale that takes absolute zero ($-273.16°$ C) as its starting point zero K; each degree on the kelvin scale is equal to one degree Celsius.

kenapophyte A plant colonizing cleared land.

Kendall's rank correlation coefficient A non-parametric method for determining the significance of association between two variables, the values of which have been replaced by ranks within their respective samples; usually denoted by the Greek letter tau (τ); $\tau = 1 - (4_s / n(n - 1))$, where s is the smallest number of interchanges required to transform the ranks on the second variable to be the same as the ranks on the first, and n is the number of observations.

keratinophilic Thriving on horny (keratin rich) substrates; **keratinophile, keratinophily**.

key A tabulation of character states facilitating identification.

key character Diagnostic character.

key classification Artificial classification *q.v.*

key fossil Index fossil *q.v.*

key gene Oligogene *q.v.*

key species Any species that may be used to assess the degree of utilization, or the environmental balance, of a habitat or area.

keystone predator The dominant predator, often the top predator in a given food web; a predator having a major influence upon community structure often in excess of that expected from its relative abundance; keystone species.

kilo- (k) Prefix used to denote unit $\times\ 10^3$.

kilocalorie (Kcal.; Cal.) A derived cgs unit of heat equal to 1000 calories (4.1855×10^3 joules at $15°$ C); see Appendix 13.

kilogram (kg) The standard SI base unit of mass arbitrarily defined as the mass of a prototype platinum/iridium cylinder kept at Sèvres in France; **kilogramme**; see Appendix 13.

kin selection Selection acting on one or more individuals and favouring or disfavouring the survival and reproduction of relatives (other than offspring) that possess the same genes by common descent, as in the case of selection for altruistic behaviour between genetically related individuals; kinship selection; *cf.* group selection.

kinematic viscosity Of a fluid, the viscosity of the fluid divided by its density.

-kinesis Suffix meaning movement; akinesis, barokinesis, chemokinesis, hygrokinesis, hygroklinokinesis, hygroorthokinesis, klinokinesis, orthokinesis, photokinesis, statokinesis, stereokinesis, thigmokinesis.

kinesis A change in rate of random movement of an organism as a result of a stimulus; the linear or angular velocity changes in response to an alteration in the intensity of the stimulus, but no orientation of movement occurs with respect to the direction of the stimulus; **kinetic**; *cf.* orthokinesis, klinokinesis.

kinetophilous Opportunistic species *q.v.*

kingdom The highest category in the hierarchy of classification; the five kingdom classification of living organisms proposed by Whittaker and currently widely used is, Monera, Protista, Fungi, Plantae, and Animalia.

kinopsis Attraction within a group of social organisms resulting from visual perception of movement alone.

kinship Possession of a recent common ancestor; the condition of being related.

kinship, coefficient of The probability that a gene taken at random from each of two individuals will be identical by common descent; coefficient of consanguinity; coefficient of parentage.

kinship selection Kin selection *q.v.*

kipuka An island of vegetation isolated by lava flows.

kite diagram A two dimensional representation of a frequency distribution with samples typically arranged symmetrically about a common vertical axis; used to depict the vertical stratification of a population or community.

klado- Clado- *q.v.*

kleisto- Cleisto- *q.v.*

kleistogamy Cleistogamy *q.v.*; **kleistogamous**, **kleistogamic**.

klendusity The capacity of an otherwise susceptible variety of a species to escape infection because of its growth pattern.

kleptobiosis An interspecific association found in some social organisms in which one species steals food from the stores of another species but does not live or nest in close proximity to it; cleptobiosis.

kleptoparasitism A form of parasitism found in some social organisms in which a female of one species steals the prey or food stores of the female of another species to feed her own progeny; cleptoparasitism; **kleptoparasite**.

klinogeotropism The drooping tendency displayed by the tip of a climbing plant during nutation *q.v.*; **klinogeotropic**.

klinokinesis A change in rate of random movement of an organism (kinesis) in which the rate of change of direction (frequency of turning movements) varies with the intensity of the stimulus; *cf.* orthokinesis.

klinophyte A terrestrial plant that is dependent upon other plants for physical support.

klinotaxis An orientation response of a motile organism (taxis) in which the direction of movement results from a comparison of the intensity of the stimulus on each side of the body by successive deviations from side to side; **klinotactic**; *cf.* tropotaxis, telotaxis.

klinotropism Clinotropism *q.v.*; **klinotropic**.

knephopelagic Pertaining to the middle pelagic zone of the sea extending from about 30 metres to the lower limit of the photic zone; **knephopelagile**.

knephoplankton Planktonic organisms of the twilight zone between about 30–500 m depth; *cf.* phaoplankton, skotoplankton.

Knight–Darwin law That nature abhors perpetual self-fertilization.

knot (kn) A unit of velocity equal to one nautical mile (6080 ft, 1.8532 km) per hour; see Appendix 13.

koino- Coeno- *q.v.*

kollaplankton Planktonic organisms rendered buoyant by encasement in gelatinous envelopes; collaplankton.

koprophagous Coprophagous *q.v.*; **koprophage**, **koprophagy**.

kormogene association A colony in which the different individuals are attached morphologically to each other, but without organic connection.

kormogene society A colony in which the different individuals are organically connected to each other.

K – r spectrum A linear system of reproductive strategies with K-selected species *q.v.* and r-selected species *q.v.* representing the two extremes.

kremastoplankton Planktonic organisms that possess modified appendages or surface structures to reduce rates of sinking.

Krogh's law That the rate of a biological process or event is directly correlated with temperature.

kronismus Cronism *q.v.*

krotovina A former animal burrow in one soil horizon that has been filled with organic material or material from another horizon; crotovina.

krummholz A discontinuous belt of stunted forest or scrub typical of windswept alpine regions close to the tree line; *cf.* elfin forest.

Kruskal–Wallis test A non-parametric method comprising a one-way analysis of variance by rank, used to test for differences in mean level between several samples where the number of counts per sample may be different.

K-selected species A species characteristic of a relatively constant or predictable environment, typically with slow development, relatively high competitive ability, late reproduction, large body size and iteroparity; K strategist; *cf.* r-selected species.

Kulezinski coefficient An index of similarity in species composition between two communities (A and B) which is relatively little influenced by dissimilar sample sizes; calculated as $S_k = c/(a+b)$ where c is the number of species common to both, and a and b are the numbers occurring in communities A and B respectively; Kulezynski's coefficient.

Kuroshio Current A warm surface ocean current that flows north and east in the North Pacific, fed from the North Equatorial Current and forming the western limb of the North Pacific Gyre; Kuro Shio; see Appendix 6.

Kuroshio Extension A warm surface ocean current that flows east in the North Pacific forming the northern limb of the North Pacific Gyre.

kurtosis A measure of the departure of a frequency distribution from a normal distribution, in terms of its relative peakedness or flatness; *cf.* leptokurtic, mesokurtic, platykurtic.

Ky Abbreviation denoting 1000 years.

kybernetics Cybernetics *q.v.*

kymatology The study of waves and wave motion.

L

l.c. Abbreviation of the Latin *loco citato*, meaning at the place cited; *loc. cit.*

labile Plastic; readily modified.

lability, evolutionary The ease and speed with which particular categories of traits evolve.

Labrador Current A cold surface ocean current that flows south off the east coast of Canada; see Appendix 6.

lacustrine Pertaining to, or living in, lakes or ponds; used of plants floating on or immersed in fresh water; **lacustral**.

LAD Leaf area duration *q.v.*

ladder of nature The classification scheme devised by Aristotle in which nature is regarded as a continuum leading from inanimate matter via plants, lower animals, and higher animals, to mankind.

lag growth phase The period in which little or no population growth occurs, preceding the exponential growth phase *q.v.*

Lagerberg–Raunkiaer cover scale A scale for estimating cover of a plant species, comprising 4 categories of percentage ground cover; 1 (0–10%), 2 (11–30%), 3 (31–50%), 4 (51–100%).

LAI Leaf area index *q.v.*

lair-flora The plants inhabiting ground manured by animals.

lake A large body of fresh or saline standing water with negligible current, having a narrow peripheral beach largely devoid of vegetation as a result of wave action.

Lamarckism Inheritance of acquired characters; the theory that changes in use and disuse of an organ result in changes in size and functional capacity and that these modified characters acquired by organisms in response to environmental factors are transmitted to the offspring; theory of use and disuse.

lambda link A dichotomously branching link in a food web involving one predator feeding on two prey species; *cf.* gamma link, iota link.

lambert A unit measure of surface luminance equal to 1 lumen per square centimetre; see Appendix 13.

lamiation A stratal or layer society of an association, based on life form and environmental differences.

lamies A stratal or layer socies of an associes *q.v.*

lamina A thin layer; the smallest recognizable unit in a bedded sediment.

laminar flow The streamlined movement of water, in which water particles appear to move in smooth paths, and one layer of water slides over another; *cf.* turbulence.

land breeze A light wind blowing off the land at night when the sea surface is warmer than the land; *cf.* sea breeze.

land-bridge A more or less continuous connection between adjacent land masses, forming a potential route for migration and dispersal; *cf.* filter bridge.

land-bridge theory That land-bridges of continental size existed during the Cenozoic but have since disappeared leaving no trace of their existence.

langley (ly) A derived cgs unit of solar radiation density equal to 1 calorie per square centimetre per minute, or 697.8 watt per square metre per second; see Appendix 13.

langmuir circulation A wind generated pattern of water circulation, in which water tends to move in vortices resulting in zones of upwelling and downwelling.

lapidicolous Living under or among stones; saxicolous; **lapidicole**.

lapsus calami Latin, meaning a slip of the pen; in nomenclature a misspelling made by the author.

LAR Leaf area ratio *q.v.*

larvicidal Lethal to larvae.

larviparous Producing eggs that are hatched internally with release of free-living larvae; **larvipary**; *cf.* oviparous, ovoviviparous, viviparous.

larvivorous Feeding on larvae; **larvivore**, **larvivory**.

lasion Fouling; a dense growth of aquatic organisms, particularly on submerged objects projecting into free water.

lat. Abbreviation of the Latin *latus*, meaning broad, wide.

latebricolous Living in holes; **latebricole**.

latent homology The condition in which two or more related lineages share a character which is absent from the ancestral stock, although the ancestor is presumed to have had the potential to develop the structure.

latent learning The association of indifferent stimuli and situations without explicit reinforcement *q.v.*

latent period Reaction time *q.v.*

later synonym Junior synonym *q.v.*

lateral planation Erosion of the two sides of a river or stream by undercutting.

Laterite soil A zonal soil, highly weathered, having a reddish brown clay subsoil, low in minerals and soluble constituents, formed in forested warm temperate to tropical regions; Latosol.

laterization A soil forming process occurring in tropical regions in which the minerals are converted to clay and pronounced leaching results in the accumulation of hydroxides of aluminium and iron in the deeper layers; **laterite**.

Latin square A square array of *n* symbols or blocks arranged so that each symbol or block appears only once in every row and column; used in designing the layout of replicate plots in field experiments; Roman square.

latinization In taxonomy, the treatment of a name, not of Latin origin, according to the grammatical rules of Latin; **latinized**.

latiphenic Used of a population exhibiting an extremely broad range of continuous phenotypic variation; *cf.* monopheny.

Latosol Laterite soil *q.v.*

laurad A sewer or drain plant; laurophyte.

Laurasia The northern supercontinent formed by the break up of Pangaea in the Mesozoic (*ca.* 150 million years B.P.), and comprising North America, Greenland, Europe, and Asia, excluding India; Eurasia; *cf.* Gondwanaland.

laurilignosa Subtropical and warm temperate evergreen forest vegetation.

laurium A sewer or drain community.

laurophilus Thriving in sewers and drains; **laurophile, laurophily**.

laurophyte A sewer or drain plant; laurad; **laurophyta**.

law An empirical generalization; a statement of a biological principle that appears to be without exception at the time it is made, and has become consolidated by repeated successful testing; rule.

law of acceleration Acceleration, law of *q.v.*

law of constant extinction An empirical law stating that for any group of related organisms sharing a common ecology there is a constant probability of extinction of any taxon per unit time; Van Valen's law.

law of faunal assemblages Faunal assemblages, law of *q.v.*

law of faunal succession Faunal succession, law of *q.v.*

law of independent assortment Independent assortment, law of *q.v.*

law of integration Oligomerization *q.v.*

law of minimum Minimum, law of *q.v.*

law of relativity of ecological valency Ecological valency, law of relativity of *q.v.*

law of segregation Segregation, law of *q.v.*

law of substitution Substitution, law of *q.v.*

law of superposition Superposition, law of *q.v.*

law of tolerance Tolerance, law of *q.v.*

layer Stratum; a horizontal zone of vegetation; four such zones are recognized by plant ecologists – tree layer, shrub (bush) layer, field (herb) layer, and ground (moss) layer.

layer society A plant community characterized by a horizontal layering of foliage at a more or less well defined height; a horizontal stratum of vegetation usually determined by levels of illumination.

LC 50 Lethal concentration, 50%; the concentration of a toxin or pollutant that kills half the organisms in a test population per unit time.

LD 50 Lethal dose, 50%; the dose level required to kill half the organisms in a test population per unit time.

leaching The removal of readily soluble components, such as chlorides, sulphates and carbonates, from soil by percolating water; *cf.* flushing.

leaf area The entire surface area of a leaf, including both upper and lower surfaces.

leaf area duration (LAD, or D) An integrated value of the leaf area index over a given period, usually a growing season.

leaf area index (LAI, or L) A measure of the area of photosynthetic surface expanded over a given area of ground.

leaf area ratio (LAR) A measure of the leaf area exposed by a plant, expressed as a fraction of the plant mass, in square centimetres per gram.

league An itinerant unit of distance equal to about 3 nautical miles.

leaky gene Hypomorphic allele *q.v.*

learning The production and accumulation of adaptive changes in individual behaviour as a result of experience.

least squares estimate In statistics, the estimate obtained by the method of least squares, used in regression analysis and analysis of variance.

least squares, method of In statistics, a technique of curve fitting and of estimating the parameters of the corresponding equation, involving choosing the parameters to minimize the sum of the squared differences between the observed values and their respective expected values.

Lebensort-Typen Life forms that are thought to have originated in, and are characteristic of, a particular habitat.

lebensspur Trace fossil *q.v.*; trace.

lecithotrophic Pertaining to developmental stages that feed upon yolk, and to eggs rich in yolk.

lectoallotype Allolectotype *q.v.*

lectoholotype Lectotype *q.v.*

lectoparatype Paralectotype *q.v.*

lectotype In taxonomy, the type specimen selected from the syntype series by a subsequent author in the absence of a holotype; hololectotype; lectoholotype.

leeward Pertaining to the side facing away from a wind, ice, or water current.

legitimate name *Nomen legitimatum q.v.*

leimocolous Living in moist grassland or meadowland; leimicolous; **leimicole, leimocole**.

leimonapophyte A plant introduced into grassland or meadowland.

leimophyte A wet-meadow plant.

lek 1: An assembly area for communal courtship display. 2: A communal prenuptial display. specifically those displays involving ritualized contests between competitors, often for the use of symbolic sites; **lekking**.

Lemuria A continental land-bridge thought to have connected South Africa, Madagascar and India up to the end of the Cretaceous.

Lemurian realm The zoogeographical region comprising the island of Madagascar; Malagasy region.

lentic Pertaining to static, calm or slow-moving aquatic habitats; lenitic; *cf.* lotic.

lepidophagous Feeding on scales; used of fishes that feed on the external scales of other fishes; **lepidophage, lepidophagy**.

lepidopterid Used of flowers adapted for pollination by butterflies and moths.

lepidopterophilous Pollinated by butterflies and moths; **lepidopterophily**.

leptokurtic Used of a frequency distribution that is more peaked than the corresponding normal distribution; **leptokurtosis**; *cf.* kurtosis.

leptology The study of minute structures or particles.

leptopel Large organic molecules or aggregates of colloidal proportions suspended in water.

leptophenic Used of a population exhibiting a narrow range of continuous phenotypic variation; *cf.* monopheny.

leptophyll A Raunkiaerian leaf size class *q.v.* for leaves having a surface area less than 25 mm^2.

Lessepsian migration Migration between the Red Sea and the Mediterranean by way of the Suez canal.

lessivage The translocation of clay particles by percolating water from the upper layer of a soil profile to deeper layers, thus producing marked textural changes within the profile; the clay enriched deeper layers are denoted by suffix *t* in soil profile.

lestobiosis A symbiosis between social insects in which colonies of a small species inhabit the nests of a larger species and prey on its brood or rob the food stores; synclopia.

lethal Pertaining to, or causing, death by direct action.

lethal dose LD 50 *q.v.*

lethal equivalent value The mean number of harmful genes carried by each member of a population multiplied by the mean probability that each gene will cause genetic death when homozygous.

lethal trait Any inherited character that causes the death of an organism before it can reproduce.

lethargic Used of an animal exhibiting a slow or relatively slow response to an external stimulus; **lethargy**.

leucism Paleness, as of coloration.

levée A sediment bank.

level of significance In statistics, the probability of erroneously rejecting a true null hypothesis.

Lias A geological epoch of the early Jurassic period (*ca.* 210–175 million years B.P.); see Appendix 1.

lichenicolous Living in or on a lichen; **lichenicole**.

lichenism The symbiotic relationship between a fungus and an alga that comprise a lichen.

lichenology The study of lichens.

lichenometry A method for dating rocks based on a knowledge of the growth rates of encrusting lichens; **lichenometric**.

lichenophilous Thriving on, or having an affinity for, lichens or lichen-rich habitats; **lichenophile, lichenophily**.

Liebig's law That the growth of an organism is limited by the essential element which is present in the lowest concentration in proportion to the requirement of that organism; law of minimum.

life assemblage Biocoenosis *q.v.*

life cycle 1: The sequence of events from the origin as a zygote, to the death of an individual. 2: Those stages through which an organism passes between the production of gametes by one generation and the production of gametes by the next.

life expectancy The average period of time that an organism is expected to survive at any given point in its life cycle.

life form The characteristic structural features and method of perennation of a plant species; the result of the interaction of all life processes, both genetic and environmental; *cf.* Raunkiaerian life forms.

life history The significant features of the life cycle

through which an organism passes, with particular reference to strategies influencing survival and reproduction.

life span Longevity; the maximum or mean duration of life of an individual or group.

life table A tabulation of the complete mortality data of a population or cohort, with respect to age.

life zone A biogeographical region having a characteristic fauna and flora.

light climate The general condition of illumination in a body of water, determined by the level of incident light, absorption, scattering and turbidity.

light flux density The flux density *q.v.* of light, usually measured in units of joules per square metre per second.

light saturation Illumination intensity beyond the level at which photosynthetic yield continues to increase as other non-photochemical processes are limiting.

lignicolous Growing on or in wood; **lignicole**.

ligniperdous Used of organisms that destroy wood.

lignivorous Feeding on wood; dendrophagous; hylophagous; zylophagous; **lignivore, lignivory**.

lignophilic Thriving on or in wood; **lignophile, lignophily**.

lignosa Woody vegetation.

likelihood In statistics, an estimate of the relative support for each alternative hypothesis; the chance that a given event will occur.

limacology The study of slugs.

limen Threshold *q.v.*

limicolous Inhabiting mud; shore-dwelling; **limicole**.

limiphagous Limophagous *q.v.*; **limiphage, limiphagy**.

limiting factor Any environmental factor, or group of related factors, which exist at suboptimal level and thereby prevent an organism from reaching its full biotic potential.

limivorous Feeding on mud; limophagous; **limivore, limivory**.

limn- Prefix meaning lake; *cf.* limno-.

limnad A lake plant.

limnaen Submersed wet meadowland.

limnetic Pertaining to lakes or to other bodies of standing fresh water; often used with reference only to the open water of a lake away from the bottom; **limnal, limnic**.

limnetic zone The surface zone of a lake above the compensation depth; see Appendix 8.

limnicolous Living in lakes; limniicolous; **limnicole, limniicole**.

limnion All lakes, ponds, pools and other bodies of standing fresh water; **limnial**.

limnium A lake community.

limno- Prefix meaning marshy; *cf.* limn-.

limnobiology The study of organisms living in lakes, ponds and other standing freshwater bodies; limnology.

limnobiont A freshwater organism; limnobion; **limnobiontic**.

limnobios The total life of fresh waters; that part of the Earth's surface occupied by freshwater organisms; limnobiontic; *cf.* geobios, holobios, hydrobios.

limnocrene A pool of spring-water with no outflow; limnokrene.

limnocryptophyte A marsh or pond plant with perennating organs below the surface; helophyte *q.v.*

limnodad A salt-marsh plant; limnodophyte.

limnodic Pertaining to salt marshes.

limnodium A salt-marsh community.

limnodophilous Thriving in salt marshes; **limnodophile, limnodophily**.

limnodophyte A salt-marsh plant; limnodad; **limnodophyta**.

limnology The study of lakes, ponds, and other standing waters, and their associated biota; **limnological**.

limnonereid A freshwater plant.

limnoneuston Freshwater neuston *q.v.*

limnophilous Thriving in lakes or ponds; **limnophile, limnophily**.

limnophyte A marsh or pond plant; helophyte.

limnoplankton Planktonic organisms of freshwater lakes and ponds.

limonile Far beneath the ground surface.

limophagous Feeding on mud; limivorous; limiphagous; **limophage, limophagy**.

Lincoln index An estimate of population size N obtained after release and recapture of marked animals; population size is calculated from the number of marked animals released (N_m), the number captured in a sample (N_s) after release, and the number of marked individuals in the sample (N_{ms}), using the formula $N = N_m \times N_s/N_{ms}$; catch-mark-recatch method.

line A unit of length equal to 1/12 inch (2.1167 mm); denoted by ′′′.

line plot survey A method of sampling vegetation by recording the plants present in a series of plots of equal area, taken at regular intervals along a given line.

line precedence In taxonomy, the precedence of a name, for the purposes of priority, that occurs on an earlier line of the same page than another name for the same taxon.

line transect method A method of sampling vegetation by recording the plants that fall on or above a measured line or transect placed close to the ground; line intercept method; line survey.

lineage A line of common descent; line.

lineage group A group of species sharing descent from a common ancestor.

linear regression analysis In statistics, the analysis of paired data $(x_1y_1, x_2y_2 \ldots \ldots)$ where the x's are measured without error and the y's are values of random variables dependent on the x's.

linkage 1: The presence of specific genes on the same chromosome such that the traits are not independently assorted; the greater the proximity of these genes the smaller the chance of their separation by crossing over, and the stronger the linkage. 2: In numerical taxonomy, the connection between two or more branches of a dendrogram.

linkage group A group of genes associated to a greater or lesser extent by linkage; the maximum number of such groups is equal to the number of chromosomes.

linkage map A chromosome map showing the linear sequence of genes within a given linkage group.

linked character Dependent character *q.v.*

Linnaean classification The system of hierarchical classification and binomial nomenclature established by Linnaeus; Linnean classification.

Linnaean species A broad concept of a species often comprising many varieties; macrospecies; taxonomic species; Linnaeon.

Linnaeon Linnaean species *q.v.*; Linneon.

lipogenesis 1: Accelerated development in evolution by the loss of certain developmental stages. 2: The formation of fat within the body.

lipopalingenesis The loss of one or more developmental stages in a descendant that were present in the ancestral form.

lipoxenous Used of a parasite that leaves its host after feeding; lipoxeny.

liptocoenosis An assemblage of dead organisms and their biogenic products; necrocoenosis.

List, Official Official List *q.v.*

list quadrat A method of sampling vegetation in which a record is made of the species occurring in a given sample area (quadrat).

litharch An ecological succession on a hard rock substratum.

lithic Pertaining to rock.

lithification The physicochemical process that produces rock from sedimentary deposits.

lithocarp A fossil fruit.

lithodomous Used of an organism that lives in holes in rock, or that bores into rock; lithotomous.

lithology The study of rocks and rock forming processes; petrology.

lithophagic 1: Used of organisms that erode or bore into rock; further subdivided into epilithophagic, mesolithophagic and endolithophagic according to the degree of penetration; lithophagous, lithophage, lithophagy. 2: Eating small stones, as in some birds.

lithophilous Thriving in stony or rocky habitats; lithophilus; lithophile, lithophily.

lithophyl A fossil leaf.

lithophyte A plant growing on rocks or stones; lithophyta, lithophytia.

lithosequence A group of related soils that differ in certain properties as a result of differences in parent rock material.

lithosere An ecological succession originating on an exposed rock surface; *cf.* sere.

Lithosol A very stony shallow azonal soil without pronounced horizons, and with incomplete surface layers comprising imperfectly weathered rock fragments; skeletal soil.

lithosphere The rigid crustal plates of the Earth.

lithostratigraphy The organization and classification of rock strata according to their lithological character; lithostratigraphic.

lithotomous Used of an organism that burrows into rock; lithodomous.

lithotrophic Pertaining to organisms that use inorganic compounds as electron donors in their energetic processes; lithotroph; *cf.* organotrophic.

lithoxyle Fossilized wood.

litoral Littoral *q.v.*

litre A derived metric unit of volume equal to the volume of 1 kilogram of water at its temperature of maximum density.

litter 1: Recently fallen plant material which is only partially decomposed and in which the organs of the plant are still descernible, forming a surface layer on some soils; L-layer. 2: Those animals produced at a multiple birth.

litter size The number of eggs or offspring produced at one time in a single batch or cocoon.

Little Ice Age An interval characterized by expanding mountain glaciers (*ca.* 4000–2000 years B.P.); about equivalent to the Subboreal period; Medithermal period; Katathermal period; Hypothermal period.

littoral 1: Pertaining to the shore. 2: The shore of a lake to a depth of about 10 m. 3: The intertidal zone of the seashore; eulittoral; sometimes used to refer to both the intertidal zone on the seashore and the adjacent continental shelf to a depth of about 200 m;

litoral; littorine; subdivided into supralittoral, eulittoral (intertidal zone), infralittoral and circalittoral; see Appendix 7.

littoriprofundal Pertaining to a transition zone of light penetration in a lake in which only scattered photosynthetic organisms are present, typically having moneran and algal herpobenthos; see Appendix 8.

litus A strand community on the shore.

living fossil A species that has persisted to the present time with little or no change over a long period of geological time.

L-layer The surface litter layer of a soil profile comprising loose fragmented material in which the original plant structures are still readily discernible; Ol-layer; see Appendix 12.

loam A friable soil comprising a mixture of clay, silt, sand and organic matter.

loc. cit. Abbreviation of the Latin *loco citato*, meaning in the place cited; used to avoid the repetition of a bibliographic reference; *l.c.*

local character species 1: Used for a species that is characteristic *q.v.* over only part of the total range in which the vegetation unit and the species occur in common; the relative distributions of the vegetation unit and the species have been classified, conlocal, intralocal, circumlocal, paralocal. 2: A faithful *q.v.* species to a single association occurring in only part of the range of the association.

local type Chorotype *q.v.*

locality The geographical position of an individual, population or collection.

lochmad A thicket plant; lochmophyte.

lochmium A thicket community.

lochmocolous Inhabiting thickets; **lochmocole**.

lochmodium A dry-thicket community.

lochmodophilous Thriving in dry thickets; **lochmodophile, lochmodophily**.

lochmodophyte A plant inhabiting dry thickets; **lochmodophyta**.

lochmophilus Thriving in thickets; **lochmophile, lochmophily**.

lochmophyte A plant inhabiting thickets; lochmad; **lochmophyta**.

lociation A ranked category in the classification of vegetation; a local variation of a faciation *q.v.* differing in the subdominant species largely as a result of local edaphic factors.

loco citato (*loc. cit., l.c.*) Latin, meaning in the place cited; used to avoid the repetition of a reference.

locotype Topotype *q.v.*

loculicidal dehiscence Spontaneous opening of a ripe fruit along the centre line.

locus The position of a given gene on a chromosome; **loci**.

loess A fine unconsolidated wind-blown sediment; löss.

log. growth phase The period in which maximum population growth occurs, with the number of individuals doubling per unit time; logarithmic growth phase; exponential growth phase; *cf.* lag growth phase.

logistic curve A sigmoid curve representing an exponential function such as the logistic equation *q.v.*; used in models of population growth in limited environments.

logistic equation A simple mathematical model of a population (N) growing from a very small initial number of individuals to an upper limiting population (K) in a finite environment; the differential form is $dN/dt = rN[(K - N)/K]$; Verhulst logistic equation.

logistic growth Growth that follows a sigmoid curve in which growth rate is gradual initially (lag growth phase), then rapid (exponential growth phase) and then slow again (stationary growth phase) as the population or system reaches its maximum size or carrying capacity *q.v.*

log-normal distribution A frequency distribution of a variable, the logarithm of which is normally distributed; used as a model of species abundances in which the x-axis is the logarithm of the number of individuals represented in a sample, and the y-axis is the number of species; logarithmic normal distribution.

logotype 1: A type determined from a written description in the absence of both a specimen and an illustration. 2: Genolectotype *q.v.*

Long Drought Xerothermal period *q.v.*

long-day plant A plant which has a flowering period initiated and accelerated by exposure to a light regime with more than 12 h illumination daily; *cf.* short-day plant.

longevity The average life span of the individuals of a population under a given set of conditions.

longshore Used of currents or movement parallel to the coastline.

lophad A hill-top plant; lophophyte.

lophium A hill-top community.

lophophilus Thriving on hill tops; **lophophile, lophophily**.

lophophyte A hill-top plant; lophad; **lophophyta**.

löss Loess *q.v.*

lotic Pertaining to fast running-water habitats, such as rivers and streams; *cf.* lentic.

Lotka-Volterra equations 1: A simple model for a predator-prey system; $dN/dt = aN - \alpha NP$ and $dP/dt = -bP + \beta NP$, where the prey population

(N) has a propensity for exponential growth (aN), which is limited by predation: the effect of the predator on the prey population is measured by the functional response $q.v.$ term αNP. The predator population (P) has an intrinsic death rate ($-bP$) and a growth rate which depends on prey abundance, as βNP. 2: A simple model for two competing populations (N_1 and N_2) based on the logistic equation: $dN_1/dt = r_1 N_1 [1-(N_1 + \alpha_{12}N_2)/K_1]$ and $dN_2/dt = r_2N_2[1-(N_2 + \alpha_{21}N_1)/K_2]$, where K_1 and K_2 are the carrying capacities $q.v.$ of species 1 and 2 respectively, r_1 and r_2 are their intrinsic growth rates, α_{12} is the competition coefficient measuring the effect of species 2 on species 1 and α_{21} measures the effect of species 1 on species 2.

low tide Low water; the mimimum height of the falling tide.

low water (LW) The lowest surface water level reached by the falling tide; low tide.

lower high water (LHW) The lower of two high waters during any tidal day where there are marked inequalities of tidal height.

lower low water (LLW) The lower of two low waters during any tidal day where there are marked inequalities of tidal height.

lowerunderstorey Understorey $q.v.$

lowest astronomical tide (LAT) The lowest low water produced only by the gravitational effects of the sun and moon.

Lowestoft glaciation Anglian glaciation $q.v.$; see Appendix 2.

lucicolous Living in open habitats with ample light; **lucicole**; $cf.$ umbraticolous.

luciferous Light-producing; bioluminescent.

lucifugous Intolerant of light; photophobic; **lucifugal**; $cf.$ luciphilous.

lucipetal Luciphilous $q.v.$

luciphilous Thriving in open, well-lit habitats; photophilic; lucipetal; **luciphile, luciphily**; $cf.$ lucifugous.

Ludwig's theory That new genotypes can be added to a gene pool if they can utilize new components of the environment and thereby occupy a new subniche, even if they are competitively inferior in the ancestral niche.

lumachelle An accumulation of shells in a sediment.

lumen (lm) A derived SI unit of luminous flux defined as the light energy emitted per second within a unit solid angle of 1 steradian by a uniform point source of 1 candela luminous intensity; see Appendix 13.

luminance A measure of light emitted (luminous intensity) per unit projected area of surface, perpendicular to the direction of view, measured as candelas per square metre.

luminescence The emission of light by biochemical processes without the production of heat; $cf.$ phosphorescence.

lumping The taxonomic practice of ignoring minor variation in the definition or recognition of species; **lumper**; $cf.$ splitting.

lunar Pertaining to the moon.

lunar day The time interval between consecutive moonrises; 24.8 h; $cf.$ solar day.

Lusitanian Having affinities with the Iberian peninsula.

lutaceous Pertaining to mud.

luticolous Inhabiting mud; **luticole**.

Lutz phytograph A polygonal configuration used to symbolize the structural features of a plant community plotted from four structural characters – percentage density, percentage frequency, size class and basal area percentage.

lux (lx) A derived SI unit of illumination, defined as 1 lumen per square metre; see Appendix 13.

luxuriance That part of hybrid vigour that is non-adaptive and does not increase the fitness of the organism.

lygophilous Thriving in dark or shaded habitats; **lygophile, lygophily**.

Lyon hypothesis That dosage compensation $q.v.$ is achieved by the random inactivation of an X-chromosome in female somatic cells.

lyophilization Freeze drying $q.v.$

Lysenkoism The doctrine that rejects the gene theory of inheritance in favour of the inheritance of acquired characters $q.v.$; neo-Lamarckism.

lysimeter An apparatus for measuring the actual uptake of water by a plant.

lysocline The oceanic depth zone between the carbonate dissolution depth (about 4000 m) and the carbonate compensation depth (about 5000 m), representing the major facies change between well-preserved and poorly preserved calcareous elements of the ocean floor sediment.

lysogeny The condition in which a bacteriophage is latent, integrated with the genetic apparatus of its host bacterium, yet capable of being activated by certain stimuli.

M

m. Abbreviation of the Latin *mihi*, meaning to me, of me.

Macallum's hypothesis That the inorganic composition of an animal's blood resembles the composition of the ocean at an earlier geological period.

Macaronesia A biogeographical area encompassing the islands off the coast of N.W. Africa and Europe, including the Azores, Canaries, Cape Verde Islands, and Madeira.

Macaronesian region A subdivision of the Boreal kingdom; see Appendix 4.

MacArthur's 'broken-stick' model 'Broken-stick' distribution *q.v.*

MacArthur–Wilson theory of island biogeography That the number of species inhabiting an island is a function of island area and distance from the mainland, and is determined by the relationship between extinction and immigration rates.

macro- Prefix meaning large, long, great.

macrobenthos The larger organisms of the benthos, exceeding 1 mm in length; *cf.* meiobenthos, microbenthos.

macrobiota Large soil organisms, exceeding about 40–50 mm in length; *cf.* mesobiota, microbiota.

macrobiotic Long-lived.

macroclimate 1: The climate of a major geographical region. 2: The conditions of temperature, precipitation, relative humidity, sunshine and other meteorological factors, recorded about 1.5 m above ground level to avoid topological, vegetational and edaphic influences; *cf.* microclimate.

macrodichopatric Dichopatric *q.v.*; **macrodichopatry**; *cf.* microdichopatric.

macroevolution Major evolutionary events or trends usually viewed through the perspective of geological time; the origin of higher taxonomic categories; transpecific evolution; macrophylogenesis; megaevolution; *cf.* microevolution.

macrogamete The larger of two anisogametes, usually regarded as the ovum or female gamete; megagamete; *cf.* microgamete.

macrogamy Hologamy *q.v.*

macrogenesis The evolution of new types by drastic and sudden mutational changes (saltation).

macromonophyletic A term proposed to describe a higher taxon such as a class derived from two or more ancestral taxa of lower rank such as

subclass or order, all belonging to an ancestral taxon of the same higher taxonomic rank (i.e. class); *cf.* micromonophyletic, micropolyphyletic, macropolyphyletic.

macromorphology Gross external morphology; *cf.* micromorphology.

macromutation The simultaneous mutation of a number of different characters or traits.

macromutationism The theory that new taxa evolve essentially instantaneously via a major genetic mutation that establishes both reproductive isolation and new adaptations simultaneously.

macronutrients Those nutrients required in relatively large amounts for optimal growth; *cf.* micronutrients.

macroparapatric Parapatric *q.v.*; **macroparapatry**; *cf.* microparapatric.

macrophagous Feeding on relatively large food particles or prey; **macrophage, macrophagy**; *cf.* microphagous.

macrophanerophyte Megaphanerophyte *q.v.*

macrophyll A Raunkiaerian leaf size class *q.v.* for leaves having a surface area between 18 225 and 164 025 mm^2.

macrophylogenesis Macroevolution *q.v.*

macrophyte A large macroscopic plant, used especially of aquatic forms such as kelp; **macrophytic**.

macrophytophagous Feeding on higher plant material only; **macrophytophage, macrophytophagy**.

macroplankton Large planktonic organisms 20–200 mm in diameter; see Appendix 9.

macropolyphyletic A term proposed to describe a higher taxon such as a class derived from two or more ancestral taxa of lower taxonomic rank such as order or family belonging to two or more ancestral higher taxa of the same taxonomic rank (i.e. class); *cf.* macromonophyletic, micromonophyletic, micropolyphyletic.

macroporous Used of a sediment having large interstitial spaces; *cf.* microporous.

macroscopic Of relatively large size; used of a structure or occurrence visible to the naked eye, or with the aid of a hand lens; megascopic; *cf.* microscopic.

macrosmatic Used of an organism possessing a highly developed sense of smell; *cf.* microsmatic.

macrospecies Linnaean species *q.v.*; a large polymorphic species.

macrospore The larger of two kinds of haploid spores produced by vascular plants, regarded as the female spore; megaspore; *cf.* microspore.

macrosymbiont The larger of two partners in a symbiotic association; *cf.* microsymbiont.

macrotaxonomy That branch of taxonomy concerned with classification of supraspecific taxa; beta taxonomy; *cf.* microtaxonomy.

macrotherm Megatherm *q.v.*

macrothermophilus Thriving in the tropics; thriving in warm habitats; **macrothermophile, macrothermophily.**

macrothermophyte Megatherm *q.v.*; macrothermophyta.

macrozoogeography The study of the dispersion of higher taxa over wide areas; *cf.* microzoogeography.

Madagascan region A subdivision of the African subkingdom of the Palaeotropical kingdom; see Appendix 4.

Madagascan subregion Malagasy subregion *q.v.*

madescent Becoming moist.

madid Wet or moist.

mafic Pertaining to rocks rich in magnesium and iron.

magnafacies Used in palaeontology for the major biogeographical regions characterized by distinctive plant and animal fossils; the palaeontological equivalent of biome *q.v.*

magnetic anomaly A reversal in the magnetic polarity in rocks or sediments indicating a reversal of the Earth's magnetic field.

magnetic stripes Alternating bands of magnetic polarity found across the ocean floor, typically symmetrical around an ocean spreading centre or mid-ocean ridge.

magnetotropism Orientation in response to a magnetic field; **magnetotropic.**

Mahalanobis' D² A measure of generalized distance between community samples, involving the computation of means, variances and covariances of various properties of replicate samples as a type of multivariate analysis; the larger the value of D^2 the greater the distance between the samples.

maintenance In taxonomy, the continued use of, or support for, a correct name in nomenclature; retention.

maintenance ration In energetics, the level of consumption *q.v.* that provides for maintenance of physiological processes but allows for no increase in biomass.

maiosis Meiosis *q.v.*; **maiotic.**

major gene Oligogene *q.v.*

malacogamous Malacophilous *q.v.*; **malacogamy.**

malacology The study of molluscs.

malacophilous Pollinated by snails; **malacophily.**

Malagasy subregion A zoogeographical area comprising Madagascar, the Archipelago of the Comores, the Mascarenes and the Seychelles; regarded either as a component part of the Ethiopian region or as a separate region; Malgashan subregion; Madagascan subregion; see Appendix 3.

Malayan subregion Indo-Malayan subregion of the Oriental region; see Appendix 3.

Malayasian region A subdivision of the Indo-Malayasian subkingdom of the Palaeotropical kingdom; see Appendix 4.

malaxation The act of chewing; mastication; manduction.

male The sperm producing form of a bisexual or dioecious organism; symbolized ♂; *cf.* female.

Malgashan subregion Malagasy subregion *q.v.*

Malm A geological epoch of the late Jurassic period (*ca.* 160–140 million years B.P.).

malnutrition Undernourishment; a deficiency condition in which one or more necessary nutrients is available in an insufficient amount for normal growth and maintenance.

Malthusian model An expression of exponential population increase, in which the rate of increase in numbers dN/dt equals rN, where r is a constant rate of increase and N is the original population size.

Malthusian parameter The intrinsic rate of natural increase of a population with a given age distribution, birth rate and death rate.

mammalology The study of mammals.

Manchurian subregion A subdivision of the Palaearctic region; see Appendix 3.

mandatory change In taxonomy, an alteration in spelling of the ending of a family-group name or epithet as required by the rules of nomenclature.

manduction Mastication *q.v.*

mangrove A tidal salt marsh community dominated by trees and shrubs, particularly of the genus *Rhizophora*, many of which produce adventitious aerial roots.

Manhattan distance The taxonomic distance *q.v.* between two OTUs plotted not as a straight line but as the moves of a rook on a chess board; city block distance.

Mann–Whitney U-test A non-parametric method in which sample counts are ranked in a simple sequence; used to determine whether two independent random samples are drawn from populations having the same distributions.

manometabolous Used of a pattern of development in some insects in which a minor or very gradual metamorphosis occurs without a resting stage; **manometaboly.**

manuscript A text (MS) or texts (MSS), typewritten or handwritten, not reproduced as multiple copies; typically that copy submitted for publication; *cf.* autograph.

manuscript name In taxonomy, an unpublished scientific name.

map unit The distance along a chromatid that has an average of one crossing over per meiosis, or a recombination frequency of 1%; used as a measure of relative distance between linked genes for purposes of chromosome mapping; Morgan unit.

marginal Pertaining to the periphery, edge or margin.

marginal species A plant species that occurs on the edge of a habitat or community.

mariculture Cultivation, management and harvesting of marine organisms in their natural habitat or in specially constructed channels or tanks.

marine Pertaining to the sea.

marine biome A major ecological community type characterized by distinctive animal species; *cf.* biome.

marine humus The products of organic decomposition that accumulate in solution or suspension in the sea.

maritime Living by the sea; having a special affinity for the sea.

Markov process A stochastic process where future development is determined only by the present state and is independent of the mode of development of the present state.

marsh An ecosystem of more or less continuously waterlogged soil dominated by emersed herbaceous plants, but without a surface accumulation of peat.

masculus Latin, meaning male; *masc.*

masking gene A gene that modifies or inhibits the expression of another gene; epistatic gene.

mass provisioning The storage, at the time of egg-laying, of a food supply adequate for the entire larval development; *cf.* progressive provisioning.

mass selection Selection by man of desired phenotypes from a population; the culling out of inferior organisms from a breeding stock.

mast year A year in which seed production is exceptionally high, usually followed by a period of several years during which relatively few seeds are produced.

master limiting factor The single ecological factor that actually prevents a population from realizing its biotic potential; master factor; *cf.* paired limiting factor.

master sample A sample used repeatedly for subsequent subsampling.

mastication The act of chewing; manduction, malaxation.

material In taxonomy, the sample available for study.

maternal Pertaining to, or derived from, the female parent; *cf.* paternal.

maternal inheritance Cytoplasmic inheritance; inheritance in which characters or traits in the offspring are determined by cytoplasmic factors, such as mitochondria or chloroplasts, that are carried by the female gamete; matrilinear inheritance; maternal transmission.

maternal sex determination The condition in which the sex of the offspring is determined by the idiotype of the female gamete.

mathematical model The symbolic representation of a number of hypotheses or assumptions about a system in the form of an equation or set of equations; used to describe or predict the behaviour of a system.

mating Pairing of unisexual individuals for purposes of reproduction; **mate**.

mating continuum An aggregate of interbreeding individuals that systematically exchange genes or chromosomes.

mating group A group of haploid or diploid individuals having genetic or environmental characteristics that favour mating within the group at the expense of mating outside the group.

mating system The pattern of matings between individuals of a population, including such factors as extent of inbreeding, pair-bonding, and number of simultaneous mates.

mating type A group of individuals having a genotypically controlled mating capacity, such that they do not usually mate amongst themselves but with individuals of a complementary type.

matric potential The force with which water is held in a matrix such as a soil by capillary action and adsorption to colloids; matric pressure.

matric pressure Matric potential *q.v.*

matriclinous Used of offspring having a greater inherited resemblance to the female parent than to the male; metroclinal; matrilinear; **matriclinic**; **matriclinal**; **matricliny**; *cf.* patriclinous.

matrifocal Pertaining to a social group in which most of the activities and personal relationships are centred on the female parents.

matrilineal Pertaining to the maternal line; passed from mother to offspring; maternal; *cf.* patrilineal.

matrilinear inheritance Maternal inheritance *q.v.*; matrilineal inheritance.

matrix A rectangular array of *m* rows each

containing n numbers or elements arranged in columns (an $m \times n$ matrix); trellis diagram; **matrices**.

matrix transpose A matrix derived from another by the switching of rows and columns so that the rows of one matrix are the columns of the other.

matroclinous Matriclinous *q.v.*; **matroclinal**, **matroclinic**, **matrocliny**.

matromorphic Resembling the female parent; metromorphic; *cf.* patromorphic.

maturation The attainment of sexual maturity; differentiation of the gametes; the increasing complexity or precision of behaviour patterns during growth to sexual maturity, not learned from prior experience.

maturation division Meiosis *q.v.*

matutinal Pertaining to the morning; used of plants that flower in the morning.

maximum entropy inference A quantitative generalization of Occam's razor *q.v.*

maximum homology hypothesis That the optimum reconstruction of a phylogenetic tree is one which maximizes identity due to common ancestry, as indicated by homologous genetic coding sites; red king hypothesis.

maximum illumination The level of illumination on a recording surface set at an angle of maximum illumination at a given depth.

maximum parsimony hypothesis That the optimum reconstruction of ancestral character states is the one which requires the fewest mutations in the phylogenetic tree to account for contemporary character states.

maximum sustainable yield The maximum yield or crop which may be harvested year after year without damage to the system.

meadow An area of closed herbaceous vegetation dominated by grasses.

mean Average; equal to the sum of the observations divided by the number of observations (denoted as a statistic of a sample by $\bar{x}$, and as a population parameter by μ); arithmetic mean.

mean area A quantitative measure of density used in vegetation sampling, calculated as the mean distance between individuals; equivalent to the reciprocal of density.

mean character difference (MCD) A measure of phenetic resemblance, calculated from the absolute values of the differences between OTUs for each character.

mean deviation A measure of variation within a set of data given by the arithmetic mean of the deviation from the mean; average deviation.

mean high water (MHW) The average height of all high waters recorded at a given place over a 19-year period or a computed equivalent period of time.

mean high water neap (MHWN) The average height of all high waters recorded at a given place during quadrature *q.v.* over a 19-year period or a computed equivalent period of time.

mean high water spring (MHWS) The average height of all high waters recorded at a given place during syzygy *q.v.* over a 19-year period or a computed equivalent period of time.

mean higher high water (MHHW) The average height of all higher high waters *q.v.* recorded at a given place over a 19-year period or a computed equivalent period of time.

mean low water (MLW) The average height of all low waters recorded at a given place over a 19-year period or a computed equivalent period of time.

mean low water neap (MLWN) The average height of all low waters recorded at a given place during quadrature *q.v.* over a 19-year period or a computed equivalent period of time.

mean low water spring (MLWS) The average height of all low waters recorded at a given place during syzygy *q.v.* over a 19-year period or a computed equivalent period of time.

mean lower low water (MLLW) The average height of all lower low waters *q.v.* recorded at a given place over a 19-year period or a computed equivalent period of time.

mean phenotypic value The average value of any quantitative character within a population.

mean range The difference in height between mean low water and mean high water.

mean sea level (MSL) The average height of the surface of the sea determined from all stages of the tide over a 19-year period or a computed equivalent period of time; sea-level datum; sea level.

mean square In an analysis of variance, the sum of squares divided by the corresponding number of degrees of freedom.

mean tide level (MTL) The average of the observed heights of high water and low water; ordinary tide level; half tide level.

mean water level (MWL) The mean surface level determined by averaging the height of the water at a given place at hourly intervals over a prolonged period of time.

mechanical isolation Reproductive isolation resulting from the structural incompatibility of male and female reproductive organs that prevents pollination or copulation; a premating isolation *q.v.* mechanism; mechanical incompatibility.

mechanical weathering The breakdown of parent

rock material without chemical changes, as part of the soil forming process.

mechanistic theory That natural processes are mechanistically determined and can be fully explained by the laws of physics and chemistry.

mechanotropism Orientation in response to a mechanical stimulus; **mechanotropic**.

median That value of a variable in an ordered array that has an equal number of observations or items above and below it; second quartile.

Mediocolumbian region Sonoran region *q.v.*

mediterranean 1: Pertaining to the region of the Mediterranean Sea. 2: Living far from the sea.

Mediterranean climate A climate characterized by hot dry summers and mild wet winters.

Mediterranean region A subdivision of the Boreal kingdom; see Appendix 4.

Mediterranean subregion A subdivision of the Palaearctic region; see Appendix 3.

Medithermal period An interval characterized by expanding mountain glaciers (*ca.* 4000–2000 years B.P.); about equivalent to the Subboreal period; Little Ice Age; Hypothermal period; Katathermal period; *cf.* Altithermal period, Anathermal period.

mega- (M) Prefix meaning large, great, greater than usual; used to denote unit $\times$ 10^6.

megaevolution The evolution of higher taxonomic categories; macroevolution.

megafauna 1: Large animals visible to the naked eye. 2: A widespread group of animals. 3: The animals of a large uniform habitat.

megaflora 1: Large plants visible to the naked eye. 2: A widespread group of plants. 3: The plants of a large uniform habitat.

Megagaea A zoogeographical area comprising the Palaearctic, Nearctic, Ethiopian and Oriental regions; Arctogaea; Megagea; *cf.* Neogaea, Notogaea, Palaeogaea.

megagamete The larger of two anisogametes, usually regarded as the ovum or female gamete; macrogamete; *cf.* microgamete.

megagametogenesis The development of megagametes (ova).

megaloplankton Largest of the planktonic organisms, typically greater than 10 mm in diameter.

megaphanerophyte A tall phanerophyte *q.v.* with renewal buds more than 30 m above ground level; macrophanerophyte.

megaphyll A Raunkiaerian leaf size class *q.v.* for leaves having a surface area greater than 164 025 mm².

megaplankton Planktonic organisms of the largest size group, 200 mm–2 m; see Appendix 9.

megascopic Used of a structure or occurrence observable with the naked eye, or with the aid of a hand lens; macroscopic; *cf.* microscopic.

megaspore The larger of two kinds of haploid spores produced by vascular plants; regarded as the female spore; macrospore; *cf.* microspore.

megasporogenesis The production of megaspores in vascular plants.

megatherm A plant favouring warm habitats, and with a requirement for a minimum temperature of 18° C in the coldest month; macrotherm; macrothermophyte; megistotherm; **megathermic**; *cf.* hekistotherm, mesotherm, microtherm.

Megathermal period Hypsithermal period *q.v.*

megecad A group of ecads *q.v.*

megistotherm Megatherm *q.v.*; **megistothermic**.

meio- Prefix meaning smaller, less than.

meiobenthos Small benthic organisms that pass through a 1 mm mesh sieve but are retained by a 0.1 mm mesh; *cf.* macrobenthos, microbenthos.

meiofauna The small interstitial animals that pass through a 1 mm mesh sieve but are retained by a 0.1 mm mesh.

meiosis Reduction division; two successive divisions of a diploid nucleus preceding the formation of the haploid gametes or meiospores, each of which contains one of each pair of the homologous chromosomes of the parent cell; maturation division; gametogenesis; maiosis; miosis; prokaryogamety; **meiotic**.

meiospore A haploid cell produced by meiosis that forms the gametophyte by mitotic division.

meiotaxis The suppression of an entire whorl in phyllotaxis.

meiotherm A plant thriving in cool-temperate habitats; **meiothermous**.

meiotic Pertaining to meiosis *q.v.*

meiotic apogamy The development of a sporophyte plant from the oosphere *q.v.*; haploid apogamy.

meiotic drive Any mechanism operating differentially during meiosis in a heterozygote to produce the two kinds of gametes with unequal frequencies.

meiotic parthenogenesis Parthenogenesis *q.v.* in which meiosis is preserved and diploidy is reinstated either by fusion of haploid nuclei within a single gamete or by the formation of a restitution nucleus; automictic parthenogenesis.

Melanesia A geographical area and an ethnological unit, sometimes considered as a distinct biogeographical unit separate from Polynesia; comprising the continental islands of the West Pacific south of the equator, and some islands of volcanic origin; includes New Guinea, New Caledonia, Fiji, New Hebrides, Bismarck archipelago and the Solomon islands.

Melanesian and Micronesian region A subdivision of the Polynesian subkingdom of the Palaeotropical kingdom; see Appendix 4.

melangeophilous Thriving in or on black loam; **melangeophile, melangeophily.**

melangeophyte A plant growing on black loam; **melangeophyta.**

melanism An increase in the amount of black or dark pigment in an organism, population or group; melanic, melanistic; *cf.* industrial melanism.

meliphagous Feeding on honey; mellivorous; mellisugous; **meliphagus, meliphage, meliphagy.**

melittology The study of bees.

melittophilous Thriving in association with bees; used of an organism that spends part of its life cycle in association with bees; pollinated by bees; mellittophilous; **melittophile, melittophily.**

melliferous Honey-producing.

mellisugous Feeding on honey; honey sucking; meliphagous; mellivorous; **mellisugent.**

mellivorous Feeding on honey; meliphagous; mellisugous; **mellivore, mellivory.**

Mendelian character Any character that is inherited in accordance with Mendel's second law of independent assortment *q.v.*; Mendelian trait.

Mendelian inheritance The inheritance of characters through nuclear chromosomes in contrast to cytoplasmic inheritance *q.v.*

Mendelian population A group of interbreeding individuals sharing a common gene pool; a panmictic unit.

Mendelian ratio The expected ratio of alleles in offspring generations according to Mendel's law of segregation *q.v.*

Mendelian trait Mendelian character *q.v.*

Mendelism Inheritance in accordance with the chromosome theory of heredity; particulate inheritance.

mendelize To undergo Mendelian segregation.

Mendel's laws First law: that in sexual organisms the two members of an allele pair or pair of homologous chromosomes separate during gamete formation and that each gamete receives only one member of the pair; law of segregation. Second law: that the random distribution of alleles to the gametes results from the random orientation of the chromosomes during meiosis; law of independent assortment.

menotaxis A directed response of a motile organism at a constant angle to the stimulus source; **menotactic.**

mensuration The act of measuring; quantitative estimation.

Mercalli scale A numerical scale, indicated by the Roman numerals I to XII, for expressing the intensity of an earthquake.

mercy call Fright cry *q.v.*

merdicolous Living on or in dung; stercoraceous; **merdicole.**

merdivorous Feeding on dung and faecal matter; coprophagous; scatophagous; **merdivore, merdivory.**

meridian A north–south reference line passing through the geographical poles of the Earth.

meridional Southern; pertaining to, or situated in, the South.

meristic variation Variation in the number of structures or parts.

mero- Prefix meaning part, incomplete.

meroepineuston Organisms that spend only part of their life cycle in the epineuston *q.v.*; **meroepineustont, meroepineustonic.**

merogamodeme A part of a relatively isolated, naturally interbreeding, population (gamodeme); a phenodeme *q.v.* of intrapopulation variants.

merogamy A mode of reproduction in protistans involving gametes that are smaller than the vegetative cell and are produced by multiple fission; microgamy; merogony; merogeny; **merogamete;** *cf.* hologamy.

merohermaphroditic population One comprising equal numbers of hermaphrodite and unisexual individuals; metahermaphroditic population.

merohyponeuston Organisms that spend only part of their life cycle in the hyponeuston *q.v.*; **merohyponeustont, merohyponeustonic.**

merological method A study of the component parts of a system in order to derive a concept of the operation of the whole; merological; *cf.* holological method.

meromictic Used of a lake that is permanently stratified due to the presence of a density gradient (pycnocline) resulting from chemical stratification; *cf.* mictic.

meromorphosis Partial regeneration of a lost part or structure.

meroneuston Organisms that spend only part of their life cycle in the neuston *q.v.*; **meroneustont, meroneustonic.**

meroparasite A partial or facultative parasite that can survive in the absence of the host; hemiparasite; *cf.* holoparasite.

meropelagic Used of aquatic organisms that are only temporary members of the pelagic community; *cf.* holopelagic.

meroplankton Temporary members of the planktonic community; hemiplankton; **meroplanktonic;** *cf.* holoplankton.

merostasis Partial paedomorphosis *q.v.*; the

possession of a small number of juvenile characters in an otherwise adult organism.

merotype In taxonomy, a part of the organism furnishing the type specimen.

Mertensian mimicry Mimicry in which a moderately offensive form is the model and a fatally offensive form is the mimic; the fatally offensive species cannot be the model because its lethality does not allow the operator to learn; *cf.* mimicry.

mesad Mesophyte *q.v.*

mesarch succession An ecological succession beginning in a habitat with a moderate amount of water; mesosere; *cf.* hydrarch succession, xerarch succession.

mesic Pertaining to conditions of moderate moisture or water supply; used of organisms occupying moist habitats.

mesic-supralittoral The wettest part of the supralittoral *q.v.* zone of the seashore, occasionally washed by splashing waves; splash zone.

meso- Prefix meaning middle, intermediate.

Meso-Atlantic bridge The proposed land-bridge *q.v.* between the Mediterranean region and the West Indies; Atlantis: Archatlantis.

mesobenthos Those organisms inhabiting the sea bed in the archibenthal zone, between 200–1000 m depth.

mesobiota Soil organisms of intermediate size, from about 40–50 mm in length to a size just visible with the aid of a hand lens; *cf.* macrobiota, microbiota.

mesochthonophilus Thriving in midlands; **mesochthonophile, mesochthonophily.**

mesohabitum Part of an otherwise uniform habitat that differs to a limited extent in one or more environmental factors.

mesohaline Pertaining to brackish water having a salinity between 3–10 parts per thousand or sea water having a salinity between 30–34 parts per thousand; *cf.* oligohaline, polyhaline.

mesohalobous Pertaining to planktonic organisms living in brackish water with a salinity between 5–20 parts per thousand; meshalobous.

mesohydrophyte A plant thriving under damper conditions than a true mesophyte; **mesohydrophytic.**

mesohylile Pertaining to moist forest habitats.

mesokurtic Used of a frequency distribution curve having a close resemblance in peakedness to a normal distribution; **mesokurtosis;** *cf.* kurtosis.

Mesolithic An archaeological period from about 10 000–4000 years B.P.; Middle Stone Age.

mesolithion Organisms inhabiting cavities in rock.

mesolithophagic Lithophagic *q.v.*

mesology Ecology *q.v.*

mesomorphic Possessing intermediate characters or traits; mesoplastic.

mesomorphous Used of plants requiring moderate conditions of moisture, or only partially protected against desiccation.

Meso-Pacific bridge The proposed land-bridge *q.v.* connecting S.E. Asia, Polynesia and Central America during the Palaeogene.

mesopelagic The pelagic zone of intermediate depth, 200–1000 m; see Appendix 7.

mesophanerophyte A medium-sized phanerophyte *q.v.* with renewal buds 8–30 m above ground level.

mesophenic Used of a population exhibiting only a moderately broad range of continuous phenotypic variation; *cf.* monopheny.

mesophilic 1: Thriving under intermediate or moderate environmental conditions; sometimes restricted to conditions of moderate moisture, or to moderate temperature; noterophilous; mesophilous; mesophilus; **mesophil, mesophile, mesophily.** 2: Used of microorganisms having an optimum for growth between 20–45° C.

mesophorbium An alpine meadow community.

mesophyll A Raunkiaerian leaf size class *q.v.* for leaves having a surface area between 2025–18 225 mm²; sometimes subdivided into notophyll (2025–4500 mm²) and mesophyll (4500–18 225 mm²).

mesophyte A plant thriving under intermediate environmental conditions of moderate moisture and temperature, without major seasonal fluctuations; noterophyte; mesad; **mesophyta, mesophytic;** *cf.* hydrophyte, hygrophyte, xerophyte.

mesophythmile Pertaining to the floor of a lake at depths between 6–25 m.

Mesophytic The period of geological time during the first appearance of the angiosperms; *cf.* Aphytic, Archaeophytic, Caenophytic, Eophytic, Palaeophytic.

Mesophyticum The plant life of the Mesozoic; *cf.* Caenophyticum, Palaeophyticum.

mesophytium A plant community able to tolerate arid conditions to only a limited extent; **mesophytism.**

mesoplankton Planktonic organisms of intermediate body size, 0.2–20 mm; see Appendix 9.

mesoplastic Mesomorphic *q.v.*

mesopleustophyte Any large plant floating freely between the surface and the floor of a lake.

mesopsammon Those organisms living in the interstitial spaces of a sandy sediment; psammon.

mesosaprobic Pertaining to a polluted aquatic

habitat having reduced oxygen concentration and a moderately high level of organic decomposition; **mesosaprobe**; *cf.* saprobity system.

mesosere 1: Mesarch succession *q.v.* 2: An eosere *q.v.* of the Mesozoic period.

mesosphere The outer layer of the atmosphere above the stratosphere, 50–85 km above the Earth's surface in which temperature decreases with increasing altitude, typically with efficient vertical convection currents.

mesostatic Used of an ecological succession that reaches a climax under mesic conditions.

mesotherm A plant favouring intermediate temperature conditions, with a minimum of 22° C in the warmest month and a range between 6–18° C in the coldest month; **mesothermophyte**; **mesothermic**; *cf.* hekistotherm, megatherm, microtherm.

mesothermophilous Thriving in temperate regions; **mesothermophile, mesothermophilic, mesothermophily**.

mesothermophyte Mesotherm *q.v.*; **mesothermophyta; mesothermophytia**.

mesothermy The condition in which a maximum occurs in a temperature curve of the water column of a lake; *cf.* dichothermy, poikilothermy.

mesotraphent Used of an aquatic plant characteristic of water bodies with intermediate nutrient concentrations; *cf.* eutraphent, oligotraphent.

mesotrophic 1: Having intermediate levels of primary productivity; pertaining to waters having intermediate levels of the minerals required by green plants; *cf.* dystrophic, eutrophic, oligotrophic. 2: Sometimes used of organisms that are incompletely autotrophic; **mesotrophy**,

mesotropic 1: Used of a mesic ecological succession that tends towards hydric. 2: Exhibiting mesotropism *q.v.*

mesotropism Orientation towards the midline or median plane; **mesotropic**.

mesoxerophytic Used of plants that prefer conditions intermediate between mesic and xeric; **mesoxerophyte**.

Mesozoic A geological era (*ca.* 245–65 million years B.P.) comprising Cretaceous, Jurassic and Triassic periods; see Appendix 1.

messenger RNA RNA that carries the code for a protein from the DNA in the nucleus to the ribosomes, and acts as a template for the synthesis of that protein; mRNA.

meta- Prefix meaning after, change, between, among.

metabiosis A symbiosis between two organisms,

one of which inhabits an environment prepared by the other; **metabios**.

metabolic water That water formed as a product of cellular metabolism.

metabolism 1: The totality of the synthetic and degradative biochemical processes of living organisms. 2: In ecological energetics, sometimes used as equivalent to respiration *q.v.*

metabolizable energy Assimilation *q.v.*

metabolous Used of a pattern of development that includes a metamorphosis.

metacentric Used of a chromosome having the centromere located at or near the middle; *cf.* acrocentric, telocentric.

metachrosis The ability to change colour.

metacommunication A form of communication that conveys information about the interpretation of subsequent communications or signals.

metacyclic Used of a stage in the life cycle of a parasite that infects the definitive host *q.v.*

metafemale A female having an additional female sex chromosome.

metagamic Pertaining to reproductive cycles that alternate between sexual and asexual phases.

metagamic sex determination The condition in which the sex of the offspring is not fixed genetically but is largely the result of external environmental influences.

metagenesis The regular alternation of sexual and asexual diploid generations.

metagynic Used of a hermaphroditic organism that matures first as a male, followed by the transition to a functional female condition; protandrous; **metagyny**.

metahermaphroditic population Merohermaphroditic population *q.v.*

metalimnetic Pertaining to the metalimnion.

metalimnion The zone of steep temperature gradient (thermocline) between the epilimnion and the hypolimnion in a lake.

metallophyte A plant confined to substrates with very high levels of heavy metals; *cf.* pseudometallophyte.

metamerism The division of the body along an anteroposterior axis into a series of segments, each segment containing representatives of all the organ systems of the body; **metameric**.

metamorphic rock Rock formed by the restructuring of pre-existing rocks under the effects of high temperature and pressure; *cf.* igneous rock, sedimentary rock.

metamorphosis A marked structural transformation during the development of an organism, often representing a change from larval stage to adult; **metamorphic**.

metamps Different forms of the same species.

metandric Used of a hermaphroditic organism that matures first as a female, followed by the transition to a functional male condition; protogynous; **metandry**.

metanimic Pertaining to a temporary stream.

metaphoetesis The situation in which a species or individual shifts its position in a food chain or web during its life cycle.

metaphyte 1: A multicellular plant; *cf.* metazoan. 2: A plant with completely differentiated tissues; *cf.* protophyte.

metaplasia 1: An evolutionary state characterized by maximum vigour and diversification of organisms; metaplasis; *cf.* anaplasia, cataplasia. 2: The change of one type of tissue into another; **metaplastic**.

metapopulation A group of conspecific populations coexisting in time but not in space.

metasitism Cannibalism.

metastasis 1: A change of state, form, position, or function. 2: The transportation of pathogenic organisms around the host body.

metathetely The retention of some juvenile characters by an insect adult that has developed through the normal number of moults; a type of neoteny *q.v.*; **metathetelic**.

metatopotype In taxonomy, a metatype *q.v.* from the type locality.

metatrophic Used of organisms that utilize organic nutrients as the source of both carbon and nitrogen.

metatype 1: In taxonomy, a specimen from the type locality determined by the original author subsequent to the original description of the species. 2: A specimen subsequently determined by the author after comparison with the type.

metaxenia The physiological influence of pollen on maternal tissue.

metazoan A multicellular animal.

meteorological tide The change in water level resulting from meteorological factors, largely barometric pressure and wind; **meteorologic tide**.

meteorology The study of weather or local atmospheric conditions; **meteorological**; *cf.* climatology.

methanogenic Methane producing, as of certain autotrophic and chemolithotrophic bacteria.

metochy A neutral symbiosis *q.v.* in which a guest organism freely inhabits the nest of a host species without demands upon the resources of the colony.

metoecious Used of a parasite that is not host specific; heteroecious; **metoecius**.

metonym A taxonomic synonym; a name given to a taxon that already has a valid name.

metonymous homonyms In taxonomy, homonyms based on different nomenclatural types but which are considered to belong to a single taxon.

metoxenous Used of a parasite that occupies different hosts at different stages of the life cycle; heteroecious; heteroxenous.

metre (m) The standard SI base unit of length, defined as 1 650 763.73 of the wavelength emitted by krypton *in vacuo* in the transition $2p_{10}$ to $5d_5$ (orange).

-metric Suffix meaning measure.

metric scaling A technique for ordination *q.v.*; principal coordinate analysis.

metroclinal Matriclinous *q.v.*; **metrocliny**.

metromorphic Resembling the female parent; matromorphic; **metromorphy**.

Mexican subregion A subdivision of the Neotropical region; Tropical North American subregion; see Appendix 3.

micelle The individual particles of the disperse phase of a colloid in which the dispersion medium is a liquid.

Michurinism The theory that the genetic constitution of the scion can be modified by grafting.

micro- (μ) 1: Prefix meaning small, short. 2: Used to denote unit $\times 10^{-6}$.

microaerophilic 1: Thriving in a free oxygen concentration significantly less than that of the atmosphere; micraerophilic; microaerophilous; **microaerophile, microaerophily**. 2: Used of an environment in which the partial pressure of oxygen is significantly below normal atmospheric levels but which is not fully anaerobic *q.v.*

micro-association Sociation *q.v.*

microbenthos Microscopic benthic organisms less than 0.1 mm in length; *cf.* macrobenthos, meiobenthos.

microbiology The study of microorganisms.

microbiota Microscopic soil organisms not visible with the aid of a hand lens; *cf.* macrobiota, mesobiota.

microbivorous Feeding on microorganisms, in particular bacteria; **microbivore, microbivory**.

microclimate The climate of the immediate surroundings or habitat, differing from the macroclimate *q.v.* as a result of the influences of local topography, vegetation and soil; bioclimate.

microcosm A small world; a small community that is taken as representative of a much larger ecological system.

microdichopatric Used of populations that coexist in the same locality but occupy different microhabitats; allotopic; **microdichopatry**; *cf.* dichopatric.

microendemic Used of an organism restricted to a very small area.

microenvironment Microhabitat *q.v.*

microevolution Minor evolutionary events usually viewed over a short period of time, consisting of changes in gene frequencies, chromosome structure or number within a population over a few generations; microphylogenesis; *cf.* macroevolution.

microfauna 1: Small animals not visible to the naked eye. 2: A localized group of animals. 3: The animals of a microhabitat; *cf.* microflora.

microflora 1: Small plants not visible to the naked eye. 2: A localized flora. 3: The plants of a microhabitat; *cf.* microfauna.

microfossil A microscopic fossil.

microgamete The smaller of two anisogametes, usually regarded as the sperm or male gamete; *cf.* macrogamete.

microgametogenesis The development of microgametes (spermatozoa).

microgamy Merogamy *q.v.*

microgeographic races Local races restricted to very small areas, often with considerable migration between adjacent races; components of a subspecies *q.v.*; microsubspecies.

microhabitat A small specialized habitat; microenvironment.

micromelittophilous Pollinated by small bees; micromelittophily.

micromonophyletic A term proposed to describe a higher taxon such as a family derived from a single ancestral taxon of lower rank such as a genus or species; *cf.* macromonophyletic, macropolyphyletic, micropolyphyletic.

micromorphology Microscopic surface topography or structure, typically studied with the aid of a scanning electron microscope; *cf.* macromorphology.

micromutation Point mutation *q.v.*

micromyiophilous Pollinated by small flies; micromyiophily.

micron A derived metric unit of length, equal to 10^{-6} m; (μm).

Micronesia An assemblage of many small oceanic islands within the warmer part of the western Pacific Ocean; a recognized geographical and ethnological area which is sometimes considered to be a distinct biogeographical unit, separate from Polynesia.

micronutrients Trace elements; nutrients required in minute quantities for optimal growth; *cf.* macronutrients.

microorganisms Organisms of microscopic or ultramicroscopic size; commonly including bacteria, blue-green algae, yeasts, some lichens and fungi, protistans, viroids and viruses.

micropalaeontology The study of microscopic

fossils – those fossils ranging in size from one or two centimetres to a few microns, that must be studied by light or electron-microscopy; micropaleontology.

microparapatric Used of populations that coexist in the same locality and in adjacent but not overlapping macrohabitats so that gene flow between them is possible; **microparapatry**; *cf.* parapatric.

microparasite A parasite of microscopic size.

microphagous Feeding on relatively minute particles or on very small prey; microphagic; **microphage, microphagy**; *cf.* macrophagous.

microphanerophyte A small phanerophyte *q.v.* with renewal buds 2–8 m above ground level.

microphilic Thriving only within a narrow range of temperature; **microphil, microphile, microphily**.

microphyll A Raunkiaerian leaf size class *q.v.* for leaves having a surface area between 225–2025 mm².

microphylogenesis Microevolution *q.v.*

microphyte A very small or microscopic plant; **microphytic**.

microphytic Used of a plant community comprising only lichens or algae.

microplankton Small planktonic organisms, 20–200 μm in diameter; see Appendix 9.

micropolyphyletic A term used to describe a higher taxon such as a family derived from two or more ancestral taxa of lower rank such as a genus or species all belonging to an ancestral taxon of the same higher taxonomic rank (i.e. family); *cf.* macromonophyletic, macropolyphyletic, micromonophyletic.

microporous Used of a sediment having small interstitial spaces; *cf.* macroporous.

micropredator An organism that feeds on a prey organism which is larger than itself and to which it attaches temporarily.

microscopic Used of an organism or occurrence that cannot be observed without the use of a microscope; *cf.* macroscopic.

microsere An ecological succession within a microhabitat, often failing to reach a stable climax; serule.

microsmatic Used of an organism possessing a poorly developed sense of smell; *cf.* macrosmatic.

microsomia Dwarfism; nanism.

microspecies 1: A microgeographic race or other infrasubspecific group; jordanon. 2: A component of a species aggregate *q.v.*

microspore The smaller of two kinds of haploid spores produced by vascular plants, regarded as the male spore; *cf.* megaspore.

microsporogenesis The production of microspores

in vascular plants; the production of the male gametophyte (pollen).

microsubspecies Microgeographic races *q.v.*; the small local populations that comprise a subspecies *q.v.*

microsymbiont The smaller of two partners in a symbiotic association; *cf.* macrosymbiont.

microtaxonomy That branch of taxonomy concerned with the classification of species, varieties and populations; *cf.* macrotaxonomy.

microtherm A plant favouring relatively cold habitats with a minimum temperature of 6° C in the coldest month and a range between 10–22° C in the warmest month; **microthermic**; *cf.* hekistotherm, megatherm, mesotherm.

microthermic Used of organisms requiring only a minimum of heat, or favouring relatively low temperatures; **microtherm**.

microthermophilus Thriving in boreal regions; **microthermophile, microthermophily**.

microthermophyte A plant of boreal regions; **microthermophyta**.

microzoogeography The study of the dispersion of geographical races, species, or genera over small areas; *cf.* macrozoogeography.

microzoon A microscopic animal.

microzoophilous Pollinated by small animals, **microzoophily**.

microzoophobous Used of organisms that repel small animals.

mictic 1: Pertaining to patterns of water circulation in a lake; *cf.* amictic, dimictic, holomictic, meromictic, monomictic, oligomictic, polymictic. 2: In genetics, used of females that produce offspring of both sexes by apomixis *q.v.* or amphigony *q.v.*

mictium A heterogenous assemblage of species that often occurs in a transition zone between different vegetation stands, often resulting from the interspersion of two or more seres and having the dominant species of each equally represented.

micton A widely distributed hybrid population which is fully fertile with the parent species.

midden A refuse heap; used especially in archaeology.

Middle Stone Age Mesolithic *q.v.*

middlestorey Understorey *q.v.*

mid-ocean ridge A topographical feature of the ocean floor comprising rifts and mountain ridges assumed to be the sites of ocean floor upwelling and spreading.

midseral Used of communities that have developed beyond the initial stage of a succession but have not reached a subclimax condition.

migrant selection Selection according to the migratory ability of individuals of differing genotypes.

migrarc An arc or circle of migration.

migration 1: Movement of an organism or group from one habitat or location to another; periodic or seasonal movement, typically of relatively long distance, from one area, stratum or climate to another; any general movement that affects the range of distribution of a population or individual; **migrate** *cf.* dispersion. 2: Movement of a pathogen within the host body. 3: Gene flow; exchange of genetic information between populations.

migrule A unit step or agent of migration.

mihi (**m**) Latin, meaning to me; used in taxonomy after a name to indicate that the author was responsible for its proposal.

mile (**mi**) A secondary fps unit of length equal to 1760 yards or $1.609\,344 \times 10^3$ metres; statute mile; see Appendix 13.

milieu The characteristic environs or surroundings of an organism or population.

millennial Pertaining to periods of thousands of years; *cf.* decennial, secular.

millennium A unit of time equal to 1000 years or 1 millicron.

milli (**m**) Prefix meaning thousand, thousandth; used to denote unit $\times 10^{-3}$.

millimetre (**mm**) A metric unit of length equal to 10^{-3} metre.

millimicron A metric unit of length equal to 10^{-9} metre or 10 ångstrom; nanometre.

mimesis Mimicry *q.v.*; **mimetic**.

mimetic polymorphism Polymorphism in which the various morphs resemble other unpalatable or venomous species in order to deceive the operator (predator); often restricted to females.

mimic 1: To imitate; any organism that imitates another. 2: A gene having similar phenotypic effect to a non-allelic gene; *cf.* genocopy.

mimicry The close resemblance of one organism (the mimic) to another (the model) to deceive a third (the operator); mimesis; *cf.* aggressive mimicry, Batesian mimicry, Mertensian mimicry, Müllerian mimicry.

mimotype A form that resembles another phenotypically but not genotypically, and which occupies a similar ecological niche in a different geographical area.

min. parte Abbreviation of the Latin, *pro minore parte*, meaning for the smaller part.

Mindel glaciation A glaciation of the Quaternary Ice Age in the Alpine area, with an estimated duration of 90 thousand years; see Appendix 2.

Mindel-Riss interglacial An interglacial period in

the middle of the Quaternary Ice Age in the Alpine area; see Appendix 2.

mineral replacement A mode of fossilization in which organic material is removed and replaced by minerals such as silica or calcium carbonate.

minerotrophic Used of organisms nourished by minerals.

minimal area The smallest area in which a particular community develops the combination of species characteristic of that community; minimalraum; minimiareal; minimum quadrat area.

minimalraum Minimal area *q.v.*

minimiareal Minimal area *q.v.*

minimum, law of That the minimal requirement in respect to any given factor is the ultimate determinate in controlling the distribution or survival of a species; Liebig's law.

minimum lethal dose (MLD) The minimum dose of an agent sufficient to cause 100% mortality of the test population.

minimum quadrat area Minimal area *q.v.*

minimum spanning tree A two-dimensional graphical representation of all the OTUs linked together through their nearest neighbours.

Miocene A geological epoch within the Tertiary period (*ca.* 26–5 million years B.P.); see Appendix 1.

miosis Meiosis *q.v.*

miothermic Used of geological periods characterized by climatic temperatures similar to those found at present, as opposed to warmer (pliothermic) or colder periods; pertaining to present day temperature conditions.

mire An ecosystem in which the vegetation is rooted in wet peat; a collective term for various bogs and fens.

misapplication In taxonomy, the use of a scientific name for a taxon which does not include the type within its range of variation.

misapplied name In taxonomy, a name given to a taxon which excludes the nomenclatural type of the name.

miscible Used of liquids that can be mixed together in any ratio without separation; miscibility.

misidentification An erroneous assignment of a specimen or group to a particular taxon.

misogamy Reproductive isolation *q.v.*

misquotation A reference to the wrong paper; *cf.* bibliographic error.

missense mutation A point mutation *q.v.* within a codon resulting in the incorporation of a different amino acid into the polypeptide chain, producing an inactive or unstable enzyme; *cf.* nonsense mutation, samesense mutation.

mitoschisis Mitosis *q.v.*

mitosis The process of chromosome division and separation that takes place in a dividing cell, producing daughter cells of equivalent chromosomal composition to the parent cell; typically four main stages are recognized: prophase, metaphase, anaphase and telophase; cytodiaeresis; karyokinesis; karyomitosis; mitoschisis; **mitotic**.

mitospore A diploid spore produced by mitosis.

mitotic index The proportion of cells within a tissue that are undergoing mitosis at any given time.

mitotic poison A substance that blocks mitosis.

mixed association An association in which several species compete for dominance.

mixed layer A layer of water well mixed by wave action or thermocline convection.

mixed nest A nest containing two or more social groups in which both adults and larvae intermingle; *cf.* compound nest.

mixis The fusion of gametes; karyogamy and karyomixis.

mixolimnion The upper low density region of periodic free circulation above the monimolimnion *q.v.* in a meromictic lake.

mixoploidy The condition in which the different cells or tissues of a mosaic *q.v.* or chimaera *q.v.* differ in chromosome number; **mixoploid**.

mixotrophic Used of organisms that are both autotrophic and heterotrophic; mesotrophic; **mixotroph, mixotrophy**.

mks system The system of scientific units based on the metre-kilogram-second as in the SI units *q.v.*

MLD Minimum lethal dose *q.v.*

mnemotaxis A directed response of a motile organism to a memory stimulus; **mnemotactic**.

mobbing A collective attack by a group of animals, on a predator that is too large or aggressive to be repelled by individual effort.

mobility The tendency of an organism or population to change its location or distribution with time; vagility; **mobile**.

modality 1: The state or quality of a stimulus or sensation. 2: In statistics, the state of having one (unimodal), two (bimodal) or more (polymodal) modes or peaks in a frequency distribution.

mode The value which occurs most frequently in a given set of data or distribution.

model 1: A mathematical formulation intended to represent a natural phenomenon or system. 2: A noxious or unpalatable organism displaying distinctive signals for recognition and which is imitated by a mimic.

modification Any non-heritable environmentally induced change in a phenotype.

modified mean In statistics, the mean of the remaining values after the omission of one or more items judged to be atypical of the data.

modifier Any gene that alters the expression of another gene at a different locus; modifying factor.

modular unit Each of the components of a society *q.v.*; member.

moisture equivalent The percentage weight of water that is retained by an oven dried soil sample subjected to a centrifugal force equal to 1000 times that of gravity for a specified period; ranges from about 2% for coarse sand to about 40% for heavy clay.

moisture stress The tension or force with which water is retained in the soil.

mole (mol) The standard SI base unit of amount of substance, defined as the amount of substance of a system which contains as many elementary entities as there are atoms in 0.012 kg of carbon-12 isotope.

molecular clock model of evolution The hypothesis that point mutations *q.v.* occur at a sufficiently regular interval to permit the dating of phylogenetic dichotomies; it assumes a direct relationship between the extent of molecular divergence and the time of ancestral separation of the two branches.

molecular fossil Biogenic molecules comprising the more stable remnants of larger organic molecules from organisms that have been degraded during fossilization; chemicofossil.

molecular systematics The study of organisms and their interrelationships using biochemical characters, and employing techniques such as electrophoresis, chromatography, serology and DNA hybridization; the study of evolutionary relationships using comparative molecular data.

molluscicide A chemical used to kill snails, slugs and other molluscs.

moltinism Moultinism *q.v.*

monacmic Exhibiting one abundance peak per year; *cf.* diacmic, polyacmic.

monandrous Used of a female that mates with a single male; **monandry**; *cf.* polyandrous.

monarchistic dominance The dominance of an intraspecific group by one individual which inhibits or represses the interactions between other members of the group; monarchy.

monarchy Monarchistic dominance *q.v.*

monarsenous Polygamous *q.v.*

monaxenic Used of a mixed culture of an organism with one prey species; *cf.* axenic, dixenic.

monecious Monoecious *q.v.*

moneran Pertaining to the prokaryotic organisms of the kingdom Monera; *cf.* protistan.

monimolimnion The perennially stagnant high density, usually saline, deep water layer in a meromictic lake *q.v.*; *cf.* mixolimnion.

monistic evolution, theory of That evolutionary changes can be explained by single factors or events; *cf.* synthetic evolution, theory of.

mono- Prefix meaning one, single.

monoallelic Used of a polyploid in which all the alleles at a given locus are similar.

monobasic In taxonomy, used of a genus established on, or comprising, a single species; monotypic; *cf.* polybasic.

monocarpic Producing a single fruit, or having only one fruiting period, during the life cycle; monotokous; *cf.* polycarpic.

monochronic Occurring only once.

monoclimax The single major climatic climax of a geographical region; *cf.* polyclimax.

monoclinous Used of a flower having both male and female organs; hermaphrodite; perfect; **monocliny**; *cf.* diclinous.

monoculture The cultivation or culture of a single crop or species to the exclusion of others.

monocyclic Used of a species having two (bivoltine) or more (multivoltine) broods or generations per year, that changes its mode of reproduction once a year, typically to produce resting stages.

monocyclic parthenogenesis Reproduction by a series of parthenogenetic generations alternating with a single sexually reproducing generation in each annual cycle; cyclic parthenogenesis; holocyclic parthenogenesis; *cf.* parthenogenesis.

monodelphous homonyms In taxonomy, devalidated names taken up by different post-starting point authors independently but validated with different nomenclatural types.

monodomic Used of colonies of social insects that occupy only one nest; monodomous; *cf.* polydomic.

monoecious 1: Used of individuals having both male and female reproductive organs; hermaphrodite; bisexual. 2: Used of a bisexual plant species having separate male and female flowers (unisexual) on the same individual; autoicous; ambisexual; monecious; **monoecism**, **monoecy**; *cf.* dioecious, trimonoecious, trioecious.

monoecious polygamy The possession of both single sexed and perfect flowers on the same individual; *cf.* dioeciopolygamy.

monoestrous Having a single breeding period in a sexual season; *cf.* anoestrus, dioestrus, polyoestrus.

monofactorial Pertaining to phenotypic characters controlled by, or to sex determined by, a single gene; monogenic, unifactorial; *cf.* polyfactorial.

monogamous Pertaining to the condition in which a single male and female form a prolonged and more or less exclusive breeding relationship; **monogamic, monogamy**; *cf.* polygamous.

monogenesis 1: Asexual reproduction; monogony; monogenetic, monogenous. 2: The development of organisms from a single cell; unigenesis.

monogenetic 1: Pertaining to a symbiont having only one host throughout its life cycle; *cf.* digenetic, trigenetic. 2: Used of asexually reproducing organisms, or asexual reproduction; monogenous.

monogenic 1: Used of characters or traits controlled by a single gene; monofactorial, unifactorial; *cf.* digenic, oligogenic, polygenic, trigenic. 2: Producing only male or only female offspring; **monogeny**; *cf.* amphogenic, arrhenogenic, thelygenic, allelogenic.

monogenous Pertaining to asexual reproduction or asexually reproducing organisms.

monogoneutic Producing only one brood per year or season; univoltine; **monogoneutism**; *cf.* digoneutic, trigoneutic, polygoneutic.

monogony Asexual reproduction; monogenesis.

monograph A comprehensive published work on a single group or subject; **monographic**.

monogynopaedium An assemblage of one female parent with her immediate progeny; *cf.* polygynopaedium.

monogyny 1: The mating of a male with only one female. 2: The presence of a single queen in a colony of social insects; **monogynous**; *cf.* oligogyny, polygyny.

monohybrid cross A cross between two individuals heterozygous for only one pair of alleles; *cf.* dihybrid cross.

monolepsis The condition in which the characters of only one parent are transmitted to the progeny.

monometrosis The founding of a colony of social organisms by a single fertile female, as from a queen in some social insects; haplometrosis; **monometrotic**; *cf.* pleometrosis.

monomial Uninominal name *q.v.*; mononym.

monomictic Used of a lake having a single period of free circulation or overturn per year, with consequent disruption of the thermocline; may be either cold monomictic or warm monomictic; *cf.* mictic.

monomorphic 1: Pertaining to a population or taxon showing no genetically fixed discontinuous variation, therefore comprising a single discrete morph; continuous (unimodal) variation may

occur within the population with an extremely broad (latiphenic), moderately broad (mesophenic) or narrow (leptophenic) range; *cf.* dimorphic, polymorphic. 2: Having only one type of flower on an individual plant; *cf.* dimorphic. 3: Used of a population, with respect to a particular allele that is present in all members of the population.

monomorphism 1: In social insects, the presence of a single worker caste in a colony. 2: The state of being monomorphic *q.v.*

monomorphous life cycle A life cycle in which successive generations comprise only one type of individual and a single reproductive strategy; *cf.* dimorphous life cycle.

monophagous Utilizing only one kind of food; feeding upon a single species or food plant; host specific; monotrophic; univorous; **monophage, monophagy**; *cf.* oligophagous, polyphagous.

monopheny The occurrence of a single phenotype in a population; continuous (unimodal) variation may occur within a population with an extremely broad (latiphenic), moderately broad (mesophenic) or narrow (leptophenic) range; **monophenic, monophenism**; *cf.* dipheny, polypheny.

monophyletic Derived from the same ancestral taxon; used of a group sharing the same common ancestor; **monophyly, monophyletism**; *cf.* diphyletic, polyphyletic.

monophyletic group 1: A group of taxa descended from a single ancestral taxon. 2: Sometimes used in a restricted sense to include the ancestral species and all descendant species; a category based on the common possession of apomorphic *q.v.* characters; holophyletic group; monophyletic assemblage; *cf.* paraphyletic group, polyphyletic group.

monophylistic Pertaining to phylogenetic relationships between taxa based on both monophyly and genotypic similarity; **monophylist, monophylistics**.

monoplanetic Having a single motile stage during a life cycle; **monoplanetism**; *cf.* diplanetic, polyplanetic.

monoploid Having a single set of chromosomes with a basic chromosome number; equivalent to the lowest haploid number of a polyploid series; **monoploidy**; *cf.* diploid, haploid, polyploid.

monosomic Pertaining to an aneuploid *q.v.* having an abnormal chromosome complement in which one chromosome of one homologous pair is absent; **monosomy, monosome**.

monospermy The penetration of an ovum by a single sperm during normal fertilization; **monospermic**; *cf.* disLaspermy, polyspermy.

monotaxic Belonging to the same taxonomic group.

monothelious Polyandrous *q.v.*; **monothely**.

monothetic key A dichotomous key in which each couplet has single contrasting statements requiring a simple either/or answer; *cf.* polythetic key.

monothetic taxon A taxon specified by a unique combination of diagnostic characters which comprise both necessary and sufficient criteria for identification; *cf.* polythetic taxon.

monotokous Having only one offspring per brood; fruiting only once during a life cycle; monocarpic; uniparous; monotocous; *cf.* ditokous, oligotokous, polytokous.

monoton plankton Plankton very heavily dominated by one particular species; *cf.* pantomictic plankton, polymictic plankton, prevalent plankton.

monotopic Occurring or arising in a single locality or geographical area; **monotopy**; *cf.* polytopic.

monotrophic Monophagous *q.v.*; **monotrophy**.

monotropism Orientation movement in one direction only; **monotropic**.

monotypic Used of a taxon comprising a single immediately subordinate principal taxon, as in a family or genus comprising a single genus or species respectively; monobasic; *cf.* bitypic, polytypic.

monotypy In taxonomy, the establishment of a genus or subgenus upon a single species; or of a species upon a single specimen; **monotype, monotypic**; *cf.* polytypy.

monovoltine Univoltine *q.v.*

monovular Monozygotic *q.v.*

monoxenous Used of a parasite utilizing a single host species during its life cycle; **monoxeny**; *cf.* dixenous, heteroxenous, oligoxenous, trixenous.

monozygotic Used of twins derived from a single fertilized egg; such twins are thus genetically identical and of the same sex; identical (twins); monovular; uniovular; *cf.* dizygotic.

monsoon A seasonal climatic pattern in which a cool dry period alternates with a hot wet period; also used of the vegetation of such a climatic region.

Monsoon current A warm surface ocean current that flows east in the northern part of the Indian Ocean produced by the south-westerly winds of the summer monsoon; North East Monsoon Drift; see Appendix 6.

monstrosity Any abnormal, malformed, or markedly aberrant individual.

montane Pertaining to mountainous regions; the cool moist upland habitat below the tree line, dominated by evergreen trees.

Monte Carlo method A method for the solution of statistical problems involving random sampling, typically from a simulated population on a computer.

monticolous Living in mountainous habitats; **monticole**.

mood The preliminary state of charge or readiness for action, necessary to the performance of a given course of instinctive behaviour.

moor An open elevated region of wet acidic peat typically dominated by heathers, sedges and some grasses; moorland.

mor A forest soil type characterized by a discrete humus layer separated from the underlying mineral soil, typically highly acidic; a raw humus produced under cool moist conditions; mar; *cf.* mull.

mores Groups of organisms having similar ecological requirements and behavioural attributes; mune.

Morgan unit The distance along a chromatid that has an average of one crossing over per meiosis or a recombination frequency of 1%; used as a measure of relative distance between linked genes; map unit.

moribund Dying; close to death.

Morisita's index of dispersion An index of dispersion that is independent of the sample mean, calculated as $I = (\Sigma(x^2) - \Sigma x)/((\Sigma x)^2 - \Sigma x)$. The index has a value equal to, greater than, or less than, unity for random, contagious and uniform distributions respectively.

-morph- Combining affix meaning form, shape.

morph A form; any of the individuals of a polymorphic group; any phenotypic or genetic variant; any local population of a polymorphic species exhibiting distinctive morphology or behaviour.

morphallaxis The process of limb or tissue regeneration in which a novel or alternative tissue or appendage is developed.

morphism Polymorphism *q.v.*

morphocline A graded series of character states of a homologous character; morphological transformation series.

morphocline polarity The primitive/derived orientation of a graded series of character states; can be unidirectional or multidirectional.

morphogenesis 1: The totality of the process of embryological development and growth. 2: The evolution of morphological structures; **morphogenetic, morphogeny**.

morphological correspondence Structural or functional similarity in two or more organisms inferred from morphological, anatomical and embryonic considerations.

morphological series A graded series of fossils showing continuous variation in a given character or suite of characters.

morphological species concept The concept of a species (a morphospecies) based on morphological characters alone, without consideration of other biological factors; *cf.* biological species concept, genetic species concept.

morphological transformation series Morphocline *q.v.*

morphology 1: The study of form and structure of organisms. 2: The form and structure of an organism, with special emphasis on external features; **morphologic, morphological**. 3: The structural features of rocks and sediments.

morphometry The measurement of external form and structure; the characterization of form for quantitative analysis; **morphometric**.

morphon A definitively formed individual.

morphoplankton Planktonic organisms rendered buoyant by anatomical specializations such as oil droplets or gas vesicles, or in which the rate of sinking is reduced by structural features or diminutive body size.

morphoplasy The morphological potential or plasticity of a developing organ or organism; **morphoplastic**.

morphosis Variation in morphogenesis of an individual induced by environmental changes.

morphospecies A species established solely on morphological characters.

morphotype 1: In taxonomy, a specimen selected to represent a given intrapopulation variant (morph). 2: A list of the morphological characters presumed present in an ancestral species.

mortality Death rate as a proportion of the population expressed as a percentage or as a fraction; mortality rate; often used in a general sense as equivalent to death; *cf.* natality.

mos A community or assemblage of species living together but without mutual interdependence.

mosaic An organism comprising tissues of two or more genetic types; usually used with reference to plants; chimaera.

mosaic chart A map of background vegetation types plotted as a graph, the two axes of which are selected environmental gradients.

mosaic evolution The change of different adaptive structures, traits or other components of the phenotype at different times or at different rates in an evolutionary sequence; non-harmonious character transformation.

motile With the capacity for movement; exhibiting movement; motility.

motivation A behavioural drive directed towards a specific goal or need.

mould A cavity or space in a sediment remaining after the organic material of an organism has been removed by the action of soil water; mold.

mouldering Decomposition of organic matter under conditions of low oxygen resulting in the formation of carbon-rich residues; *cf.* decay, putrefaction.

moultinism A polymorphism in which different strains have a different number of larval stages; moltinism.

Mozambique Current A warm surface ocean current that flows south through the Mozambique Channel; see Appendix 6.

mRNA Messenger RNA *q.v.*

MS Abbreviation of the Latin *manuscripta*, meaning manuscript *q.v.* (MSS, manuscripts).

mucilaginous Comprising or resembling mucilage; muculent.

mucivorous Feeding on plant juices or mucilage; **mucivore, mucivory**.

muck Highly decomposed plant material typically darker and with higher mineral content than peat.

mud Fine sediment particles of diameter less than 0.002 mm; fine grained, slimy, marine or lacustrine detrital sediment; see Appendix 11.

muddy shore biome A shore biome characterized by mobile fine sediments, typically having relatively rich infauna; *cf.* oceanic biome, rocky shore biome, sandy shore biome.

mull A forest soil type in which the organic matter is intermixed with mineral soil without a discrete humus layer; *cf.* mor.

Müllerian association An assemblage of different species at a particular site, all of which display similar aposematic coloration *q.v.*

Müllerian mimicry Imitative similarity, typically based on warning coloration amongst a number of mimic species all of which are unpalatable or otherwise offensive to a predator (the operator); Muellerian mimicry; *cf.* mimicry.

Muller's rachet A mechanism incorporated in an asexual population that prevents any lines of the population from attaining a mutation load smaller than that already existing in its least-loaded lines.

multi- Prefix meaning many.

multi-access key Multiple-entry key *q.v.*

multidimensional species concept The concept of a species as a multi-population system; although several distinct morphospecies may be recognized, without phenotypic intermediates, this concept regards all such populations as members of a single dispersed species network in which morphological variants are replacing each other geographically; *cf.* nondimensional species concept, typological species concept.

multifactorial Pertaining to phenotypic characters controlled by the integrated action of multiple independent genes; multigenic; polyfactorial; polygenic; *cf.* unifactorial.

multiform Polymorphic.

multigeneric Pertaining to a cross involving three or more different genera; *cf.* bigeneric.

multigenic Used of characters or traits controlled by the integrated action of several independent genes; multifactorial; polygenic; polyfactorial; *cf.* digenic, monogenic.

multihybrid An individual heterozygous at more than one locus.

multiparous 1: Producing two or more offspring in a single brood; multiparity; *cf.* biparous, uniparous. 2: Having produced more than one previous brood.

multiple alleles Two or more different forms of a gene occupying a specific locus; multiple allelomorphs.

multiple factors Polygene *q.v.*

multiple fission A variant of binary fission *q.v.* found in some protistans in which a subsequent division commences before the earlier one is complete.

multiple gene inheritance Inheritance of a character or trait controlled by more than one gene.

multiple regression A method of relating one dependent variable to several independent variables.

multiple-entry key An identification key, often utilizing some form of punch card system, which allows the operator freedom to select any character in any sequence from a given list; multi-access key; random-access key; polyclave.

multistate character A character having three or more different states; *cf.* binary character.

multistratal Pertaining to vegetation composed of several horizontal layers; *cf.* unistratal.

multivalent An association of three or more chromosomes synapsed during the first meiotic division.

multivariate analysis In statistics, a group of techniques for the simultaneous analysis of more than one independent variable, including analysis of variance and of covariance, regression and correlation methods.

multivariate data Data in which an observation is characterized by more than one independent variable.

multivoltine Having several broods or generations per year; polyvoltine; polygoneutic; *cf.* voltine.

mune A group of organisms having similar ecological requirements and a characteristic behavioural response; mores.

mural Pertaining to, or growing on, walls; rupestral; murarius.

Murphy's law That bills of birds inhabiting islands are longer than those in populations of the same species living on the mainland.

mus. Abbreviation of the Latin *museum*, meaning museum.

muscarian Used of flowers that attract flies by a putrid odour.

muscicolous Living on or in mosses or moss-rich communities; **muscicole, muscicoline**.

muscology The study of mosses; bryology.

muskeg A grassy bog habitat, with scattered, stunted conifers.

mut. char. Abbreviation of the Latin *mutatis characteribus*, meaning with the characters changed (by); equivalent to emend *q.v.*

mutable gene An unstable gene that mutates spontaneously during the course of development with a relatively high frequency.

mutagen Any agent that produces a mutation or enhances the rate of mutation; **mutagenic, mutagenesis**.

mutant Any organism, gene, or character that has undergone a mutational change.

mutant type Any type derived from the wild type by a mutational change.

mutation 1: A sudden heritable change in the genetic material, most often an alteration of a single gene by duplication, replacement or deletion of a number of DNA base pairs. 2: An individual that has undergone such a mutational change; mutant.

mutation frequency The proportion of mutant alleles in a population.

mutation pressure A measure of the effect of mutation in changing the gene frequency in a population.

mutation rate The probability of a particular mutation occurring per organism per population.

mutational load The reduction in population fitness *q.v.* due to the accumulation of deleterious genes resulting from recurrent mutation; *cf.* genetic load.

mutationism De Vriesianism *q.v.*

mutatis characteribus Mut. char. *q.v.*

mutilous Harmless; without defensive structures.

muton The mutational site within a gene; the smallest part of DNA that will produce a mutation when transformed.

mutualism A symbiosis in which both organisms benefit, frequently a relationship of complete dependence; interdependent association; synergism.

M-V linkage Linkage between genes determining morphological characters and genes controlling viability.

-mycetes The ending of a name of a class of fungi.

-mycetidae The ending of a name of a subclass of fungi.

mycetogenic Produced by fungi.

mycetology The study of fungi; mycology.

mycetophagous Feeding on fungi; fungivorous; mycophagous; **mycetophage, mycetophagy**.

mychogamy Self-fertilization *q.v.*

mycobiont The fungal partner of a lichen (algal/fungal symbiosis); *cf.* phycobiont.

mycobiota The fungal flora of an area or habitat.

mycocriny The decomposition of organic material by the action of fungi.

mycology The study of fungi; mycetology.

mycophagous Feeding on fungi; fungivorous; mycetophagous; **mycophage, mycophagy**.

mycophthorous Used of organisms that destroy fungi.

mycorrhiza The association between a fungus and the root system of a plant.

-mycota The ending of a name of a division of fungi.

mycotic Produced by the agency of fungi; **mycosis**.

-mycotina The ending of a name of a subdivision of fungi.

mycotrophic Used of plants living in a symbiosis with a fungus, and nutritionally dependent upon the fungus, as in those plants with mycorrhizal associations; **mycotrophy**.

myiophilous Pollinated by flies; **myiophily**.

myria- 1: Prefix used to denote unit $\times 10^4$.
2: Prefix meaning many, innumerable.

myrmeco- Combining form meaning ant; often used also to include termites.

myrmecobromous Myrmecotrophic *q.v.*

myrmecochorous Dispersed by the agency of ants; **myrmecochore, myrmecochory**.

myrmecoclepty A symbiosis between ant species in which the guest species steals food directly from the host species.

myrmecodomatia Plant structures inhabited by ants or termites.

myrmecodomus Used of a plant affording shelter to ants.

myrmecology The study of ants.

myrmecolous Living in ant or termite nests; **myrmecocole**.

myrmecophagous Feeding on ants or termites; **myrmecophage, myrmecophagy**.

myrmecophilous 1: Thriving in association with ants; used of organisms that spend part of their life cycle in ant or termite nests; **myrmecophile, myrmecophily**; *cf.* myrmecophobous.
2: Pollinated by ants or termites.

myrmecophobous Used of organisms that repel ants or termites; **myrmecophobe, myrmecophobic, myrmecophoby**; *cf.* myrmecophilous.

myrmecophyte A plant having specialized structures for sheltering ants or termites, or having a mutual interdependence with ants or termites; myrmecoxenous plant.

myrmecosymbiosis A symbiosis between an ant and its host plant; **myrmecosymbiotic**.

myrmecotrophic Pertaining to plants and animals that provide food for ants; myrmecobromous.

myrmecoxenous Used of plants that provide both food and shelter for ants and termites; **myrmecoxeny**.

myxotrophic Used of an organism obtaining nutrients through the ingestion of particles; **myxotrophy**.

N

n. 1: Abbreviation of the Latin *nobis*, meaning to (of) us. 2: Abbreviation of the Latin *nomen*, meaning name. 3: Abbreviation of the Latin *novus*, meaning new. 4: Abbreviation of the prefix nano, denoting unit $\times$ 10^{-9}.

n.n. Abbreviation of the Latin *nomen novum q.v.* or *nomen nudum q.v.*

n.nov. Abbreviation of the Latin *nomen novum*, meaning new name.

n.sp. Abbreviation of the Latin *species nova*, meaning new species.

n.v. Abbreviation of the Latin *non visus*, meaning not seen.

nadir The point on the celestial sphere opposite the zenith *q.v.*

naive Used of an animal with no previous experience of infection with a particular parasite.

naked flower A flower lacking both stamens and pistils; sterile flower.

namatad A brook or stream plant; namatophyte.

namatium A brook or stream community.

namatophilus Thriving in brooks and streams; **namatophile, namatophily**.

namatophyte A plant inhabiting a stream or brook; namatad; **namatophyta**.

name-group In zoological nomenclature, an assemblage of coordinate categories.

nanander A dwarf male; nannander.

nanism The condition of being stunted or smaller than normal, or of having restricted growth; dwarfism; microsomia; nanoid.

nanno- Nano- *q.v.*; dwarf.

nano- (n) Prefix denoting unit $\times$ 10^{-9}; nanno-.

nanofossils Very small marine fossils, usually algae, typically close to the resolution limit of a light microscope and studied by electron microscopy; nannofossil.

nanometre (nm) A unit of length equal to 1×10^{-9} metre, or 10 ångstrom; millimicron.

nanophanerophyte A dwarf phanerophyte *q.v.* with renewal buds less than 2 m above ground level; nannophanerophyte; **nanophanerophytium**.

nanophyll A Raunkiaerian leaf size class *q.v.* for leaves having a surface area between 25–225 mm^2.

nanoplankton Minute planktonic organisms with a body diameter between 2–20 μm; nannoplankton; formerly used for those organisms that passed through a 0.03–0.04 mm mesh silk bolting cloth; see Appendix 9.

NAR Nett assimilation rate *q.v.*

narcotropism Orientation movements resulting from the effects of narcotics; **narcotropic**.

nascence The origin or commencement of a community in a previously barren area.

nastic Pertaining to plant movement in response to a diffuse stimulus, and to structural curvature resulting from differential growth of opposite surfaces; **nasty**; *cf*. autonasty, chemonasty, epinasty, geonasty, hyponasty, nyctinasty, photoepinasty, photohyponasty, photonasty, seismonasty, thermonasty, traumatonasty.

natality Birth rate; the number of offspring produced per female or head of the population per unit time; the production of new individuals by birth, germination or fission; natality rate; *cf*. mortality.

natatorial Adapted for swimming; natatory; natatorious; **natant**.

native Indigenous; living naturally within a given area; used of a plant species that occurs at least partly in natural habitats and is consistently associated with certain other species in these habitats.

natural classification A hierarchical classification based on hypothetical phylogenetic relationships such that the members of each category in the classification share a single common ancestor; phylogenetic classification; *cf*. artificial classification.

natural erosion Geological erosion *q.v.*

natural group An assemblage of organisms related by descent from a common ancestor; a monophyletic group.

natural history The study of nature, natural objects and natural phenomena.

natural key An identification key constructed from a natural classification and indicating the supposed evolutionary relationships of the group within the branching sequences of the key; *cf*. artificial key.

natural period The fundamental periodicity of a biological clock when not entrained to a zeitgeber *q.v.*; free-running period.

natural selection The non-random and differential reproduction of different genotypes acting to preserve favourable variants and to eliminate less favourable variants; viewed as the creative force that directs the course of evolution by preserving those variants or traits best adapted in the face of natural competition.

naturalized Used of an alien or introduced species that has become successfully established.

nautical mile (International) A secondary SI unit of length equal to 1852 metres; the average distance on the Earth's surface subtended by one minute of latitude; see Appendix 13.

nautical mile (UK) A secondary fps unit of length equal to 6080 feet (1.853184 km) (British Hydrographic Office) or 6080.27 feet (1.853266 km) (US Coastal Survey); the average distance on the Earth's surface subtended by one minute of latitude; see Appendix 13.

neallotype In taxonomy, a newly designated type specimen selected in the absence of extant type material, of opposite sex to the neotype.

nealogy The study of young and immature organisms.

neanic 1: Adolescent; pertaining to the larval phase that precedes the adult condition. 2: Used of characters which first appeared, in evolution, in early ontogenetic stages; *cf.* ephebic.

neanotype In taxonomy, the type of a pupal stage.

neap tide The tide of minimum range occurring at the time of first and third quarters of the moon, when the gravitational attraction of the sun and the moon act at right angles to each other during quadrature *q.v.*; *cf.* spring tide.

Nearctic region A zoogeographical region comprising North America, Greenland and northern Mexico; subdivided into Alleghany, Californian, Canadian and Rocky Mountain subregions; see Appendix 3.

nearest neighbour method A method of plotless sampling of populations in which the distance is measured from an individual (chosen on the basis of its proximity to a randomly selected point) to its nearest neighbour, and the procedure repeated; population density estimates can be based on these measurements as for example, by the formula $m = 1/4 \, r^2$ where m is the mean density per unit area and r is the mean distance between nearest neighbours.

Nebraskan glaciation A glaciation of the Quaternary Ice Age in North America with an estimated duration of 100 thousand years; see Appendix 2.

nec Latin, meaning not (of); nor (of).

necrocoenosis An assemblage of dead organisms; liptocoenosis.

necrocoleopterophilous Pollinated by carrion beetles; **necrocoleopterophily**.

necrogenic abortion The rapid death of the tissue of a plant immediately proximal to the point of attack of a pathogen, thus preventing further spread of the pathogen.

necrogenous 1: Growing on, or inhabiting, dead bodies. 2: Used of organisms or factors that premote decay.

necrogeography The study of the distribution of dead organisms; thanatogeography.

necrology The study of decomposition, fossilization and other processes affecting plant and animal remains after death.

necron Dead plant material not yet decomposed to form humus; dead organisms.

necrophagous Feeding on dead material; saprophytic; necrophilous; **necrophage**, **necrophagy**.

necrophilous Necrophagous *q.v.*; **necrophile**, **necrophily**.

necrophoresis The transport of dead individuals away from a colony, as in some social insects.

necrophytophagous Feeding on dead plant material; nekrophytophagous; **necrophytophage**, **necrophytophagy**.

necrotrophic symbiosis A symbiosis established between two living organisms in which one symbiont continues to use the other as a food source even after complete or partial death has occurred; *cf.* biotrophic symbiosis.

necrotype A fossil; an extinct organism.

nectariferous Nectar producing.

nectarivorous Feeding on nectar; **nectarivore**, **nectarivory**.

nectism Swimming by means of cilia.

nectobenthic Swimming off the sea bed; nektobenthic.

nectobenthos Nektobenthos *q.v.*; nectobenthic.

necton Nekton *q.v.*

negative assortative mating Disassortative mating *q.v.*

negative binomial distribution A mathematical distribution used as a model of an aggregated or contagiously distributed population in which the presence of an individual at any given point increases the probability of another individual occurring nearby and in which the variance is greater than the mean.

negative eugenics Improvement of the genetic constitution of a population by the prevention of reproduction in individuals with undesirable heritable characteristics; *cf.* positive eugenics.

neglected name *Nomen neglectum q.v.*

neidioplankton Planktonic organisms possessing some form of swimming apparatus.

neighbourhood Used in population genetics for the area around an individual within which the gametes that produced the individual can be considered to have been drawn at random.

neighbouringly sympatric Used of populations that inhabit different habitats within the same geographical area; **neighbouring sympatry**; *cf.* biotically sympatric.

nekrophytophagous Feeding on dead plant material; necrophytophagous; **nekrophytophage, nekrophytophagy**.

nektobenthic Swimming off the sea bed; nectobenthic.

nektobenthos Organisms typically associated with the benthos that swim actively in the water column at certain periods; nectobenthos; **nektobenthont, nektobenthonic, nektobenthic**.

nektogenic Derived from the nekton *q.v.*

nekton Those actively swimming pelagic organisms able to move independently of water currents; typically within the size range 20 mm–20 m, divided into three categories on the basis of body size – centimetre nekton, decimetre nekton, metre nekton; necton; **nektonic**; *cf.* plankton.

nematicide A chemical used to kill nematode worms.

nematodology Nematology *q.v.*

nematology The study of nematodes; **nematodology**.

nemoricolous Living in open woodland; nemoral; nemorose; **nemoricole**.

nemorose Living in open woodland; nemoricolous.

neo- Prefix meaning new.

neo-Adansonism Phenetic method *q.v.*; neo-Adansonian.

neobiogenesis The theory that life originated repeatedly from inorganic substrates.

neobiogeography The study of the geographical distribution of modern plants and animals; *cf.* palaeobiogeography.

neobiont A primordial life form arising independently from non-living matter.

neocotype In taxonomy, a replacement syntype *q.v.* designated in the absence of the original type or type series; neosyntype.

neo-Darwinism The modern theory of evolution that combines both natural selection and population genetics, in which the Darwinian concept of spontaneous variation is explained in terms of mutation and genetic recombination; neo-Darwinian evolution.

neoendemic An endemic species having a limited geographical range attributable only to its recency of origin; a recent endemic; *cf.* palaeoendemic.

neofemale A female arising by sex reversal from a male.

Neogaea A zoogeographical area originally comprising both Nearctic and Neotropical regions; now generally used to refer to the Neotropical region only; Neogea; *cf.* Arctogaea, Notogaea, Palaeogaea.

neogaean Pertaining to the New World; neogeic, neontogeic; *cf.* amphigeic, gerontogeic.

neogamy Precocious syngamy *q.v.*

Neogene A division of the Tertiary period comprising the Pliocene and Miocene epochs (*ca.* 26–1.6 million years B.P.); see Appendix 1.

neoholotype In taxonomy, a new type specimen selected in the absence of the original type; neotype.

neoichnology The study of tracks, burrows and other trace structures made by living organisms.

neo-Lamarckism Inheritance of acquired characters; the theory that characters acquired by organisms as a response to environmental factors are assimilated into the genome and transmitted to the offspring.

Neolithic An archaeological period from about 4000–2500 years B.P.; a period of human history characterized by the use of polished stone tools; the late Stone Age; New Stone Age.

neologism A new term or expression often disapproved of because of its novelty or barbarousness.

neomorphic allele A mutant allele producing a phenotypic effect quantitatively different from that of the wild type allele; neomorph; *cf.* amorphic allele, hypermorphic allele, hypomorphic allele.

neomorphosis The regeneration of a structure that differs morphologically from the original part.

neonatal Newborn; recently hatched; **neonate**.

neontogeic Pertaining to the New World; neogeic; neogaean; *cf.* amphigeic, gerontogeic.

neontology The study of living organisms; *cf.* palaeontology.

neoparatype In taxonomy, a figured specimen used in addition to the neotype.

neophyte A recently introduced plant.

Neophytic Caenophytic *q.v.*

neoplasm A new growth of tissue in some part of the body where it serves no physiological function, as in the case of a tumour; **neoplasia, neoplastic**.

neosyntype Neocotype *q.v.*

neotectonic Pertaining to crustal plate movements of the late Tertiary and Quaternary periods.

neotene A neotenic *q.v.* organism.

neoteny Paedomorphosis *q.v.* produced by retardation of somatic development, such that sexual maturity is attained in an organism retaining juvenile characters; palaeogenesis; neoteinia; neotenia; neoteiny; **neotene, neotenic, neotenous**; *cf.* progenesis.

Neotropical kingdom One of six major phytogeographical areas characterized by floristic composition; comprising Andean, Caribbean,

Juan Fernandez, Venezuela, and Guianan and Pampas regions; see Appendix 4.

Neotropical region A zoogeographical region comprising South America, West Indies and Central America south of the Mexican plateau; subdivided into Antillean, Brazilian, Chilean, and Mexican subregions; see Appendix 3.

neotype In taxonomy, a newly designated type specimen selected in the absence of extant type material (holotype, paratype or syntype); proxy type.

neotypology A method of classification involving the use of numerical taxonomy within typology as a means of assessing ranges of variation about the type.

Neozoic The period of geological time from the end of the Mesozoic (*ca.* 65 million years B.P.) to the present.

NEP Nett ecosystem production *q.v.*

nepheloid layer The turbid layer of bottom ocean water carrying very fine suspended particulate matter.

nepionic Used of the post-embryonic developmental stage which does not exhibit distinctive specific characters.

nepionotype In taxonomy, the type of a larval stage of a species.

nepotism Showing favouritism to a relative.

nereid A water plant.

nereidion A water-plant community.

neritic Pertaining to the shallow waters overlying the continental shelf; *cf.* oceanic.

neritopelagic Inhabiting shallow coastal waters over the continental shelf.

neritoplankton Plankton of shallow continental shelf waters.

nervicolous Living on or in the veins of leaves; **nervicole**.

nest epiphyte An epiphyte with a twisted growth form which accumulates humus within the tangle.

nest odour The distinctive odour of a nest by which its inhabitants are able to distinguish the nest from others, and from the surrounding environment.

nest parasitism Symbiosis between two termite species in which colonies of one species live in, and feed on, the walls of the nests of a second (host) species.

nest provisioning A form of progressive provisioning *q.v.* in which parents return to the nest regularly to feed the developing offspring.

nested hierarchy A system of consecutively subordinate categories.

net See nett.

nett assimilation rate (NAR) The increase in dry matter (dW) of a single plant per unit time (dt)

referred to the assimilation area A; individual production rate; calculated as NAR = dW/dt × 1/A.

nett ecosystem production (NEP) The total assimilation of organic matter by an ecosystem, less that consumed by the catabolic processes of respiration (ecosystem respiration).

nett photosynthesis Apparent photosynthesis measured as the nett uptake of carbon dioxide into the leaf, and equal to gross photosynthesis less respiration.

nett precipitation That part of total precipitation that actually reaches the ground; calculated as total precipitation less interception (that part of precipitated water intercepted by plant surfaces).

nett primary production (NPP) The total assimilation of organic matter by an autotrophic individual or population per unit time per unit area or volume, less that consumed by the catabolic processes of respiration; net primary production; *cf.* gross primary production.

nett production The total assimilation of organic matter by an individual, population or trophic unit per unit time per unit area or volume, less that consumed by the catabolic processes of respiration; often referred to simply as production *q.v.*; net production; visible production; realized production; observable production.

nett secondary production The total assimilation of organic matter by a primary consumer *q.v.* individual or population per unit time per unit area or volume, less that consumed by the catabolic processes of respiration; net secondary production.

neurobiotaxis Migration or growth of embryonic nervous tissue towards the source of most frequent stimulation.

neurotropic Affecting or having an affinity for the nervous system.

neuston Small to medium size organisms that live on (epineuston) or under (hyponeuston) the surface film of water bodies; **neustont, neustonic**; *cf.* pleuston.

neustonology The study of neuston – those organisms living at the air/water interface of water bodies.

neustophagous Feeding on organisms living at the surface film of a water body (neuston); **neustophage, neustophagy**.

neuter Of neither sex; with imperfectly developed or non-functional reproductive organs.

neutral allele Neutral mutation *q.v.*

neutral mutation A gene mutation that gives no selective advantage or disadvantage to the organism; neutral allele.

neutral term A taxonomic term of convenience

having no nomenclatural significance or hierarchical rank, such as complex, group.

neutralism An association between two species in which the dynamics of each population is unaffected by the presence of the other; the absence of any interaction between two associated populations; neutral association.

New Caledonian region A subdivision of the Polynesian subkingdom of the Palaeotropical kingdom; see Appendix 4.

new name *Nomen novum q.v.*

new replacement name *Nomen novum q.v.*

New Stone Age Neolithic *q.v.*

New Zealand region A subdivision of the Antarctic kingdom; see Appendix 4.

New Zealand subregion A subdivision of the Australian region; see Appendix 3.

Newer glaciation Devensian glaciation *q.v.*

newton (N) A derived SI unit of force, defined as the force required to give a mass of 1 kilogram an acceleration of 1 metre per second per second; see Appendix 13.

nexus hypothesis That each phenotypic character is likely to be influenced by more than one gene and, conversely, that most genes affect more than one character.

niche The ecological role of a species in a community; conceptualized as the multidimensional space, of which the coordinates are the various parameters representing the condition of existence of the species, to which it is restricted by the presence of competitor species; sometimes used loosely as an equivalent of microhabitat in the sense of the physical space occupied by a species; *cf.* fundamental niche, realized niche.

niche breadth Used in the context of a niche as a multidimensional space to indicate the upper and lower limits of a given parameter (one axis of the hyperspace), that is the range of any factor within which the species can function.

niche diversification Divergence of structure or behaviour between species competing for the same resource; character displacement.

niche overlap Direct competition for a given resource by two or more species; in the context of the niche as a multidimensional space, the co-occurrence of two or more niches along all or part of the same resource axis.

niche size A measure of the relative size of the realized niche *q.v.* of a species in a group, community or habitat, usually estimated in terms of distribution and tolerance; habitat tolerance.

nidation Implantation *q.v.*

nidicolous Living in a nest; used also of young animals, especially birds, that remain in the nest

for a prolonged period after birth; **nidicole**; *cf.* nidifugous.

nidification Nest building.

nidifugous Used of young animals, especially birds, that leave the nest soon after birth; **nidifuge**; *cf.* nidicolous.

nidus 1: A nest, domicile, breeding place, or point of origin. 2: The focus or primary site of an infection.

niphic Pertaining to snow; niveal.

nitration The oxidation of nitrite to nitrate, especially by aerobic soil microorganisms.

nitrification The oxidation of ammonia to nitrite, and nitrite to nitrate, by aerobic microorganisms.

nitrogen cycle The biogeochemical cycle of nitrogen involving fixation, nitrification, decomposition by putrefaction and denitrification.

nitrogen fixation The reduction of gaseous nitrogen to ammonia or other inorganic or organic compound by microorganisms or lightning.

nitrophilous Thriving in soil rich in nitrogenous compounds; **nitrophile, nitrophily**.

nitrophyte A plant thriving in soil rich in nitrogenous compounds.

nitrozation The oxidation of ammonia to nitrite by aerobic soil microorganisms.

nival Pertaining to snow; niphic; niveal.

nivalflora The plants occurring above the snowline.

nivation Erosion of the land surface by the action of snow.

niveal Nival *q.v.*

niveoglacial Pertaining to the combined action of snow and ice.

nivicolous Living in snow or snow-covered habitats; **nivicole**.

nm. Abbreviation of the Latin *nothomorphus*, meaning nothomorph *q.v.*

no. Abbreviation of the Latin *numero*, meaning number.

nobis (n.; nob.) Latin, meaning to (of) us; used after a scientific name to denote the authors' responsibility.

noctilucent Used of bioluminescent organisms emitting light during darkness; noctilucous.

nocturnae Pertaining to night-flying insects; *cf.* diurnae.

nocturnal 1: Active during the hours of darkness. 2: Lasting only one night; *cf.* crepuscular, diurnal.

node A branching point in a phylogenetic tree.

nodum 1: A maximum point in a continuum. 2: A ranked category in the classification of plant communities below that of association *q.v.*; **noda**. 3: A neutral term for a plant community of any rank.

nom. Abbreviation of the Latin *nomen* *q.v.*, meaning name.

nomad 1: A wandering organism. 2: A pasture plant.

nomadic Pertaining to the habit of wandering from place to place, usually within a well defined territory; **nomadism**.

nomen Latin, meaning name; *nomina.*

nomen abortivum (*nom. abort.*) A name which at the date of publication was contrary to the Code in operation at that time.

nomen ambiguum (*nom. ambig.*) Ambiguous name; a name consistently used by different authors for different taxa.

nomen alternativum (*nom. alt.*) Alternative name.

nomen anamorphosis (*nom. anam.*) A name based on an imperfect state of a pleomorphic fungus.

nomen confusum (*nom. conf.*; *nom. confus.*) Confused name; a name based on heterogeneous elements from which it is impossible to select a lectotype.

nomen conservandum (*nom. cons.*; *nom. conserv.*) Conserved name; a generic or family name that is retained by authorization of the ICZN *q.v.* or by an International Botanical Congress, although the name strictly contravenes one or more provisions of the Code.

nomen conservandum propositum (*nom. cons. prop.*) A family or genus name proposed for conservation but whose retention has yet to be authorized by an International Botanical Congress; *cf. nomen rejiciendum propositum.*

nomen dubium (*nom. dub.*) Dubious name; a name having uncertain application because it is impossible to establish the taxon to which it should be referred.

nomen hybridum (*nom. hybr.*) Hybrid name; a name formed by combining words from different languages.

nomen illegitimum (*nom. illeg.*; *nom. illegit.*) Illegitimate name; a validly published name which must be rejected for the purposes of priority *q.v.* in accordance with the provisions of the Code; impriorable name; *cf. nomen legitimatum.*

nomen inquirendum (*nom. inq.*) A name that requires investigation.

nomen invalidum (*nom. inval.*) Invalid name; a name that is not validly published or is unavailable.

nomen legitimatum (*nom. legit.*) Legitimate name; a valid name published in accordance with the provisions of the Code; priorable name; *cf. nomen illegitimum.*

nomen monstrositatum (*nom. monstr.*) A scientific name based on a monstrosity *q.v.*

nomen neglectum Neglected name; a validly published name that has been subsequently overlooked.

nomen non rite publicatum (*nom. non rite public.*) Not properly published name; used to indicate that a name has not been validly published according to the provisions of the Code.

nomen novum (*n.n.*; *n. nov.*; *nom. nov.*) New name; a new name proposed to replace an earlier available, but preoccupied, name; new replacement name; substitute name; *nomina nova.*

nomen nudum (*n.n.*; *nom. nud.*) Naked name; a name published without sufficient descriptive information to satisfy the criteria of availability *q.v.* or valid publication *q.v.*; bare name; *nomina nuda.*

nomen oblitum (*nom. oblit.*) Forgotten name; an unused senior synonym rejected under the provisions of the Code; formerly used to denote a name that had not been used in zoological literature for a period of at least 50 years.

nomen provisorium (*nom. provis.*) Provisional name; a name proposed provisionally in anticipation of the acceptance of the taxon concerned at a future date; provisional name.

nomen rejiciendum (*nom. rejic.*) 1: Rejected name; any name, other than the valid name of a taxon, which has been officially rejected under the provisions of the Code. 2: A family or genus name that should replace another well-known name, according to the Code, but which is rejected in favour of a *nomen conservandum q.v.*

nomen rejiciendum propositum (*nom. rejic. prop.*) A family or genus name proposed for rejection in favour of a name proposed for conservation, but whose rejection has yet to be adopted by an International Botanical Congress; *cf. nomen conservandum propositum.*

nomen sed non planta (*nom. non planta*) In taxonomy, used to indicate that an author has applied the name to a taxon other than that to which the type of the basionym *q.v.* belongs.

nomen superfluum (*nom. superfl.*) Superfluous name; a name incorrectly applied to a taxon when another should have been adopted according to the provisions of the Code.

nomen triviale Trivial name; the specific name or epithet.

nomenclator A list of scientific names assembled for nomenclatural rather than taxonomic purposes, such as Sherborn's *Index Animalium*; nomenclatorial.

nomenclatorial Pertaining to a nomenclator.

nomenclatural Pertaining to nomenclature; nomenclatural act.

nomenclatural status The status of a scientific name according to the provisions of the Code.

nomenclatural synonyms Homotypic synonyms *q.v.*

nomenclatural type In taxonomy, the single element of a taxon to which its name is permanently attached; type; nomenifer; onomatophore.

nomenclature The system of scientific names applied to taxa, or the application of these names; **nomenclatural**.

nomenifer In taxonomy, the name bearer; the selected type specimen to which the name of the taxon is permanently attached; nomenclatural type; onomatophore.

nomenspecies Typological species *q.v.*

nomina The plural of *nomen q.v.*; names.

nominal Used of a named taxon based on a type; nominal taxon.

nominalism The theory that there are no universal essences, and that only individuals exist.

nominalistic species concept The concept of a species as a man-made abstraction formulated as a convenient way of referring to large numbers of individuals, but without any real existence in nature.

nominate In taxonomy, used of a subordinate taxon (subspecies or subgenus) containing the type of the higher taxon and bearing the same name; nominotypical.

nominotypical Nominate *q.v.*

nomium A pasture community.

nomocolous Living in pasture; **nomocole**.

nomogenesis 1: The concept that life arose as a result of natural rather than supernatural processes; abiogenesis. 2: The concept that external environmental factors can produce heritable adaptations in all individuals of the same species experiencing these factors.

nomophilous Thriving in pastures; **nomophile**, **nomophily**.

nomophyte A pasture plant; **nomophyta**.

non Latin, meaning not (of); used in a synonymy to indicate that the name that follows is a homonym of an earlier author.

non aliorum (*non al.*) Latin, meaning not of other authors.

non . . . nec Latin, meaning neither (of) . . . nor (of); used in citing a homonym.

non vide (*n.v.*) Latin, meaning not seen; used in a synonymy to indicate that a particular reference has not been checked; *non viso*.

non viso (*n.v.*) Latin, meaning not seen; *non vide*.

nona- Prefix meaning nine, ninefold.

nonadaptive zone An intermediate area between two adaptive zones *q.v.* which can be regarded as a relatively unstable ecological zone that must be crossed when a group evolves from one adaptive zone to another.

non-adiabatic Pertaining to a thermodynamic process in which heat enters or leaves the system; diabatic; *cf.* adiabatic.

non-available water The amount of water still remaining in the soil when a plant reaches the permanent wilting point.

non-biodegradable Not decomposable by natural processes.

Non-calcic Brown soil A zonal soil with a light red or reddish brown acidic A-horizon over a light brown to red B-horizon, formed in subhumid climates under a mixture of forest prairie transitional vegetation; Shantung soil.

nondimensional species concept The concept of a species in terms of the non-interbreeding of coexisting local populations; those sympatric populations that do not interbreed are regarded as distinct species, and those populations that do interbreed and as a result have morphologically intermediate forms are treated as belonging to the same species; *cf.* multidimensional species concept, typological species concept.

nondisjunction The failure of paired chromosomes to separate at meiosis, producing an unequal chromosome distribution in the gametes.

non-harmonious character transformation Mosaic evolution *q.v.*

nonhomologous 1: Used of structures, traits or properties that do not share common ancestry; proposed as a replacement term for analogous *q.v.* 2: Used of chromosomes or chromosome segments comprising dissimilar gene sequences that do not become intimately associated during meiosis; *cf.* homologous chromosomes.

non-Mendelian Used of any extrachromosomal hereditary factor.

nonparametric test A statistical test that makes no hypothesis about, and is therefore independent of, the values of the parameters in a statistical density function or the precise form of the parent distribution from which the data were derived.

non-pelagic sediment A marine sediment containing a high proportion of terrigenous material.

nonrandom With an *a priori* probability of occurrence of 0 (zero) or 1 (unity); *cf.* random.

nonsense mutation A point mutation *q.v.* within a codon which converts an amino acid specifying codon into a chain terminating codon, resulting in the premature interruption of polypeptide synthesis; *cf.* missense mutation, samesense mutation.

non-viable Incapable of normal development, or survival.

noosphere That part of the biosphere altered or influenced by the activities of man.

Nordatlantis North Atlantic bridge *q.v.*

normal curve In statistics, the typically symmetrical bell-shaped graph of a normal distribution *q.v.*; Gaussian curve.

normal distribution A probability distribution in which the density of observations is given by $(2\pi)^{-\frac{1}{2}} \exp(-x^2/2)$. If x has this standard normal distribution then $y = \mu + \sigma x$ is said to have a normal distribution with mean μ and standard deviation σ; Gaussian distribution.

normal erosion Erosion of land used by man at a rate that is not significantly greater than naturally occurring erosion.

normal soil Zonal soil *q.v.*

normalizing selection Stabilizing selection; selection for the mean or normal phenotype.

normoxic Used of a habitat having the normal atmospheric level of oxygen; **normoxicity**; *cf.* anoxic, oligoxic.

North African-Indian desert region A subdivision of the African subkingdom of the Palaeotropical kingdom; see Appendix 4.

North and East Australian region A subdivision of the Australian kingdom; see Appendix 4.

North Atlantic bridge The proposed land-bridge connecting North America, Greenland, Iceland and Europe, thought to have existed a number of times during the Palaeozoic, Mesozoic and up to the beginning of the Cenozoic; Nordatlantis.

North Atlantic Drift A north-easterly continuation of the Gulf Stream current into the eastern North Atlantic; see Appendix 6.

North Atlantic Gyre The major clockwise circulation of surface ocean water in the North Atlantic Ocean; see Appendix 6.

North Equatorial Countercurrent A warm surface ocean current that flows east in the tropical Pacific Ocean; see Appendix 6.

North Equatorial Current A major warm surface ocean current that flows west in the North Pacific and forms the southern limb of the North Pacific Gyre; see Appendix 6.

North Pacific bridge The proposed land-bridge *q.v.* between Siberia and Alaska thought to have existed several times, including much of the Tertiary and Quaternary.

North Pacific Gyre The major clockwise circulation of surface ocean water in the North Pacific; see Appendix 6.

Northeast African Highland and Steppe region A subdivision of the African subkingdom of the Palaeotropical kingdom; see Appendix 4.

Northern coniferous forest biome The circumboreal forest belt across North America and Eurasia bordering to the north with the Tundra biome *q.v.*, also found on high mountains at lower latitudes; characterized by long cold winters with deep snow, short warm summers, podsol soils; dominated by evergreen needle-leaved trees and dense mats of lichens, mosses and low shrubs, deer, rodents and seasonal swarms of insects; taiga; see Appendix 5.

nosogenic Pathogenic; disease producing.

nosology The study and classification of disease; nosography.

nosophyte A pathogenic plant.

notalian Antiboreal *q.v.*

noterophilous Thriving in mesic habitats; mesophilic; **noterophile, noterophily.**

noterophyte Mesophyte *q.v.*; **noterophytic.**

nothocline A graded series of characters or forms produced by hybridization; hybrid cline.

nothogamy Heteromorphic xenogamy *q.v.*

nothomorph (nm.) Any interspecific hybrid variant derived from the same parent stock; equivalent to variety and subordinate to collective *q.v.*

Notogaea A zoogeographical area originally comprising both Australian and Neotropical regions, now generally used to refer to the Australian region only; Notogea; *cf.* Arctogaea, Neogaea, Palaeogaea.

notophyll A leaf size category added to the Raunkiaerian leaf size class *q.v.* for leaves having a surface area between 2025–4500 mm^2.

novus (nov.) Latin, meaning new; **nova.**

NPP Nett primary production *q.v.*

NS quotient A measure of the relative humidity of a climate; determined as N/S where N is the mean annual rainfall (mm) and S is the mean saturation deficit (mm of mercury).

nucellar embryony Asexual reproduction in plants in which an embryo is produced directly from the nucellus.

nucivorous Feeding on nuts; **nucivore, nucivory.**

nuclear division Mitosis *q.v.*; Meiosis *q.v.*; Amitosis *q.v.*; karyokinesis.

nuclear gymnogamy Fertilization by nuclear fusion.

nucleomixis The formation of internuclear connections during meiosis.

nucleotide A subunit of the DNA and RNA molecules, comprising phosphoric acid, a purine or pyrimidine base and a sugar.

nudation The production of an area of bare ground by natural or artificial means.

null hypothesis In statistics, the hypothesis that no real difference or association exists between two populations; that is, that an observed difference is due to chance alone; if the calculated probability of an observed difference is equal to or less than a selected level of significance the null hypothesis is rejected in favour of the alternative hypothesis and the difference is considered to be significant; denoted by H_0.

nulliplex A polyploid that is homozygous for a given recessive allele.

nullisomic A type of aneuploid *q.v.* in which both members of a chromosome pair are missing from the chromosome complement.

numerical character Quantitative character *q.v.*

numerical hybrid A hybrid derived from parental gametes having differing chromosome numbers.

numerical mutation Any change in the number of chromosomes, either by polyploidy or aneuploidy.

numerical phenetics Numerical taxonomy *q.v.*

numerical response A change in the number of predators in a predator population as a result of a change in prey density; *cf.* functional response.

numerical taxonomy Classification based on the numerical comparison of large numbers of equally-weighted characters, scored consistently for all the groups under consideration, and in which individuals are grouped solely on the basis of observable similarities; taxometrics; taximetrics; taxonometrics; numerical phenetics.

numericlature The designation of taxa by numbers rather than names.

Nummultic Palaeogene *q.v.*

nunatak A mountain peak or rocky outcrop projecting above an ice cap.

nunatak hypothesis The theory that apparent anomalies of modern plant distribution patterns can be explained in terms of survival through periods of glaciation in ice-free areas surrounded by ice.

nuptial flight The mating flight of social insects.

nutation A rhythmic or rotational movement of the growing tip of a plant.

nutricism Symbiosis that benefits only one member of the partnership.

nutrition Ingestion, digestion and/or assimilation of food by plants and animals; **nutritive**.

nutritional plant An auxotrophic microorganism.

nyctanthous Flowering only during the night; **nyctanthic, nyctanthy**; *cf.* hemeranthous.

nyctigamous Used of flowers that open at night and close during the day; **nyctigamy**.

nyctinasty Orientation movements of plants during the night; nyctitropism; **nyctinastism, nyctinastic**.

nyctipelagic Pertaining to organisms that migrate into surface waters at night.

nyctiperiod A period of darkness; a dark phase in a light-dark cycle; *cf.* photoperiod.

nyctitropism An orientation response occurring at night; **nyctitropic**.

nymphiparous Giving birth to offspring at the nymphal stage of development; pupiparous.

O

obduction zone A zone of overthrust of crustal masses along tectonic plate boundaries; *cf.* subduction zone.

obex Any barrier separating populations; **obices**.

Obik sea An epicontinental sea joining the Arctic Sea and the Tethys Sea during the Eocene epoch (*ca.* 54–38 million years B.P.).

objective Substantive; factual; not a matter of opinion; having real status; *cf.* subjective.

objective synonyms Homotypic synonyms *q.v.*

obligate Essential; necessary; unable to exist in any other state, mode, or relationship; *cf.* facultative.

obligate anaerobe An organism surviving only in the absence of molecular oxygen.

obligate apogamy Parthenoapogamy *q.v.*

obligate gamete A gamete that is unable to develop parthenogenetically.

obligate synonyms Homotypic synonyms *q.v.*

obligate thermophile An organism requiring a temperature between 65–70° C for optimum growth, and usually unable to grow at temperatures below 40° C; *cf.* facultative thermophile.

obliterative shading A graded coloration of a body which neutralizes relief giving the appearance of a flat surface; obliterative coloration.

observable production Nett production *q.v.*

observational learning Unrewarded learning that occurs when one animal watches the activities of another; empathic learning.

observed frequency The actual number of sample values belonging to a class of a frequency distribution.

obsolescence A phase or stage of reduction leading to the extinction of a species or the loss of a character; obsolete.

Occam's razor States that the simplest sufficient hypothesis is to be preferred even though others are possible; Ockham's razor.

occasional species A species that may be found from time to time in a particular habitat or community, but is not a permanent member of the association.

occidental Western; westerly.

oceanad An oceanic plant; oceanophyte.

Oceania The islands of the Pacific and surrounding seas.

oceanic Pertaining to the open ocean waters beyond the edge of the continental shelf; *cf.* neritic.

oceanic biome Open water biome away from the influence of the shore; has been divided into planktonic, nektonic and benthic sub-biomes; *cf.* muddy shore biome, rocky shore biome, sandy shore biome.

oceanic climate A climate characteristic of continental margins and islands in which the annual temperature range is less than the average for that latitude because of the proximity of an ocean or sea.

oceanic eddy A body of water rotating within the main current system, or deflected from the main current into adjacent areas.

oceanic island A volcanic island formed independently of continental land masses; *cf.* continental island.

oceanics Oceanography *q.v.*

oceanium An oceanic community.

oceanodromous Used of organisms that migrate only within the oceanic province; *cf.* anadromous, catadromous, potamodromous.

oceanography The study of the oceans.

oceanophilus Thriving in oceanic habitats; **oceanophile, oceanophily**.

oceanophyte An oceanic plant; oceanad; **oceanophyta**.

ochetium An ecological succession associated with artificial drainage.

ochthad A bank or ridge plant; ochthophyte.

ochthium A bank community.

ochthohydrophyte A plant which may be fully aquatic during development but normally occurs with stems and leaves projecting well above the surface.

ochthophilus Thriving on banks; **ochthophile, ochthophily**.

ochthophyte A plant living on banks; ochthad; **ochthophyta**.

Ockham's razor Occam's razor *q.v.*

octa- Prefix meaning eight, eightfold.

odour trail A chemical trace (trail pheromone) deposited by one animal for another to follow.

oecad Ecad *q.v.*

oecesis Ecesis *q.v.*; oecisis.

oecology Ecology *q.v.*; **oecological**.

oecoparasite Ecoparasite *q.v.*

oecophene Ecophene *q.v.*

oecostratigraphy Ecostratigraphy *q.v.*

oecotrophobiosis Trophallaxis *q.v.*

oekiophyte A plant cultivated for ornamental purposes.

oekochronology Ecochronology *q.v.*

oekospecies Ecospecies *q.v.*

oekostratigraphy Ecostratigraphy *q.v.*

oestrus The period of maximum sexual receptivity, or heat, in female mammals, usually also the time of release of the eggs; estrus.

Official Index A list published by the International Commission on Zoological Nomenclature of names or works ruled as rejected or invalid by the Commission; the full titles are, Official Index of Rejected and Invalid Specific Names in Zoology; Official Index of Rejected and Invalid Generic Names in Zoology; Official Index of Rejected and Invalid Family-group Names in Zoology; Official Index of Rejected and Invalid Works in Zoological Nomenclature.

Official List A list published by the International Commission on Zoological Nomenclature citing available names and works; the full titles are, Official List of Specific Names in Zoology; Official List of Generic Names in Zoology; Official List of Family-group Names in Zoology; Official List of Works approved as Available for Zoological Nomenclature.

Official opinion Opinion *q.v.*

offprint Reprint *q.v.*

O-horizon The upper soil horizon having more than 20% organic matter; subdivided into litter layer (L- or Ol-layer), fermentation layer (F- or Of-layer), and humified layer (H- or Oh-layer); see Appendix 12.

-oidea The ending of a name of a superfamily in zoological nomenclature; see Appendix 10.

-oideae The ending of a name of a subfamily in botanical nomenclature; see Appendix 10.

oike Habitat *q.v.*

oikesis Ecesis *q.v.*

oikology Ecology *q.v.*

oikos Habitat *q.v.*; ece.

oikosite An attached commensal or parasitic organism.

Oiluvium Pleistocene *q.v.*

older synonym An earlier synonym that cannot be used as the name for the taxon under the provisions of the Code.

oleaginous Producing or containing oil.

oligo- Prefix meaning few, little.

oligoaerobic Pertaining to organisms that grow in conditions of low oxygen concentration.

Oligocene A geological epoch within the Tertiary period (*ca.* 38–26 million years B.P.); see Appendix 1.

oligogene A gene with a major obvious phenotypic effect; switch gene; key gene; major gene.

oligogenic Used of characters or traits controlled by only a few genes; *cf.* digenic, monogenic, polygenic, trigenic.

oligogyny 1: The mating of a single male with a few females. 2: The presence of a few queens within a single colony of social insects; **oligogynous**; *cf.* monogyny, polygyny.

oligohalobous Used of planktonic organisms living in sea water of less than 5 parts per thousand salinity.

oligohaline 1: Used of organisms that are tolerant of only a moderate range of salinities; *cf.* euryhaline, holeuryhaline, polystenohaline, stenohaline. 2: Pertaining to brackish water having a salinity between 0.5–3.0 parts per thousand, or sea water having a salinity between 17–30 parts per thousand; *cf.* mesohaline, polyhaline.

oligolectic Used of insects that collect pollen from only a few different types of flowers.

oligomerization An evolutionary trend manifesting a reduction in the number of segments of a body or structure; law of integration.

oligomictic Used of a lake having a relatively stable stratification with only rare periods of circulation; *cf.* mictic.

oligonitrophilous Thriving in habitats having a low nitrogen content; **oligonitrophile, oligonitrophily**.

oligopelic Used of an area, substrate or habitat comprising or containing only small amounts of clay.

oligopetric Pertaining to plants that prefer rocks that produce only small amounts of clay detritus.

oligophagous Utilizing only a small variety of food species, typically restricted to a single genus, family or order of food species; oligotrophic; **oligophage, oligophagy**; *cf.* monophagous, polyphagous.

oligophyletic Pertaining to a taxon derived from a small number of ancestral forms.

oligopsammic Used of an area, substrate or habitat comprising or containing only a small amount of sand.

oligorhizous Used of plants that develop only a few roots.

oligosaprobic 1: Pertaining to polluted habitats having a high oxygen concentration, low levels of dissolved organic matter, and a low level of organic decomposition; **oligosaprobe**; *cf.* saprobity system. 2: Used of a zone of clear water in a body of fresh water; *cf.* septic.

oligotaxis The suppression of a whorl or whorls in a flower.

oligothermic Tolerating relatively low temperatures; oligothermal; *cf.* polythermic.

oligotokous Having only a few offspring per

brood; oligotocous; *cf.* ditokous, monotokous, polytokous.

oligotraphent Used of an aquatic plant characteristic of water bodies having low nutrient concentration; *cf.* eutraphent, mesotraphent.

oligotrophic 1: Having low primary productivity; pertaining to waters having low levels of the mineral nutrients required by green plants; used of substrates low in nutrients; oligotropic. 2: Used of any organism requiring only a small nutrient supply, or restricted to a narrow range of nutrients; oligophagous. 3: Used of a lake in which the hypolimnion does not become depleted of oxygen during the summer; *cf.* dystrophic, eutrophic, mesotrophic. 4. Pertaining to insects that visit only a small variety of plant species.

oligotrophophyte A plant growing in soil poor in mineral nutrients.

oligotypic Used of a group or assemblage largely comprising a single type of organism or species.

oligoxenous Used of a parasite utilizing a few host species during the life cycle; **oligoxeny**; *cf.* dixenous, heteroxenous, monoxenous, trixenous.

oligoxic Used of a habitat having reduced levels of molecular oxygen; **oligoxicity**; *cf.* anoxic, normoxic.

olisthium A landslip community; olisthion.

-ology Suffix meaning discourse, study.

ombratropism An orientation response to rain; **ombratropic**.

ombrocleistogamic Used of a flower that remains closed bescause of rain and within which self-pollination occurs; **ombrocleistogamy**.

ombrophilous Thriving in habitats having abundant rain; ombrophilus; **ombrophil, ombrophile, ombrophily**; *cf.* ombrophobous.

ombrophobous Intolerant of prolonged rain; ombrophobic; **ombrophobe, ombrophoby**; *cf.* ombrophilous.

ombrophyte 1: A plant growing in situations exposed to direct rainfall. 2: An epiphyte that absorbs water by aerial assimilation through specialized structures.

ombrotiphic Pertaining to temporary pools of water formed from melting snow or rain.

ombrotrophic Pertaining to organisms that obtain nutrients largely through precipitation; **ombrotrophy**.

omni- Prefix meaning all, universally.

omnicolous Living on a wide variety of substrates; **omnicole**.

omnispective method A method of classification based on intuitive and pragmatic considerations using weighted phenotypic similarity as the criterion of relationship, with evolutionary history taken into consideration, but without full phylogenetic analysis; *cf.* cladistic method, evolutionary method, phenetic method.

omnivorous Feeding on a mixed diet of plant and animal material; pantophagous; **omnivore, omnivory**.

oncogenic Causing tumours.

one-tailed test In statistics, a test of a hypothesis in which the critical region comprises values on only one side of the mean (the left or right hand tail) of the sampling distribution; single tail test; one-sided test.

onomatography The practice of the correct writing of plant or animal names.

onomatophore A nomenclatural type; a specimen acting as the name-bearer; nomenifer.

ontocline A gradation of phenotypic characters during the ontogenic development of an organism.

ontogenetic drift Those developmental changes in form and function of higher plants that are inseparable from growth.

ontogenetic migration The occupation by an animal of different habitats at different stages of development.

ontogeny The course of growth and development of an individual to maturity; ontogenesis; **ontogenetic**.

ooapogamy The parthenogenetic development of an individual from an unfertilized female gamete.

oocyte The cell that produces eggs (ova) by meiotic division; *cf.* spermatocyte.

oogamy The fertilization of large non-motile eggs by small motile sperms; usually applied to algae.

oogenesis Female gametogenesis; the formation, development, and maturation of female gametes (ova); ovogenesis.

oology The study of eggs.

oophagous Feeding on eggs; **oophage, oophagy**.

oosphere The female gamete prior to fertilization.

oospore The fertilized female gamete.

ooze A fine grained pelagic deposit comprising at least 30% undissolved sand or silt-sized skeletal remains of marine organisms, the remainder being amorphous clay-sized material; *cf.* coccolith ooze, diatomaceous ooze, foraminiferal ooze, globigerina ooze, pteropod ooze, radiolarian ooze.

op. cit. Abbreviation of the Latin *opere citato q.v.*

open community 1: A plant community which shows incomplete group cover; typically with the plants or tufts discrete rather than touching, but separated by less than their diameter. 2: A community that is open to invasion by immigrant species.

open pollination Outbreeding *q.v.*

open population A population freely exposed to gene exchange with other populations.

operational taxonomic unit (OTU) In numerical taxonomy, any item, individual, or convenient group used for comparison or analysis.

operator The target organism of a signal, as in the case of a predator deceived by mimicry; signal receiver.

operator gene The gene within an operon responsible for switching on or off the structural gene *q.v.*; regulator gene.

opere citato (**op. cit.**) Latin, meaning in the publication cited; used in the text of a work to avoid repetition of a bibliographic reference.

operon A group of closely linked genes comprising the operator gene *q.v.* and related structural genes *q.v.*

ophiology The study of snakes.

ophiophagous Feeding on snakes; **ophiophage**, **ophiophagy**.

ophiotoxicology The study of snake venoms.

Opinion In taxonomy, a decision of the ICZN applying, interpreting or suspending the provisions of the Code, and stating the action to be taken; Official Opinion.

opium A parasite community.

opophilus Thriving or feeding on sap; **opophile**, **opophily**.

opophyte A parasitic plant; **opophyta**.

opportunistic Having the ability to exploit newly available habitats or resources.

opportunistic species A species adapted for utilizing variable, unpredictable or transient environments, typically with a high dispersal ability and a rapid rate of population growth; r-selected species; kinetophilous species.

-opsida The ending of the name of a class in botanical nomenclature; see Appendix 10.

opsonization The modification of the surface properties of an invading virus particle by binding with an antibody or a nonspecific molecule in order to facilitate phagocytosis by host cells.

optimal Most favourable; used of the levels of environmental factors best suited for growth and reproduction of organisms; *cf.* pessimal.

optimal diet theory That predators select prey type in a manner that maximizes nett rate of energy intake.

optimal foraging theory That selection favours a strategy in which a predator utilizes prey in a manner that optimizes nett energy gain per unit feeding time.

optimal yield The maximum sustainable rate of increase of a population under a given set of environmental conditions.

optimum Those conditions most favourable for growth and reproduction of an organism or the maintenance of a system; *cf.* pessimum.

order 1: A rank within a hierarchy of classification; the principal category between class and family; in botanical nomenclature takes the ending *-ales*; see Appendix 10. 2: A ranked category in the classification of vegetation comprising one or more alliances *q.v.*; has the termination *-etalia*. 3: Of a matrix, the number of rows and columns.

ordinary tide level Mean tide level *q.v.*

ordinate The vertical axis, or *y*-axis, of a graph; *cf.* abscissa.

ordination 1: A process by which plant or animal communities are ordered along a gradient. 2: A method of numerical analysis used for summarizing similarities between communities or between OTUs, by representing the communities or OTUs as points in a multidimensional space in such a way that the inter-point distances are inversely related to the similarities.

ordo naturalis (**ord. nat.**; **nat. ord.**) Latin, meaning natural order, sometimes used by nineteenth-century taxonomists for a group equivalent to the modern family.

Ordovician A geological period within the Palaeozoic (*ca.* 504–441 million years B.P.); see Appendix 1.

oread Heliophyte *q.v.*; sun-loving plant.

orgadium An open woodland community.

orgadocolous Living in open woodland; **orgadocole**.

orgadophilous Thriving in open woodlands; **orgadophile**, **orgadophily**.

orgadophyte A plant inhabiting open woodland; **orgadophyta**.

organ-genus A fossil plant genus, assigned to a family, established on the basis of the characters of a single organ or structure such as a fruit or stem; equivalent to ordinary genus names for the purposes of nomenclature.

organic Pertaining to or derived from living organisms, or to compounds containing carbon as an essential component; *cf.* inorganic.

organism Any living entity; plant, animal, fungus, protistan, or prokaryote; **organismal**, **organismic**.

organization effect Interaction among adjacent loci due to the structural organization of the chromosome.

organogenesis The formation of organs during development; **organogeny**.

organogenic Used of sediments and other structures derived from organic matter.

organology The study of organ systems and structures.

organonymy The nomenclature of organs and structures.

organophyly The phylogeny of organs and structures.

organotrophic Pertaining to organisms that use organic compounds as electron donors in their energy producing processes; heterotrophic; **organotroph**; *cf.* lithotrophic.

orgasm Intense sexual excitement or climax.

oriental Eastern; easterly.

Oriental region A zoogeographical region comprising India, Sri Lanka, Malay peninsula, Sumatra, Philippines, Borneo, Djawa (Java) and Bali; subdivided into Indian, Ceylonese, Indo-Chinese, and Indo-Malayan subregions; Sino-Indian region; see Appendix 3.

orientation The position, or change of position, of an organism or part of an organism in relation to a given axis, stimulus, or direction of stimulus.

original date The date of publication of an available work containing a nomenclatural act, a name, or a nominal taxon.

original description In taxonomy, the description of a taxon when first established.

original designation In taxonomy, the designation of the type of a taxon when first established; *cf.* subsequent designation.

original fixation In taxonomy, the fixation of the type by original designation *q.v.* or by monotypy *q.v.*

original material All material used by an author in the preparation of the original description of a taxon, including specimens, illustrations and descriptions.

original spelling In taxonomy, the spelling of an available name when first published.

ornithic Pertaining to birds.

ornithichite A fossilized track or foot-print of a bird.

ornithocoprophilous Thriving in habitats rich in bird droppings; ornithocoprophilic; **ornithocoprophile, ornithocoprophily**.

Ornithogaea The zoogeographic region comprising New Zealand and Polynesia.

ornithogamous Ornithophilous *q.v.*; **ornithogamy**.

ornithogenic Pertaining to sediments rich in bird droppings.

ornithology The study of birds; **ornithological**.

ornithophilous Pollinated by birds; ornithogamous; **ornithophily**.

oroarctic Used of a treeless zone of vegetation north of the orohemiarctic *q.v.* zone; *cf.* oroboreal.

oroboreal Used of an altitudinal zone in mountainous regions having a vegetation that closely corresponds with that of the latitudinal boreal *q.v.* zone; *cf.* oroarctic, orohemiarctic.

orogeny The process of mountain formation; **orogenic, orogenesis, orogenetic**.

orographic Pertaining to relief factors such as hills, mountains, plateaux, valleys and slopes.

orographic desert A rain shadow desert on the leeward side of a mountain range.

orographic rain Rainfall produced by supersaturation of an air mass forced upwards by a mountain range; *cf.* convective rain, frontal-cyclonic rain.

orohemiarctic Used of a latitudinal zone of vegetation having relatively few trees, north of the boreal *q.v.* zone; also known as subarctic; *cf.* oroarctic, oroboreal.

orohylile Pertaining to an alpine or subalpine forest community or habitat.

orophilous Thriving in subalpine, or in mountainous regions; **orophile, orophily**.

orophyte A subalpine plant; **orophytic, orophyta**.

orophytium A subalpine plant community.

orothamnic Pertaining to alpine heath communities.

orphan Of, or relating to, a virus that has no known association with disease.

ortet The original organism from which a clone was derived; *cf.* ramet, genet.

orth. mut. Abbreviation of the Latin ***orthographia mutata***, meaning with an altered spelling (by).

ortho- Prefix meaning straight, upright, at right angles (to).

orthogamy Autogamy *q.v.*

orthogenesis Evolution continuously in one direction over a considerable period of time; commonly used with the implication that the direction is determined by a factor internal to the organism or, at least, not determined by natural selection; aristogenesis; entelechy; **orthogenetic**; *cf.* rectilinear evolution.

orthograde In limnology, the type of distribution in which concentration increases with depth, as in the oxygen concentration in the deepest and most unproductive lakes; *cf.* clinograde, heterograde.

orthographia mutata (***orth. mut.***) Latin, meaning with an altered spelling (by); used after a name, and followed by the author responsible for the changed spelling.

orthographic error An unintentional misspelling.

orthographic variants 1: In taxonomy, two or more different spellings of the same name. 2: The names of two or more taxa that have spellings so similar that they are likely to be confused; paranyms.

orthography The spelling of words; **orthographic**.

orthoheliotropism An orientation response towards sunlight; **orthoheliotropic**.

orthokinesis A change in the rate of random movement of an organism (kinesis) in which the rate of locomotion (linear velocity) varies with the intensity of the stimulus; *cf.* klinokinesis.

orthologous Used of homologous structures determined by a common ancestral gene present in the most recent common ancestor; *cf.* paralogous.

orthophototropism An orientation response towards light; **orthophototropic**.

orthophyte The totality of a plant's existence from zygote to zygote through any alternation of generations.

orthoselection Natural selection acting continuously in the same direction over long periods of time.

orthotopic Used of an individual or group in its normal habitat; **orthotopy**.

orthotropism Orientation or growth in a straight line; parallelotropism; **orthotropic**.

orthotype In taxonomy, a type by original designation *q.v.*

oryctocoenosis 1: An outcrop community or assemblage of fossils. 2: That part of a thanatocoenosis *q.v.* preserved as fossils; **oryctocoenose**.

oryctology The study of fossils; palaeontology; oryctological; **oryctics**.

osmoconformer An organism having a body fluid of the same osmotic concentration as the surrounding medium; *cf.* osmoregulator.

osmophilic Thriving in a medium of high osmotic concentration; **osmophile**, **osmophily**.

osmoregulation Control of the volume and composition of body fluids.

osmoregulator An organism that maintains the osmotic concentration of its body fluid at a level independent of the surrounding medium; *cf.* osmoconformer.

osmosis Diffusion of a solvent through a semipermeable membrane from a region of low solute concentration to one of high solute concentration; **osmotic**.

osmotaxis A directed reaction of a motile organism to a change or gradient of osmotic pressure; tonotaxis; **osmotactic**.

osmotic water Water held in close contact by clay particles that is less available to soil organisms than capillary water *q.v.*

osmotrophic Used of organisms that are capable of absorbing organic nutrients directly from the external medium; **osmotroph**, **osmotrophy**.

osmotropism An orientation response to an osmotic stimulus; tonotropism; **osmotropic**.

osphresiology Study of the sense of smell.

osteology Study of the structure and development of bones.

OTU Operational taxonomic unit *q.v.*

outbreeding Mating or crossing of individuals that are either less closely related than average pairs in the population, or from different populations; outcrossing; open-pollination; crossbreeding; *cf.* inbreeding.

outcrossing Outbreeding *q.v.*

ovariicolous Living in ovaries; **ovariicole**.

overall similarity The proportion of shared character states.

overdispersed distribution Contagious distribution *q.v.*; some confusion exists over the use of this term as some authors have employed it with the opposite meaning, that is to describe uniform distributions *q.v.*

overdominance Superior phenotypic expression of the heterozygote over both homozygotes.

overgrazing Grazing which exceeds the recovery capacity of the community and thus reduces the available forage crop or causes undesirable changes in the community composition.

overstepping The development of a descendant organism through the ontogenetic stages, and beyond the final adult stage, of its ancestor.

overstorey The topmost layer of a forest community, formed by the tallest trees of the canopy; overstory stratum; overwood.

overturn Thorough water circulation in the sea or a lake, often occurring seasonally and caused by density differentials resulting from changes of temperature.

ovigerous Egg-bearing; used of a female carrying eggs; berried.

oviparous Egg-laying; producing eggs that are laid and hatch externally; **oviparity**; *cf.* larviparous, ovoviviparous, viviparous.

oviposition The act or process of depositing eggs; **oviposit**.

ovism The belief that the embryo is contained solely within the egg cell; encasement theory *q.v.*

ovogenesis Oogenesis *q.v.*

ovoviviparous Producing fully formed eggs that are retained and hatched inside the maternal body with the release of live offspring; **ovoviviparity**; *cf.* larviparous, oviparous, viviparous.

ovum An unfertilized egg cell; a female gamete; **ova**.

oxodad An acid-marsh or peat bog plant.

oxodic Pertaining to a peat bog community.

oxodium A humus marsh community; a peat bog community; oxodion.

oxybiotic Aerobic *q.v.*; oxybiontic.

oxygen isotope ratio ($^{18}O/^{16}O$ ratio) A method for estimating the environmental temperatures that existed at particular periods of geological history, using the ratio of oxygen isotopes ^{18}O and ^{16}O in calcium carbonate taken from fossil marine shells; the ratio of these isotopes is determined by the ambient temperature at the time of deposition of the carbonate skeleton by the live animal.

oxygenated With adequate available oxygen; aerobic; *cf.* deoxygenated.

oxygenotaxis A directed response of a motile organism towards (positive) or away from (negative) an oxygen stimulus; oxytaxis; **oxygenotactic**.

oxygenotropism An orientation movement induced by an oxygen gradient stimulus; oxytropism; **oxygenotropic**.

oxygeophilus Thriving in humus-rich habitats; **oxygeophile, oxygeophily**.

oxygeophyte A plant growing in humus; **oxygeophyta**.

oxygeophytic Pertaining to plant communities in humus-rich habitats.

oxylium A humus-marsh community.

oxylophilous Thriving in humus or humus-marsh habitats; oxylophilus; oxylyphilus; **oxylophile, oxylophily**.

oxylophyte A plant growing in a humus-rich habitat; **oxylophytic, oxylophyta**.

oxylyphilus Oxylophilous *q.v.*

oxyphilous Thriving in acidic habitats; acidophilic; **oxyphile, oxyphily**; *cf.* oxyphobous.

oxyphobous Thriving in alkaline habitats; intolerant of acidic conditions; basophilous; **oxyphobe, oxyphoby**; *cf.* oxyphilous.

oxyphyte A plant growing under acidic conditions.

oxysere An ecological succession commencing in an acidic habitat.

oxytaxis A directed response of a motile organism to an oxygen stimulus; oxygenotaxis; **oxytactic**.

oxytropism 1: An orientation response to an oxygen gradient stimulus; oxygenotropism; **oxytropic**. 2: An orientation response to an acid stimulus.

Oyashio Current A cold surface ocean current that flows south in the north west Pacific; Oya Shio; Kamchatka current; see Appendix 6.

ozonosphere The lower layer of the stratosphere *q.v.* about 15–30 km above the Earth's surface in which the absorption of ultraviolet solar radiation produces ozone.

P

p Abbreviation of pico, denoting unit $\times 10^{-12}$
P Abbreviation of peta, denoting unit $\times 10^{15}$.
p. Abbreviation of the Latin *pagina*, meaning page.
P₁ The parental generation.
P₂ The second parental generation; grandparents.
p.p. Abbreviation of the Latin *pro parte*, meaning in part, partly.
pachy- Prefix meaning thick.
Pacific Islands subregion Polynesian subregion *q.v.*
Pacific North American region A subdivision of the Boreal kingdom; see Appendix 4.
paedium An assemblage of young animals.
paedogamous autogamy The condition in which the nucleus but not the cytoplasm is involved in the formation of the zygote.
paedogamy The formation of a zygote by the fusion of two gametes produced by one individual; a form of automixis *q.v.*; pedogamy.
paedogenesis Progenesis *q.v.*; paedomorphosis *q.v.* produced by precocious sexual maturation in an organism that is still at a morphologically juvenile stage; *cf.* neoteny.
paedomorphosis The retention of juvenile characters of ancestral forms by adults, or later ontogenetic stages, of their descendants; superlarvation; **paedomorphic, paedomorphism, pedomorphic, pedomorphism**; *cf.* neoteny, progenesis.
paedoparthenogenesis Parthenogenetic reproduction occurring in larvae.
paedophagous Feeding on embryos and the young stages of other species; **paedophage, paedophagus, paedophagy**.
pagium An ecological succession on a glacial deposit.
pagophilous Thriving in foothills; pagophilus; **pagophile, pagophily**.
pagophyte A plant inhabiting foothills; **pagophyta**.
pain cry Fright cry *q.v.*
pair bonding The formation of a close and lasting association between a male and female animal, used particularly with reference to the cooperative rearing of young; pair formation.
paired limiting factors Two or more ecological factors acting synergistically to prevent a population realizing its biotic potential; *cf.* master limiting factor.
PAL Present atmospheric level.
Palaeantarctis Antarctic bridge *q.v.*
Palaearctic region A zoogeographical region

comprising Europe, North Africa, western Asia, Siberia, northern China, Japan; Palearctic region; subdivided into European, Mediterranean, Siberian and Manchurian subregions; see Appendix 3.
palaeo- Prefix meaning old, ancient; paleo-.
palaeoagrostology The study of fossil grasses; paleoagrostology.
palaeoalgology The study of fossil algae; paleoalgology.
palaeoautecology Study of the ecology of individual fossil species or groups; paleoautecology; *cf.* palaeosynecology.
palaeobiocoenosis An assemblage of fossil organisms that had existed in geological history as an integrated community; palaeocoenosis.
palaeobiogeography Study of the geographical distributions of fossil faunas and floras; *cf.* neobiogeography.
palaeobiology Study of the biology of extinct organisms; paleobiology.
palaeobiotope In palaeontology, a region characterized by more or less uniform environmental conditions and by correspondingly uniform plant and animal associations.
palaeobotany Study of the plant life of the geological past; phytopalaeontology; palaeophytology; paleobotany.
Palaeocene A geological epoch within the Tertiary period (*ca.* 65–54 million years B.P.); see Appendix 1.
palaeoclimatology Study of the climate during periods of past geological time; paleoclimatology; **palaeoclimate**.
palaeocommunity All preservable taxa comprising a fossil community; paleocommunity.
palaeodendrology The study of fossil trees; paleodendrology.
palaeoecology Study of the ecology of fossil communities; paleoecology.
palaeoendemic A species with a limited geographical range but of considerable evolutionary age; an ancient endemic; paleoendemic; *cf.* neoendemic.
Palaeogaea A zoogeographical area originally comprising the Palaearctic, Ethiopian, Oriental and Australian regions; Palaeogea; *cf.* Arctogaea, Neogaea, Notogaea.
Palaeogene A division of the Tertiary period (*ca.* 65–26 million years B.P.) comprising the Palaeocene, Eocene and Oligocene epochs;

Paleogene; Eogene; Nummultic;
see Appendix 1.

palaeogenesis Neoteny *q.v.*

palaeogenetics Study of the genetics of fossil life forms; paleogenetics.

palaeoichnology The study of trace fossils; paleoichnology, palichnology; *cf.* neoichnology.

Palaeolaurentian Archaeozoic *q.v.*

palaeolimnology Study of the geological history and development of inland waters; paleolimnology.

Palaeolithic An archaeological period from about 10 thousand to 3.5 million years B.P.; a period of human history characterized by the use of stone tools; the early, older or chipped Stone Age; Paleolithic.

palaeontography The systematic description of fossil taxa; paleontography.

palaeontology The study of fossils; oryctology; paleontology; *cf.* neontology.

palaeopalynology The study of fossil spores and pollen; paleopalynology.

Palaeophytic The period of geological time between the development of the algae and the first appearance of gymnosperms, during which pteridophytes were abundant; Pteridophytic; *cf.* Aphytic, Archaeophytic, Caenophytic, Eophytic, Mesophytic.

Palaeophyticum The plant life of the Palaeozoic; Paleophyticum; *cf.* Caenophyticum, Mesophyticum.

palaeophytogeography The study of plant distributions through geological time; paleophytogeography.

palaeophytology Palaeobotany *q.v.*; paleophytology.

palaeornithology The study of fossil birds; paleornithology.

palaeosere An ecological succession during geological history; an eosere of the Palaeozoic era.

palaeosol A soil formed under past environmental conditions, distinguished by the morphology of the soil profile; a fossil soil.

palaeospecies Chronospecies *q.v.*

palaeosynecology Study of the ecology of fossil populations or communities; paleosynecology; *cf.* palaeoautecology.

palaeothanatocoenosis An assemblage of organisms buried together in the geological past; paleothanatocoenosis.

palaeothermal Pertaining to warm climates of the geological past; paleothermal; palaeothermic; paleothermic.

Palaeotropical kingdom One of six major phytogeographical areas characterized by its

floristic composition; comprising African, Indo-Malaysian and Polynesian subkingdoms; see Appendix 4.

Palaeotropical region Ethiopian region *q.v.*; Paleotropical region.

Palaeozoic A geological era (*ca.* 570–245 million years B.P.) comprising the Cambrian, Ordovician, Silurian, Devonian, Carboniferous and Permian periods; Paleozoic; see Appendix 1.

palaeozoology Study of the animal life of the geological past; paleozoology.

paleo- Palaeo- *q.v.*

palichnology Palaeoichnology *q.v.*

palingenesis 1: The development of individuals already preformed within the egg. 2: The true repetition or recapitulation of past phylogenetic stages in the ontogeny of descendants; **palingenetic**.

palingenetic Ancestral; of remote or ancient origin.

palmigrade Plantigrade *q.v.*

paludal Pertaining to marshes; uliginous; helobius; palustrine; **paludine, paludinous, paludose**.

paludicolous Inhabiting marshy habitats; **paludicole**.

paludification The process of bog expansion resulting from the rising water table produced as drainage is impeded by the accumulation of peat.

palustrine Pertaining to wet or marshy habitats; helobius; uliginous; paludal; **palustral**.

palynology The study of pollen and spores, both living and fossil; **palynologic**.

pan- Prefix meaning all, whole, completely.

pan A soil horizon of highly compacted material, or one of high clay content.

PAN Peroxi-acetyl-nitrate, an atmospheric pollutant produced from automotive and industrial fumes under strong radiation.

panalloploid A true allopolyploid *q.v.*

panautoploid A true autopolyploid *q.v.*; *cf.* hemiautoploid.

panbiogeographic model Vicariance model *q.v.*

panclimax Two or more related climaxes having similar dominants, life forms and climatic factors; panformation.

pandemic 1: Very widely distributed, ubiquitous, cosmopolitan. 2: A disease reaching epidemic proportions simultaneously in many parts of the world.

panformation Panclimax *q.v.*

Pangaea The single supercontinent formed about 240 million years B.P. comprising the present continental land-masses joined together, which began to break up about 150 million years B.P. (some authorities date the break up closer to 200 million years B.P.); Pangea; *cf.* Gondwanaland, Laurasia, Tethys.

pangamy Unrestricted random mating; panmixis; **pangamic**.

pangenesis Theory of heredity, that somatic cells contain particles (gemmules) that can be influenced by the environment and the activity of the organs containing them; these particles can move to the sex cells to influence the course of heredity; gemmule theory.

panmictic Pertaining to a randomly interbreeding population (a panmictic unit or deme).

panmictic index Relationship, coefficient of *q.v.*

panmixis Unrestricted random mating; the free interchange of genes within an interbreeding population; pangamy; **panmictic, panmixia**.

panphenetic Pertaining to overall similarity between taxa based on a large number of characters.

panphotometric Pertaining to leaves orientated to avoid maximum light; sometimes used of all leaves that vary their orientation according to the type of illumination; *cf.* euphotometric.

panphytophagous Feeding on both higher and lower plants; **panphytophage, panphytophagy**.

panspermia The theory that life did not originate on Earth but was introduced from elsewhere in the Universe.

Panthalassa The single large ocean surrounding Pangaea *q.v.*

panthalassic Pertaining to marine organisms or species living in both neritic and oceanic waters.

pantogenous Used of symbiotic fungi not confined to a single host.

pantomictic plankton Plankton not dominated by any particular species; *cf.* monoton plankton, polymictic plankton, prevalent plankton.

pantophagous Omnivorous *q.v.*; **pantophage, pantophagy**.

pantropic Extending or occurring throughout the tropics and subtropics, or at least widespread in tropical regions; **pantropical**.

pantropism An orientation movement that has no fixed direction; **pantropic**.

para- Prefix meaning by, beside, along, beyond.

parabiosis 1: The temporary suspension of physiological activity. 2: Utilization of the same nest by colonies of different species of social insects which nevertheless keep their broods separate; **parabiont, parabiotic**. 3: Anatomical and physiological union as in the case of siamese twins; parabiotic twins; phylacobiosis.

parabiosphere The space surrounding the eubiosphere *q.v.* in which conditions are too extreme for active metabolism, but in which resistant life cycle stages may exist.

parachoric Used of organisms that live within the body of another organism without being parasitic.

parachrosis The process of colour change.

paraclimax Subclimax *q.v.*

paraclonal Used of a society *q.v.* comprising genotypically identical modular units which cannot regenerate or follow an independent existence if separated from the parent organism; *cf.* euclonal, pseudoclonal.

paracme In the phylogeny of a group, the period of decline after the point of maximum vigour or peak of development; phylogerontic period; *cf.* acme, epacme.

paradigm An archetype or outstanding example of type.

paraedeotype A paratype specimen of which the genitalia have been examined.

parafluvial nappe The moist region of a river bank fed by permeating river water and sometimes by adjacent ground water; may extend up to 3 m from water edge.

paragamy The formation of a zygote nucleus by the fusion of nuclei within a syncytium *q.v.*

paragenesis Reproduction induced by a cross between a parent and an otherwise sterile hybrid.

paragenetic Pertaining to a chromosomal change that affects the expression of a gene rather than its constitution, such as a position effect.

paraheliotropism Movement of leaves to avoid or minimize exposure to sunlight; paraphototropism; **paraheliotropic**.

paralectotype Any specimen of an original syntype series remaining after the designation of a lectotype *q.v.*; lectoparatype.

paralimnion The entire spaciotemporal collection of factors constituting the surroundings of a lake; the littoral zone of a lake from the water margin to the deepest level of rooted vegetation; **paralimnetic, paralimnic**.

parallel evolution 1: The independent acquisition in two or more related descendant species of similar derived character states evolved from a common ancestral condition; parallelism; *cf.* convergent evolution. 2: The maintenance of constant differences in the evolution of characters between two unrelated lines.

parallel key Bracketed key *q.v.*

parallel polymorphism Tautomorphism *q.v.*

parallelism Parallel evolution *q.v.*

parallelogeotropism An orientation response that brings the main axis of the body to the vertical (in line with the force of gravity); **parallelogeotropic**.

parallelotropism 1: Orientation or growth in a

straight line; orthotropism. 2: An orientation response towards a light source, parallel to the light rays; **parallelotropic**.

parallelotype Paratype *q.v.*

parallotype In taxonomy, a paratype of the same sex as the allotype *q.v.*; alloparatype.

paralocal Used of a local character species *q.v.* that has a geographical range only partly overlapping the range of the vegetation unit of which it is a characteristic species.

paralogous Used of homologous structures that arose by gene duplication and evolved in parallel within a single line of descent; *cf.* orthologous.

parameter A characteristic of a distribution of a variable or population, such as the mean; usually denoted by Greek letters; *cf.* statistic.

parametric test A test that makes a hypothesis about the form of the underlying distribution.

paramorph Any phenotypic variant lacking accurate definition.

paramorphic homology The condition of two distinct species having polypeptide variants whose ancestral forms were polymorphic alleles in the ancestral species; polymorphism-dependent homology.

paramutualism A facultative symbiosis in which both species benefit.

paranym In taxonomy, a name having a spelling so similar to another name that confusion is likely to occur; orthographic variant.

parapatric Used of populations whose geographical ranges are contiguous but not overlapping, so that gene flow between them is possible; macroparapatric; **parapatry**; *cf.* allopatric, dichopatric, sympatric.

parapatric speciation Speciation in which initial segregation and differentiation of diverging populations take place in disjunction but complete reproductive isolation is attained after range adjustment so that populations become separate but contiguous.

paraphototropism 1: Diaphototropism *q.v.*; **paraphototropic**. 2: Paraheliotropism *q.v.*

paraphyletic group A group of taxa derived from a single ancestral taxon, but one which does not contain all the descendants of the most recent common ancestor; a category based on the common possession of plesiomorph characters (symplesiomorphy).

pararegional Used of a regional character species *q.v.* that has a geographical range only partly overlapping the range of the vegetation unit of which it is a characteristic species.

parasematic Pertaining to markings, structures or behaviour intended to deter, distract or mislead a predator; **paraseme**.

parasexual Pertaining to organisms that achieve genetic recombination by means other than the regular alternation of meiosis and karyogamy *q.v.*; parasexual reproduction; *cf.* eusexual, protosexual.

parasite Any organism that is intimately associated with, and metabolically dependent upon another living organism (the host) for completion of its life cycle, and which is typically detrimental to the host to a greater or lesser extent.

parasite-saprophyte A plant parasite that kills its host then lives saprophytically on its decomposing remains.

parasitic castration Reproductive death of a host organism resulting from a parasitic infection.

parasitic male A dwarf male that lives a parasitic existence attached to the body of its female, typically having well developed reproductive organs but an otherwise degenerate body form; complemental male.

parasitism An obligatory symbiosis between individuals of two different species, in which the parasite is metabolically dependent on the host, and in which the host is typically adversely affected but rarely killed; parasitosis; **parasitic**.

parasitocoenosis The total parasite load of any given host organism.

parasitoid An organism with a mode of life intermediate between parasitism and predation; usually a species of hymenopteran in which the larva feeds within the living body of another organism eventually causing the death of the host.

parasitology The study of parasites and parasitism.

parasocial Used of a social group in which at least one of the following characteristics of a eusocial group is absent; cooperation in care of young; reproductive division of labour; overlap of generations contributing to colony labour; presocial.

parasymbiosis 1: Symbiosis without benefit or detriment to either partner; neutralism. 2: The association of an organism (the parasymbiont) with an existing symbiosis, as in a lichenicolous fungus.

parataxon A taxonomic category for fossilized remains that form only part of the whole organism and are commonly widely separated from other such parts.

paratelic Used of superficially similar characters or traits derived by convergent evolution; **parately**.

paratenic host A host that is not essential for the completion of a parasite's life cycle but is used as a temporary habitat or as a means of reaching the definitive host *q.v.*; transfer host; transport host; *cf.* intermediate host.

parathermotropism Movement of leaves to avoid or minimize the heating effects resulting from exposure to sunlight; **parathermotropic**.

paratonic Used of movement or growth stimulated, enhanced or retarded by an external stimulus; aitiogenic; *cf.* autonomic.

paratopotype A paratype from the type locality having the same collection data as the holotype.

paratrepsis Distraction display *q.v.*

paratrophic Feeding in the manner of a parasite; parasitic; **paratroph, paratrophism, paratrophy**.

paratype 1: In taxonomy, any specimen from the type series remaining after the designation of the holotype or isotype *q.v.*; parallelotype. 2: Any abnormal type of a species.

parent material The inorganic deposit which has been modified by weathering, translocation or other pedogenic process, to give rise to different soil types.

parentage, coefficient of Kinship, coefficient of *q.v.*

parental generation (P$_1$) The original generation in an experimental study.

parental investment Any behaviour toward offspring that increases the probability of survival of the offspring to the detriment of possible investment in other offspring.

parentheses ( ) Round brackets; pairs of printed marks placed around a word or words; used in taxonomy to enclose the name of a subgenus or of an original author when the epithet is placed in a combination other than the original combination; *cf.* brackets.

parhomology Apparent similarity of structure or function.

parity The number of times a particular female has given birth, or the number of offspring produced at each birth; *cf.* biparous, multiparous, uniparous.

-parous Combining suffix meaning producing.

parsimonious The condition of being economic or frugal in use or application; **parsimony**.

parsimony, law of That the simplest sufficient hypothesis is to be preferred even though others are possible; Occam's razor.

part. Abbreviation of the Latin *partim*, meaning part.

parthenapogamy Diploid parthenogenesis; parthenogenesis of diploid individuals in which meiosis has been suppressed so that neither chromosome reduction nor any corresponding phenomenon occurs; obligate apogamy; somatic parthenogenesis.

parthenocarpy The development of fruit without seeds, typically resulting from lack of fertilization; **parthenocarpic**.

parthenocaryogamy Parthenokaryogamy *q.v.*

parthenogamete A gamete capable of parthenogenetic development.

parthenogamy The parthenogenetic development of a diploid organism, arising from the fusion of two nuclei within a single gamete; automictic parthenogenesis.

parthenogenesis The development of an individual from a female gamete without fertilization by a male gamete; geneagenesis; gynecogeny; *cf.* ameiotic parthenogenesis, androcyclic parthenogenesis, anholocyclic parthenogenesis, arrhenotoky, automictic parthenogenesis, cyclic parthenogenesis, haploid parthenogenesis, hemizygoid parthenogenesis, heteroparthenogenesis, holocyclic parthenogenesis, meiotic parthenogenesis, monocyclic parthenogenesis, paedoparthenogenesis, polycyclic parthenogenesis, thelytoky, tychoparthenogenesis, zygoid parthenogenesis.

parthenogenome An organism produced by parthenogenesis; parthenote.

parthenokaryogamy The fusion of two female haploid nuclei; parthenocaryogamy.

parthenomixis The fusion of two female nuclei produced within a single gamete or gametangium.

parthenospore A spore produced from an unfertilized female gamete; aboospore; azygospore.

parthenote An organism produced by parthenogenesis; parthenogenome.

partial Secondary; incomplete.

partial habitat A habitat occupied by a plant during only one phase of its life cycle.

partial parasite Semiparasite *q.v.*

partial selection Selection in which the value of the selection coefficient *q.v.* is greater than zero but less than one.

particle size analysis The classification of sediment particles according to size; three grade scales commonly used are phi, Wentworth and Udden; see Appendix 11.

particulate inheritance The theory that maternal and paternal genetic factors do not blend but retain their integrity from generation to generation; Mendel's theory of particulate inheritance; *cf.* blending inheritance.

parturition The act of giving birth.

pascal (Pa) A derived SI unit of pressure or stress defined as a pressure of 1 newton per square metre; see Appendix 13.

pascual Pertaining to, or inhabiting, pasture.

pasculomorphosis Change in the structure of plants as a result of grazing by animals.

passive chamaephyte A chamaephyte *q.v.* subtype

in which the erect vegetation shoots persist through unfavourable seasons in a procumbent position.

passive immunity An immune state created by inoculation with serum antibodies or lymphocytes from an immune animal rather than by exposure to the antigen.

passive process A system or process not requiring the expenditure of metabolic energy; *cf.* active process.

passive uptake The absorption of ions by diffusion or other physical process not involving the expenditure of metabolic energy; *cf.* active uptake.

Pasteur point The transition point at which a facultative anaerobe no longer utilizes oxygen for respiration but obtains energy through fermentation.

Pastonian interglacial An interglacial period of the Quaternary Ice Age *q.v.* in the British Isles; see Appendix 2.

patabiontic Used of an organism permanently inhabiting the forest litter layer; patobiontic; **patabiont**.

patacolous Patocolous *q.v.*; **patacole**.

Patagonian region A subdivision of the Antarctic kingdom; see Appendix 4.

patchy distribution Contagious distribution *q.v.*

paternal Pertaining to, or derived from, the male parent; *cf.* maternal.

paternal sex determination The condition in which the sex of the offspring is determined by the idiotype of the male parent.

patho-, pathy- Prefix meaning suffering, feeling, associated with disease.

pathogenesis The course or development of a disease.

pathogenic Producing or capable of producing disease; nosogenic; **pathogen, pathogenicity**.

pathology The study of disease; nosology.

patobiontic Used of an organism that spends its entire life cycle in the forest litter; cryptozoic; patabiontic; **patobiont**.

patocolous Used of an organism that inhabits the forest litter for only part of its normal life cycle; patacolous; **patocole**.

patoxenous Used of an organism that occurs in the forest litter only by accident; **patoxene**.

patriclinous Used of offspring having a greater hereditary resemblance to the male parent than to the female; patroclinous; patriclinic; patriclinal; **patricliny**; *cf.* matriclinous.

patrilineal Pertaining to the paternal line; passed from male parent to offspring; paternal; *cf.* matrilineal.

patristic affinity A degree of similarity due to

common ancestry, including parallel evolution but not convergent evolution; patristic similarity; *cf.* cladistic affinity.

patristic character A character or trait inherited by all members of a group from their most recent common ancestor.

patristic distance A measure of the amount of genetically determined change that has occurred between any two points of a phylogenetic tree; divergence index; *cf.* cladistic distance, phenetic distance.

patroclinous Patriclinous *q.v.*; **patroclinal, patroclinic, patrocliny**.

patroendemic A diploid plant having a restricted geographical distribution that has given rise to a more widespread polyploid; a type of neoendemic.

patrogenesis The parthenogenetic development of an organism from an enucleated egg following fusion with a normal male gamete; male parthenogenesis; androgenesis.

patrogynopaedium A family group in which both parents remain with the offspring for some time; *cf.* gynopaedium, patropaedium.

patromorphic Resembling the male parent; *cf.* matromorphic.

patronymic A scientific name based on the name of a person or persons.

patropaedium A family group in which the male parent remains with the offspring for some time; *cf.* gynopaedium, patrogynopaedium.

paurometabolous Used of a pattern of development displaying a gradual metamorphosis in which changes are inconspicuous.

PBB Polybrominated biphenyl; a class of compounds related to PCB.

PCB Polychlorinated biphenyl; a chlorinated hydrocarbon used in industry which occurs as a persistent pollutant, particularly in aquatic ecosystems.

peat An accumulation of unconsolidated partially decomposed plant material found in more or less waterlogged habitats of fen (alkaline peat), bog (acid peat) and swamp.

pebble A sediment particle between 4 and 64 mm in diameter; medium gravel; see Appendix 11.

peck A unit of volume equal to 2 gallons (9.09218 dm³).

peck order A dominance hierarchy *q.v.*, especially in birds; peck dominance.

Peckhammian mimicry Aggressive mimicry *q.v.*

peculiar Diagnostic *q.v.*

ped A unit of soil structure formed by natural processes, such as an aggregate, block, crumb.

pedalfer An acid soil type found in humid regions

characterized by aluminium and iron compounds, and the absence of carbonates; *cf.* pedocal.

pedigree A formal list or register of ancestry or genealogy of an individual or family group.

pediophilous Thriving in uplands; pediophilus; **pediophile, pediophily.**

pediophyte A plant of an upland community; **pediophyta.**

pedo- Paedo- *q.v.*

pedocal An alkaline soil type found in arid regions characterized by a high calcium carbonate content; *cf.* pedalfer.

pedogamy Paedogamy *q.v.*

pedogenesis 1: The formation of soils; pedogenics. 2: Paedogenesis *q.v.*

pedogenic factor Any of the factors contributing to the ultimate profile of a soil.

pedology Study of the structure and formation of soils; soil science.

pedon 1: Those organisms that live on or in the substrate of an aquatic habitat. 2: The smallest vertical column of soil containing all the soil horizons at a given location.

pedonic Pertaining to an organism of the bottom community of a freshwater lake; **pedon.**

pedosphere That component of the biosphere comprising the soil and soil organisms.

pegmatypic Used of a mating system in which imprinting occurs during a particular period of juvenile development.

pelagad A plant of the sea surface; **pelagophyte.**

pelagic Pertaining to the water column of the sea or lake; used of organisms inhabiting the open waters of an ocean or lake; **pelagial;** *cf.* benthic.

pelagic sediment A deep ocean sediment containing no significant terrigenous component.

pelagium A sea-surface community.

pelagophilus Thriving in the open surface waters of the sea; **pelagophile, pelagophily.**

pelagophyte A plant living at the sea surface; pelagad; **pelagophyta.**

pelasgic Moving from place to place.

pelochthium A mud-bank community.

pelochthophilous Thriving on mud-banks; **pelochthophile, pelochthophily.**

pelochthophyte A mud-bank plant; **pelochthophyta.**

pelogenous Mud-producing or clay-producing.

pelophilous Thriving in habitats rich in clay; pelophilus; **pelophile, pelophily.**

pelophyte A plant living in clay or muddy soil; **pelophyta.**

pelopsammic Pertaining to or comprising both clay and sand.

pelopsammogenous Producing both clay and sand.

pene- Prefix meaning almost, nearly.

penecontemporary Almost contemporary; occurring at more or less the same time.

penetrance The percentage frequency with which a dominant or homozygous recessive gene is manifest in the phenotype.

penta- Prefix meaning five, fivefold.

per- Prefix meaning through.

per mille Per thousand parts.

per se As such; intrinsically; of or by itself.

percentage plant aeration (PPA) The proportion of root oxygen derived by transport down from the shoot, calculated as PPA $= 100(ft - rl)/Re$, where ft is the flux into the top of the root, rl is the rate of radial loss and Re is the total respiration rate.

percentage similarity (PS) A measure of the similarity of communities, expressed as $Ps = 100 - 0.5 \Sigma(a - b)$ in which a and b are the percentages of importance values, such as density, for given species in samples A and B.

perch A unit of length equal to 16.5 feet (5.0292 m); rod; pole.

percolation Downward movement of water through porous sediments.

perdominant A species that is present in all, or nearly all, of the associations of a formation.

perennation The survival of plants from year to year with an intervening period of reduced activity; **perennial, perennate.**

perennial 1: Used of plants that persist for several years with a period of growth each year; restibilic. 2: Occurring throughout the year; *cf.* annual, biennial.

perfect 1: Pertaining to a hermaphrodite or bisexual flower; monoclinous. 2: Pertaining to fungi or fungal life cycle stages that produce sexual spores.

pergelicolous Living in geloid soils *q.v.* having a crystalloid content below 0.2 parts per thousand; **pergelicole;** *cf.* gelicolous, halicolous, perhalicolous.

perhalicolous Living in haloid soils *q.v.* having a crystalloid content above 2.0 parts per thousand; **perhalicole;** *cf.* gelicolous, halicolous, pergelicolous.

peri- Prefix meaning around.

pericontinental sea A shallow sea covering a modern continental shelf, typically less than 200 m in depth; *cf.* epicontinental sea.

perigean tides The tides of increasing amplitude occurring at the time the moon is nearest the Earth.

perigee The point in the orbit of the moon or other satellite nearest the Earth; *cf.* apogee.

perigenesis The theory of heredity, that somatic

cells are modified by the environment and by the activity of the organs containing them, and that these changes are passed as wave motions to the sex cells and influence the course of heredity.

perihelion The point in the Earth's orbit nearest to the sun; *cf.* aphelion.

period 1: A division of geological time; see Appendix 1. 2: The time interval of one complete cycle; **periodicity**.

periodic plankton Organisms that enter the plankton only at certain periods of the day, month or season.

periodicity The periodic or rhythmic occurrence of an event; also the duration of a single phase of such an oscillation.

peripheral isolate A population isolated at or beyond the outer boundary of the species' range.

periphyton A community of plants, animals and associated detritus adhering to and forming a surface coating on stones, plants and other submerged objects; aufwuchs; **periphyte**, **periphytic**.

perittogamy Random cytoplasmic fusion between undifferentiated cells of a gametophyte generation, without nuclear fusion.

permafrost Permanently frozen subsoil.

permanence of the continents, theory of That the continents have always existed and have occupied the same relative positions as they do today; fixed continent theory.

permanent heterozygosity Permanent hybrid *q.v.*

permanent hybrid An organism in which heterozygosity is fixed by the action of balanced or partially balanced lethal factors; permanent heterozygosity.

permanent wilting The condition in which the leaves of a plant are unable to recover turgidity even in saturated air without the addition of water to the soil.

permanent wilting percentage The wilting coefficient *q.v.* expressed as a percentage.

permanent wilting point (PWP) The soil water level at which permanent wilting occurs, so that a plant cannot drive its water potential below that of the soil; a soil water potential of -15 bar $(-15 \times 10^3$ N m$^{-2})$ is taken as the conventional norm for the PWP.

permeability 1: The condition of having pores or spaces that permit the passage of fluid molecules; measured in soils as the rate of passage of water through a given cylindrical section of a core sediment. 2: A measure of the freedom of entry of new members into a community or society.

permeant An organism that frequently moves from one community to another.

Permian A geological period of the Palaeozoic (*ca.* 290–245 million years B.P.); see Appendix 1.

permineralization A process of fossilization by the infiltration and subsequent precipitation of dissolved minerals into the pore spaces of bones, shells and other skeletal parts.

perpelic Used of plants that live in areas of pure or abundant clay.

perpsammic Used of plants that live in areas of pure or abundant sand.

pers. comm. Abbreviation of personal communication.

persistence The number of generations that an allele persists in a population before elimination.

perthophyte A plant living on dead, or partially decayed, tissues of a living plant; **perthophytic**.

perturbation A disturbance; any departure of a biological system from a steady state.

Peru Current The northerly extension of the cold Humboldt Current off the Peruvian coast, divided into an inner coastal current and an outer oceanic current by a tongue of warm water from the South Equatorial Countercurrent; see Appendix 6.

perversum confusa An aggregation during the breeding season of individuals of the same sex but of different species.

perversum simplex The attempted mating of one male with another male.

pessimal Least favourable; used of the levels of environmental factors that are close to the tolerance limits of an organism; *cf.* optimal.

pessimum Those conditions outside the optimum for growth and development of an organism, or for the maintenance of a system; *cf.* optimum.

pesticide A chemical agent that kills insects and other animal pests.

pestilent Destructive; devastating; epidemic; pernicious; infectious; contagious; **pestilence**, **pestilential**.

peta (P) Prefix used to denote unit $\times 10^{15}$.

petrad A rock plant; petrophyte.

petric Pertaining to rock-living communities or species.

petricolous Growing within rock; endolithic; **petricole**.

petrification A process of fossilization in which tissues are preserved by impregnation with carbonate or silicate minerals; **petrified**, **petrifaction**.

petrimadicolous Used of an organism living in the surface film of water on rocks; hygropetric; **petrimadicole**.

petrium A rock community.

petrobiontic Living on or amongst rocks or stones; **petrobiont**.

petrochthium A rock-bank community.

petrochthophilous Thriving on rock-banks; **petrochthophile, petrochthophily**.

petrochthophyte A rock-bank plant; **petrochthophyta**.

petrocolous Living in rocky habitats; **petrocole**.

petrodad A plant inhabiting rocky or stony habitats; petrodophyte.

petrodium A boulder-field, or ravine community.

petrodophilus Thriving in boulder fields; **petrodophile, petrodophily**.

petrodophyte A plant inhabiting rocky or stony habitats; petrodad.

petrophilous Thriving on rocks or in rocky habitats; petrophilus; **petrophile, petrophily**.

petrophyte A rock plant; petrad; **petrophyta**.

pF Back-pull; the logarithm of the suction force of a soil, measured in centimetres of water, with which the soil water is in equilibrium.

pH The negative logarithm of the hydrogen ion (H^+) concentration, giving a measure of acidity on a scale from 0 (acid) through 7 (neutral) to 14 (alkaline).

phaenogamy Phanerogamy *q.v.*

phaenology Phenology *q.v.*

phaenotype Phenotype *q.v.*

phage Bacteriophage; a bacterial virus.

phagocytosis The ingestion of solid particulate matter by a cell.

phagotrophic Feeding by ingesting organic particulate matter; used of cells in the blood or body fluid that ingest foreign particles; **phagotroph, phagotrophy**.

-phagous Suffix meaning feeding on, eating; **-phagic, -phage, -phagy**; *cf.* acridophagous, algophagous, autocoprophagous, autophagous, bacteriophagous, biophagous, carpophagous, cerophagous, coprophagous, creophagous, cytophagous, dendrophagous, detriophagous, ectoophagous, ectophagous, endolithophagic, endophagous, entomophagous, entophagous, epilithophagic, euryphagous, galliphagous, geophagous, haematophagous, haemophagous, heterophagous, hylophagous, ichthyophagous, isophagous, lepidophagous, limophagous, lithophagic, macrophagous, macrophytophagous, meliphagous, mesolithophagic, microphagous, microphytophagous, monophagous, mycetophagous, mycophagous, myrmecophagous, necrophagous, nekrophytophagous, neustophagous, oligophagous, oophagous, ophiophagous, paedophagous, pantophagous, panphytophagous, phycophagous, phyllophagous, phytophagous, plasmophagous, pleophagous, pollenophagous, polyphagous, rhizophagous, rypophagous, saprophagous, saprophytophagous, sarcophagous, scatophagous, stenophagous, sycophagous, tecnophagous, telmophagous, xylophagous, zoophagous, zoosaprophagous.

phaneric Pertaining to conspicuous coloration; *cf.* cryptic.

phanerogamous Reproducing by conspicuous means, used in particular of a plant having conspicuous flowers; phaenogamy; **phanerogam, phanerogamy**.

phanerogenic Of known descent; used of a fossil species having an established phylogeny from species occurring in earlier geological formations; *cf.* cryptogenic.

phanerophyta scandentia Woody climbing plants with renewal buds high above ground level.

phanerophytes Tall aerial perennial plants, mostly trees or shrubs with their renewal buds at least 250 mm above ground level; subdivided on the basis of bud height above the ground into – nanophanerophyte (up to 2 m), microphanerophyte (2–8 m), mesophanerophyte (8–30 m), megaphanerophyte (greater than 30 m); *cf.* Raunkiaerian life forms.

Phanerozoic That part of geological history in which there is abundant evidence of past life in fossil remains; the aeon comprising the Palaeozoic, Mesozoic and Cenozoic eras (*ca.* 570–0 million years B.P.); see Appendix 1.

phaopelagic Pertaining to the upper 30 m of the oceanic water column.

phaoplankton The surface plankton of the upper photic zone; phaeoplankton; *cf.* knephoplankton, skotoplankton.

PhAR Photosynthetically active radiation *q.v.*

pharotaxis 1: Navigation by means of landmarks. 2: Movement towards a specific place in response to a learned or conditioned stimulus.

phase 1: A characteristic stage. 2: The basic unit in cytology *q.v.*

phase type In taxonomy, a type specimen of a different life-history stage or sex from the holotype.

phellad A plant inhabiting stony soil; phellophyte.

phellium A rock-field community.

phellophilous Thriving on stony or rocky ground; **phellophile, phellophily**.

phellophyte A plant living on gravel or a loose stony substratum; phellad; **phellophyta**.

phene Any phenotypic character that is genetically determined.

phenetic Pertaining to overall similarity based on many characters selected without regard to evolutionary history, and thus including character states arising from common ancestry (patristic affinity), parallel evolution and convergent evolution.

phenetic distance A measure of the difference in phenotype between any two points on a phylogenetic tree; *cf.* cladistic distance, patristic distance.

phenetic gap A morphological discontinuity manifested as a gap between proximate clusters of OTUs.

phenetic method A method of classification based on the criteria of overall morphological, anatomical, physiological or biochemical similarity or difference, with all characters equally weighted and without regard to phylogenetic history; neo-Adansonism; phenetics; **pheneticist**; *cf.* cladistic method, evolutionary method, omnispective method.

phengophilous Thriving in, or having an affinity for, light; phengophilic; **phengophil, phengophile, phengophily**; *cf.* phengophobous.

phengophobous Intolerant of light; phengophobic; **phengophobe, phengophoby**; *cf.* phengophilous.

phenocline A graded series of phenotype frequencies within the geographical range of a species.

phenocontour Isophene *q.v.*

phenocopy An environmentally induced phenotypic variation that resembles the effect of a known gene mutation.

phenocritic Used of that phase during development at which two different genotypes begin to diverge morphologically and physiologically.

phenocritical period That phase during development at which the expression of a gene is most easily affected by externally applied factors.

phenodeme A local interbreeding population characterized by observed structural and functional properties (phenotype).

phenodeviant An individual that deviates phenotypically from the population norm due to special gene combinations, such as excessive homozygosity.

phenogenesis The development of the genetically determined phenotypic characters or traits.

phenogenetics A branch of genetics; the study of genotypic expression in the development of the phenotype.

phenogram A branching diagram representing degrees of overall phenetic similarity of taxa from which phylogenetic relationships may be inferred; *cf.* cladogram.

phenology Study of the temporal aspects of recurrent natural phenomena, and their relation to weather and climate; phaenology; phenomenology.

phenome Phenotype *q.v.*

phenomenology Phenology *q.v.*

phenometry The quantitative measurement of plant growth, mass and leaf area; **phenometric**.

phenon A sample or group of phenotypically similar organisms, used in numerical taxonomy to replace 'taxon'; males and females of sexually dimorphic species belong to different phenons.

phenon line A horizontal line on a dendrogram or phenogram indicating group affiliation.

phenophases Those externally observable phases of the life cycle of a plant.

phenotype The sum total of observable structural and functional properties of an organism; the product of the interaction between the genotype and the environment; reaction type; phaenotype; phenome.

phenotypic expression The manifestation of a particular gene.

phenotypic plasticity The capacity for marked variation in the phenotype as a result of environmental influences on the genotype during development.

phenotypic polymorphism The presence of two or more phenotypes in a population.

phenotypic sex determination The condition in which the sex of an individual is determined by external environmental factors, and not directly by syngamy or karyogamy; environmental sex determination.

phenotypic variance The total variation in a given phenotypic character or trait, comprising both environmental variance and genetic variance.

pheromone A chemical messenger secreted by an organism that conveys information to another individual, often eliciting a specific response; ectohormone; sociohormone.

phi grade scale A logarithmic transformation of the Udden grade scale of sediment particle size categories based on the negative log to the base 2 of the particle diameter in millimetres; phi; phi unit; phi notation; Φ; see Appendix 11.

phi notation Phi grade scale *q.v.*

phi quartile deviation (QDΦ) A measure of the particle size distribution in a sediment sample, calculated as $QD\Phi = Q3\Phi - Q1\Phi/2$ where $Q1\Phi$ and $Q3\Phi$ are the first and third quartiles: a value of zero is given by a sample having particles of one size only, the $QD\Phi$ value increasing with increasing heterogeneity of particle sizes.

phi quartile skewness (SkqΦ) A measure of the departure from a normal distribution of the particle size distribution within a sediment, calculated as $Skq\Phi = (Q_3\Phi + Q_1\Phi - 2Q_2\Phi) \times 0.5$, where $Q_1\Phi$, $Q_2\Phi$ and $Q_3\Phi$ are the first, second and third quartiles: negative values indicate a greater heterogeneity amongst the larger particles, positive values indicating greater heterogeneity amongst smaller particles.

-philic -Philous *q.v.*

philopatric Exhibiting a tendency to remain in the native locality; used of species or groups that show little capacity to spread or disperse and of individuals that tend to remain in, or return to, their home areas or domiciles; **philopatry.**

philoprogenetive Prolific; producing large numbers of offspring.

philothermic Thriving in a warm climate; **philotherm, philothermy.**

-philous 1: Suffix meaning loving, thriving in; **-phil; -phile; -philic; -philus; -phily;** *cf.* acarophilous, actophilous, acidophilous, acrodendrophilous, acrohygrophilous, acrophilous, agrophilous, aigialophilus, aiphyllophilous, aithalophilus, aletophilous, alkaliphilic, alsophilus, amathophilus, ammochthophilous, ammophilous, ancophilous, androphilous, anemophilous, anheliophilous, anthophilous, anthropophilous, argillophilous, barophilic, basophilous, bathophilus, bathyphilous, biophilous, bryophilous, calciphilic, carboxyphilic, chalicophilous, chasmophilous, cheradophilous, chersophilous, chianophilous, chimonophilous, chionophilous, chledophilous, chomophilous, circumneutrophilous, coniophilous, conophorophilous, coprophilous, coryphilous, cremnophilous, crenophilous, crymophilous, cryophilic, dendrophilous, drimophilous, enaulophilous, eremophilous, euhydrophilous, eurotophilous, frigophilic, geophilic, gypsophilous, halophilous, heliophilous, helioxerophilous, helohylophilous, helolochmophilous, helophilous, helorgadophilous, hemerophilous, hemichimonophilous, hydrophilous, hydrotribophilus, hylodophilus, hylophilous, keratinophilic, laurophilus, lichenophilous, lignophilic, limnodophilous, limnophilous, lithophilous, lochmodophilous, lochmophilus, lophophilus, lygophilous, macrothermophilus, melangeophilous, melittophilous, mesochthonophilus, mesophilic, mesothermophilous, microaerophilic, microphilic, microthermophilus, myrmecophilous, namatophilus, necrophilous, nitrophilous, nomophilous, noterophilous, oceanophilus, ochthophilus, oligonitrophilous, ombrophilous, opophilous, orgadophilous, ornithocoprophilous, orophilous, osmophilic, oxygeophilus, oxylophilous, pagophilous, pediophilous, pelagophilus, pelochthophilous, pelophilous, petrochthophilous, petrodophilus, petrophilous, phellophilous, phengophilous, photophilic, phretophilus, phycophilic, planktophilous, pluviophilous, poleophilous, polyhalophilic, pontophilous, poophilous, potamophilous, psamathophilous, psammophilous, pseudoxerophilous, psilophilus, psychrophilic, ptenophyllophilus, ptenothalophilus, pyrophilous, pyroxylophilous, rheophilous, rhizophilous, rhoophilous, rhyacophilus, rugophilic, saprophilous, sathrophilus, sciophilous, scotophilic, serpentinophilous, siderophilous, skiophilous, skotophilic, spermatophilic, sphagnophilous, spiladophilus, stasophilous, stenothermophilic, sterrhophilus, stygophilic, subhydrophilous, substratohygrophilous, subxerophilous, syrtidophilus, taphrophilus, telmatophilous, termitophilic, thalassophilus, thamnophilic, thermophilic, thinophilous, thiophilic, tiphophilous, troglophilic, tropophilous, turfophilous, umbrophilic, urophilic, xerohylophilous, xerophilous, xeropoophilous, xylophilous.

2: Suffix meaning pollinated by; *cf.* allophilous, aerophilous, anemophilous, autophilous, cantharophilous, chiropterophilous, dientomophilous, drosophilus, entomophilous, euphilous, hydrophilus, lepidopterophilous, malacophilous, melittophilous, micromelittophilous, micromyiophilous, microzoophilous, myiophilous, myrmecophilous, necrocoleopterophilous, ornithophilous, protozoophilous, psychophilous, sapromyiophilus, sphingophilous, zoidiophilous, zoophilous.

-phobic Suffix meaning intolerant of, lacking affinity for; **-phobous, -phobe, -phoby.**

phobo- Combining affix meaning manifest fear; sometimes used to negate the operative prefix as in phobophototropism that means negative phototropism.

phobotaxis An avoidance reaction of a motile organism; **phobotactic.**

phobophototropism Negative phototropism *q.v.*

pholadophyte A plant living in hollows, intolerant of high light intensity.

phoresy A symbiosis in which one organism is merely transported on the body of an individual of a different species.

phorophyte The host plant of an epiphyte *q.v.*

phosphorescence The emission of light in darkness by the release of absorbed radiation; also the light so produced; sometimes incorrectly applied to bioluminescence; **phosphorescent.**

phot The cgs unit of surface illumination, equal to 1 lumen per square centimetre; see Appendix 13.

photic zone The surface zone of the sea or a lake having sufficient light penetration for photosynthesis; *cf.* dysphotic zone, euphotic zone.

photo- Prefix meaning light.

photoautotrophic Used of organisms that obtain metabolic energy from light by a photochemical process; phototrophic; **photoautotroph, photoautotrophy;** *cf.* chemoautotrophic.

photocleistogamic Used of a plant in which self-pollination occurs within flowers that remain closed either because of insufficient illumination or as a result of unilateral photohyponastic growth; **photocleistogamy.**

photocliny A response of an organism to the direction of incident light.

photocybernetic effects The role of solar radiation as a stimulus regulating plant development.

photodestructive effects The role of solar radiation as a cause of injury to plants.

photodynamics Study of the effects of light on living organisms and on metabolic processes.

photoenergetic effects The role of solar radiation as a source of energy.

photoepinasty Upward curvature induced by light; **photoepinastic;** *cf.* photohyponasty.

photogenic Light-producing; bioluminescent; **photogenesis, photogenetic.**

photogrammetric Pertaining to the use of photography as a field technique for quantitative ecological analysis, as in measurements of density taken from photographs.

photographotype Prototype *q.v.*

photoharmosis A response of an organism to light; **photoharmose.**

photoheterotrophic Photoorganotrophic *q.v.*

photohoramotaxis The directed reaction of a motile organism to colour, or to a light-pattern stimulus; photohorotaxis; **photohoramotactic.**

photohyponasty Downward curvature induced by light; **photohyponastic;** *cf.* photoepinasty.

photoinhibition The cessation of growth or activity due to light.

photokinesis A change in random movement of an organism in response to a light stimulus; **photokinetic.**

photolithotrophic Used of organisms that utilize radiant energy and inorganic electron donors; **photolithotroph;** *cf.* photoorganotrophic.

photolytic Pertaining to chemical decomposition by the action of radiant energy; **photolysis.**

photometric Used of leaves that assume a definite orientation relative to the direction of incident light and to light intensity.

photomorphogenesis 1: The influence of solar radiation on the structure and development of a plant at the subcellular, cellular and individual levels; **photomorphogenetic.** 2: The production by plants of particular form or structure in response to level of illumination or light regime.

photomorphosis The influence of light on the growth and development of plants, independent of photosynthesis; **photomorphic.**

photonasty Growth curvature in response to a diffuse light stimulus; **photonastic.**

photoorganotrophic Used of organisms that utilize radiant energy and organic electron donors; photoheterotrophic; **photoorganotroph;** *cf.* photolithotrophic.

photopathic Photophobic *q.v.*

photoperiod The light phase of a light–dark cycle; *cf.* nyctiperiod.

photoperiodic Pertaining to the response of an organism to changes in day-length or to a light–dark cycle; photoperiodism.

photophilic Thriving in conditions of full light; luciphilous; photophilous; **photophile, photophily;** *cf.* photophobous.

photophobic Intolerant of, or avoiding, conditions of full light; lucifugous; photopathic; photophobous; photophygous; **photophobe, photophoby;** *cf.* photophilic.

photophobotaxis A response of an organism to a temporal change in light intensity; positive when stimulated by a decrease in light intensity and negative when stimulated by an increase; **photophobotactic.**

photophygous Photophobic *q.v.*

photoplagiotropism Orientation at an oblique angle to incident light; **photoplagiotropic.**

photorespiration A metabolic process involving the uptake of oxygen and release of carbon dioxide, exhibited by many green plants when exposed to light, and which ceases in the dark; occurs only in chlorophyllous plant tissue and in part utilizes the photosynthetic mechanism in the chloroplasts.

photosynthates The products of carbon dioxide assimilation during photosynthesis.

photosynthesis The biochemical process that utilizes radiant energy from sunlight to synthesize carbohydrates from carbon dioxide and water in the presence of chlorophyll; **photosynthetic;** *cf.* chemosynthesis.

photosynthetic capacity The maximum rate of nett photosynthesis by a plant at a particular stage of development under natural conditions of atmospheric carbon dioxide content and optimal conditions with respect to other external factors.

photosynthetic efficiency coefficient (K_F) The ratio of gross photosynthesis to total respiration; commonly approximated by the ratio of the sum of nett photosynthesis (F_n) and respiration (R) to respiration: $K_F = (F_n + R)/R$

photosynthetic energy utilization efficiency coefficient The proportion of absorbed energy fixed by a plant in the form of chemical bonds by the conversion of carbon dioxide to carbohydrate, expressed as a percentage.

photosynthetic quotient The ratio of the amount of oxygen produced to the amount of carbon dioxide absorbed during photosynthesis.

photosynthetically active radiation (PhAR) Radiation capable of driving the primary processes of photosynthesis, in the range of wavelengths between 380–710 nm.

phototaxis A directed response of a motile organism towards (positive) or away from (negative) a light stimulus; phototopotaxis; **phototactic**.

phototonic Sensitive to light; phototonus; **phototonicity**.

phototopotaxis Phototaxis *q.v.*

phototrophic Photoautotrophic *q.v.*

phototropism An orientation response to light; **phototropic**.

phototype A photograph of a type in the absence of the type specimen; not usually given nomenclatural status; fototype; photographotype.

phragmosis The action of closing the entrance to a nest or burrow with the body.

phratry A neutral classificatory term for a subtribe or clan.

phreatic Pertaining to ground-water.

phreaticolous Inhabiting ground-water (phreatic) habitats; phreatobie; **phreaticole**.

phreatobiology The study of ground-water organisms.

phreatophyte A plant that absorbs water from the permanent water table.

phretium A water tank community.

phretophilous Thriving in water tanks; phretophile, phretophily.

phretophyte A plant inhabiting a water tank; phretophyta.

-phyceae The ending of a name of a class in botanical nomenclature; see Appendix 10.

-phycidae The ending of a name of a subclass in botanical nomenclature; see Appendix 10.

phycobiont The algal partner of a lichen (an algal/fungal symbiosis); *cf.* mycobiont.

phycocoenology The study of algal communities.

phycology The study of algae.

phycophagous Feeding on algae; phycophage, phycophagy.

phycophilic Thriving in algae-rich habitats; living on algae; phycophile, phycophily.

phylacobiosis Parabiosis *q.v.*; phylacobiotic.

phylesis The course of evolution or phylogenetic development; **phyletic**.

phyletic Pertaining to a line of direct descent, or a course of evolution.

phyletic evolution A sequence of changes occurring more or less uniformly throughout the entire geographical range of a species over an extended period of time; often regarded as a type of speciation since it may establish a sequence of distinct species (chronospecies) within a single line of descent through geological time, although it does not increase the number of species in existence at any time.

phyletic extinction Used in a phylogenetic context for the loss or extinction of an ancestral species which, since the species lineage continues, does not actually result in a nett loss of species; pseudoextinction; *cf.* terminal extinction.

phyletic gradualism A model of evolution in which species change gradually through time, by slow directional transformation within a lineage, producing a long and graded series of differing forms; gradualism; *cf.* punctuated evolution.

phyletic lineage An ancestor–descendant sequence between two successive branching points of a phylogenetic tree, which may comprise two or more successive chronospecies.

phyletic speciation The gradual evolution within a phyletic lineage of one species from another, eventually leading to its replacement by that species; successional speciation.

phyletic tree Phylogenetic tree *q.v.*

phyletic trogloxene An organism only occasionally found in caves that does reproduce in the cave habitat; *cf.* aphyletic trogloxene.

phylistic Pertaining to concepts of phylogenetic relationship based on both cladogenesis and evolutionary divergence; **phylist, phylistics**.

phylktioplankton Planktonic organisms rendered buoyant by hydrostatic means.

phyllobiology The study of leaves.

phyllophagous 1: Feeding on leaves; phyllophagic; **phyllophage, phyllophagy**. 2: Pertaining to plants that obtain nourishment from their leaves; phyllophyte.

phyllosphere The microhabitat of a leaf; the immediate surroundings of plant leaves that are influenced by the presence of the leaves; *cf.* rhizosphere.

phyllotaxis The arrangement of leaves on a stem; **phyllotaxy**.

phylogenesis The evolutionary history of a taxon; phylogeny; **phylogenetic**.

phylogenetic Pertaining to evolutionary relationships within and between groups; phylogeny.

phylogenetic classification Natural classification *q.v.*

phylogenetic inertia Those underlying genetic mechanisms and preadaptations that determine the direction and rate of evolutionary change; *cf.* prime mover.

phylogenetic noise Misinterpretation of phylogenetic relationships due to convergent evolution.

phylogenetic relationship Evolutionary relationship; affinity based on recency of common ancestry.

phylogenetic systematics A method of classification based on the study of evolutionary relationships between species in which the criterion of recency of common ancestry is fundamental and is assessed primarily by recognition of shared derived character states.

phylogenetic tree A branching diagram in the form of a tree representing inferred lines of descent; evolutionary tree; phyletic tree; phylogram.

phylogeny 1: The evolutionary history of a group or lineage. 2: The origin and evolution of higher taxa; phylogenesis.

phylogerontic In the phylogeny of a group, pertaining to the senescent stages following the point of maximum vigour or peak of development; paracmic; *cf.* phyloneanic.

phylogram Phylogenetic tree *q.v.*

phylon 1: A line of descent. 2: Phylum *q.v.*

phyloneanic An adolescent or youthful stage in the phylogeny of a group; *cf.* phylogerontic.

phylum (phyla) A rank within the zoological hierarchy of classification; the principal category immediately below kingdom; see Appendix 10.

physical environment The abiotic component of an ecosystem; all structural and chemical environmental factors.

physically controlled communities Communities formed under conditions of high physiological stress and controlled largely by physical fluctuations and unfavourable physical conditions of the environment; stability-time hypothesis *q.v.*

physiogenesis 1: The origin and development of physiological processes. 2: Cellular differentiation during ontogeny.

physiognomic dominant The dominant species of a consociation having the life form characteristic of the formation; consociation dominant.

physiognomy The characteristic features or appearance of a plant community or vegetation; **physiognomic**.

physiographic Pertaining to geographical features of the Earth's surface; **physiography**.

physiographic climax A climax community determined largely by topographic and edaphic factors; *cf.* climax.

physiological drought A condition of drought in plants caused by factors affecting the uptake of water rather than by lack of soil water.

physiological longevity The maximum life span of an organism that dies of old age.

physiological race A race characterized by physiological properties; physiological form.

physiological specialization The adaptive radiation of a species into two or more physiological races.

physiology Study of the normal processes and metabolic functions of living organisms; **physiological**.

physogastry A condition of excessive enlargement of the abdomen, as found in some insects.

-phyta The ending of a name of a division in botanical nomenclature; see Appendix 10.

phytal zone That part of a shallow lake bottom supporting rooted vegetation.

phytalfauna The fauna of protected or concealed plant microhabitats.

-phytina The ending of a name of a subdivision in botanical nomenclature; see Appendix 10.

-phytium Suffix denoting a plant community.

phyto- (-phyt-) Combining affix meaning plant.

phytobenthos 1: A bottom-living plant community; phytobenthon. 2: That part of the bottom of a stream or lake covered by vegetation; *cf.* geobenthos.

phytobiology Plant biology.

phytobiontic Pertaining to organisms that spend most of their active life on or within plants; planticolous; phytobiotic; **phytobiont**.

phytochemistry Chemotaxonomy of plants.

phytocidal Lethal to plants.

phytocide A chemical used to destroy and control plants.

phytocoenology Phytosociology; the study of plant communities.

phytocoenosis 1: The total plant life of a given habitat or community. 2: A syntaxon in the classification of plant communities comprising one or more synusiae *q.v.*

phytodyte A plant living above ground level.

phytoecology Study of the relationships between plants and their environment; plant ecology.

phytoedaphon Soil flora; the plant community of the soil.

phytogamy Cross fertilization in plants.

phytogenesis 1: The origin of plants on Earth; phytogeogenesis. 2: The evolution and development of a plant species; **phytogeny**.

phytogenic Arising from or caused by, plants.

phytogeocoenosis The assemblage of plants in a given habitat or community and its physical environment.

phytogeogenesis Phytogenesis *q.v.*

phytogeographical kingdoms The major geographical divisions of the world characterized by floristic composition; the six commonly recognized kingdoms are – Antarctic, Australian, Boreal, Neotropical, Palaeotropical and South African; see Appendix 4.

phytogeography Study of the biogeography of plants; geobotany.

phytography Descriptive botany.

phytoliths Plant fossils; small particles of opaline silica found in the cell walls of some plants and studied as plant trace fossils.

phytology Botany.

phytomass Plant biomass; any quantitative estimate of the total mass of plants in a stand, population, or within a given area, at a given time.

phytome 1: A plant community; phytoma; phytomata. 2: The totality of an individual plant.

phyton The smallest detached fragment that can develop into a new plant; propagation unit.

phytoneuston The plant component of neuston *q.v.*; **phytoneustonic, phytoneustont**.

phytopalaeontology Palaeobotany *q.v.*; phytopaleontology.

phytoparasite A parasitic plant.

phytopathogen Any pathogenic organism occurring on a plant host.

phytopathology The study of plant disease.

phytophagous Feeding on plants or on plant material; herbivorous; phytophilous; **phytophage, phytophagy**.

phytophenology Study of the periodic phenomena of plants, such as flowering and leafing.

phytophilous 1: Phytophagous *q.v.* 2: Planticolous *q.v.*; **phytophile, phytophily**.

phytoplankton Planktonic plant-life; **phytoplankter, phytoplanktont**; *cf.* zooplankton.

phytopleuston Plants free floating in aquatic habitats.

phytosis A disease caused by a plant parasite.

phytosociology The study of vegetation, including the organization, interdependence, development, geographical distribution and classification of plant communities; plant sociology.

phytostrote An organism that migrates or disperses through the agency of plants.

phytosuccivorous Feeding on sap; sap-sucking; **phytosuccivore, phytosuccivory**.

phytotelmic Used of organisms that inhabit small pools of water within or upon plants.

phytoteratology The study of malformations and monstrosities in plants; plant teratology.

phytotopography Study of the vegetation of a given region.

phytotoxic Poisonous to plants; **phytotoxicity**.

phytotoxin A substance toxic to plants.

phytotron A laboratory for the large-scale culture of plants under controlled conditions.

phytotrophic Autotrophic *q.v.*

pico- (p) Prefix used to denote unit $\times$ 10^{-12}.

picoplankton Planktonic organisms between 0.2 and 2.0 μm in diameter.

pictograph A graphic display of biological information using pictorial symbols.

pie chart A circle divided radially into sectors whose sizes are proportional to the quantities they represent; often used for representing data arranged in percentage categories.

piezo- Prefix meaning pressure.

piezoelectric Pertaining to a crystalline substance the electrical property of which is changed by pressure.

piezotropism A growth movement in response to a compression stimulus; **piezotropic**.

pigmentation rule Gloger's rule *q.v.*

pinnigrade A form of locomotion by swimming using flippers as paddles.

pinocytosis Active ingestion of fluid by a cell, by invagination of the cell membrane to form vesicles; **pinocytotic**.

pint A measure of volume equal to 1/8 gallon (0.56826 dm³); see Appendix 13.

pioneer The first species or community to colonize or recolonize a barren or disturbed area, thereby commencing a new ecological succession.

piscicolous Living on or within fish; **piscicole**.

piscivorous Feeding on fish; ichthyophagous; **piscivore, piscivory**.

pladobolous Used of a plant dispersed by the action of moisture.

pladopetric Pertaining to moist or wet rocks.

plagio- Prefix meaning oblique.

plagioclimax A climax community formed following the deflection of an ecological succession by the activity or influence of man.

plagiogeotropism Geoplagiotropism *q.v.*; **plagiogeotropic**.

plagiosere An ecological succession deflected from its natural course by continuous human interference; **plagioseral**.

plagiotropism An orientation response at an oblique angle to the vertical; **plagiotropous, plagiotropic**.

planetic Motile; possessing motile or swarming stages; **planetism**.

plankter An individual planktonic organism; phytoplankter; zooplankter; **plankt, planktont**.

planktobiontic Living solely in the plankton; holoplanktonic; **planktobiont**.

planktohyponeuston Planktonic organisms that

congregate just below the water surface at night but migrate downwards during the day.

plankton Those organisms that are unable to maintain their position or distribution independent of the movement of water or air masses; see Appendix 9 for size categories; *cf.* nekton.

plankton pulse A periodic fluctuation in the abundance of plankton, or of a particular species, in a given area.

planktont An individual planktonic organism; phytoplanktont; zooplanktont; plankt; plankter.

planktophilous Living or thriving in the plankton; **planktophile, planktophily.**

planktophyte A planktonic plant; an individual of the phytoplankton *q.v.*

planktotrophic Feeding on plankton; **planktotroph, planktotrophy.**

planktoxenic Occurring only occasionally in the plankton; **planktoxene.**

planogamete A motile gamete; zoogamete; *cf.* aplanogamete.

planomenon All free living organisms; those organisms not rooted or attached to a substrate; *cf.* ephaptomenon, rhizomenon.

planont A motile gamete, spore or zygote; planospore.

planophyte A free floating freshwater plant; either a large pleustophyte *q.v.* or small planktophyte *q.v.*; planophyton.

Planosol An intrazonal soil with an eluviated surface layer over a strongly compacted eluviated claypan, formed in humid to subhumid climates.

plant 1: Any member of the kingdom Plantae. 2: A multicellular eukaryote organism typically exhibiting holophytic nutrition, lacking locomotion, lacking obvious nervous or sensory organs, and possessing cellulose cell walls.

plant sociology Phytosociology *q.v.*

planticolous Used of organisms that spend most of their active life on or within plants; phytobiontic; phytophilous; **planticole.**

plantigrade Walking with the entire sole of the foot in contact with the ground; palmigrade.

plasmagene Any cytoplasmic hereditary particle.

plasmagynogamous Used of fertilization in which the cytoplasm of the zygote is derived only from the female gamete.

plasmaheterogamous Used of fertilization in which the cytoplasm of the zygote is derived unequally from both male and female gamete.

plasmaisogamous Used of fertilization in which the cytoplasm of the zygote is derived equally from both male and female gamete.

plasmatogamy Plasmogamy *q.v.*

plasmogamy Fusion of the cytoplasm of two or more cells without nuclear fusion; plasmatogamy.

plasmon The totality of all extrachromosomal hereditary factors (plasmagenes); plasmotype.

plasmophagous Feeding on body fluid; used of a parasite that freely absorbs the cellular fluid of the host; **plasmophage, plasmophagy.**

plasmotomy The division of a multinucleated cell into multinucleated daughter cells, without accompanying mitosis.

plasmotype The sum total of extrachromosomal hereditary factors; plasmon.

plastic strain An irreversible physical or chemical change in a living organism produced by stress.

plasticity The capacity of an organism to vary morphologically, physiologically or behaviourally as a result of environmental fluctuations.

plastidiotype That part of the total hereditary determinants of a plant located within the plastids; plastom.

plastidule A subcellular building block, conceived by Haeckel as the basic unit of life.

plastocotype In taxonomy, a mould taken from a cotype *q.v.*

plastodeme A local interbreeding population characterized by environmentally induced phenotypic features.

plastoecodeme An ecodeme *q.v.* characterized by environmentally induced phenotypic features.

plastogamy The fusion of the cytoplasm of unicellular organisms to form a plasmodium, without fusion of the nuclei.

plastogene A hereditary factor contained within a plastid.

plastoholotype In taxonomy, a mould taken from a holotype *q.v.*

plastom Plastidiotype *q.v.*

plastotype In taxonomy, a mould taken from a primary type.

plate tectonics The concept that the Earth's crust is divided into a number of rigid plates that are in motion relative to each other; the plates are formed at the midocean ridges (seafloor spreading) and typically destroyed in deep sea trenches.

platykurtic Used of a frequency distribution that is less peaked than the corresponding normal distribution curve; **platykurtosis;** *cf.* kurtosis.

play Trial and error learning especially in young animals.

pleio- Prefix meaning more.

pleiocyclic Existing through more than one cycle of activity.

pleiomorphic Exhibiting polymorphism at different stages of the life cycle; pleomorphic;

polymorphic; **pleiomorph**, **pleiomorphism**, **pleiomorphous**, **pleiomorphy**.

pleiotropic Used of a gene that has more than one, apparently independent, phenotypic effect; **pleiotropism**, **pleiotropy**.

pleioxenous Used of a parasite that is not host specific, or a parasite that has several hosts during its life cycle; heteroxenous; polyxenic; **pleioxeny**.

pleisiotypic state The ancestral character state in a transformation series *q.v.*; plesiotypic state; *cf.* apotypic state.

Pleistocene A geological epoch of the Quaternary period (*ca.* 1.6–0.01 million years B.P.); Great Ice Age; Oiluvium; see Appendix 1.

Pleistocene glaciation Quaternary Ice Age *q.v.*

Pleistocene refuge An area peripheral to a major ice sheet, or an elevated, ice-free area within an ice sheet (a nunatak) that has been relatively unaltered by glaciation during the Pleistocene and which has a biota once typical of the region as a whole; glacial refugium.

plenary power In taxonomy, the authority of the ICZN to suspend the provisions of the Code *q.v.*

pleogamy The maturation and pollination of different flowers on an individual plant at different times; **pleogamic**.

pleometrosis The founding of a colony of social organisms by more than one original female; **pleometrotic**; *cf.* monometrosis.

pleomorphic Polymorphic; assuming various shapes or forms within a species or group, or during a single life cycle; pleiomorphic; **pleomorphism**, **pleomorph**, **pleomorphous**.

pleonasm In taxonomy, the use of a specific epithet with a generic name in which both have the same or similar meaning, but not identical orthography (tautology); equivalent in zoology to virtual tautonym *q.v.*

pleophagous 1: Feeding on a variety of food substances or food species; polyphagous; euryphagous; plurivorous; pleotrophic; **pleophage**, **pleophagy**. 2: Used of a parasite associated with a variety of hosts.

pleophyletic Polyphyletic *q.v.*

pleotrophic Pleophagous *q.v.*

plesiobiotic Used of organisms living in close proximity to one another; **plesiobiosis**.

plesiomorphic 1: Used of ancestral or primitive characters or character states; pleisiomorphic; *cf.* apomorphic. 2: Having similar shape or structure; **plesiomorph**, **plesiomorphous**.

plesiotype In taxonomy, a specimen identified by a subsequent author as belonging to a particular species.

pleuriilignosa A rain forest or rain bush community; pluviilignosa.

pleustohelophyte A plant which floats at the surface of a water body but which also has emergent structures.

pleuston 1: Aquatic organisms that remain permanently at the water surface by their own buoyancy, normally positioned partly in the water and partly in the air; **pleustonic**, **pleustont**; *cf.* neuston. 2: All free-floating macroscopic plants.

pleustophyte A macroscopic free-floating plant.

plexus A network; applied to a reticulation in the phylogeny of a taxon that results from sexual reproduction or hybridization.

Pliocene A geological epoch within the Tertiary period (*ca.* 5.0–1.6 million years B.P.); see Appendix 1.

pliothermic Pertaining to the geological periods characterized by warmer than average climates; *cf.* miothermic.

ploidy The number of sets of chromosomes present; *cf.* monoploid, haploid, diploid, triploid, tetraploid, polyploid, euploid, aneuploid, allopolyploid, autopolyploid.

plot A circumscribed sampling area for vegetation.

plotophyte A floating plant, usually having special flotation structures.

pluriparous 1: Producing several offspring within a single brood; multiparous; *cf.* biparous, uniparous. 2: Having produced more than one previous brood.

plurivorous Feeding on a variety of different food sources or food species; euryphagous; polyphagous; pleophagous; **plurivore**, **plurivory**.

pluvial 1: Pertaining to, or resulting from, the action of rain or precipitation. 2: Used of a geological period, or the climate, characterized by abundant rainfall.

pluvial periods Periods during the Quaternary *q.v.* characterized by excessive rainfall and the formation of great lakes in tropical and subtropical regions.

pluviilignosa Tropical and subtropical evergreen rain forest vegetation; pleuriilignosa.

pluviofluvial Pertaining to the combined action or effects of rainfall and streams.

pluviophilous Thriving in conditions of abundant rainfall; **pluviophile**, **pluviophily**; *cf.* pluviophobous.

pluviophobous Intolerant of conditions of abundant rainfall; **pluviophobe**, **pluviophoby**; *cf.* pluviophilous.

pluviotherophyte A short lived vascular plant that germinates after heavy rainfall and rapidly completes the life cycle whilst moisture levels remain adequate.

pneumotaxis The directed response of a motile organism towards (positive) or away from (negative) a stimulus of dissolved carbon dioxide or other gas; pneumatotaxis; **pneumotactic**.

pneumotropism An orientation response to a stimulus of dissolved carbon dioxide or other gas; **pneumotropic**.

pnoium An ecological succession on an aeolian soil.

poad A meadow plant; poophyte.

POC Particulate organic matter, usually expressed in grams carbon per litre.

Podsol Podzol soil *q.v.*

Podzol soil A zonal soil with thin acidic layer of litter and duff over a highly acidic leached A-horizon and dark brown illuvial acidic lower horizon with translocated deposits of humus and iron oxide, formed in cool temperate to temperate, humid climates, under coniferous or mixed coniferous/deciduous forest or heath; Podsol.

podzolization A soil forming process in regions of high precipitation, good internal soil drainage, and low calcium carbonate levels; the soil becomes more acidic and iron and aluminium sesquioxides, as well as humus colloids, are translocated down the soil profile by percolating water; sesquioxide deposits in the profile are denoted by suffix *s* or *fe*.

poecilandric Used of organisms having more than one form of male; **poecilandry**; *cf.* poecilogynic.

poecilo- Poikilo- *q.v.*

poecilogony 1: The developmental condition in which adult stages of related taxa are virtually identical but juvenile stages are highly divergent; **poecilogonic**. 2: Intraspecific variation in the duration of ontogenetic stages induced by environmental factors; poikilogony.

poecilogynic Used of organisms having more than one form of female; **poecilogyny**; *cf.* poecilandric.

poic Pertaining to meadows.

poikilo- Prefix meaning various, variable; poecilo-.

poikilodynamic Used of a hybrid (heterodynamic hybrid) that displays the phenotype of one parent more or less exclusively.

poikilogony Poecilogony *q.v.*

poikilohaline Used of organisms having body fluids that conform to external changes in salinity.

poikilohydric Used of protistans, fungi and plants that cannot compensate for short-term fluctuations in water supply and rate of evaporation so that their water content varies with ambient humidity levels; *cf.* homoiohydric.

poikilosmotic Used of organisms that have an internal osmotic pressure that conforms to the osmotic pressure of the external medium; *cf.* homoiosmotic.

poikilothermic Used of an organism having no mechanism or only a poorly developed mechanism for internal temperature regulation; cold-blooded; having a body temperature that fluctuates with that of the immediate environment; ectothermic; haematocryal; heterothermic; **poikilotherm, poikilothermal, poikilothermy, poikilothermous**; *cf.* homoiothermic.

poikilothermy 1: The condition of being poikilothermic *q.v.* 2: The condition in which one or more minima coexist with one or more maxima in a temperature curve of a water column of a lake; *cf.* dichothermy, mesothermy.

point estimate In statistics, a single number estimating a population parameter; point estimation; *cf.* interval estimate.

point mutation A mutation within a single gene; infragenic mutation; genovariation; micromutation; transgenation.

point-centred quarter method A method of plotless sampling of vegetation in which lines are erected at right angles from the sampling point to produce 4 quarters in each of which a measure is taken of the distance from the sampling point to the nearest neighbour.

poise A derived cgs unit of dynamic viscosity, with the dimensions gram per centimetre per second; see Appendix 13.

Poisson distribution A mathematical distribution used to describe or test for a random distribution, or process, and as a model of randomly distributed populations in which the presence of an individual at any given point does not increase or decrease the probability of another individual occurring nearby and in which the variance is approximately equal to the mean; Poisson series.

Poisson process A random event in time.

poium A meadow community.

polar Pertaining to the areas within the Arctic and Antarctic circles; characteristic of regions around the geographical poles; *cf.* temperate, tropical.

polar lake A lake having a surface temperature that does not exceed 4° C at any time of the year.

polar zone A latitudinal zone between 72° and 90° in either hemisphere.

polarity 1: The directional differentiation or trend within a system. 2: The direction of evolution or change within a morphocline or transformation series. 3: The direction of magnetism; polarity of rocks is termed normal if it matches the Earth's present magnetic field and reversed if in the opposite direction.

pole A unit of length equal to 16.5 feet (5.03 m); perch, rod.

poleophilous Thriving in urban habitats; **poleophile, poleophily;** *cf.* poleophobous.

poleophobous Intolerant of urban habitats; **poleophobe, poleophoby;** *cf.* poleophilous.

poleotolerance, index of A qualitative scale ranking species according to their tolerance of urban habitats.

poleotolerant Tolerant of urban habitats.

poles of pendulation, theory of A biogeographical theory according to which the Earth, as well as having the North–South axis, also has an axis of pendulation the poles of which are situated in Sumatra and Ecuador; regions in the vicinity of these poles have a permanently equatorial climate while those more distant show greater climatic variation reflected in the distribution of the taxa.

pollakanthic Having several flowering periods during the life cycle; *cf.* hapaxanthic.

pollen The microspores of flowering plants containing the male gametophyte.

pollen analysis The study of pollen in Quaternary sediments; especially the qualitative and quantitative determination of pollen as an aid to archaeological, geochronological or palaeoecological research.

pollen profile A representation of a vertical section of a sediment illustrating the sequence of contained pollen, fossil pollen and spores; pollen diagram.

pollen rain The total amount of pollen and spores deposited in a given area over a specified period of time.

pollen spectrum The frequency distribution of the different kinds of pollen in a sample.

pollen statistics Pollen analysis *q.v.*; palynology *q.v.*

pollenophagous Feeding on pollen; **pollenophage, pollenophagy.**

pollination Transference of pollen from the anther to the receptive area of a flower; used loosely to mean fertilization of a seed plant.

polliniferous Adapted for the transport of pollen; pollen-bearing; pollinigerous.

pollution The contamination of a natural ecosystem, especially with reference to the activity of man; **pollute, pollutant.**

poloicous Having fertile and sterile flowers on the same and on different plants.

polotropism An orientation response tending to direct the growth of both proximal and distal

apices towards the same point; as in two growing points coming together in the conjugation of some algal cells; **polotropic.**

poly- Prefix meaning many.

polyacmic Exhibiting many abundance peaks per year; *cf.* diacmic, monacmic.

polyandrous Used of a female that mates with several males; monothelious; **polyandry;** *cf.* monandrous.

polyaxenic Comprising several axenic cultures *q.v.*

polybasic In taxonomy, used of a genus established on, or comprising, two or more species; polytypic; *cf.* monobasic.

polybrominated biphenyl PBB *q.v.*

polycarpic Producing fruit or spores more than once during a life cycle; polytokous; *cf.* monocarpic.

polychlorinated biphenyl PCB *q.v.*

polyclave Multiple-entry key *q.v.*

polyclimax A compound climax occurring within a given climatic region, that comprises a number of local edaphic climaxes.

polycyclic parthenogenesis Reproduction involving two or more cycles of alternating parthenogenetic and sexually reproducing generations per year; *cf.* parthenogenesis.

polydemic Used of species occurring in several different areas.

polydomic Used of colonies of social insects that occupy more than one nest; polydomous; *cf.* monodomic.

polyembryony The formation of multiple embryos from a single zygote, ovule or other cell.

polyergistic Used of characters determined by two or more interacting genes; polyfactorial; polygenic; multifactorial; multigenic.

polyestrous Polyoestrous *q.v.*

polyethism The division of labour within a society on the basis of morphological castes (caste polyethism) or age (age polyethism).

polyfactorial Pertaining to phenotypic characters controlled by, or to sex determined by, the integrated action of multiple independent genes; multifactorial; multigenic; polyergistic; polygenetic; polygenic; *cf.* monofactorial.

polygamous 1: Pertaining to the condition in which a single male has many female mates at any one time; monarsenous; **polygamy;** *cf.* monogamous. 2: Used of a plant having perfect and imperfect flowers together on the same individual.

polygene 1: One of an integrated group of independent genes which collectively control the expression of a character or trait; multigene; multiple factors. 2: A gene controlling quantitative characters.

polygenetic 1: Polyphyletic *q.v.* 2: Having a multiple origin either in time or space. 3: Polygenic *q.v.*

polygenic Used of characters or traits controlled by the integrated action of multiple independent genes; multifactorial; multigenic; polyergistic; polyfactorial; polygenetic; *cf.* digenic, monogenic, oligogenic, trigenic.

polygoneutic Producing several broods per year or season; multivoltine; **polygoneutism**; *cf.* monogoneutic, digoneutic, trigoneutic.

polygynopaedium An assemblage comprising one female with her offspring plus the parthenogenetic progeny of the latter; *cf.* monogynopaedium.

polygyny 1: The mating of a single male with several females. 2: The presence of many queens within a single colony of social insects; *cf.* monogyny, oligogyny.

polyhaline Pertaining to brackish water having a salinity between 10 and 17 parts per thousand; or to sea water having a salinity greater than 34 parts per thousand; *cf.* mesohaline, oligohaline.

polyhalophilic Thriving in a wide range of salinities; eurycladous; euryhaline; **polyhalophile**, **polyhalophily**.

polymictic Used of a lake having no persistent thermal stratification, which is continually circulating with only brief periods of stability; *cf.* mictic.

polymictic plankton Plankton having several dominant or abundant species; *cf.* monoton plankton, pantomictic plankton, prevalent plankton.

polymodal Used of a population or frequency distribution having three or more modes or peaks; *cf.* modality.

polymorph One of the two or more forms (morphs) of a polymorphic species.

polymorphic behaviour hypothesis A hypothesis of population self-regulation; that spacing behaviour limits population density and that individual differences in spacing behaviour have a genetic basis and respond to rapid natural selection; Chitty hypothesis; genetic-behavioural polymorphism hypothesis.

polymorphism 1: The co-occurrence of several different forms. 2: The co-existence of two or more discontinuous, genetically determined, segregating forms in a population, where the frequency of the rarest type is not maintained by mutation alone; different manifestations of polymorphism include – balanced polymorphism, chromosomal polymorphism, cryptic polymorphism, genetic polymorphism, phenotypic polymorphism, pseudoseasonal polymorphism, transient polymorphism; pleiomorphism; pleomorphism; morphism; **polymorphic**, **polymorphous**; *cf.* dimorphic, monomorphic.

polymorphism-dependent homology Paramorphic homology *q.v.*

Polynesia An assemblage of a large number of small oceanic islands scattered over the eastern Pacific Ocean, extending North to include Hawaii and East to Easter Island, sometimes also including New Zealand; used loosely as a geographical area and ethnological unit to include both eastern Melanesia *q.v.* and Micronesia *q.v.*

Polynesian region A subdivision of the Polynesian subkingdom of the Palaeotropical kingdom; see Appendix 4.

Polynesian subkingdom A subdivision of the Palaeotropical kingdom; see Appendix 4.

Polynesian subregion A zoogeographical area comprising the Polynesian and Micronesian Archipelagos from Fiji to the Marquesas Isles; regarded either as a component of the Australian region or as a separate region; originally including also the Hawaiian Archipelago which is now generally regarded as a separate region; The Pacific Islands subregion; see Appendix 3.

polynomial Comprising many words; used of a family or genus group name consisting of more than one word, a species name of more than two words, and a subspecies name of more than three words.

polynya An expanse of open water in the middle of sea ice.

polyoestrous Having a succession of breeding periods in one sexual season; polyestrous; *cf.* anoestrus, dioestrus, monoestrous.

polyoxybiotic Used of organisms requiring abundant free oxygen; polyoxybiont, polyoxybiontic.

polyphagous Feeding on a wide variety of different foods or food species; euryphagous; pleophagous; plurivorous; **polyphage, polyphagia, polyphagy**; *cf.* monophagous, oligophagous.

polyphasic Used of a species showing marked morphological variation within a given habitat; **polyphasy**.

polyphasic taxonomy Taxonomic investigation of a given set of organisms by a number of different techniques, applied successively or simultaneously.

polyphenic 1: Used of populations exhibiting polypheny *q.v.* 2: Used of experimental procedures which yield many characters.

polypheny The occurrence of extensive phenotypic variation lacking genetic fixation; commonly used in a more restricted sense for the occurrence in a population of several discontinuous phenotypes (phenes) that lack genetic foundation; polyphenism; **polyphenic**; *cf.* dipheny, monopheny.

polyphyletic Derived from two or more distinct ancestral lineages; used of a group comprising taxa derived from two or more different ancestors; pleophyletic; polyphylesis; **polyphyly**; *cf.* diphyletic, monophyletic.

polyphyletic group A group of taxa derived from two or more distinct ancestral taxa; a category based on convergence rather than common ancestry; *cf.* monophyletic group, paraphyletic group.

polyplanetic Having several motile stages and intervening resting phases during the life cycle; **polyplanetism**; *cf.* diplanetic, monoplanetic.

polyploid Having more than two sets of homologous chromosomes; **polyploidy**; *cf.* diploid, haploid, monoploid.

polyploid hybrid Allopolyploid *q.v.*

polyploid variety Autopolyploid *q.v.*

polyploidy The state of being polyploid *q.v.*

polysaprobic Pertaining to a polluted habitat having a very high concentration of decomposable organic matter and a low concentration or absence of oxygen; *cf.* saprobity system.

polysomic Used of an aneuploid *q.v.* having one or more, but not all, chromosomes in the polyploid state.

polyspermy The penetration of two or more sperms into a single ovum at the time of fertilization; **polyspermic**; *cf.* dispermy, monospermy.

polystenohaline Used of organisms that only inhabit oceanic waters of relatively constant high salinity; *cf.* euryhaline, holeuryhaline, oligohaline, stenohaline.

polytaxis Discontinuous change.

polythermic Tolerating relatively high temperatures; polythermal; *cf.* oligothermic.

polythetic key A key in which multiple contrasting characters are used simultaneously in at least some of the couplets; *cf.* monothetic key.

polythetic taxon A taxon specified by the combination of a large number of characters none of which is diagnostic alone; *cf.* monothetic.

polytokous Having many offspring per brood; fruiting many times during the life cycle; caulocarpous; multiparous; polytocous; *cf.* ditokous, monotokous, oligotokous.

polytopic Occurring or arising independently in two or more separate localities or geographical areas; **polytopism, polytopy**; *cf.* monotopic.

polytopic subspecies A subspecies comprising two or more widely separated but phenotypically identical populations.

polytrophic Feeding on a variety of different food substances or food species; **polytrophy**.

polytypic Used of a taxon comprising more than two subordinate principal taxa, or of a family or genus comprising two or more genera or species respectively; *cf.* bitypic, monotypic.

polytypic species A species comprising several subspecies or geographical variants; composite species; Rassenkreise; rheogameon; *cf.* monotypic species.

polytypic species concept Multidimensional species concept *q.v.*

polytypy In taxonomy, the establishment of a taxon based upon more than one type; **polytype**; *cf.* monotypy.

polyvoltine Multivoltine *q.v.*

polyxenic Pleioxenous *q.v.*; **polyxeny**.

polyzoic Used of a habitat that contains a wide variety of animal life.

Pomeranian glaciation A subdivision of the Weichselian glaciation *q.v.*; see Appendix 2.

pomeridianus Pertaining to the afternoon.

pomology The study and practice of fruit cultivation.

pond A small natural or artificial body of standing fresh water, intermediate in size between a pool and a lake, usually with negligible current and having more or less continuous vegetation from the marginal land areas into the water, in the absence of significant wave action; *cf.* lake.

pontic Pertaining to the deep sea.

pontium A deep sea community.

pontohalicolous Living in salt marshes; **pontohalicole**.

pontophilous Thriving in the deep sea; **pontophile, pontophily**.

pontophyte A deep sea plant; **pontophyta**.

poocolous Living in meadows; **poocole**.

pool feeder Telmophagous *q.v.*

poophilous Thriving in meadows; **poophile, poophily**.

poophyte A meadow plant; poad; **poophyta**.

Popperian method A scientific method involving the formulation of theories or hypotheses from which singular statements (predictions) are deduced that can be tested; a theory or hypothesis may thus be falsified or corroborated, but not proven; deductive method.

population 1: All individuals of one or more species within a prescribed area. 2: A group of organisms of one species, occupying a defined area and usually isolated to some degree from other similar groups. 3: In statistics, the whole group of items or individuals under investigation.

population biology Study of the spatial and temporal distributions of organisms.

population bottleneck A sharp decrease in the size of a population with consequent reduction in the size of the gene pool; *cf.* bottleneck effect.

population cycle The more or less regular fluctuation in abundance of a species or population.

population density The number of conspecific individuals in a given area or volume of a habitat.

population dynamics The study of changes within populations and of the factors that cause or influence those changes; the study of populations as functioning systems.

population genetics The study of gene frequencies and selection pressures in populations.

population growth The change in population size with time as a nett result of natality, mortality, immigration and emigration.

population regulation The control of population density by density dependent factors *q.v.*

population size The number of items or individuals in a finite population; denoted by the letter *N*.

population structure The composition of a population according to age and sex of the individuals; *cf.* life table.

pore volume The volume of air-filled space in a soil.

poricidal dehiscence Spontaneous release of the contents of a ripe fruit through pores.

porosity The ratio of the volume of pore space in a sediment to its total volume, usually expressed as a percentage.

position effect A change in the phenotypic expression of a gene as a result of a change in its spatial relationship to other genes on the chromosome.

positive assortative mating Assortative mating *q.v.*

positive binomial distribution A mathematical distribution used as a model when each individual in a sample may exhibit a character in either of two alternative states; also used as an approximate model of a uniformly distributed population in which the presence of one individual at any given point decreases the probability of another individual occurring nearby and in which the variance is less than the mean.

positive eugenics Improvement of the genetic constitution of a population by selective breeding of individuals with desirable heritable characteristics; *cf.* negative eugenics.

post- Prefix meaning after, behind, succeeding.

post coitum Occurring after mating.

postadaptation An evolutionary change occurring in response to a particular environmental selection pressure; postadaption; *cf.* preadaptation.

postclimax A climax community occurring in more mesic conditions than those obtaining at present, often considered to be a remnant of an earlier climatic climax.

postclisere An ecological succession arising in more mesic conditions than those obtaining at present.

postinteractive niche Realized niche *q.v.*

postmating isolation The condition in which interbreeding between two or more populations is prevented by intrinsic factors effective after mating; includes all postzygotic isolating mechanisms *q.v.* and prezygotic isolating mechanisms *q.v.* operative after copulation; *cf.* premating isolation.

postpartum Occurring after parturition *q.v.*

postventitious Delayed in development.

postzygotic isolating mechanism A mechanism preventing interbreeding between two or more populations that is effective after zygote formation.

potable water Water that is palatable and safe for human consumption, in which any toxic substances, pathogenic organisms and factors have been reduced to safe or acceptable levels.

potamad A river plant; potamophyte.

potamic Pertaining to rivers, or transport by river currents.

potamicolous Living in rivers; **potamicole**.

potamium A river community.

potamodromous Used of an organism that migrates only within fresh water; *cf.* anadromous, catadromous, oceanodromous.

potamology The study of rivers.

potamophilous Thriving in rivers; **potamophile**, **potamophily**.

potamophyte A river plant; potamad; **potamophyta**.

potamoplankton Planktonic organisms of slow-moving streams and rivers.

potamous Pertaining to the lower reaches of rivers and streams; **potamon**; *cf.* rhithrous.

potential evapotranspiration The potential water loss from the ground or water body surface by evaporation and from vegetation by transpiration given a surfeit of available soil water; *cf.* evapotranspiration.

potential soil water deficit (**PWD**) A measure of the amount of drying which may affect a soil in a given period computed as the excess of evapotranspiration over precipitation (negative precipitation effectiveness) for that period.

pound The fps unit of mass defined as 0.453 592 37 kilogram; see Appendix 13.

ppm 1: Parts per million. 2: Parts per mille; ppt.

ppt Parts per thousand; ‰; parts per mille.

praecocial Precocial *q.v.*; **praecoces**.

praeform The actual or postulated ancestor of a taxon; archetype.

Prairie soil A zonal soil with a dark greyish brown to black moderately acidic A-horizon grading through brown into a light coloured parent material, formed in temperate to cool–temperate humid climates under tall grass vegetation; brunizem.

pratal Pertaining to grassland and meadowland.

pratinicolous Living in grassland and meadowland; **pratinicole**.

pratum A meadow or grassy field community.

pre- Prefix meaning before, in front of, prior to, earlier than.

preadaptation The possession by an organism of characters or traits that would favour its survival in a new or changed environment.

prebiotic Pertaining to conditions existing before the origin of life; prebiological; **prebios**.

Preboreal period A period of the Blytt-Sernander classification *q.v.* (*ca.* 9000–10 000 years B.P.), characterized by birch and pine vegetation and by relatively colder and wetter conditions than in the Boreal period; Subarctic period.

Precambrian A geological era ending about 570 million years B.P. comprising the Proterozoic, Cryptozoic, Archaeozoic and Azoic eons; see Appendix 1.

precedence In taxonomy, the order of seniority of available names or nomenclatural acts.

precipitation Rainfall; also including snow, hail and sleet.

precipitation effectiveness A measure of soil wetness calculated as precipitation minus evapotranspiration *q.v.* for a given period.

precision In statistics, the closeness of repeated measurements of the same quantity; *cf.* accuracy.

preclimateric Used of a period or phase prior to a climateric *q.v.*

preclimax A climax community occurring in more xeric conditions than those obtaining at present, often considered a seral stage preceding the attainment of a climatic climax; *cf.* postclimax.

preclisere An ecological succession arising under more xeric conditions than those obtaining at present.

precocial Used of offspring that exhibit a high level of independent activity from birth; praecocial; *cf.* altricial.

precocious Occurring particularly early in development.

predation The consumption of one animal (the prey) by another animal (the predator); also used to include the consumption of plants by animals, and the partial consumption of a large prey organism by a smaller predator (micropredation).

predation pressure The effects of predation on the dynamics of a prey population.

predator An organism that feeds by preying on other organisms, killing them for food; **predaceous, predacious, predation**.

predictive classification A classification having a high information content from which inferences can be drawn concerning structural and functional properties of the taxa; the more predictive the classification the more natural it is considered to be.

predominant Of outstanding abundance or importance in a community.

preferential species A plant species found in a variety of communities, but one showing greater abundance in one particular community; a species of relatively low fidelity *q.v.*

prefix An appended beginning of a compound word; *cf.* suffix.

preformation theory That germ cells contain preformed miniature adults which unfold during development; encasement theory; incasement theory; preformism; *cf.* epigenesis.

pre-Fresian In taxonomy, pertaining to names or works published before 1 January 1821.

prehensile Adapted for gripping.

preinteractive niche Fundamental niche *q.v.*

pre-Linnaean In taxonomy, pertaining to names or works published before 1 January 1758 (in Zoology) or 1 May 1753 (in Botany).

premating isolation The condition in which interbreeding between two or more populations is prevented by factors operating outside the organism (extrinsic) and effective before mating or fertilization, including ecological, ethological, geographical, mechanical and temporal isolating mechanisms; *cf.* postmating isolation.

premunition A resistance to reinfection or superinfection conferred by an existing infection, but one that does not destroy the organisms of the infection already present; complete clearance of all parasites will render the host susceptible to reinfection; concomitant immunity.

Prenebraskan glaciation An early glaciation of the Quaternary Ice Age in North America; see Appendix 2.

preoccupied name In zoological nomenclature, a

name already in use for another taxon based on a different type; junior homonym.

prepotency The preferential fertilization of a flower by pollen from another flower rather than its own, when both are present.

preprint A work distributed in advance of a book or periodical in which it is to be published.

prepubertal Pertaining to a condition occurring or existing before puberty; prepuberal.

presaturation dispersal The emigration of individuals from a population that remains at low density, or is in an early phase of growth; typically individuals in good condition and of either sex; *cf.* saturation dispersal.

presentation time The minimum time that an organism must be exposed to a stimulus in order to elicit a response.

preservation The maintenance of individual organisms, populations or species by planned management and breeding programmes; *cf.* conservation.

preservational association An association of fossils occurring together because of a common method of preservation.

presocial Pertaining to a social group, or the condition, in which members display some degree of social behaviour but fall short of full eusocial *q.v.* behaviour; includes both parasocial and subsocial.

pre-starting point In taxonomy, pertaining to names or works published before the starting point of the group; *cf.* pre-Linnaean; pre-Fresian.

Preston's octaves of frequency A natural series of groups of rarity/commonness of species forming a sequence of octaves of frequency which may be plotted on a logarithmic base.

prevalent plankton Plankton dominated by one particular species that comprises at least half the total plankton of that area; *cf.* monoton plankton, pantomictic plankton, polymictic plankton.

prevernal Pertaining to early spring.

prey An animal or animals killed and consumed by a predator.

prey set The total range of prey items utilized by a predator.

prezygotic isolating mechanism A mechanism preventing interbreeding between two or more populations that is effective before fertilization and zygote formation; *cf.* postzygotic isolating mechanism.

primary consumer A heterotrophic organism that feeds directly on a primary producer.

primary homonym In taxonomy, each of two or more identical specific or subspecific names of

different taxa established in the same genus; *cf.* secondary homonym.

primary host Definitive host *q.v.*

primary intergradation The occurrence of intermediates developed *in situ* in the zone of contact between two phenotypically distinct populations; *cf.* secondary intergradation.

primary producer An organism that synthesizes complex organic substances from simple inorganic substrates; autotroph.

primary production The assimilation of organic matter by autotrophs; *cf.* gross primary production, nett primary production.

primary productivity The productivity *q.v.* of autotrophic organisms.

primary sere Prisere *q.v.*

primary sex ratio The number of males per 100 females at the time of zygote formation; *cf.* secondary sex ratio, tertiary sex ratio.

primary sexual characters Those differences between the sexes that relate to the reproductive organs and gametes; *cf.* secondary sexual characters.

primary speciation The splitting of a single lineage into two or more separate evolutionary lineages; *cf.* secondary speciation.

primary succession An ecological succession commencing in a habitat or on a substrate that has never previously been inhabited; prisere; primary sere.

primary type The original specimen on which a species was based; proterotype; prototype; type.

prime mover The ultimate factor that determines the direction and rate of evolutionary change, comprising phylogenetic inertia *q.v.*, and ecological pressure *q.v.*

primer pheromone A pheromone that affects the physiology of the recipient organism eventually eliciting a response; *cf.* releaser pheromone.

primeval Pertaining to earliest times, or the beginning of the Universe.

primitive 1: Early; simple; poorly developed; unspecialized. 2: Used of characters or character states present in an ancestral species or taxon; plesiomorphic; *cf.* derived.

primitive similarity Symplesiomorphy; the common possession of homologous primitive characters or character states.

Primordial Cambrian *q.v.*

primordial Primitive; primary; used of the earliest stage in ontogeny or development of an organ or system; original.

principal component analysis A method of transforming the axes of multidimensional space in which observations occur so that the first axis explains the maximum amount of variance, the second (orthogonal to the first) explains the maximum of the remaining variance and so on; for *p*-dimensional space there is a maximum of *p* principal components (normalized eigenvectors) but the first two or three explain most of the variance.

principal coordinate analysis A technique for ordination *q.v.*; metric scaling.

principle of commonality Commonality, principle of *q.v.*

priorable name *Nomen legitimatum q.v.*

priority In taxonomy, the principle that seniority is fixed by the date of valid publication or availability.

prisere An ecological succession in a habitat or on a substrate that has never previously been inhabited; primary sere; primary succession.

pro- Prefix meaning before, in front of, forward.

pro hybrida (**pro hybr.**) Latin, meaning as a hybrid; used in nomenclature to indicate that the name of a species was first published as the name of a hybrid.

pro minore parte (**pro min. parte**) Latin, meaning for the smaller part.

pro parte (**p. p.**) Latin, meaning in part; used in nomenclature to indicate that only part of a taxon is being referred to.

pro speciei (**pro spec.**) Latin, meaning as a species; used in nomenclature to indicate that the name of a hybrid was first published as the name of a species.

pro synonymo (**pro syn.**) Latin, meaning as a synonym; used in nomenclature to indicate that a name was first published in a synonymy as a synonym.

proaposematic coloration Warning coloration as exhibited by many venomous or unpalatable animals; proaposematism; *cf.* aposematic coloration.

probabilistic identification Any identification in which a measure of the probability that the identification is correct is also given; *cf.* Bayesian method.

probability In statistics, the chance that a given event will occur; the probability of an impossible event is zero, and that of an inevitable event unity; likelihood.

problematic fossil Dubiofossil *q.v.*

problematicum Dubiofossil *q.v.*

procaryotic Prokaryotic *q.v.*; **procaryote**.

prochosium An ecological succession occurring on an alluvial soil.

proclimax A stage in an ecological succession replacing the typical climatic climax; a stable plant community reflecting favourable local variations in edaphic or biotic factors; subclimax.

procryptic Pertaining to coloration and behaviour that affords protection against enemies; **procrypsis**; *cf.* anticryptic.

prodophytium The predecessors of a plant formation.

producer An organism that synthesizes complex organic substances from simple inorganic substrates; primary producer.

production (P) 1: Gross production, the actual rate of incorporation of energy or organic matter by an individual, population or trophic unit per unit area or volume. 2: Nett production, that part of assimilated energy converted into biomass through growth and reproduction by an individual, population or trophic unit, per unit time per unit area or volume; that is the balance between assimilation (A) and respiration (R), calculated as $P = A - R$. 3: The biomass, organic matter or energy accumulated by a population or trophic unit plus that lost by elimination (E) *q.v.* per unit time per unit area or volume, calculated as $P = B + E$.

productivity 1: The potential rate of incorporation or generation of energy or organic matter by an individual, population or trophic unit per unit time per unit area or volume; rate of carbon fixation; *cf.* production. 2: Often used loosely for the organic fertility or capacity of a given area or habitat.

productivity coefficient A dimensionless quantity derived as the ratio of gross community production (estimated as nett community production, PP_n, plus respiration, R) to respiration, calculated as:
$$Kpp = (PP_n + R)/R.$$

proepisematic Pertaining to a character or trait that aids social recognition; *cf.* episematic.

profile The vertical sequence of distinct horizontal layers of a soil, sediment or vegetation.

profile transect Stratum transect *q.v.*

profundal Pertaining to the deep zone of a lake below the level of effective light penetration, and hence of vegetation.

progamete The cell which gives rise directly to gametes, either by a single division or by intracellular metamorphosis.

progamic sex determination The condition in which the sex of the offspring is determined in the egg prior to fertilization.

progamous Occurring or existing before fertilization.

progenesis Paedomorphosis *q.v.* produced by precocious sexual maturation of an organism that is still at a morphologically juvenile stage.

progenitor An ancestor.

progeny The offspring of a single mating or of an asexually reproducing individual.

progeotropism Positive geotropism; the tendency of a plant organ to bend downwards; **progeotropic**.

prognosis A forecast of the likely course of a disease.

progressive provisioning Repeated feeding of larvae during development; *cf.* provisioning.

progressive selection Directional selection *q.v.*

progressive succession An ecological succession leading towards a climax; *cf.* retrogressive succession.

progynous Protogynous *q.v.*; **progyny**.

prohydrotropism Positive hydrotropism *q.v.*; **prohydrotropic**.

proiospory Prospory *q.v.*

prokaryogamete The nucleus of a progamete *q.v.*

prokaryogamety Meiosis *q.v.*

prokaryotic Used of organisms lacking a discrete nucleus separated from the cytoplasm by a membrane, as in bacteria and blue-green algae; procaryotic; **prokaryote, protokaryotic, protocaryotic**; *cf.* eukaryotic.

proliferation Multiplication; growth; cell division; reproduction.

prometatropic Pertaining to obligatory cross pollination.

promiscuous Pertaining to the mating system in which neither sex is restricted to a single mate.

promunturium A rocky seashore community.

prop. Abbreviation of the Latin *propositus*, meaning proposed.

propagation 1: Vegetative increase. 2: Sexual or asexual multiplication; **propagate**.

propagule 1: Any part of an organism, produced sexually or asexually, that is capable of giving rise to a new individual; diaspore; **propagulum**. 2: The minimum number of individuals of a species required for colonization of a new or isolated habitat.

property Any attribute, feature or character.

prophototaxis Positive phototaxis *q.v.*; **prophototactic**.

prophototropism Positive phototropism *q.v.*; **prophototropic**.

prophylactic Pertaining to the prevention of a disease; **prophylaxis**.

proportion rule Allen's law *q.v.*

proposal In taxonomy, any attempt to establish a name or nominal taxon.

propositus Latin, meaning proposed.

pros- (pro-) Prefix denoting the positive condition.

prosaerotaxis Positive aerotaxis *q.v.*; **prosaerotactic**.

proschairlimnetic Used of organisms occasionally found in the plankton of fresh water bodies; **proschairlimnic**.

proschemotaxis Positive chemotaxis *q.v.*; **proschemotactic**.

prosgalvanotaxis Positive galvanotaxis *q.v.*; **prosgalvanotactic**.

prosgeotropism Positive geotropism *q.v.*; progeotropism; **prosgeotropic**.

prosheliotropism Positive heliotropism *q.v.*; **prosheliotropic**.

proshydrotaxis Positive hydrotaxis *q.v.*; **proshydrotactic**.

prososmotaxis Positive osmotaxis *q.v.*; **prososmotactic**.

prospective Potential; possible; probable; *cf.* realized.

prospective ecospace Fundamental niche *q.v.*

prosphototaxis Positive phototaxis *q.v.*; **prosphototactic**.

prospory Seed production by plants that are not fully developed; proiospory.

protandrous Pertaining to a hermaphrodite organism that assumes a functional male condition during development before reversal to a functional female state; proterandrous; **protandrism, protandry**; *cf.* protogynous.

protective call A sound produced by an animal when danger threatens but is still at a distance; alarm call.

protective coloration Cryptic coloration *q.v.*

protelean Used of an organism that is parasitic when immature and free-living as an adult.

proterandrous Protandrous *q.v.*; **proterandric, proterandry**.

proterogenesis The theory that new evolutionary features appear suddenly in ancestral juvenile stages and are displaced slowly forwards by neoteny towards the adult stage of descendants; *cf.* recapitulation.

proterogynous Protogynous *q.v.*; **proterogyny**.

Proterophytic Archaeophytic *q.v.*

proterotype 1: Primary type *q.v.* 2: All the original specimens collected at one time from which the types were selected.

Proterozoic A geological period (*ca.* 2400–570 million years B.P.); the late Precambrian era; Agnotozoic; see Appendix 1.

prothetely Accelerated sexual maturation in insects resulting in the production of an imago exhibiting larval or pupal characters; a type of progenesis *q.v.*; **prothetelic**.

protistan Pertaining to the eukaryotic organisms of the kingdom Protista, including some algae, fungi and protozoans; formerly used only of unicellular eukaryotic organisms; protobiontic; *cf.* moneran.

protistology The study of protistans *q.v.*

proto- Prefix meaning first or original.

protobiogenesis The origin of life; **protobiogenetic**.

protobiontic Protistan *q.v.*

protobios All ultramicroscopic life forms.

protocaryotic Prokaryotic *q.v.*

protocooperation A facultative symbiosis in which both symbionts benefit more or less equally; cooperation.

protoepiphyte A plant that is epiphytic throughout its life cycle, having aerial roots only.

protogamy The fusion of gametes to produce a binucleate zygote.

protogene A dominant allele; *cf.* allogene.

protogenesis 1: Early embryonic development to gastrulation; *cf.* deuterogenesis. 2: Reproduction by budding.

protogenic Pertaining to structures that persist from the earliest stages of development.

protograph All the original figures of a holotype.

protogynous Pertaining to a hermaphroditic organism that assumes a functional female condition first during development before changing to a functional male state; metagyny; progynous; proterogynous; **protogynic**, **protogyny**; *cf.* protandrous.

protokaryotic Prokaryotic *q.v.*

protologue In taxonomy, all information relating to a name when first published; **protolog**, **protologist**.

protomorphic Primordial; primitive; original.

protonym In taxonomy, an effectively but not validly published name, or devalid name *q.v.* that is reused and validly published subsequently.

protophyte 1: A plant without well differentiated tissues; protophyton; *cf.* metaphyte. 2: A plant of the sexual generation; protophyt.

protoplastotype In taxonomy, a cast of a primary type.

protosexual Pertaining to organisms that achieve genetic recombination by conjugation *q.v.*; transduction *q.v.* or lysogenation *q.v.*; *cf.* eusexual, parasexual.

protospecies An ancestral species.

prototrophic 1: Autotrophic *q.v.*; protrophic. 2: Obtaining nourishment from a single source only; **prototroph**, **prototrophism**. 3: Used of microorganisms capable of growth on simple sugars and salts without specific additional nutrient requirements.

prototype 1: Primary type *q.v.* 2: A hypothetical ancestral stage.

protozoology The study of Protozoa.

protozoophilous Pollinated by Protozoa or other

microscopic animals, as in some aquatic plants; **protozoophily**.

protrophic Prototrophic *q.v.*

protype In taxonomy, an entire specimen that replaces a fragmentary holotype; neotype.

prov. Abbreviation of the Latin *provisorius*, meaning provisional.

provenance The place of origin.

province 1: A major division of the biosphere *q.v.*; a phytogeographical subdivision of a Region characterized by dominant plant species of similar past history. 2: A biogeographical zone characterized by a 25–50% endemic flora or fauna.

provisional name *Nomen provisorium q.v.*

provisioning Providing food for young; *cf.* mass provisioning, nest provisioning, progressive provisioning.

***provisorius* (*prov.*)** Latin, meaning provisional.

proximate factor An environmental factor that acts as the immediate stimulus for periodic biological activity; zeitgeber; proximate causation; *cf.* ultimate factor.

proxy type Neotype *q.v.*

psamathad A strandline plant of a sandy seashore; psamathophyte.

psamathium A strandline community of a sandy seashore.

psamathophilus Thriving in the strandline of a sandy seashore; **psamathophile, psamathophily**.

psamathophyte A strandline plant of a sandy seashore; psamathad; **psamathophyta**.

psammobiontic 1: Living in sand; arenicolous; psammobiontic; sabulicolous. 2: Living interstitially between or attached to sand particles.

psammofauna Those animals associated with a sandy substrate, or living within a sandy area.

psammogenous Sand producing; forming sandy soils.

psammolittoral Pertaining to a sandy shore.

psammon Those organisms living in, or moving through, sand; the interstitial flora and fauna; mesopsammon; *cf.* benthos.

psammophilous Thriving in sandy habitats; **psammophile, psammophilic, psammophily**; *cf.* cheradophilous, thinophilous.

psammophyte A plant growing or moving in unconsolidated sand; **psammophyta**.

psammosere An ecological succession commencing on an area of unconsolidated sand.

P-SD ratio The ratio of precipitation to saturation deficit; used as a representation of regional climatic conditions.

psephonecrocoenosis A necrocoenosis *q.v.* of dwarf individuals.

pseudapogamy The fusion of two vegetative cells, or an ovum with a vegetative cell, to form a sporophyte; pseudoapogamy.

pseudaposematic coloration Warning coloration which takes the form of a bluff, as in Batesian mimicry *q.v.*; pseudoaposematic coloration; pseudaposematism; *cf.* aposematic coloration.

pseudaspory Spore formation without meiosis giving rise to a diploid spore which produces the gametophyte.

pseudepisematic Pertaining to a character or trait aiding recognition but involving deception; pseudoepisematic; pseudosematic; *cf.* episematic.

pseudo- Prefix meaning false; **pseud-**.

pseudoalleles Any of two or more mutations that have the same effect but are not structurally allelic.

pseudoapogamy Pseudapogamy *q.v.*

pseudoaquatic Used of organisms that thrive in wet ground.

pseudoclonal Used of a society *q.v.* comprising non-genotypically identical modular units which are so closely related and coordinated that they can be regarded as ecologically and functionally equivalent to a clone; *cf.* euclonal, paraclonal.

pseudocopulation A mode of pollination in which a male insect attempts to copulate with a flower that resembles a female insect, and in the process transfers pollen from one flower to another.

pseudoephemer A flower that remains open for about one day.

pseudoepiphyte A plant which germinates in the soil but subsequently becomes an epiphyte following loss of soil roots and other basal structures.

pseudoextinction Phyletic extinction *q.v.*

pseudofossil A structure, object or mineral of inorganic origin that has a marked resemblance to a particular type of fossil and may be mistakenly identified as such.

pseudogamy The activation of an egg by a male gamete that degenerates' without its nucleus fusing with that of the egg; gynogenesis; pseudomixis; somatogamy.

pseudogonochorism The condition of a consecutive hermaphrodite *q.v.* in which either male or female gonads are normally found and a hermaphrodite gonad occurs at sex reversal; pseudohermaphroditism.

pseudohermaphroditism Pseudogonochorism *q.v.*

pseudohybrid A hybrid that exhibits the phenotypic characters of one parent only.

pseudohydrophyte An aquatic plant that produces persistent submerged clonal populations asexually, often never producing flowers.

pseudometallophyte A plant able to tolerate high levels of heavy metals but one not confined to such substrates; facultative metallophyte; *cf.* metallophyte.

pseudomixis Pseudogamy *q.v.*

pseudomycorrhizal association A symbiosis between a fungus and a higher plant in which the fungus is parasitic.

pseudoparasitism The chance entry and survival of a free-living organism in the body of another; false parasitism; **pseudoparasite**.

pseudoperiphyton Vagile or motile aquatic organisms belonging to the community of plants and animals adhering to and forming a surface coating on stones, plants and other submerged objects; *cf.* euperiphyton.

pseudoplankton Organisms that are not normally planktonic occurring accidentally in the plankton; tychoplankton.

pseudopolyploidy 1: The condition in which the chromosome number is doubled without a corresponding increase in the genetic material. 2: A numerical relationship of chromosome sets in groups of related species that leads to the erroneous inference that they are polyploids.

pseudoseasonal polymorphism The condition of plants that have different phenotypes according to the time of year at which they germinate and flower, caused by local fluctuations in temperature, lengths of season and other climatic factors.

pseudosematic Used of a character or trait that serves as a false danger signal or warning; pseudepisematic.

pseudosymphilic Used of predators or parasites of social insects that obtain nourishment from the trophallactic secretions of the host larvae; **pseudosymphile**.

pseudotripton Tripton *q.v.* transported into a system or region from elsewhere.

pseudotroglobionte An organism that occurs with some regularity in cave habitats; a regular trogloxene *q.v.*

pseudotrophic Pertaining to the mode of nutrition of the parasitic fungus in a pseudomycorrhizal association *q.v.*

pseudotype In taxonomy, a false type; a type designated in error.

pseudovicarism The existence of unrelated or distantly related, but ecologically equivalent, forms (pseudovicars) occupying different geographical areas; *cf.* vicarism.

pseudovicars Unrelated or distantly related, but ecologically equivalent, species that tend to be mutually exclusive occupying separate geographical areas.

pseudoxerophilous Thriving in subxeric habitats; used of a plant exhibiting less moisture sensitivity than a true xerophyte; **pseudoxerophile, pseudoxerophily.**

psicolous Living in prairie or savannah habitats; psilicolous; **psicole.**

psilad A prairie or savannah plant; psilophyte.

psilic Pertaining to savannah or prairie communities or habitats; psilile.

psilicolous Living in savannah or prairie habitats; psicolous; **psilicole.**

psilium A prairie or savannah community.

psilopaedic Used of birds that are naked when hatched; *cf.* ptilopaedic.

psilophilus Thriving in prairie or savannah habitats; **psilophile, psilophily.**

psilophyte A prairie or savannah plant; psilad; **psilophyta.**

psolidomeiomology The study of scale row reductions in snakes.

psychogenetic Pertaining to the origin and development of social instincts and behaviour; **psychogenesis.**

psychogenic Generated by mental activity; originating in the mind.

psychophilous Pollinated by diurnal lepidoptera; **psychophile, psychophily.**

Psychozoa A grade of organization represented by man and characterized by a unique method of evolution by cultural transformation.

Psychozoic The geological era dominated by man; Anthropozoic.

psychric Pertaining to low temperatures or cold habitats.

psychro- Prefix meaning cold.

psychrocleistogamic Used of plants in which self-pollination takes place within flowers that remain closed because of abnormally cold conditions.

psychrokliny Growth behaviour influenced by low temperatures.

psychrophilic 1: Thriving at low temperatures; psychrophilous, psychrophilus, **psychrophil, psychrophile, psychrophily.** 2: Used of microorganisms having an optimum for growth below 20° C.

psychrophyte A plant living in relatively cold arctic or alpine habitats or on a cold substrate.

ptenophyllium A deciduous forest community.

ptenophyllophilous Thriving in deciduous forest; **ptenophyllophile, ptenophyllophily.**

ptenophyllophyte A deciduous woodland plant; **ptenophyllophyta.**

ptenothalium A deciduous thicket community.

ptenothalophilus Thriving in deciduous thickets; **ptenothalophile, ptenothalophily.**

ptenothalophyte A deciduous thicket plant; **ptenothalophyta.**

pteridology The study of ferns.

Pteridophytic Palaeophytic *q.v.*; the age of ferns.

pteropod ooze A calcareous pelagic sediment containing at least 30% calcium carbonate in the form of skeletal remains, predominantly of pteropods; *cf.* ooze.

ptilopaedic Used of birds that are covered with down when hatched; *cf.* psilopaedic.

puberty The onset of sexual maturity.

publication In taxonomy, the process of distributing graphic material in a manner that satisfies the criteria of effective publication of the appropriate Code; any item so published; valid publication.

pulse A sudden increase in abundance of organisms or species, often occurring at regular intervals.

punctuated evolution A model of evolution in which species are relatively stable and long-lived and in which new species appear during concentrated outbursts of rapid speciation, followed by differential success of certain species; punctuated equilibrium model of evolution; *cf.* phyletic gradualism.

Punnett square A matrix used to derive the genotypic and phenotypic ratios of zygotes produced by the union of gametes of known parentage.

pupate To change from larva to pupa.

pupigerous Containing a pupa.

pupiparous Producing offspring already at the pupal stage, or larvae ready to pupate; nymphiparous.

pupivorous Feeding on pupae; **pupivore, pupivory.**

pure association A plant association dominated by a single species.

pure line The descendants of a single homozygous parent produced by inbreeding or self-fertilization; pure strain.

pure stand A plant community in which a single species is predominant.

putrefaction The decomposition of organic matter with incomplete oxidation, producing methane and other gaseous products subject to further oxidation; *cf.* decay, mouldering.

pycnocline A zone of marked density gradient; a density discontinuity.

pyramid of biomass A graphic representation of the biomass relationships of a community, in which total biomass at each successive trophic level is proportional to the width of the pyramid at the appropriate height.

pyramid of energy A graphic representation of the energy relationships of a community in which the total energy available per unit time at each successive trophic level is proportional to the width of the pyramid at the appropriate height.

pyramid of numbers A graphic representation of the numerical relationships of a community in which the number of organisms at each successive trophic level is proportional to the width of the pyramid at the appropriate height.

pyric Pertaining to fire.

pyrium An ecological succession following fire or burn-off.

pyrophilous Thriving on ground that has recently been scorched by fire; **pyrophile, pyrophily**.

pyrophobic Intolerant of the soil conditions produced by fire; used of a plant that is unable to re-establish itself following a fire; **pyrophobe, pyrophoby**.

pyrophyte A plant resistant to permanent damage by fire.

pyroxylophilous Thriving on burnt wood; **pyroxylophile, pyroxylophily**.

pythmic Pertaining to lake bottoms.

Q

Q technique In numerical taxonomy, a technique of multivariate analysis based on comparison between OTUs rather than between the character states; *cf.* R technique.

Q_{10} Temperature coefficient *q.v.*

q.e. Abbreviation of the Latin *quod est*, meaning which is.

q.v. Abbreviation of the Latin *quod vide*, meaning which see.

QO_2 The oxygen uptake of an organism, expressed in microlitres per milligram (dry weight) per hour.

quadrat 1: A delimited area for sampling flora or fauna; usually taken randomly within the study area and typically consisting of a 1 metre square frame; a larger area of 4 square metres is often referred to as a major quadrat. 2: The sampling frame itself.

quadrat chart A diagram showing the position of each organism within a quadrat.

quadrature The time at which the sun and moon are approximately at right angles with respect to the Earth; *cf.* syzygy.

quadri- Prefix meaning four, fourfold, square, at right angles.

quadrigenomic Containing four independently derived genomes.

quadrillion A million billion, 10^{15}.

quadrivalent An association of four chromosomes synapsed during the first meiotic division.

quadrivoltine Having four broods per year or per season; *cf.* voltine.

qualitative Descriptive; non-numerical.

qualitative character A character having discrete states that are not numerical or morphometric; discrete character; qualitative trait; *cf.* quantitative character.

qualitative inheritance The inheritance of phenotypic characters showing discontinuous variation, that are mainly controlled by only one or a few genes; *cf.* quantitative inheritance.

qualitative multistate character A qualitative character exhibiting several states that cannot be arranged linearly along a single axis; *cf.* quantitative multistate character.

quantitative Numerical; based on counts, measurements, ratios or other values.

quantitative character 1: A character based on counts, measurements, ratios or other numerical values; numerical character; quantitative trait; *cf.* qualitative character. 2: A character showing continuous variation and controlled by several interacting genes.

quantitative inheritance The inheritance of phenotypic characters showing continuous variation, that are controlled by several interacting genes; *cf.* qualitative inheritance.

quantitative multistate character A quantitative character exhibiting several states that can be arranged linearly along a single axis; *cf.* qualitative multistate character.

quantum evolution A rapid evolutionary shift to a new adaptive zone under strong selection pressure.

quantum speciation 1: A rapid evolutionary shift from one adaptive zone to another, across a nonadaptive zone. 2: Rapid speciation brought about by a combination of natural selection and genetic drift *q.v.* and involving some degree of spatial isolation.

quantum yield (Φ) Of photosynthesis, the photochemical work performed by the absorption of light quanta, expressed in mol oxygen liberated per mole absorbed photons.

quaquaversal Sloping out in all directions; directed or bending in every direction.

quart A unit of volume equal to 1/4 Imperial gallon, or 1.13650 litre.

quartiles The three points that divide a distribution into first, second, third and fourth quarters by area or frequency, cutting the distribution at the 25% (first quartile), 50% (second quartile) and 75% points (third quartile); the second quartile is also the median.

quasi-normal A mutation that causes the death of less than 10% of the mutants in the population; *cf.* semilethal, subvital.

quasisocial Pertaining to a group in which members of the same generation use the same composite nest and cooperate in brood care.

Quaternary A geological period of the Cenozoic (*ca.* 1.6 million years B.P. to present), comprising the Pleistocene and the Holocene; Anthropogene; see Appendix 1.

Quaternary Ice Age A period in the Quaternary when glaciation was much more widespread than at present; Pleistocene glaciation; see Appendix 2.

queen substance A pheromone or combination of pheromones produced by a queen to control the reproductive behaviour of workers and other queens; trans-9-keto-2-decenoic acid is the most potent of the set of such pheromones used by a queen.

queenright Used of a colony of honeybees that contains a functional queen.

Quenouille's test A non-parametric method for testing differences in mean level between several samples, independent of any assumption about population variance.

quiescence 1: A temporary resting phase characterized by reduced activity, inactivity or cessation of development; **quiescent**. 2: The condition of seeds that have not germinated because one or more environmental variables is unfavourable; *cf.* dormancy.

quin- Prefix meaning five, fivefold.

quintillion A billion billion, 10^{18}.

quirotype Cheirotype *q.v.*

quoad Latin, meaning as to (the); as regards (the); with respect to (the); used in citation to indicate which part of a taxon as circumscribed by an earlier author is being referred to by the present writer.

quod vide (*q.v.*) Latin, meaning which see; used in a text as a cross reference without additional information.

quotient The value obtained by dividing one quantity by another.

R

r. Abbreviation of the Latin *rarus*, meaning rare.

r extinction Extinction, r *q.v.*

r selection Selection favouring a rapid rate of population increase, typical of species that colonize short-lived environments or of species that undergo large fluctuations in population size; *cf. K* selection.

r strategist *r*-selected species *q.v.*

R technique In numerical taxonomy, a technique of multivariate analysis based on direct comparisons between character states rather than between the OTUs; *cf.* Q technique.

r value The intrinsic rate of increase of a population.

race An intraspecific category characterized by conspicuous physiological (physiological race), biological (biological race), geographical (geographical race) or ecological (ecological race) properties.

rachion The zone on the shore of a lake subject to pronounced wave action.

raciation The formation of new races within a species.

rad A derived SI unit of absorbed radiation, defined as the energy absorption of 10^{-2} joules per kilogram of irradiated material; see Appendix 13.

radiational aridity index (RAI) A measure of the degree of aridity over large areas based on the assumption that, on average during the course of a year, the radiation input is used to evaporate water:

$$RAI = \lambda \, Pr/Q_R,$$ where λ is the specific heat of evaporation of water, Pr is the annual precipitation and Q_R is the positive radiation balance.

radicicolous Growing on or in roots; **radicolous**; **radicicole**.

radicivorous Feeding on roots, **radicivore**, **radicivory**.

radiolarian ooze A pelagic siliceous sediment comprising at least 30% siliceous material, predominantly in the form of radiolarian tests; *cf.* ooze.

radiotropism A growth effect produced by radiation; **radiotropic**.

rafting Dispersal of terrestrial organisms, sediment, rock or other material, across water on floating objects or matter.

RAI Radiational aridity index.

rain factor A measure of humidity expressed as the ratio of mean annual rainfall (in millimetres) to mean annual temperature (in degrees Celsius).

rain forest A tropical forest having an annual rainfall of at least 254 centimetres (100 inches).

rain shadow An area on the leeward side of a mountain or range of mountains having a significantly lower rainfall than an area on the windward side.

rain shadow desert An area of desert or arid land in the rain shadow on the leeward side of a mountain range; orographic desert.

rainfall The amount of water precipitated from the atmosphere over a given period of time; rain.

ramet A member or modular unit of a clone, that may follow an independent existence if separated from the parent organism; *cf.* genet, ortet.

ramicolous Living on twigs and branches; **ramicole**.

ramification A branch or dichotomy; branching.

random In statistics, with an *a priori* probability of occurrence other than zero or unity; used also for haphazard, having no recognizable pattern.

random distribution 1: A distribution in which the outcome is the result of pure chance. 2: A spatial distribution pattern in which the presence of one individual has no influence on the distribution of other individuals; random dispersion pattern; *cf.* contagious distribution, uniform distribution.

random error Any deviation, the magnitude and direction of which cannot be predicted.

random fixation The chance loss of one allele and fixation of the other allele in a population.

random mating Mating within a population irrespective of genotype or phenotype; panmixis.

random numbers In statistics, a table of numbers in which the probability of any number occurring at any one time is constant and independent of all preceding numbers.

random pairs method A method of plotless sampling of vegetation in which the individual closest to the sampling point is used and a 90° exclusion angle erected on either side of it; a measure is taken of the distance from the individual to the nearest neighbour outside the exclusion angle.

random process A process governed in part by a chance or random mechanism; stochastic process.

random sample In statistics, a subset in which every member of the population has an equal and independent chance of being included in the sample.

random variable In statistics, a variable that can assume any of a given set of values with assigned probability.

random-access key Multiple-entry key *q.v.*

random-entry key Multiple-entry key *q.v.*

range 1: The limits of the geographical distribution of a species or group. 2: In statistics, a measure of variation within a set of data, given by the difference between the maximum and minimum values. 3: Home range *q.v.*

ranivorous Feeding on frogs; **ranivore, ranivory**.

rank The position of a taxon in a hierarchy of classification; taxonomic rank; taxonomic level.

rank correlation In statistics, the degree of association between two random variables, the values of which have been replaced by ranks within their respective samples; *cf.* Kendall's rank correlation coefficient, Spearman's rank correlation coefficient.

rapacious Feeding on prey; voracious; ravenous.

raptorial Adapted for seizing prey; preying; raptatory.

rare 1: Very seldomly occurring. 2: Used of a plant species that occurs only once within 20 samples taken from an area of 64 square metres; or of a species known to exist in a community that is absent from a sample series from that community.

rarefaction A method of scaling down measurements of species diversity to a common sample size, achieved by randomly subsampling the actual samples by computer simulation.

rasorial Adapted for scratching or scraping the ground.

Rassenkreis A polytypic species *q.v.*

rate of natural increase The percentage increase in a population per unit time; calculated as the birth rate less the death rate.

ratio cline A clinal variation in a polymorphic species, in which successive populations show progressive change in the relative frequency of the morphs.

Raunkiaerian leaf size classes A system of classifying leaf sizes according to surface area, as a physiognomic character of vegetation; the six categories are leptophyll (up to 25 mm²), nanophyll (25–225 mm²), microphyll (225–2025 mm²), mesophyll (2025–18 225 mm²), macrophyll (18 225–164 025 mm²), megaphyll (over 164 025 mm²); the category notophyll has been added for a small mesophyll (2025–4500 mm²).

Raunkiaerian life forms A system of classification of the life forms of plants based on the type and position of the renewal buds with respect to ground level; categories include chamaephyte, cryptophyte, geophyte, helophyte, hemicryptophyte, hydrophyte, phanerophyte, therophyte.

raw data Collected data which have not been organized numerically or treated statistically; uncooked data.

re- Prefix meaning again, repeat.

reaction Any change in the activity of an organism in response to a stimulus.

reaction time The time interval between a stimulus and the response to that stimulus; latent period.

realized Actual; *cf.* prospective.

realized ecospace Realized niche *q.v.*

realized niche That part of the fundamental niche *q.v.* actually occupied by a species in the presence of competitive or interactive species; realized ecospace; postinteractive niche.

realized production Nett production *q.v.*

realm The largest of the biogeographical units, encompassing major climatic or physiographic zones.

recapitulation The repetition of a sequence of ancestral adult stages in the embryonic or juvenile stages of descendants; *cf.* biogenetic law.

Recent Holocene *q.v.*

recent Still existing; extant; *cf.* fossil.

recent. Abbreviation of the Latin *recentiorum*, meaning of recent authors.

recessive An allele that is not expressed in the phenotype except when homozygous; recessive trait; *cf.* dominant.

recessiveness The failure of one allele (recessive) to be expressed over another (dominant) in the phenotype of the heterozygote; *cf.* dominance.

reciprocal altruism Any mutually beneficial behaviour.

reciprocal crosses Crosses between two species in which the sources of the gametes are interchanged as in A♀B♂, A♂B♀.

reciprocal feeding Trophallaxis *q.v.*

reciprocal genes Complementary genes *q.v.*

reciprocal hybrids Hybrid offspring derived from reciprocal crosses, one from the male of one species with the female of another, the other *vice versa*.

reciprocal yield law That there is a linear relationship between the reciprocal of mean plant weight of a plant population and the density of that population.

recombination Any process that gives rise to a new combination of hereditary determinants, such as the reassortment of parental genes during meiosis through crossing over.

recombination frequency The number of recombinants as a proportion of the total progeny; the average frequency with which two given alleles are separated by recombination; recombination fraction.

recommendation In taxonomy, a non-mandatory provision of the Code giving preferred actions or practices; *cf.* Article.

recon The smallest unit of a chromosome the interchange of which can be detected by fine-structure techniques.

record A list of species in a plant community, in some cases including a measure of the frequency of occurrence.

recretion The elimination of mineral salts from plants in the same form in which they were taken up.

recruitment The influx of new members into a population by reproduction or immigration.

rectilinear Used of growth or movement that proceeds in a straight line.

rectilinear evolution Continued change in one direction within a line of descent over a considerable period of time, with no implication as to how the direction of evolution is maintained; *cf.* orthogenesis.

recurrent selection Repeated selection of superior phenotypes from a heterozygous population and their subsequent propagation by selfing; artificial selection.

recycling coefficient That fraction of the total quantity of minerals absorbed (M_{abs}) by vegetation per area per year lost during the same year by recretion (M_r) and in the form of detritus (M_L):
$$K_m = M_L + M_r/M_{abs}$$

red clay A pelagic sediment comprising an accumulation of volcanic and aeolian dust with less than 30% biogenic material; brown clay.

Red Desert soil A zonal soil with a light reddish brown A-horizon overlying a heavy reddish alkaline B-horizon above calcareous material formed under warm temperate to dry tropical climates.

red/far red ratio A component of the light climate characterizing a plant habitat, calculated as the proportion of red light to far red light in the available radiation spectrum.

red king hypothesis Maximum homology hypothesis *q.v.*

red light Solar radiation in the spectral range 620–680 nm.

red mud A well oxidized terrigenous marine sediment having a red coloration produced by ferric oxide.

red queen hypothesis That each evolutionary advance by any one species represents a deterioration of the environment of other species so that each species must evolve as fast as it can merely to survive.

red tide A reddish brown coloration of sea water caused by a bloom of dinoflagellates.

Reddish Brown lateritic soil A zonal soil, highly weathered, with a thin litter and duff layer, reddish brown slightly acidic A-horizon over a red or reddish brown clayey, strongly acidic, B-horizon, formed in forested warm temperate and tropical regions.

Reddish Brown soil A zonal soil with a reddish brown surface layer over a darker red horizon and a light calcareous lower horizon, formed in warm, temperate to tropical, semi-arid climates beneath short grass and shrub vegetation.

Reddish Chestnut soil A zonal soil with a thick red to dark brown A-horizon over a heavy reddish brown B-horizon and a calcareous lower horizon, formed in warm temperate, semi-arid climates beneath mixed grasses and shrub.

Reddish Prairie soil A zonal soil with a dark reddish brown acidic A-horizon grading through a reddish brown soil to the parent material, formed in warm to temperate, humid to subhumid, climates under tall grasses.

redeposition In taxonomy, the transfer of a taxon to a new position, with or without a change in the name or epithet.

redirected activity The direction of behaviour away from the primary target toward another less appropriate object.

redox potential (Eh) Oxidation-reduction potential; a measure of the tendency of a given system to act as an oxidizing (electron acceptor) or reducing (electron donor) agent.

reducer Any heterotrophic organism responsible for degrading or mineralizing organic matter; decomposer.

reduction The process of producing or evolving simpler structures from more complex ones; *cf.* elaboration.

reduction division The first division of meiosis in which the chromosome complement is halved.

reductionism The theory that all complex phenomena can be explained by reducing them to the simplest possible terms, and that all biological phenomena can be explained by simple physical laws; *cf.* holism.

redundant character Any character that does not contribute useful information to an analysis.

Red-Yellow podzolic soil An acidic zonal soil with little or no litter, a light coloured leached A-horizon over a red to yellow clayey B-horizon, formed in humid warm temperate or tropical climates under coniferous or deciduous forest.

reef A raised aggregation of carbonate skeletal

material existing in its original growth position having sufficient elevation from the sea bed to significantly influence local sediment deposition.

refaunation The reintroduction of an animal into a host from which it had been removed; used particularly of symbiotic microorganisms of the gut.

refection The habit of an animal that eats its own faeces; autocoprophagy.

reference The citation of the name of the author and date of publication of a work, often including title of journal, volume and page numbers; bibliographic reference.

reflection depth That depth in a body of water from which reflected light still reaches the surface.

reflex A rapid innate stereotyped response to a stimulus involving the central nervous system; reflex arc.

reforestation The process of re-establishing a forest on previously cleared land.

refuge 1: An area in which prey may escape from or avoid a predator. 2: Refugium q.v.

refugium An area that has escaped major climatic changes typical of a region as a whole and acts as a refuge for biota previously more widely distributed; an isolated habitat that retains the environmental conditions that were once widespread; **refuge, refugia**.

reg A stony desert; serir.

regeneration Renewal or restoration of structures or tissues, typically after loss or damage.

region Any major world area supporting a characteristic biota; biogeographical region. 2: A major world area and climatic zone characterized by a 50–75% endemic flora or fauna. 3: A collective term for provinces q.v. having a similar vegetational appearance or physiognomy.

regional character species 1: A plant species which is limited to the range of distribution of the association to which it belongs. 2: Used for a plant species that is characteristic q.v. over the whole range in which the vegetation unit and the character species occur in common; the relative distributions of the vegetation unit and the species have been classified conregional, intraregional, circumregional and pararegional q.v.

regional climax Climatic climax q.v.

regolith All loose earth material above the underlying solid rock; more or less equivalent to the term soil q.v.; saprolith.

Regosol An azonal soil without pronounced horizons formed on unconsolidated soft mineral deposits such as loess or sand.

regression 1: In statistics, the estimation of the relationship between one variable and one or more other variables, by expressing one in terms of a linear, or more complex, function of the other(s). 2: A reversal in the direction of evolution. 3: A retreat of the sea from a land area; cf. transgression.

regression coefficient In statistics, the rate of change of the dependent random variable with respect to the independent variable(s); a coefficient in a regression equation such as b in the linear regression equation $y = a + bx$.

regression curve In statistics, a graphic representation of a regression equation, used specifically for non-linear regression equations; regression surface.

regression equation In statistics, the equation of a regression q.v. such as $y = a + bx$, where x and y are the two variables, a is the intercept and b the slope; if the equation is linear then the regression is a linear regression, if it represents a non-linear surface or curve then the regression is a curvilinear one.

regression line In statistics, the graphic representation of a linear regression equation showing the change in the dependent variable expected from a change in the independent variable; regression plane.

regression model of evolution The evolution of groundwater and other limnic species from populations of marine littoral ancestors stranded by periodic regressions of sea level.

regressive character A character or trait showing a gradual loss of differentiation as a result of ageing, or in the course of phylogeny.

regressive evolution 1: A process of increasing simplification of structure. 2: A trend of decreasing specialization.

regular distribution Uniform distribution q.v.

regulation 1: Control; homeostasis. 2: Population regulation q.v. 3: Those processes by which an embryo maintains normal development.

reinforcement Any object, situation or activity that leads to a more complete or more stable state of an organism by the acquisition or conservation of reserves, or the release of a consummatory act q.v.; can be either positive (reward) or negative (punishment).

reinforcing selection The operation of selection pressures at two or more levels of organization, such as individual, family or population, so that certain genes are favoured at all levels and their spread through the population is accelerated; cf. counteracting selection.

reject In taxonomy, to set aside a name of a taxon in favour of another name, or to set aside a work; rejected name; rejected work.

rejecta (FU) In ecological energetics, that part of the total food or energy intake of an individual, population or trophic unit (consumption) that is not utilized for production and respiration, and is lost as egesta *q.v.* and excreta *q.v.*, per unit time per unit area or volume; expressed as rejecta (FU) =egesta (F) +excreta (U); or rejecta (FU) =consumption (C) −assimilation (A).

rejected name *Nomen rejiciendum q.v.*

rejected work A publication or work that must not be used for the purposes of zoological nomenclature.

rejungent species Ring species *q.v.*

relationship 1: The degree of overall phenotypic similarity. 2: The degree of genotypic similarity inferred from phenotypic similarity. 3: Recency of common ancestry.

relationship, coefficient of The proportion of the genes of two individuals that are identical by descent; degree of relatedness; Wright's inbreeding coefficient; panmictic index.

relative abundance The total number of individuals of one taxon compared to the total number of individuals of all other taxa combined, per unit area, volume or community; *cf.* absolute abundance.

relative affinity The relationship established when the characters of a specimen do not correspond with those of any single taxon; *cf.* absolute affinity.

relative age The geological age of a fossil or rock expressed in relation to other fossils or events rather than as an actual number of years; *cf.* absolute age.

relative crowding coefficient (RCC) A measure of the aggressiveness of one species (*a*) towards another (*b*), derived from the results of a replacement series experiment and calculated as: RCC = $(X_a/X_b) / (Y_a/Y_b)$, where X_a and X_b represent the mean yield per plant of *a* and of *b* grown in mixture respectively, and Y_a and Y_b represent the mean yield per plant of *a* and *b* grown in pure stand respectively.

relative dating A method of geological dating in which only the relative age of a rock is used, rather than the absolute age *q.v.*; *cf.* absolute dating.

relative density A measure of the density of one species (*a*) in an area or community to the total number of individuals (N_{tot}) of all other species within the same unit; usually expressed as a percentage; relative density =100 N_a/N_{tot}.

relative dominance A measure of the relative importance of one species in a habitat calculated as the species basal area as a percentage of the total basal area for all species.

relative drought index (RDI) A measure of the drought survival capacity of a plant species expressed as a percentage of the relative magnitudes of the actual water saturation deficit (WSD_{act}) at a given time and the critical water saturation deficit (WSD_{crit}) for the species concerned; RDI = $WSD_{act} \times 100/WSD_{crit}$.

relative frequency 1: A measure of the occurrence of a species, calculated as the ratio of the frequency of a given species to the sum of the frequency values of all species present; usually expressed as a percentage. 2: The number of times an event occurs in *x* trials, divided by the number of trials (*n*).

relative growth rate (RGR) A measure of growth in terms of increase in weight (usually dry weight) per unit time with respect to the initial weight (*W*) of the plant; RGR = $dW/dt \times 1/W$; *cf.* absolute growth rate.

relative humidity The ratio of the actual water vapour present to that of total saturation at the existing temperature, expressed as a percentage; *cf.* absolute humidity.

relative ranking The procedure of combining monophyletic groups with their sister groups to form a series of consecutively subordinate categories which are assigned formal rank; *cf.* absolute ranking.

relative sexuality The state or condition of a gamete that is able to act as a male or as a female when mated to different gametes.

relative water content (RWC) The actual water content (W_{act}) at a given time with respect to the water content under conditions of saturation (W_s) expressed as a percentage; RWC = $W_{act} \times 100/W_s$.

relative yield A measure of the effect of competition of one plant species on the growth of another, expressed as yield of plant A = yield of B grown in mixture A + B divided by yield of A in pure stand.

relative yield total (RYT) The sum of the relative yield *q.v.* values for both of two plant species grown in a mixture.

releaser 1: A stimulus that serves to initiate instinctive activity or mood. 2: A sign stimulus used in communication.

releaser pheromone A pheromone that is rapidly perceived producing a more or less immediate response; *cf.* primer pheromone.

relevance coefficient The ratio of the number of characters shared by two taxa to the total number of characters studied.

relevé A visual description of the vegetation of an area together with appropriate habit and habitat data; a phytosociological record.

reliability In statistics, an indication of the size of chance errors.

relict species 1: Persistent remnants of formerly widespread fauna or flora existing in certain isolated areas or habitats; relic. 2: A phylogenetic relict; the existence of an archaic form in an otherwise extinct taxon.

rem Abbreviation of röntgen equivalent man; defined as the equivalent dose of ionizing radiation having the same biological effect as a single röntgen unit of X-rays.

remanie Used of a fossil assemblage comprising fossil elements derived from earlier rocks or strata.

Rendzina soil An intrazonal soil with a dark friable upper A-horizon, a calcareous lower A-horizon over a light grey to yellow calcareous B-horizon, derived from calcium rich parent material in humid and semi-arid climates under grasses or mixed grass and forest vegetation.

Rensch's laws 1: That in cold climates races of mammals have larger litters and birds larger clutches of eggs than races of the same species in warmer climates. 2: That birds have shorter wings and mammals shorter fur in warmer climates than in colder ones; wing rule. 3: That races of land snails in colder climates have brown shells and those in warmer climates have white shells. 4: That the thickness of shell is positively associated with strong sunlight and arid conditions; Rensch's rules.

repetition The close resemblance of early developmental stages in related animals, with subsequent divergence in later stages by the acquisition of distinctive characters.

repetitious DNA DNA of eukaryotes having nucleotide sequences serially repeated many times; repeated DNA; repetitive DNA.

repium An ecological succession on soil subject to subsidence; rhepium.

replacement fertility level The fertility level required to maintain a population at the level of zero growth.

replacement name Nomen novum q.v.

replacement series An experimental design for the study of competition in mixed populations of two plant species, involving sowing the two species in varying proportions while maintaining a constant overall density.

replete Fully nourished; well fed.

replicate In statistics, any repeated experiment.

replication 1: In statistics, repetition of an experiment to obtain more information for estimating experimental error; replicate. 2: The production of identical copies of information from existing units.

representative sample A subset which is representative of the population as a whole.

reprint 1: A printed copy of an article or work distributed subsequent to the publication of the work; separate; offprint. 2: A subsequent printing of a book or other printed matter that preserves the identical text of the original printing.

reproduction The act or process of producing offspring.

reproduction curve A graphical representation of the relationship between the number of individuals at a given stage of one generation and the number of individuals at that stage in the previous generation.

reproductive effort That proportion of metabolic resources devoted to reproduction.

reproductive isolation Isolation q.v.; often used more specifically for the condition in which interbreeding between two or more populations is prevented by intrinsic factors; postmating isolation; misogamy.

reproductive potential The number of offspring produced by a single female in a population.

reproductive success The number of offspring of an individual surviving at a given time.

reproductive value A measure of the expected number of offspring of a female of a given age.

reproductivity effect The decrease in the rate of production of new individuals per colony member as colony size increases.

repugnatorial Adapted for defence or offence.

repulsion configuration The relationship between two genes in which each homologous chromosome carries one dominant and one recessive gene; cf. coupling configuration.

reservoir host A host that carries a pathogen without detriment to itself and which serves as a source of infection for other host organisms; reservoir.

residence time That time a given substance remains within a system.

resident avifauna Those bird species living in an area during the season of highest breeding activity.

residual genotype Background genotype q.v.

residual variance In statistics, that part of the variability of the dependent variable, in an analysis of variance or a regression analysis which cannot be attributed to a specific source of variation.

residual weighting The bias introduced into a numerical taxonomic study by the deliberate, imposed or accidental omission of one or more characters.

resource Any component of the environment that can be utilized by an organism.

resource axes Bionomic axes *q.v.*

resource competition The competitive interaction of two or more populations or species for a limited resource.

respiration 1: The exchange of gases between an organism and its surrounding medium; breathing. 2: Intracellular oxidation of organic molecules to release energy and typically to produce carbon dioxide and water. 3: In ecological energetics, that part of the assimilated energy of an individual, population or trophic unit that is dissipated as heat as the result of exothermic metabolic processes, per unit time; can be quantified as

respiration (R) = assimilation
(A) − production (P); or
respiration (R) = consumption
(C) − production (P) − rejecta (FU).

respiratory index The amount of carbon dioxide produced per unit dry weight per hour.

respiratory quotient (**RQ**) The ratio of the volume of oxygen consumed to the volume of carbon dioxide released; respiratory ratio.

response Any change in an organism or a behaviour pattern as the result of a stimulus.

restibilic Perennial.

resurgent Outflow of a large underground river.

retardation 1: Decrease in velocity. 2: Decrease in the speed of ontogenetic development so that a feature appears later in the ontogeny of a descendant than in that of the ancestor; *cf.* acceleration.

reticulate evolution Evolution dependent on repeated intercrossing between a number of lineages producing a network of relationships in a series of related allopolyploid species; the anastomoses in the network represent sites of hybridization.

retinue Court *q.v.*

retro- Prefix meaning backwards.

retrofection A process of reinfection in which larvae of parasites hatch on the surface of the host and re-enter the host's body.

retrogressive Pertaining to a process of dedifferentiation or reversal to a simpler state or form; regressive; **retrogression**.

retrogressive succession An ecological succession which recedes from a climax as a result of an external influence or interference, such as the activity of man; *cf.* progressive succession.

revalidated name In taxonomy, a name published before the starting date of a group (devalidated name) but subsequently validly published in a post-starting point work.

reverse mutation Back mutation *q.v.*

reversion Return towards an ancestral condition, or from a cultivated to a wild state; atavism.

revertant An allele that undergoes back mutation; or the organism bearing such an allele.

reviser, first First reviser *q.v.*

revision In taxonomy, a critical reappraisal of a taxon or group in the light of new material or a new interpretation.

revivescence Emergence from hibernation or some other quiescent state.

Reynolds number (**Re**) A dimensionless quantity relating to the forces that act when a fluid moves relative to a solid; calculated as Re = $u \times a/v$ where a is the length of the body, u is the velocity of the fluid relative to the body and v is the kinematic viscocity of the fluid calculated as viscosity/density.

rheo- Prefix meaning current, flowing.

rheocrene A flowing spring.

rheogameon A polytypic species *q.v.*

rheology The study of fluid movement; that aspect of limnology devoted to running waters.

rheophilous Thriving in or having an affinity for running water; rheophilic; **rheophile, rheophily**; *cf.* rheophobous.

rheophobous Intolerant of running water; **rheophobe, rheophoby**; *cf.* rheophilous.

rheophyte A plant inhabiting running water.

rheoplankton Planktonic organisms associated with running water.

rheopositive Sensitive to currents; used of organisms showing behavioural and other responses to air or water currents.

rheoreceptive Sensitive to water or air currents.

rheotaxis A directed response of a motile organism to a water or air current, either into the current (positive rheotaxis) or with the current (negative rheotaxis); **rheotactic**.

rheotrophic Pertaining to organisms that obtain nutrients largely from percolating or running water.

rheotropism An orientation response to a water or air current; **rheotropic**.

rheoxenous Used of organisms that occur only occasionally in running water; **rheoxene**.

rhepium Repium *q.v.*

rhexigenous Produced as a result of breakage or rupture; rhexogenous.

rhithrous Pertaining to the upper reaches of a river; rhithron; *cf.* potamous.

rhizanthous Used of plants producing flowers that appear to arise directly from the roots.

rhizo- Prefix meaning root.

rhizobenthos Those organisms rooted in the substratum; rhizobenthic; *cf.* benthos.

rhizocarpous 1: Used of plants having perennial roots and annual stems and foliage. 2: Pertaining to perennial herbs that produce flowers and fruit below ground level as well as above ground.

rhizogenic Producing roots; stimulating root formation or growth; rhizogenous; **rhizogenesis**.

rhizomenon All organisms rooted into the substrate; *cf.* ephaptomenon, planomenon.

rhizophagous 1: Feeding on roots; **rhizophage**, **rhizophagy**; 2: Used of a plant that obtains nourishment through its own roots.

rhizophilous Growing on roots; **rhizophile**, **rhizophily**.

rhizoplane The root-surface component of the rhizosphere *q.v.*

rhizopleustohelophyte A plant rooted in the bank or bottom sediment of shallow water with a floating stem and emergent leaf-bearing shoots.

rhizosphere The soil immediately surrounding plant roots that is influenced structurally or biologically by the presence of such roots; root zone; *cf.* phyllosphere.

rhizotaxis The spatial arrangement of roots; **rhizotaxy**.

rhoad A creek plant; rhoophyte.

rhoium A creek community.

rhoophilus Thriving in creeks; **rhoophile**, **rhoophilous, rhoophily**.

rhoophyte A creek plant; rhoad; **rhoophyta**.

R-horizon The bedrock underlying a soil profile; D-horizon; see Appendix 12.

rhyacad A torrent plant; rhyacophyte.

rhyacium A torrent community.

rhyacophilus Thriving in, or having an affinity for, torrents; rhyacophilous; **rhyacophile**, **rhyacophily**.

rhyacophyte A plant living in a torrent; rhyacad; **rhyacophyta**.

rhypophagous Rypophagous *q.v.*; **rhypophage**, **rhypophagy**.

rhysium An ecological succession following volcanic activity; rhysion.

rhythm Any cyclical variation in the intensity of a behavioural or physiological activity; periodicity; **rhythmicity**.

Ribonucleic acid RNA *q.v.*

ribosomal RNA The structural RNA of the ribosomes; rRNA.

richness index (d) A measure of species diversity calculated as $d = (S - 1)/\log N$ where S is the number of species in the habitat or community and N is the total number of individuals of all species.

Richter scale A logarithmic scale for measuring the magnitude of seismic disturbances; ranging from 1.5 for the smallest detectable tremor to 8.5 for a devastating earthquake; *cf.* Kanamori scale.

rill erosion The formation of many small channels due to erosion by surface water, especially on cultivated land; *cf.* erosion.

ring species The graded series of populations along an extensive cline *q.v.* that curves round on itself so that the populations at the extremes overlap but are unable to interbreed successfully; rejungent species.

riparian Pertaining to, living or situated on, the banks of rivers and streams; riparial; riparious.

ripicolous Living on the banks of rivers and streams; **ripicole**.

Riss glaciation A glaciation of the Quaternary Ice Age in the Alpine area with an estimated duration of 100 thousand years; see Appendix 2.

Riss-Würm interglacial The interglacial period preceding the Würm glaciation in the Alpine area; see Appendix 2.

rite Latin, meaning according to the rules, correctly.

ritualization The evolutionary modification of a behaviour pattern to serve a communicative function, or to enhance the efficiency of a signal.

rival Pertaining to a stream or rivulet.

river model migration Gene flow by migration along a linear series of subpopulations; *cf.* island model migration.

riverain Pertaining to a river bank or the general vicinity of a river.

riverine Pertaining to a river; formed by the action of a river.

r-K spectrum A linear system of reproductive and life history strategies with *r*-selected *q.v.* and *K*-selected species *q.v.* representing the two extremes.

RNA Ribonucleic acid; a polynucleotide that governs protein synthesis in a cell, existing in a variety of forms serving particular functions, such as messenger-RNA, ribosomal-RNA, transfer-RNA.

Robertsonian changes 1: The union of two chromosomes to form a single chromosome, such as the fusion of two acrocentric or telocentric chromosomes to produce one metacentric chromosome (Robertsonian fusion). 2: The dissociation of a metacentric chromosome to give two telocentric chromosomes.

robustness The ability of a statistical method to produce a fair result, despite deviations in the data, from the premises upon which the method is based.

Rocky Mountain subregion A subdivision of the Nearctic region; Central subregion; see Appendix 3.

rocky shore biome The shore biome characterized by solid stable substrates, typically showing zonation of attached organisms; *cf.* muddy shore biome, oceanic biome, sandy shore biome.

rod A unit of length equal to 16.5 feet; perch, pole.

rodenticide A chemical used to kill rodents.

roentgen Röntgen *q.v.*

rogue 1: To remove undesirable weeds, diseased plants or genetic variants (rogues) from a population. 2: A sport *q.v.*

Roman square Latin square *q.v.*

röntgen (R) The derived SI unit of radiation, defined as the quantity of X-rays or gamma radiation that liberates by ionization 8.38×10^{-3} joules per kilogram dry air at constant temperature and pressure; roentgen.

rookery A breeding or nesting place for some mammals and birds.

root constant The maximum moisture deficiency that can occur in soil without reducing transpiration.

root parasitism A condition found in a semi-parasitic plant the roots of which penetrate the roots of the host plant and remove elaborated food material.

rot Decay; decomposition.

round brackets Parentheses *q.v.*; *cf.* brackets.

RQ Respiratory quotient *q.v.*

rr. Abbreviation of the Latin *rarissimus*, meaning very rare.

rRNA Ribosomal RNA *q.v.*

r-selected species A species characteristic of variable or unpredictable environments; typically with rapid development, high innate capacity for increase (r), early reproduction, small body size and semelparity; opportunistic species; r strategist; *cf.* K-selected species.

R-strategist Within the C-S-R triangle *q.v.* of ecological strategies, a species typically with small body size, rapid growth, short life span, efficient dispersal and devoting a large proportion of metabolic energy to the production of offspring or propagules; a ruderal species.

rubicolous Living on brambles; rubicole.

rubrification The appearance of a red coloration in a soil due to oxidative weathering and the formation of haematite.

ruderal 1: Pertaining to or living amongst rubbish or debris, or inhabiting disturbed sites. 2: A plant inhabiting such a habitat.

rudimentary Imperfectly developed; arrested at an early stage of development or evolution; sometimes used incorrectly as a synonym of vestigial *q.v.*

rugophilic Thriving on rough surfaces or in surface depressions; **rugophile, rugophily.**

rugotropism An orientation with respect to a surface depression or roughness; **rugotropic.**

rule In taxonomy, a mandatory provision (Article) of a Code of nomenclature.

running water Any body of water exhibiting continuous unidirectional flow.

runoff That part of precipitation that is not held in the soil but drains freely away.

rupestral Pertaining to or living on walls or rocks; mural; rupestrine.

rupicolous Living on walls or rocks; rupestral; mural; **rupicole.**

rut A period of sexual excitement in a mammal.

rypophagous Feeding on putrid matter or refuse; rhypophagous; ryophagous; ryophage; ryophagy, **rypophage, rypophagy.**

S

S₁ First selfing generation; the offspring generation of a self-cross.

s. ampl. Abbreviation of the Latin *sensu amplificato*, meaning in an enlarged sense.

s.l. Abbreviation of the Latin *sensu lato*, meaning in the broad sense.

s.n. Abbreviation of the Latin *sine numero*, meaning without a number, unnumbered.

s.str. (**s.s.**) Abbreviation of the Latin *sensu stricto*, meaning in the narrow sense, in the strict sense.

Saale glaciation A glaciation of the Quaternary Ice Age in northern Germany and Poland with an estimated duration (together with the Warthe glaciation) of 100 thousand years; see Appendix 2.

sabulicolous Living in sand; arenicolous; psammobiontic; **sabulicole**.

sabuline Sandy; used especially of species growing on coarse sand; sabulose, sabulous.

saline Salty; pertaining to soil or water rich in soluble salts.

saline soil Solonochak *q.v.*

salinity A measure of the total concentration of dissolved salts in sea water; more precisely the total amount of dissolved solids in parts per thousand (ppt) by weight when all the bromide and iodide has been converted to chloride, all the carbonate to oxide and all organic matter completely oxidized; *cf.* chlorinity.

salinization The process by which soluble salts accumulate in soil.

salsuginous Pertaining to, or living in, habitats inundated by salt or brackish water; **salsugine**.

salt lake An inland water body having a high salinity due to an excess of evaporation over precipitation.

salt marsh A flat poorly drained coastal swamp inundated by most high tides.

saltant An organism exhibiting saltation *q.v.*

saltation 1: A drastic and sudden mutational change; an abrupt evolutionary change; macrogenesis. 2: To move by leaping or bounding; saltatory.

saltatorial Pertaining to, or adapted for, leaping or bounding locomotion; **saltatory**.

saltem Latin meaning at least.

saltigrade Used of a leaping or hopping form of locomotion.

samesense mutation A point mutation *q.v.* within a codon that produces no phenotypic change because the mutant codon codes for the same amino acid as the original codon; *cf.* missense mutation, nonsense mutation.

sample Any subset of a population; a representative part of a larger unit used to study the properties of the whole.

sample mean The arithmetic mean of the sample, denoted by the statistic $\bar{x}$ (x bar).

sample size The number of observations in a sample, denoted by the letter n.

sample variance In statistics, the square of the standard deviation *q.v.* of the sample, denoted by s^2.

sampling The process of taking a sample.

sampling error 1: In statistics, the difference between an observed value of a statistic and the parameter it is intended to estimate. 2: The variation between the observations in any one cell of an analysis of variance; *cf.* experimental error.

sand Cohesionless sediment particles of size range 2.0–0.0625 mm; subdivided into very coarse sand (2.0–1.0 mm), coarse sand (1.0–0.5 mm), medium sand (0.5–0.25 mm), fine sand (0.25–0.125 mm) and very fine sand (0.125–0.0625 mm); see Appendix 11.

sandbar A submerged ridge of alluvial sand in shallow water; sand bank.

sandy shore biome A shore biome characterized by mobile coarse sediments, typically having relatively poor infauna; *cf.* rocky shore biome, muddy shore biome, oceanic biome.

Sangamonian interglacial The interglacial period preceding the Iowan glaciation in North America, see Appendix 2.

sanguicolous Living in blood; **sanguicole**.

sanguinivorous Sanguivorous *q.v.*; **sanguinivore, sanguinivory**.

sanguivorous Feeding on blood; haematophagous; sanguinivorous; **sanguivore, sanguivory**.

saprium A saprophyte community.

sapro- Prefix meaning rotten, decaying.

saprobe 1: A saprophytic organism. 2: An organism thriving in water rich in organic matter.

saprobic Pertaining to water rich in decaying organic matter.

saprobiontic Saprophagous *q.v.*; saprobiotic; **saprobiont**.

saprobity system An ecological classification of a polluted aquatic system that is undergoing self-purification; based on relative levels of pollution, oxygen concentration and types of indicator microorganisms; *cf.* polysaprobic zone, mesosaprobic zone, oligosaprobic zone.

saprogenic Causing, or caused by, the decay of organic matter; saprogenous; **saprogen**.

saprolith Regolith *q.v.*

sapromyiophilus Pollinated by dung flies; **sapromyiophile, sapromyiophily**.

sapropelic Used of organisms inhabiting mud rich in decaying organic matter; **sapropel**.

saprophagous Feeding on dead or decaying organic matter; saprobiontic; saprophagic; **saprophage, saprophagy**; *cf.* biophagous.

saprophilous Thriving in humus-rich substrates, or on decaying organic matter; **saprophile, saprophily**.

saprophytic Used of a plant obtaining nutriment from dead or decaying organic matter; humicular; hysterophytic; **saprophyte**.

saprophytophagous Feeding on decomposing plant material; **saprophytophage, saprophytophagy**.

saproplankton Planktonic organisms inhabiting water rich in decaying organic matter.

saprotrophic Obtaining nourishment from dead or decaying organic matter; **saprotroph, saprotrophy**.

saproxylobios Those organisms living in or on rotting wood.

saprozoic Used of an animal feeding upon decaying plant or animal matter in the form of dissolved organic compounds; **saprozoite**.

sarco- Prefix meaning flesh, fleshy.

sarcophagous Feeding on flesh; flesh-eating; carnivorous; **sarcophage, sarcophagy**.

Sarmatian Sea The enclosed sea of the late Miocene formed by the east–west closure of the Tethys *q.v.*

satellite DNA Any DNA fraction in eukaryotes that differs sufficiently in base composition from the main DNA component to be separable by centrifugation techniques.

sathrophilous Thriving on humus or decaying organic matter; **sathrophile, sathrophily**.

sathrophyte A plant living on humus; **sathrophytic**.

saturation 1: The presence of a factor at a concentration or level equal to, or in excess of, that required for maximum response or activity. 2: The equilibrium condition of a community in which immigration is balanced by extinction.

saturation deficit A measure of humidity, derived by subtracting the actual water vapour pressure from the maximum possible vapour pressure at a given temperature, expressed as a percentage of total saturation in mm of mercury; water vapour deficit; *cf.* relative humidity.

saturation dispersal The emigration of surplus

individuals from a population at or near carrying capacity; typically individuals that are juvenile, aged or in poor condition; *cf.* presaturation dispersal.

saturation-presaturation dispersal hypothesis That individuals dispersing from a population at its peak are qualitatively different from those emigrating from a declining population or one at low density; the latter are typically in better condition with higher survivorship and greater fecundity.

saurochorous Dispersed or distributed by lizards or snakes; saurophilus; **saurochore, saurochory**.

saurophilus Saurochorous *q.v.*; **saurophile, saurophily**.

savannah The tropical and subtropical grassland biome *q.v.*, transitional in character between grassland or desert and rain forest, typically having drought resistant vegetation dominated by grasses with scattered tall trees; Tropical grassland biome; savanna; see Appendix 5.

saxatile Living or growing among rocks; saxicolous.

saxicavous Rock-boring; endolithic.

saxicolous Living or growing on or among rocks or stones; lapidicolous; saxatile; saxigenous; **saxicole, saxicoline**.

saxifragous Living in rock crevices; **saxifrage**.

saxigenous Saxicolous *q.v.*

sc. Abbreviation of the Latin *scilicet*, meaning namely.

scalar Used of something which is completely described by a single number.

scansorial Climbing or adapted for climbing; **scandent**.

scat An animal faecal dropping.

scatology The study of animal faeces; **scatological**.

scatophagous Feeding on faecal matter or dung; coprophagous; merdivorous; **scatophage, scatophagy**.

scatter diagram A graph plotting sets of paired or multiple measurements as points in order to indicate the general nature of the relationship between two or more variables.

scavenger Any organism that feeds on carrion or organic refuse.

scenopoetic axes Used in the context of a niche as a multidimensional space, for those axes that relate to factors such as light, temperature, humidity, which have upper and lower tolerance values for the species (niche breadth), but do not involve competitive interactions with neighbouring species; *cf.* bionomic axes.

scheda (*sched.*) Latin, meaning the label (of a specimen).

schizoendemic A neoendemic *q.v.* derived from or

having given rise to a more widespread taxon with the same chromosome number.

schizogenesis Reproduction by multiple fission; **schizogenetic**.

schizogony A type of multiple fission in which the parent organism (schizont) divides to form many new individuals (merozoites).

schizoholotype In taxonomy, a significant fragment or part of a holotype *q.v.*

schizolectotype In taxonomy, a significant fragment or part of a lectotype *q.v.*

schizoparalectotype In taxonomy, a significant fragment or part of a paralectotype *q.v.*

schizophyte A plant that reproduces by fission.

schizotype 1: In taxonomy, a syntype regarded by a subsequent author as the type of a taxon but not specifically designated as the type for that taxon; *cf.* lectotype. 2: A significant fragment or part of an otherwise larger part or intact type specimen.

school To aggregate and swim together in an organized manner; also the swimming group itself.

Schuchterland The proposed land-bridge in the eastern Pacific between the North Pacific bridge *q.v.* and Patagonia without connections with the eastern parts of the Americas.

Schwerdtfeger's law of compensation and moderation That the limiting factor seldom operates in isolation but is influenced by the simultaneous action of other factors.

sciad A plant growing in shady situations; a shade plant; sciophyte; skiophyte; skiaphyte.

scientific name The formal Latin or latinized name of a taxon; *cf.* vernacular name.

scion That part or structure transplanted from one plant to another during a graft.

sciophilous Thriving in shaded situations, or in habitats of low light intensity; heliophobous; skiophilous; umbrophilic; **sciophile, sciophilic, sciophily**.

sciophyllous Having leaves that can tolerate shaded situations; skiophyllous; *cf.* heliophyllous.

sciophyte A plant growing in shady situations, or in habitats of low light intensity; sciad; skiaphyte; skiarophyte; skiophyte; *cf.* heliophyte.

sciophytium A shade community.

scissiparous Used of multicellular organisms that reproduce by fission.

scolite A fossil worm-burrow.

scopulus Pertaining to crags and steep overhanging cliffs.

scotophilic Skotophilic *q.v.*; scotophilous; **scotophile, scotophily**.

scotophyte A plant living in darkness; skotophyte.

scototaxis Skototaxis *q.v.*; **scototactic**.

scototeropism Skototropism *q.v.*; **scototropic**.

scramble competition Competition for a resource which is equally partitioned between the competitors so that some or all of the individuals obtain an insufficient amount of the resource for survival or reproduction; *cf.* contest, competition.

SCUBA An acronym of self-contained underwater breathing apparatus; aqualung.

sea 1: A geographical division of an ocean. 2: An inland body of salt water.

sea breeze A light wind blowing off the sea on warm days when the land surface is warmer than the sea surface; *cf.* land breeze.

sea level Mean sea level *q.v.*

sea snow The settlement of particulate organic detritus and living organisms down the water column.

seafloor spreading The formation and lateral displacement of oceanic crust by upwelling of material from the Earth's mantle along mid-ocean ridges.

seamount An elevated area of limited extent rising 1000 m or more from the surrounding ocean floor; guyot.

search image A model for visual comparison, formed as part of the behavioural mechanism by which individual predators select certain prey types and ignore others, thereby optimizing their foraging efficiency.

season Any of the major climatic periods of the year; *cf.* aestival, autumnal, hibernal, serotinal, vernal.

seasonal Showing periodicity related to the seasons; used of vegetation exhibiting pronounced seasonal periodicity marked by conspicuous physiognomic changes.

seasonal climate Any climatic regime exhibiting alternating extremes of precipitation or temperature of sufficient magnitude to produce seasonal responses in the biotic communities.

seasonal hydroperiodism The control of vegetative processes in plants by periodic seasons of dryness; hydroperiodism.

sec. Abbreviation of the Latin *secundum*, meaning according to.

Secchi disc technique A method for estimating the transparency of water by submerging a white disc of standard size and recording the depth at which it disappears from view.

second The standard SI base unit of time defined as the total duration of 9 192 631 770 cycles of radiation from the hyperfine transition in caesium-133.

secondary consumer Any heterotrophic organism that feeds directly on a primary consumer *q.v.*

secondary contact The re-establishment of contact between populations after a period of separation.

secondary homonym In taxonomy, one of two or more identical specific or subspecific names for different taxa which were established in different genera, but which were transferred into the same genus; *cf.* primary homonym.

secondary host Intermediate host *q.v.*

secondary intergradation Hybridization occurring between two phenotypically distinct populations along a zone of secondary contact (suture zone); *cf.* primary intergradation.

secondary production The assimilation of organic matter by a primary consumer; *cf.* gross secondary production, nett secondary production.

secondary sex ratio The number of males per 100 females at birth; *cf.* primary sex ratio, tertiary sex ratio.

secondary sexual characters Those differences between the sexes that relate to structures other than the reproductive organs and gametes; *cf.* primary sexual characters.

secondary speciation Interspecific hybridization between distinct evolutionary lineages; *cf.* primary speciation.

secondary succession An ecological succession resulting from the agricultural or other activities of man, or that follows destruction of all or part of an earlier community.

secretion 1: Any substance or product elaborated and released by a cell or gland to perform a specific function; secreta. 2: The process of elaboration and release; secretory; secretitious; *cf.* excretion.

section 1: A neutral term employed as a subdivision of a taxon; frequently considered to be equivalent to a subgenus if used as a primary division of a genus. 2: Any phytogeographical division of a province *q.v.* having some measure of floristic integrity.

secular 1: Pertaining to periods of hundreds of years; *cf.* decennial, millennial. 2: Used of a process persisting for an indefinitely long period.

secundum (sec.) Latin, meaning according to.

sedentary Attached to the substrate; not free-living.

sediment Particulate matter that has been transported by wind, water or ice and subsequently deposited, or that has been precipitated from water; sedimentation.

sedimentary rock Rock formed by deposition of particulate matter transported by wind, water or ice, or by precipitation from solution in water under normal surface temperatures and pressures, or by the aggregation of inorganic

material from skeletal remains; *cf.* igneous rock, metamorphic rock.

***sedis incertae** Incertae cedis q.v.*

seed 1: A fertilized ripened ovule of a flowering plant. 2: Any propagative plant or animal structure.

seed bank The store of dormant seeds buried in soil.

seedling stratum The lowest layer of a stratified forest community comprising germinating plants that will later enter the transgressive stratum *q.v.*

segmentation The division of a body or appendage into a linear sequence of similar units (segments); metamerism.

segregate A hybrid that exhibits predominantly the phenotype of only one of its parents; heterodynamic hybrid.

segregation distortion The unequal segregation of alleles in a heterozygote due to an aberrant meiotic mechanism.

segregation, law of Mendel's first law; that in sexual organisms the two members of an allele pair or a pair of homologous chromosomes separate during gamete formation and that each gamete receives only one member of the pair.

segregational load The accumulated loss of fitness *q.v.* of a population due to the segregation of genes from favoured heterozygotes to relatively unfit homozygotes; a component of genetic load; balanced load.

seiche A standing wave oscillation of an enclosed or partially enclosed water body that continues after the cessation of the original generating force, such as wind or heavy rain.

seismonasty A growth movement of a plant in response to a non-directional shock or mechanical vibration stimulus; **seismonastic, seismonic.**

seismotaxis A directed response of a motile organism to mechanical vibration or a shock stimulus; vibrotaxis; **seismotactic.**

seismotropism An orientation response to mechanical vibration or a shock stimulus; **seismotropic.**

selection 1: Non random differential reproduction of different genotypes in a population; various modes of selection are: directional selection *q.v.*, disruptive selection *q.v.*, stabilizing selection *q.v.*, artificial selection *q.v.* 2: Natural selection *q.v.*

selection, coefficient of A measure of the intensity of natural selection calculated as the proportional reduction in gametic contribution of one genotype compared with that of a standard genotype, denoted by s; the coefficient may have any value from zero to one; selection coefficient.

selection pressure 1: Any environmental factor that results in natural selection. 2: The intensity of natural selection *q.v.* measured as the alteration of the genetic composition of a population from generation to generation.

selective advantage The increased fitness *q.v.* of one genotype relative to a competing genotype.

selective breeding The selection by man of particular genotypes in a population because they exhibit desired phenotypic characters; artificial selection.

selective differential The deviation in mean phenotypic value *q.v.* of the individuals selected as parents, from the mean phenotypic value of the whole population.

selective disadvantage The decreased fitness *q.v.* of one genotype relative to a competing genotype.

selective species A plant species most commonly found in a particular community but also occasionally in other communities; a relative measure of fidelity *q.v.*, a species of relatively high fidelity.

selenophyte A plant adapted to, or tolerating high selenium levels in the soil, frequently used as an indicator of this particular soil type.

selenotropism An orientation response to moonlight; **selenotropic**.

self-compatible Used of a plant that can be self-fertilized.

self-fertilization The union of male and female gametes produced by the same individual; idiogamy; mychogamy; *cf.* cross fertilization.

self-grooming Grooming of one's own body; *cf.* allogrooming.

self-incompatibility Self-sterility *q.v.*

selfing Self-fertilizing or self-pollinating.

selfishness Any behaviour that benefits the individual in terms of genetic fitness at the expense of other members of the same species; *cf.* altruism, spite.

self-pollination Transfer of pollen from anther to stigma of the same flower or to another flower on the same plant; *cf.* cross pollination.

self-purification A natural process of organic degradation that produces nutrients utilized by autotrophic organisms.

self-regulation hypothesis That an indefinite increase in population density is prevented by a qualitative change in the population; encompassing the stress hypothesis *q.v.*, the behaviour hypothesis *q.v.*, and the polymorphic behaviour hypothesis *q.v.*

self-sterility The condition of hermaphroditic organisms that cannot produce viable offspring by self-fertilization; self-incompatibility.

self-sterility gene A gene that prevents self-fertilization in monoecious plants by controlling the rate of growth of the pollen tube down the style.

self-thinning Density dependent mortality in plant populations.

SEM Scanning electron microscope.

semantic Pertaining to the meaning or significance of words.

semantide Any biochemical character; comprising primary semantides (DNA), secondary semantides (RNA), tertiary semantides (protein) and quaternary semantides (other molecules).

semaphoront An individual organism at a given point of the life cycle, representing the basic element of biological systematics.

sematectonic Pertaining to communication by means of constructed objects.

sematic Pertaining to or serving as a signal or danger warning.

semelparous Pertaining to organisms that have only one brood during the lifetime; semelparity (big-bang reproduction); *cf.* iteroparous.

semestrial Pertaining to periods of six months; half-yearly.

semi- Prefix meaning half, partly.

semiarid Used of a climate having a rather higher annual precipitation than a truly arid climate, between 10 and 20 inches (25.4 and 50.8 cm), typically with a vegetation of grasses.

semi-cryptic polyploidy Cryptic polyploidy *q.v.*

semidiurnal Having a period of about half a lunar day; equal to 12.42 h.

semidiurnal tide A tide having two high and two low waters each lunar day.

semidominance Codominance *q.v.*

semigamy Hemigamy *q.v.*

semigeographical speciation Speciation along lines of secondary intergradation *q.v.* in zones of marked ecological transition.

semilethal A mutation that causes the death of more than 50% but less than 100% of the mutants in a population; *cf.* quasi-normal, subvital.

seminatural Used of communities modified by the activities of man.

semiotics The study of communication.

semioviparous Used of species, or a reproductive strategy, intermediate between oviparous *q.v.* and viviparous *q.v.*

semiparasitic Pertaining to an organism (partial parasite) that derives only part of its nourishment from its host, or that lives for only part of its life cycle as a parasite; **semiparasite**.

semisaprophyte A partially saprophytic plant.

semisocial Used of a social group in which

members of the same generation cooperate in brood care, with reproductive division of labour.

semispecies 1: Incipient species; populations that are intermediate between races and biological species, exhibiting partial interbreeding and intergradation, and weak reproductive isolation. 2: A component species of a superspecies; allospecies.

senescence The biological process of ageing; **senescent**.

senility The degenerative condition of old age; post-reproductive; **senile**.

senior homonym In taxonomy, the earlier published of two homonyms; earlier homonym; *cf.* junior homonym.

senior synonym In taxonomy, the earliest established of two or more synonyms.

sens. Abbreviation of the Latin *sensu*, meaning in the sense of.

sensibility 1: The capacity to perceive a stimulus. 2: The degree of the reaction of a host to infection by a parasite.

sensitivity The capacity of an organism to respond to stimuli; irritability; **sensitive**.

sensory Pertaining to a sense organ or sensation.

sensu (sens.) Latin, meaning in the sense of; used in nomenclature for author citations of misapplied names, in front of the name of the author misapplying the name.

sensu amplificato (sens. amplo.; s. ampl.) Latin, meaning in an enlarged sense; *sensu lato*.

sensu lato (sens. lat.; s.l.) Latin, meaning in the broad sense; *sensu amplo*; *cf. sensu stricto*.

sensu stricto (sens. str.; s. str.; s.s) Latin, meaning in the strict sense or in the narrow sense; *cf. sensu lato*.

separate Reprint *q.v.*; *separatum*.

sepicolous Living in hedgerows; **sepicole**.

septentrional Northern, northerly.

septic Heavily polluted; used of a habitat or zone of fresh water rich in decomposing organic matter, high in carbon dioxide and very low in dissolved oxygen; *cf.* oligosaprobic.

seq. Abbreviation of the Latin *sequens*, meaning following.

sequence affinity Phylogenetic affinity based on similarities in amino acid sequences.

sequential hermaphrodite Consecutive hermaphrodite *q.v.*

sequential key An identification key constructed as a sequence of alternative choices; dichotomous key; *cf.* simultaneous key.

ser. Abbreviation of the Latin *series*, meaning series.

seral 1: Pertaining to a sere *q.v.* 2: Developmental.

seral stages The developmental stages of an ecological succession not including the climax community.

seration A series of plant communities within a formation or ecotone.

serclimax A stable plant community that persists at a stage before the subclimax of an ecological succession.

sere A succession of plant communities in a given habitat leading to a particular climax association; a stage in a community succession; **seral**; variously classified according to habitat or major influencing factors as: adsere, aquatosere, clisere, geosere, hydrosere, lithosere, microsere, oxysere, plagiosere, postclisere, preclisere, prisere, psammosere, subsere, xerosere.

serial endosymbiosis theory Endosymbiosis theory *q.v.*

serial homology The state or condition of having a series of homologous structures, frequently functionally differentiated, in a metamerically segmented organism.

seriation The ordering of a set of OTUs so that similar OTUs are adjacent; the rearrangement of a similarity matrix with the highest similarity values placed nearest to the principal diagonal of the matrix and with an orderly decrease in similarity values away from the diagonal.

series 1: A category of classification intermediate in rank between section and species. 2: In taxonomy, the sample available for study. 3: Any major division of a stratigraphical system.

serir A stony desert; reg.

serodeme A local interbreeding population characterized by immunological properties.

serology The study of antigens and antibodies; **serological**.

serotaxonomy Taxonomy based on serological characters.

serotinal Pertaining to the late summer.

serotinous Coming late; used particularly with reference to late flowering plants, and for behaviour that occurs late in the day, or late in the season.

serotype An infraspecific variant characterized by antigenic structure; serological type.

serozem Sierozem *q.v.*

serpentinophilous Thriving in habitats rich in the mineral serpentine; **serpentinophile**, **serpentinophily**.

serpulite A fossil polychaete tube.

serule Microsere *q.v.*

sesquireciprocal hybrid A hybrid crossed with one of the parental types.

sessile Non-motile; permanently attached at the base.

seston The total particulate matter suspended in water, comprising bioseston *q.v.* and abioseston *q.v.*

seventy-five percent rule That population *x* can be considered subspecifically distinct from population *y* if at least 75% of the individuals of *x* are different from all individuals of *y*.

sewage Liquid or solid waste matter channelled through sewers.

Sewall Wright effect Genetic drift *q.v.*

sex The sum of all structural, functional and behavioural characteristics distinguishing males, females and hermaphrodites; **sexual**.

sex chromosome A chromosome or group of chromosomes of eukaryotic organisms, represented differently in the sexes, responsible for the genetic determination of sex; idiochromosome.

sex cycle The recurrent series of processes that lead to genetic recombination, such as the alternation of karyogamy and meiosis.

sex determination Any mechanism by which the sex of an individual is determined; including genotypic sex determination, maternal sex determination, metagamic sex determination, phenotypic sex determination, progamic sex determination, syngamic sex determination.

sex differentiation The process of diverse development of characters in male and female individuals.

sex digamety The capacity of an individual of one sex (the heterogametic sex) to produce male and female-determining gametes through a sex chromosome mechanism; *cf.* sex monogamety, sex polygamety.

sex limited character A character expressed in only one sex.

sex linkage The location of a gene or character on a sex chromosome; sex-linked.

sex monogamety The phenomenon of a hermaphrodite or parthenogenetic individual, or population, producing gametes of only one kind with regard to sex determination; *cf.* sex digamety, sex polygamety.

sex mosaic A phenotype that manifests characters of both the male and the female; gynandromorph; gynander.

sex plurigamety Sex polygamety *q.v.*

sex polygamety The phenomenon of an individual producing gametes of many different types with regard to sex determination through the mechanism of polyfactorial sex determination; sex plurigamety; *cf.* sex digamety, sex monogamety.

sex ratio The relative number of males and females in a population, expressed as the number of males per 100 females, or as a simple ratio;

cf. primary sex ratio, secondary sex ratio, tertiary sex ratio.

sex reversal A change from one functional sex to the other; consecutive hermaphroditism.

sex-conditioned Sex-influenced *q.v.*

sex-influenced Used of a gene the expression of which is modified by the sex of the individual; sex-conditioned.

sexology The study of sex and sexual behaviour.

sexual Pertaining to sex, sex related characters or traits, or to the process of sexual reproduction; idiotypic.

sexual dimorphism Marked phenotypic differences between males and females of the same species; antigeny; **sexually dimorphic**.

sexual reproduction Reproduction involving the regular alternation of gamete formation by meiosis and gamete fusion (karyogamy) to form a zygote; eusexual reproduction.

sexual selection 1: Choosing a mate; the ability of individuals to differentiate and select a mate; epigamic selection; amphiclexis. 2: The differential abilities of individuals to acquire mates in competition with other individuals of the same sex; intrasexual selection.

sexuparous Used of organisms that produce male and female offspring either by sexual reproduction or by parthenogenesis.

s-g. Subgenus; s-gg (subgenera).

shallow scattering layer A more or less well defined layer present in neritic waters reflecting sound from echosounding equipment, produced by stratified populations of marine organisms.

Shannon–Wiener index of diversity (**H**) An index of diversity based loosely upon information theory, calculated as

$$H = \sum_{i=1}^{s} (p_i)(\log_2 p_i)$$

where p_i is the probability that one individual belongs to species *i* and can be derived from $p_i = N_i/N$ where N_i is the number of individuals in species *i* and *N* is the total number of individuals; its minimum value occurs if all individuals belong to the same species and its maximum value if each individual belongs to a different species; Shannon–Weaver index.

Shantung soil Noncalcic Brown soil *q.v.*

sheet erosion The removal of a more or less uniform layer of surface soil over a wide area by the action of runoff water; *cf.* erosion.

shelf break The outer margin of the continental shelf marked by a pronounced increase in the slope of the sea bed; shelf edge; see Appendix 7.

Shelford's law of tolerance That any value, quantity or factor below a critical maximum will exclude certain organisms from that area.

shift The translocation of a chromosome segment to another point on the same chromosome; internal translocation.

shingle Rounded rock fragments larger than 16 mm in diameter.

shoreface Within a typical beach profile, the zone between mean low water and the fair weather wave base (about 2 m).

short-day plant A plant which has a flowering period initiated and accelerated by exposure to a short photoperiod of less than 12 h duration; *cf.* long-day plant.

SI Units (Système International d'Unités) An internationally agreed system of metric units; comprising seven base units – metre, kilogram, second, ampere, kelvin, candela and mole; other metric units may be derived units or supplementary units; see Appendix 13.

sib mating Mating of progeny derived from the same parents; adelophogamy.

sib pollination Pollination of one plant by another plant derived from the same parents; adelphogamy.

Siberian subregion A subdivision of the Palaearctic region; see Appendix 3.

sibling species Pairs or groups of closely related and frequently sympatric species which are morphologically indistinguishable but which are reproductively isolated; aphanic species; espèces jumelles; twin species.

siblings Organisms sharing close genetic relationship, especially those derived from the same parents; **sibs**.

sibs Siblings *q.v.*

sibship The relationship between siblings.

sic Latin, meaning in this manner; used in citations in parentheses after a printed word or number to confirm that an orthographic or other error is intended, or that it represents an exact quotation of an original.

siccideserta Dry sand deserts.

siccocolus Living in dry arid habitats; siccolous; siccicolous; **siccocole, siccicole, siccole**.

sidereal day The mean time taken for one revolution of the Earth; each year comprises 365.256 sidereal days; *cf.* solar day.

siderophilous Thriving in an iron-rich medium; **siderophile, siderophily**.

siderotrophic Used of water bodies rich in dissolved iron compounds.

Sierozem A zonal soil with pale grey A-horizon, a slightly darker alkaline B-horizon merging into a calcareous lower layer, formed in temperate to cool arid climates; serozem; cerozem.

sigma A Greek letter. 1: Σ, the summation sign, Σx means the sum of all x's. 2: σ, the symbol

denoting the standard deviation of a population.

sigma-t (σt) The density of sea water at atmospheric pressure.

sigmoid curve An S-shaped curve.

sign stimulus A stimulus or one of a group of vital stimuli, by which an animal distinguishes important objects.

signal Any behaviour that conveys information from one individual to another.

signal receiver The target organism of a signal, as in the case of a predator deceived by mimicry; operator.

significance test In statistics, a test that provides a criterion for deciding whether a difference between expected (according to hypothesis) and observed values is small enough to be reasonably attributed to chance.

silent allele An allele that manifests no detectable phenotypic effect; silent mutant; silent section (of a chromosome).

silic Pertaining to rocks containing more than 66% silica (SiO_2) by weight.

siliceous ooze A fine-grained pelagic sediment containing more than 30% siliceous skeletal remains of pelagic organisms, primarily radiolarians and diatoms.

silicolous 1: Living in soil rich in silica or silicates; silicicolous; **silicole**. 2: Growing on flints.

silification A process of fossilization in which the fossil is encrusted, partially replaced or impregnated with silica derived by the degradation of silicates, from biogenic organic silica solutions or from products of vulcanism.

silt Cohesionless sediment particles between 0.004 and 0.0625 mm in diameter; coarse mud; see Appendix 11.

silt load The quantity of particulate matter carried in suspension by a stream or river; *cf.* bed load.

silting The deposition of silt; siltation.

Silurian A geological period within the Palaeozoic (*ca.* 441–413 million years B.P.); Gothlandian; see Appendix 1.

silvicolous Inhabiting woodland; sylvicolous; **silvicole**.

silvics The study of forest trees; **silvical**.

silviculture The management and exploitation of forests.

similarity coefficient A measure of the association of character states of two specimens or taxa.

similarity index A measure of the similarity in species composition of two communities, such as Jaccard's coefficient, Kulezinski's coefficient and Sørensen's coefficient.

Simpson index (D) An index of diversity based on the probability of picking two organisms at random that are different species; calculated as

$$D = 1 - \sum_{i=1}^{i} (P_i)^2$$ where P_i is the proportion of individuals of species i in a community of s species.

simulation The imitation or mimicking of the behaviour of a system by a model.

simultaneous hermaphrodite Synchronous hermaphrodite $q.v.$

simultaneous key An identification key in which the unknown is compared simultaneously with all taxa, with the aim of securing an unambiguous identification in a single step; $cf.$ sequential key.

sine Latin, meaning without.

sine numero (*s.n.*) Latin, meaning without number; used to indicate the absence of a collector's number from a list of specimens.

sine typo Latin, meaning without type.

single tail test One-tailed test $q.v.$

sink A buffering reservoir; any large reservoir that is capable of absorbing or receiving energy or matter without undergoing significant change.

sinking Downwelling; downward movement of surface water, generally caused by converging currents or by the density of a water mass increasing relative to surrounding water.

Sino-Indian region Oriental region $q.v.$

Sino-Japanese region A subdivision of the Boreal kingdom; see Appendix 4.

siotropism An orientation response to shaking; **siotropic**.

siphonogamy Fertilization by means of a pollen tube; **siphonogamous**.

sire In animal breeding, the male parent; $cf.$ dam.

sister groups The two monophyletic groups produced by a single dichotomy; each is the other's nearest relative; sister species-groups.

sister species A pair of species that have arisen from a single speciation event, each is the other's closest relative; the two descendants of a common ancestral species.

site Place; position; habitat.

site-type A group of plant communities sharing a particular set of site characteristics.

sitotropism An orientation response to a food stimulus; **sitotropic**.

size rule Bergmann's rule $q.v.$

skato- Scato- $q.v.$

skatobios Those organisms inhabiting detritus or faecal matter; **skatobiont, skatobiontic**.

skeletal soil Lithosol $q.v.$

skewness In statistics, the degree of asymmetry or departure from symmetry of a population; positive or negative skewness indicates the presence of a long thin tail on the right or left of the distribution respectively.

skiaphyte Skiophyte $q.v.$

skiophilous Thriving in shaded situations, or in habitats of low light intensity; heliophobous; sciophilous; skiophilic; umbrophilic; **skiophile**, **skiophily**.

skiophyllous Sciophyllous $q.v.$

skiophyte A plant growing in shady situations or in habitats of low light intensity; sciad; sciophyte; skiaphyte; skiarophyte; $cf.$ heliophyte.

skotophilic 1: Thriving in darkness or in darkened situations; scotophilic; scotophilous; **skotophile**, **skotophily**. 2: Pertaining to the dark phase of a light–dark cycle.

skotophyte Scotophyte $q.v.$

skotoplankton Planktonic organisms living in darkness, below the photic zone; $cf.$ knephoplankton, phaoplankton.

skototaxis The directed movement of a motile organism towards a dark area, as distinct from negative phototaxis; scototaxis; **skototactic**.

skototropism An orientation movement towards a dark area; scototropism; **skototropic**.

slack water An interval of low velocity tidal current, usually the period of reversal between ebb and flow.

slick An aggregation of floating debris resulting in reduced wave activity and a shiny water surface.

smathophyte Psamathophyte; a plant of sandy soils.

smoothing The averaging of data in time or space in order to compensate for random errors or variations on a scale smaller than that presumed significant to a given problem.

snow line The line marking the lower altitudinal limit of perpetual snow.

sociability The pattern of occurrence of a species within a community, classified on a five point scale: 1 (growing singly), 2 (grouped or tufted), 3 (small patches), 4 (extensive patches) and 5 (pure populations).

social Pertaining to a society, or to the interrelationships between organisms.

social cohesion hypothesis A hypothesis relating behaviour to dispersal patterns; that social interaction prior to emigration is the major factor determining dispersal patterns rather than agonistic interactions at the time of emigration.

social drift A random divergence of the behaviour, or the organization, of societies.

social facilitation Facilitation $q.v.$

social group An assemblage of individuals of the same species organized cooperatively.

social homeostasis Maintenence of a steady state within a society.

social parasitism A symbiosis in which a parasite

diverts to its own use the workers of the host-species society.

social releaser Any stimulus provided by an organism that elicits an instinctive activity in another individual either of the same or another species; releaser.

social subordination hypothesis That as population density increases, increased intraspecific competition leads to increased aggression which results in subordinate individuals being forced to disperse to suboptimal habitats.

sociality The tendency to form social groups.

socialization The modification of the behaviour of an organism as an adaptation to social existence.

sociation 1: A ranked category in the classification of vegetation; a plant community having a dominant species and some constant species *q.v.* but lacking differential *q.v.* or faithful species *q.v.*; micro-association; society. 2: A stable plant community with one or more dominants at each level; an aspect society.

socies A sere or seral community with at least one dominant species.

society 1: A ranked category in the classification of vegetation, a small plant climax community representing a subdivision of a consociation *q.v.* and characterized by a subdominant or subordinate species. 2: A group of organisms of the same species which has a social structure and consists of repeated members or modular units; may be either a continuous modular society *q.v.* or a discontinuous modular society *q.v.* 3: Sociation *q.v.*

sociobiology Study of the biological basis of social behaviour.

sociocline A graded series of social organization among related species, interpreted to represent stages in an evolutionary trend.

sociogram A catalogue of all the social behaviour patterns of a species.

sociohormone Pheromone *q.v.*

sociology The study of communities or of collectively organized groups of animals or plants.

sociotomy Colony fission *q.v.* in termites.

software All the programs, programming languages and operating routines used with computers; *cf.* hardware.

soil The superficial weathered layers of the Earth's crust and any intermixed organic material.

soil climax Edaphic climax *q.v.*

soil evolution The sequence of changes in a soil profile during development towards maturity.

soil granulation scale A scale of soil particle sizes: coarse gravel/rubble (greater than 20 mm diameter), gravel (20–2 mm diameter), coarse sand (2–0.2 mm diameter), fine sand (0.2–0.02 mm diameter), silt/dust (0.02–0.002 mm diameter), fine silt/clay (less than 0.002 mm diameter); *cf.* Appendix 11.

soil horizons The horizontal subdivisions of a soil profile; often given the notation O,A,E,B,C, and R-horizons, see Appendix 12.

soil moisture deficit The amount of water required to restore soil to its field capacity *q.v.*

soil profile A vertical section of a soil through the system of horizontal layers (soil horizons) more or less within the depth zone penetrated by plant roots; see Appendix 12.

soil reaction The measure of acidity or alkalinity of a soil, categorized according to pH value; extremely acid (less than 4.5), very strongly acid (4.5–5.0), strongly acid (5.1–5.5), moderately acid (5.6–6.0), slightly acid (6.1–6.5), neutral (6.6–7.3), slightly alkaline (7.4–7.8), moderately alkaline (7.9–8.4), strongly alkaline (8.5–9.0), very strongly alkaline (greater than 9.1).

soil skeleton The chemically inert quartz fraction of a soil.

soil structure The macrostructure of a soil; the aggregation of soil particles into compound units (or peds), described according to shape of units – crumb (rounded), prismatic (vertically elongated), and platy (flattened).

solar constant The mean intensity of solar radiation at the outer edge of the atmosphere.

solar day The mean time interval between consecutive sunrises, or any other given position of the sun; 24 h; *cf.* lunar day.

solar radiation The radiant energy emitted by the sun; the ecosphere receives solar radiation at wavelengths ranging from 290 nm to about 3000 nm.

solarization The inhibitory effect of extremely high light intensities on photosynthesis, resulting largely from photo-oxidation; heliosis.

solarize To expose to sunlight or solar radiation.

solifluction The gradual downhill flow of fragmented surface material, typically over a frozen substrate.

Solod soil Soloth soil *q.v.*

Solonchak An intrazonal soil with grey shallow salty surface layer over a fine mulch and a greyish salty lower horizon, formed in subhumid to arid climates under conditions of poor drainage and halophytic vegetation; Saline soil.

Solonetz An intrazonal soil with a thin friable surface layer over a dark alkaline horizon, formed in subhumid to arid climates under conditions of moderate drainage and sparse halophytic vegetation; soil formed by leaching of salts from a Solonchak soil *q.v.*

Soloth An intrazonal soil with a light brown friable leached surface layer over a dark lower horizon, formed as a desalinated, decalcified, degraded Solonetz soil *q.v.*, in subhumid to arid climates under conditions of moderate drainage; Solod soil.

solstitial tide The tide occurring near the times of the solstices when the sun reaches its greatest northerly or southerly declination; *cf.* tropic tide.

solubility The extent to which a substance (solute) mixes with a liquid (solvent) to produce a homogeneous mixture (solution).

solubilize To render soluble.

solum (sola) The upper and most weathered part of the soil profile, comprising A- and B-horizons.

Somali Current A warm surface ocean current that flows north off the coast of Somalia, produced by the south-westerly winds of the summer monsoon; South West Monsoon Drift; see Appendix 6.

somatic Pertaining to the body or any non-germinal cell, tissue, structure or process.

somatic mutation theory That ageing is the result of an accumulation of mutations in somatic tissue.

somatic parthenogenesis Parthenapogamy *q.v.*

somatic spory Sporogenesis in which the spores represent somatic cells of the sporophyte.

somatogamy A form of plasmogamy *q.v.* in which somatic cells or hyphae fuse.

somatogenesis The development of somatic structure; **somatogenetic**.

somatogenic variations The various reactions of somatic tissues, or of the body as a whole, to external influences and the effects of use or disuse.

somatropism An orientation response to the substratum; somatotropism; **somatotropic, somatropic**.

sonagram A graphic representation of animal vocalizations.

song An elaborate vocal signal.

soniferous Capable of producing sounds; **sonic**.

Sonoran region A zoogeographical region comprising northern Mexico and the southern parts of North America; the zone intermediate between Nearctic and Neotropical regions; Mediocolumbian region.

Sørensen coefficient (S) A measure of the similarity between the species composition of two communities, calculated as $S = 2c/(a + b)$, where a and b are the numbers of species in communities A and B respectively and c the number common to both; usually expressed as a percentage; Sørensen index.

South African region A subdivision of the African subkingdom of the Palaeotropical kingdom; see Appendix 4.

South African subregion A subdivision of the Ethiopian region; see Appendix 3.

South and West Australian region A subdivision of the Australian kingdom; see Appendix 4.

South Atlantic Gyre The major anticlockwise circulation of surface ocean water in the South Atlantic; see Appendix 6.

South Equatorial Countercurrent A warm surface ocean current that flows east in the equatorial Pacific; see Appendix 6.

South Equatorial Current A warm surface ocean current that flows west in the equatorial Pacific and forms the northern limb of the South Pacific Gyre; see Appendix 6.

South Indian Gyre The major anticlockwise circulation of surface ocean water in the southern Indian Ocean; see Appendix 6.

South Pacific Gyre The major anticlockwise circulation of surface ocean water in the South Pacific; see Appendix 6.

South Temperate American subregion Chilean subregion of the Neotropical region; see Appendix 3.

South Temperate Oceanic Island region A subdivision of the Antarctic kingdom; see Appendix 4.

South West Monsoon Drift Somali Current *q.v.*

sp. Abbreviation of species *q.v.*; plural **spp**.

sp.indet. (sp.ind.) Abbreviation of the Latin *species indeterminata q.v.*

sp.nov. (sp.n.) Abbreviation of the Latin *species nova q.v.*

spanandry A scarcity or progressive decrease in the number of males; *cf.* spanogyny.

spanogyny A scarcity or progressive decrease in the number of females; *cf.* spanandry.

sparse Pertaining to vegetation in which the plants are more than twice their diameter apart on average.

spatial distribution The pattern of distribution of organisms in space.

spatial isolation The spatial separation of populations leading to reproductive isolation.

spatial niche The spatial parameters of a realized niche *q.v.*

spawn The eggs of certain aquatic organisms; also the act of producing such eggs or egg masses.

spay The removal of, or interference with the activity of, the ovaries of a female; *cf.* geld.

Spearman's rank correlation coefficient A non-parametric method for determining the significance of association between two variables the values of which have been replaced by ranks within their respective samples; denoted by r, and expressed as $r = 1 - 6 \Sigma(R_1 - R_2)^2/n(n^2 - 1)$ where $R_1 - R_2$ is the difference between the ranks R_1 and R_2 of paired variables 1 and 2, and n is the number of observations.

spec. Abbreviation of specimen.

Special Creation The belief that all living organisms were specially created by God.

special form *Forma specialis q.v.*

specialist A species having a very narrow habitat range or food preference; *cf.* generalist.

specialization Evolutionary adaptation to a particular mode of life or habitat.

specialized Highly modified from the ancestral condition and no longer able to carry out the function performed when in the ancestral state; adapted to perform a particular function; specialization.

speciation The formation of new species; the splitting of a phylogenetic lineage; acquisition of reproductive isolating mechanisms producing discontinuities between populations; *cf.* allochronic speciation, allopatric speciation, parapatric speciation, stasipatric speciation, quantum speciation, sympatric speciation.

specient An individual member of a species.

species 1: A group of organisms, minerals or other entities formally recognized as distinct from other groups. 2: A taxon of the rank of species; in the hierarchy of biological classification the category below genus; the basic unit of biological classification; the lowest principal category of zoological classification; **species-group**; other species concepts include, biological species *q.v.*, chronospecies *q.v.*, evolutionary species *q.v.*, morphospecies *q.v.*, successional species *q.v.*, taxospecies *q.v.*, typological species *q.v.*

species aggregate A group of species that are morphologically similar and thus difficult to distinguish; collective group.

species equilibrium A condition in which the extinction rate equals the rate of arrival of new species by immigration.

species indeterminata (*sp. indet.; sp. ind.*) Latin, meaning indeterminate species; a species that cannot be identified from the original description.

species inquirenda Latin, meaning a species to be queried; an incompletely defined species that requires further clarification or characterization.

species niche Niche *q.v.*; the role of a species within a community; the basic functional unit of an ecosystem.

species nova (*sp. nov.; sp. n.*) Latin, meaning new species; cited after a new binomen in place of the authority; *n. sp., nov. sp.*

species odour The particular odour found on the bodies of social insects that is peculiar to a given species.

species packing In the context of the multidimensional niche, used as a measure of the number of species that a given hyperspace or hyperspace axis can contain.

species recognition The exchange of species specific stimuli and responses between individuals particularly prior to mating.

species richness The absolute number of species in an assemblage or community.

species swarm A large number of closely related species occurring together in the same general geographical area and derived by multiple splitting of an ancestral stock.

species-abundance curve A graph of species numbers against the number of individuals per species.

species-area curve A graph of the relationship between number of species found and the size of the area surveyed; can be used to determine the minimal area *q.v.*; often used with reference to island biotas where the relationship between number of species (S) and island area (A) is described by the expression $S = CA^z$, where C is a constant giving the number of species when the area is unity and z is the slope of the regression and is related to the degree of isolation; area-species curve.

species-flock A group of several ecologically diverse and closely related species that have evolved within a single macrohabitat, such as a particular lake basin.

species-group 1: A group of closely related species, usually with partially overlapping ranges; sometimes used as an equivalent of superspecies. 2: In zoology, the categories of species and subspecies.

specific Pertaining to a particular individual, situation, quality, trait, effect or species; peculiar; free from ambiguity.

specific action potential A state of the animal responsible for its readiness to perform the behaviour pattern of one instinct in preference to all other behaviour patterns.

specific character Any character diagnostic of a species.

specific epithet The second word of a binomial name of a species or of a trinomial name of a subspecies; *cf.* specific name.

specific heat capacity The amount of heat required to raise the temperature of 1 kilogram of substance by 1 kelvin.

specific leaf area (SLA) The relationship between the transpiring surface area of a leaf and the amount of water stored in its tissues; expressed as SLA = leaf surface area/fresh weight of leaf.

specific modifier A gene that has the specific function of modifying the expression of a gene at another locus.

specific name The scientific name of a taxon of species rank; binomen; formerly used as equivalent to epithet *q.v.*

specific survival time A measure of the degree to which a plant can conserve the water stored in its shoots after complete cessation of water replacement; calculated as the ratio of available water (that consumed between the time of stomatal closure and the appearance of desiccation injury) to cuticular transpiration.

specificity 1: The state of being restricted to a given species. 2: Host specificity *q.v.*

specimen An individual considered to be typical of a group, or a part typical of the whole; any portion or sample of material for scientific study.

specioid A taxonomic category in cultivated plants that corresponds to a species.

speleogenesis The formation of caves.

speleology The study of caves; speology; spelaeology; **speleological**.

speleotherm A structure within a cave formed by the deposition of secondary minerals, such as a stalactite or stalagmite.

sperm Male gamete; spermatozoa.

spermateleosis Spermiogenesis *q.v.*

spermatism Spermism *q.v.*

spermatize Fertilize; impregnate.

spermatocyte The cell that produces male gametes by meiotic division; *cf.* oocyte.

spermatogamete A male gamete; sperm; spermatozoon.

spermatogenesis The process of sperm formation; spermogenesis; **spermatogenic, spermatogenous**.

spermatosome Spermatozoon *q.v.*

spermatozoon A male gamete; sperm; spermatozoan; spermatozoid; spermatozooid; spermium; spermatosome; spermatogamete.

spermiogenesis The transformation of a spermatid into a spermatozoon; spermateleosis; spermiohistogenesis; spermioteleosis.

spermism Encasement theory *q.v.*; the belief that the embryo is contained solely within the spermatozoon; spermatism.

spermium Spermatozoon *q.v.*

spermogenesis Spermatogenesis *q.v.*

spermology The study of seeds.

spermophilic Feeding on seeds; **spermophile, spermophily**.

spermotype In taxonomy, a plant specimen grown from the seed of a type plant.

sphagnicolous Living in peat-moss; **sphagnicole**.

sphagniherbosa A plant community dominated by *Sphagnum* growing on peat; *cf.* aquiherbosa.

sphagnophilous Thriving on *Sphagnum* moss, or in *Sphagnum*-rich habitats; **sphagnophile, sphagnophily**.

sphagnophyte A *Sphagnum*-bog plant.

sphagnous Pertaining to *Sphagnum* moss, or to *Sphagnum* bogs.

sphalmate (sphalm.) Latin, meaning in error, by mistake.

sphecology The study of wasps.

sphingophilous Pollinated by hawk moths or nocturnal Lepidoptera; **sphingophile, sphingophily**.

sphyric Pertaining to rock slides.

spiladophilus Thriving on clay or in clay rich habitats; **spiladophile, spiladophily**.

spiladophyte A plant living on clay soils; **spiladophyta**.

spite Any behaviour that is detrimental in terms of genetic fitness *q.v.* to both the perpetrator and the target organism; *cf.* altruism, selfishness.

splash erosion The displacement of soil particles by impacting rain droplets; *cf.* erosion.

splash zone The region of the shore immediately above the highest water level that is subject to wetting by splash from breaking waves; see Appendix 7.

splitting The taxonomic practice of subdividing species on the basis of more or less minor differences; **splitter**; *cf.* lumping.

spongicolous Living on or in sponges; **spongicole**.

spontaneous Developing or occurring without apparent external stimulus or cause.

spontaneous generation The theory that organisms arise spontaneously from non-living components of the environment by natural processes without the intervention of supernatural powers; abiogenesis; xenogenesis.

sporadic Scattered; occasional.

sporadophytium A widely dispersed plant community.

spore A plant reproductive cell capable of developing into a new individual, directly or after fusion with another spore; produced either meiotically (meiospores) or mitotically (gonidia).

sporiferous Bearing or producing spores.

sporiparity Reproduction by spore formation.

sporiparous Sporogenous *q.v.*

sporogamy Production of spores by an organism developed from a zygote; **sporogamic**.

sporogenesis The formation of spores; reproduction by means of spores; **sporogenic, spory**.

sporogenous Spore producing; sporiparous; sporigenous.

sporogony Multiple fission of a cell (sporont) to produce many dormant spores (sporozoites); **sporogonic**, **sporogonous**.

sporologic Pertaining to palynology q.v.

sporophyte The diploid, spore producing, asexual generation in the life cycle of a plant; typically formed by fusion of haploid gametes; diplophyte; synkaryophyte.

sport Any freak, or somatic mutation, that deviates from the basic type; rogue.

sporulation The production or liberation of spores.

spory The formation of spores; reproduction by means of spores; sporogenesis; cf. aneuspory, euspory, apomeiotic spory, somatic spory.

spray zone The region of the shore immediately above the splash zone that is subject to wetting by the spray from breaking waves; see Appendix 7.

spring tide A tide of maximum range occurring at the time of new and full moon, when the gravitational attraction of the sun and the moon act together during syzygy q.v.; cf. neap tide.

square brackets Brackets q.v.

square root transformation In statistics, the transformation $y = \sqrt{x}$; used for example to make values of random variables having Poisson distributions amenable to analysis of variance.

ssp. Abbreviation of subspecies.

SSSI Site of special scientific interest.

S-strategist Within the C-S-R triangle q.v. of ecological strategies, a species with small body size, slow growth, long to very long life span, low dispersal ability, strong physiological tolerance to environmental stress and devoting a small proportion of its metabolic energy to the production of offspring or propagules; a stress tolerant species.

st. Abbreviation of the Latin *status*, meaning rank.

stability Resistance to change; tendency to remain in, or return to, an equilibrium state; the ability of populations to withstand perturbations without marked changes in composition.

stability–time hypothesis That where physiological stresses have been historically low, biologically accommodated communities have evolved; as the gradient of environmental stress increases the nature of the community gradually changes to a predominantly physically controlled community, until the stress conditions exceed the adaptive abilities of the organisms and an abiotic condition is reached; the number of species present diminishes continuously along the stress gradient.

stabilizing selection Selection for the mean or

intermediate phenotype with the consequent elimination of peripheral variants, maintaining an existing state of adaptation in a stable environment; centripetal selection; normalizing selection; cf. directional selection, disruptive selection.

stable Remaining constant or unaltered for an extended period of time, or from generation to generation.

stade A climatic episode within a glacial stage during which a secondary glacial advance occurs; **stadial**; cf. interstade.

stadial One of two or more peaks or maxima (periods of minimum temperature) within a single glaciation.

stage 1: Any distinguishable phase of growth or development of an organism; state. 2: A biostratigraphical subdivision of a system, typically based on a sequence of biostratigraphical zones.

stagnicolous Living in stagnant water; **stagnicole**.

stagnoplankton Planktonic organisms and floating vegetation of stagnant water bodies; **stagnoplanktonic**.

stand 1: A unit of vegetation comprising a single species, homogeneous in composition and age. 2: A unit of vegetation which together with similar units having similar combinations of species forms an association q.v.; sometimes used in the sense of a phytosociological record q.v.

stand table A list of species occurring in a stand.

standard deviation In statistics, a measure of variation within a set of data, calculated as the square root of the variance, denoted by σ or s.

standard error In statistics, the standard deviation of the sampling distribution of a statistic; in the case of the sample mean calculated as $SE = \sigma/\sqrt{n}$, where σ is the standard deviation of the original observations and n the number of observations making up the mean.

standard sea water A comparison standard for chlorinity measurements of sea water samples having a chlorinity between 19.30 and 19.50 parts per thousand, determined to within ± 0.001 part per thousand; prepared by the Institute of Oceanographic Sciences, Wormley, U.K.

standing crop Biomass; the total mass of organisms comprising all or part of a population or other specified group, or within a given area, measured as volume, mass (live, dead, dry, ash-free) or energy (calories); standing stock.

standing water Any body of water that does not exhibit a continuous flow in a single direction.

starting point In taxonomy, the date of a published work that for nomenclatural purposes is considered to be the first available *q.v.* or validly published *q.v.* for a particular group.

stasad A stagnant-water plant; stasophyte.

stasigenesis The persistence of a form or group showing little change over an extended period of time, irrespective of changing environmental conditions; the tendency to form well integrated gene pools with well-adapted phenotypic products persisting under the influence of stabilizing selection *q.v.*; *cf.* anagenesis, catagenesis.

stasimorphic 1: Used of a deviant form arising as a result of arrested development. 2: Used of a character retaining the ancestral condition; stasimorphy; *cf.* apomorphic.

stasimorphic speciation The formation of new species without morphological differentiation.

stasipatric speciation The formation of new species as a result of chromosomal rearrangements giving homozygotes which are adaptively superior in a particular part of the geographical range of the ancestral species.

stasis Cessation or retardation of growth or movement; static.

stasium A stagnant-water plant community.

stasophilous Thriving in stagnant water; stasophile, stasophily.

stasophyte A stagnant-water plant; stasad; stasophyta.

stat. Abbreviation of the Latin *status*, meaning rank.

statary Pertaining to a relatively quiescent phase in the activity of a colony of social insects.

state 1: The particular expression or condition of a character. 2: A stage in the life cycle of a plant. 3: A variant that arises in a bacterial culture and alters the gross appearance of the culture.

station 1: The precise location of an organism or species in a habitat; or the site at which a collection or observation was made. 2: A circumscribed area of uniform environmental conditions and vegetation; the typical habitat of a given community.

stationary growth phase A period of little or no population growth following a period of exponential growth; *cf.* lag growth phase.

statistic Any function of a sample, often used as an estimate of the corresponding parameter of the population from which the sample was drawn; usually denoted by Roman letters; *cf.* parameter.

statistical inference The process of predicting or estimating population parameters on the basis of sample data; inductive statistics.

statistics Any kind of data collection, organization and analysis; the branch of mathematics dealing with generalization, estimation and prediction in situations where uncertainty exists; deals particularly with populations, variation and reduction of data.

statokinesis Movement associated with the maintenance of equilibrium; statokinetic.

statu nascendi Latin, meaning (in) the state of being formed; nascent; incipient.

status 1: The standing of a name in nomenclature; a name that does not satisfy the criteria of availability or valid publication has no status and is considered as non-existent. 2: The position of a name in a taxonomic hierarchy; rank. 3: The hybrid or non-hybrid nature of a taxon.

status conidialis (*stat. conid.*) Latin, meaning the conidial state; an indication, used in synonymies, that a name refers to the conidial state of a pleomorphic fungus.

status imperfectus (*stat. imperf.*) Latin, meaning the imperfect state; an indication, used in synonymies, that a name refers only to the imperfect state of a pleomorphic fungus.

status novus (*stat. nov.*) Latin, meaning new status, new rank; used to indicate that a taxon has changed either in rank, without a change of name or epithet, or in its specific or hybrid status.

status pycnidialis (*stat. pycnid.*) Latin, meaning the pycnidial state; an indication, used in synonymies, that a name refers only to the pycnidial state of a pleomorphic fungus.

statute mile A unit of length equal to 5280 ft, or 1.6093 km.

statute of limitation A provision of the International Commission on Zoological Nomenclature introduced to protect universally adopted junior synonyms against the revival of senior synonyms that had not been used for a period of at least 50 years.

staurogamy Cross fertilization *q.v.*

steady state An apparently unchanging condition of a system maintained at equilibrium by antagonistic processes.

steganochamaephytium A community of dwarf shrubs under trees.

stem 1: That part of a word or name to which an appropriate prefix or suffix is added. 2: The single starting point of a dendrogram or cladogram; stem species.

stem-flow Rainwater flowing down a tree trunk or plant stem.

steno- Prefix meaning narrow; *cf.* eury-.

stenobaric Tolerant of a narrow range of atmospheric pressure or hydrostatic pressure; *cf.* eurybaric.

stenobathic Tolerant of a narrow range of depth; *cf.* eurybathic.

stenobenthic Living on the sea or lake bed within a narrow range of depth; *cf.* eurybenthic.

stenobiontic Used of an organism requiring a stable uniform habitat; *cf.* eurybiontic.

stenochoric Having a narrow range of distribution; stenochorous; **stenochore**; *cf.* eurychoric.

stenocoenose Having a restricted distribution; *cf.* eurycoenose.

stenoecious Restricted to a narrow range of habitats and environmental conditions; stenoecic; stenecious; stenoekous; *cf.* amphioecious, euryoecious.

stenohaline Used of organisms that are tolerant of only a narrow range of salinities; *cf.* euryhaline, holeuryhaline, oligohaline, polystenohaline.

stenohydric Tolerant of a narrow range of moisture levels or humidity; *cf.* euryhydric.

stenohygric Tolerant of a narrow range of atmospheric humidity; *cf.* euryhygric.

stenohygrobia A fauna tolerant of only a narrow range of humidity.

stenoionic Having or tolerating a narrow range of pH; *cf.* euryionic.

stenolume Tolerant of a narrow range of light intensity; *cf.* eurylume.

stenomorphic Pertaining to organisms that are of small size owing to cramped habitat conditions.

stenophagous Utilizing, or tolerant of, only a limited variety of foods or food species; stenophagic; **stenophage, stenophagy**; *cf.* euryphagous.

stenophotic Tolerant of a narrow range of light intensity; *cf.* euryphotic.

stenoplastic Exhibiting only a limited developmental response (phenotypic variation) to different environmental conditions; *cf.* euryplastic.

stenosubstratic Tolerant of a narrow range of substrate types; *cf.* eurysubstratic.

stenosynusic Used of plants having a restricted distribution; *cf.* eurysynusic.

stenothermic Tolerant of a narrow range of environmental temperatures; stenothermal; stenothermous; **stenotherm**; *cf.* amphithermal, eurythermal.

stenothermophilic Tolerant of only a narrow range of high temperatures; **stenothermophile, stenothermophily**; *cf.* eurythermophilic.

stenotopic 1: Tolerant of a narrow range of habitats; **stenotopy**. 2: Having a narrow geographical distribution; *cf.* amphitopic, eurytopic.

stenotropic Used of organisms exhibiting a limited response or adaptation to changing environmental conditions; **stenotropism**; *cf.* eurytropic.

stenoxenous Tolerating only a narrow range of host species; *cf.* euryxenous.

steppe Semi-arid areas of treeless grassland found in the mid-latitudes of Europe and Asia.

stercoraceous Living on or in dung; merdicolous.

stereokinesis A change in linear or angular velocity in response to a contact stimulus; thigmokinesis; **stereokinetic**.

stereotaxis A directed response of a motile organism to continuous contact with a solid surface; thigmotaxis; **stereotactic, stereotaxy**.

stereotropism An orientation response to a contact stimulus; haptotropism; thigmotropism; **stereotropic**.

sterile 1: Unable to produce viable propagules or to reproduce sexually; **sterility**. 2: Free from contamination by microorganisms.

sterrhad A moorland plant; sterrhophyte.

sterrhium A moorland community.

sterrhophilus Thriving on moorland; sterrophilous; **sterrhophile, sterrhophily**.

sterrhophyte A moorland plant; sterrhad; sterrophyte.

sterric Pertaining to a heath community or heathland habitat.

stimulus Any internal or external agent that produces a reaction or change in an organism.

stirp 1: A lineage; the sum total of inherited genetic elements. 2: A group of animals more or less equal to a superfamily. 3: A variety or race of plants.

stochastic model A mathematical model founded on the properties of probability so that a given input produces a range of possible outcomes due to chance alone; *cf.* deterministic model.

stock 1: A race, stem or lineage. 2: A population; a group of individuals of one species within a specified area or volume. 3: The stem of a plant that receives a graft.

stocking density A measure of the abundance of individuals per unit area, calculated for woodland as the number of stems per hectare.

stomatal conductance The reciprocal of stomatal resistance *q.v.*

stomatal resistance The property of the stomata in restricting the free exchange of carbon dioxide (CO_2) by a plant leaf; the major constraint on CO_2 uptake into the plant leaf governed largely by the diameter of the stomatal pores.

stomatal transpiration The loss of water vapour through the open stomata of leaves; *cf.* cuticular transpiration.

stone A sediment particle larger than 20 mm in diameter; a component of gravel; see Appendix 11.

Stone Age An archaeological period dating from

about 10 000 years B.P.; the earliest technological period of human history characterized by the use of stone tools; three periods are recognized, Palaeolithic, Mesolithic and Neolithic *q.v.*

storey A horizontal layer or stratum of a given height within a stratified plant community.

strain 1: Any physical or chemical change in a living organism produced by a stress *q.v.* 2: A group of individuals with common physiological traits and presumed common ancestry; an infraspecific group having characteristic properties.

strand plant A plant growing on the shore immediately above high tide level.

strandline A line on the shore comprising debris deposited by a receding tide; commonly used to denote the line of debris at the level of extreme high water.

strange species A species found only as a rare or accidental member of a community, probably a relic of an earlier community; a species of low fidelity *q.v.*

strategy 1: The plan, method or structure utilized by an organism or group of organisms to meet a particular set of conditions. 2: A plan, in game theory *q.v.*, which completely specifies the choices a competitor should make in all circumstances that can possibly arise.

stratification Organization into horizontal strata; the vertical structuring of a community or a habitat into superimposed horizontal layers; the grouping of individuals of a community or habitat into height classes.

stratigraphic palaeontology Biostratigraphy *q.v.*

stratigraphic range The distribution of a taxon through geological time, determined by its distribution in strata of known geological age.

stratigraphic unit A stratum or group of adjacent strata treated as a unit for the purposes of classification with reference to one or more characters.

stratigraphy Study of the origin, composition, distribution and succession of rock strata.

stratocladistic Pertaining to phylogenetic relationships inferred from weighted derived similarities between fossil taxa and selected stratigraphic data not contradicted by morphological analysis.

stratophenetic Pertaining to phylogenetic relationships inferred from overall similarity of fossil taxa and their stratigraphic sequence.

stratosphere 1: The nearly uniform cold ocean water masses in high latitudes and near-bottom waters of middle and low latitudes; ocean water below the thermocline. 2: The layer of the

atmosphere above the troposphere, 15–50 km above the Earth's surface, in which temperature ceases to fall with increasing height, typically without strong convection currents; *cf.* mesosphere, troposphere.

stratotype 1: In taxonomy, the original or subsequently designated type of a named stratigraphic unit or boundary, used as the standard for identification and definition; further qualified by the prefixes, holo-, para-, neo-, lecto- and hypo- as used in biological taxonomy. 2: The type of an ant soldier caste.

stratum (strata) 1: A layer, as of rock, possessing characters that serve to distinguish it from adjacent layers. 2: A horizontal layer of vegetation within a stratified plant community; storey; **stratose**.

stratum society A plant society occurring as a well-defined horizontal layer in a community.

stratum transect A profile of vegetation showing the heights of plants and shoots; profile transect.

stress Any environmental factor that restricts growth and reproduction of an organism or population; any factor acting to disturb the equilibrium of a system; *cf.* strain.

stress avoidance Stress resistance *q.v.* by avoiding thermodynamic equilibrium with the stress either by means of a physical barrier which insulates the living tissue from the stress, or by a steady state exclusion of the stress (a chemical or metabolic barrier); *cf.* stress tolerance.

stress hypothesis A hypothesis of population self-regulation; that mutual interactions in an increasing population lead to physiological changes which are phenotypic in origin that eventually reduce the number of births and increase deaths; *cf.* behaviour hypothesis, polymorphic behaviour hypothesis.

stress resistance The ability of a living organism to survive exposure to an unfavourable environmental factor; quantifiable as the stress necessary to produce a specific strain *q.v.*

stress tolerance Stress resistance *q.v.*, through the ability of an organism to come to thermodynamic equilibrium with the stress without injury, because the organism prevents, decreases or repairs the strain *q.v.* induced by the stress; *cf.* stress avoidance.

stridulation Production of sound by rubbing together of specially modified structures or surfaces; **stridulate**.

strobilation Reproduction by successive budding.

stromatolite A biogenic sedimentary structure, either fossil or recent; often regarded as plant trace fossils.

strophogenesis The evolution of a system of

alternation of generations in an organism which previously passed through a succession of similar generations.

strophotaxis A twisting movement in response to an external stimulus; strophism; **strophotactic**.

structural gene Any gene that determines the amino acid sequence of a polypeptide.

structure 1: The arrangement of parts or manner of construction of a given entity. 2: In phytosociology, the spatial arrangement of the components of vegetation.

struggle for existence Competition between members of a population for a limited vital resource resulting in natural selection.

Student's t-test t-test *q.v.*

stygobie An obligate cavernicole; troglobie.

stygon The groundwater biotope; *cf.* crenon, thalasson, troglon.

stygophilic Thriving in caves or subterranean passages; troglophilic; **stygophile, stygophily**.

stygoxenous Found only occasionally in caves or subterranean passages; trogloxenous; **stygoxene**.

sub- Prefix meaning under, beneath, lesser (in structure or significance); *cf.* super-.

subaerial Occurring immediately above the surface of the ground.

subalpine Alpestrine *q.v.*

subandroecious Incompletely androecious; used of a plant having mostly male flowers but with a few female or hermaphrodite flowers.

subaquatic Not completely aquatic.

subaqueous Occurring in or under water.

subarctic Pertaining to regions adjacent to the Arctic Circle, or to montane regions having similar climatic conditions or vegetation; orohemiarctic.

Subarctic period Preboreal period *q.v.*

Subarctic subregion Canadian subregion of the Nearctic region; see Appendix 3.

subarctic zone A latitudinal zone between 58° and 66.5° in either hemisphere.

subassociation A ranked category in the classification of vegetation; a non-geographical subdivision of an association *q.v.*; syntaxa of subassociation rank have the ending -*etosum*.

Subatlantic period A period of the Blytt–Sernander classification *q.v.* (ca. 2500 years B.P. to present), characterized by beech and linden vegetation and by mild and wet conditions.

subboreal Pertaining to a climate or biogeographic zone approaching frigid or boreal conditions; **sub-boreal**.

Subboreal period A period of the Blytt–Sernander classification *q.v.* (*ca.* 4500–2500 years B.P.), characterized by oak, ash and linden vegetation

and by cooler and drier conditions than in the Subatlantic period.

subclimax The stage in an ecological succession that precedes the final or climax community and that is prevented from attaining full development by one or more edaphic or biotic factors (arresting factors); paraclimax.

subdioecious Incompletely dioecious; used of a plant population that is largely unisexual but has a few bisexual individuals.

subdominant 1: A life-form subordinate to the dominant *q.v.* of a plant community. 2: A species that may appear to be more abundant at certain times of the year than the true dominant species in the climax.

subdominule A subdominant of a microcommunity.

subduction The process by which crustal material is returned to the upper mantle.

subduction zone A zone in which the edge of one tectonic plate descends below an adjacent plate; *cf.* obduction zone.

subfamily A subdivision of a family *q.v.*; a taxon at the rank of subfamily.

subforma A category subordinate to *forma q.v.*; subform.

subformation A ranked category in the classification of vegetation; a distinct geographical facies of a formation *q.v.*, equivalent in rank to an association *q.v.*

subfossil A post-Pleistocene fossil; used of plant and animal remains not strictly Recent but which are not old enough to be regarded as fossil.

subgeneric name A scientific name of a taxon of subgenus rank, placed in parentheses after the genus name.

subgenus A subdivision of a genus *q.v.*; a taxon at the rank of subgenus.

subgeocolous Living underground; **subgeocole**.

subgregarious Used of organisms forming loose aggregations or groups.

subgynoecious Incompletely gynoecious; used of a plant having mostly female flowers but with a few male or hermaphrodite flowers.

subhumid Used of a climate or habitat transitional between humid and subarid in terms of rainfall.

subhydrophilous Thriving in habitats experiencing periodic inundation by fresh water; **subhydrophile, subhydrophily**.

subjective Lacking reality; assessed as a matter of opinion or judgement; *cf.* objective.

subjective synonyms Heterotypic synonyms *q.v.*

sublethal Not causing death by direct action, but possibly modifying behaviour, physiology, reproduction, or life cycle, and showing cumulative effects.

subliminal Below threshold level; insufficient to elicit a response.

sublittoral 1: The deeper zone of a lake below the limit of rooted vegetation; 2: The marine zone extending from the lower margin of the intertidal (littoral) to the outer edge of the continental shelf at a depth of about 200 m; sometimes used for the zone between low tide and the greatest depth to which photosynthetic plants can grow; sublitoral; see Appendix 7.

sublittoral fringe Infralittoral fringe *q.v.*

submaritime Used of organisms characteristic of the seashore but also occurring inland.

submergent Pertaining to a plant or plant structure growing entirely under water; submersed; **submerged**; *cf.* emergent.

submersed Pertaining to a plant or plant structure growing entirely under water; immersed; submerged; *cf.* emersed.

submersiherbosa Submergent herbaceous vegetation of ponds and swamps; *cf.* aquiherbosa.

submesic-supralittoral That part of the supralittoral *q.v.* zone of the shore seldom splashed by waves.

submission The behaviour of an animal defeated in an aggressive contact that adopts a submissive posture to prevent further attack.

subordinate taxon A taxon of lower rank than that with which it is compared.

subprovince A subdivision of a biogeographical province *q.v.*

subsequent designation In taxonomy, the designation of the type of a taxon in a work published subsequent to the establishment of the taxon; *cf.* original designation.

subsequent fixation The fixation of the type of a taxon in a work published after the establishment of the taxon; fixation by subsequent designation *q.v.* or subsequent monotypy *q.v.*

subsequent monotypy The assumption of type status by the first species placed in a genus that was originally established without any included species.

subsequent spelling In taxonomy, any modified spelling of a name other than the original spelling.

subsere An ecological succession on a denuded area, or an ecological succession arrested by edaphic or biotic factors; secondary sere.

subsidence theory That the upward growth of coral atolls and barrier reefs matched the slow subsidence of a volcanic island over a prolonged period of time; postulated to explain the formation of coral atolls.

subsocial Pertaining to a social group, or the condition, in which parents care for their own young; presocial.

subsociation A component of an association distinguished by the presence of a subdominant *q.v.* which is itself under the influence of the dominant.

subsoil The layer of soil beneath the top-soil and overlying the bed rock; the C-horizon of a soil profile.

subspecies 1: A group of interbreeding natural populations differing taxonomically and with respect to gene pool characteristics, and often isolated geographically, from other such groups within a biological species *q.v.*, and interbreeding successfully with these groups where their ranges overlap. 2: A taxon at the rank of subspecies.

subspecific epithet The third word of a trinomial name of a subspecies; *cf.* subspecific name.

subspecific name The scientific name of a taxon of subspecies rank; trinomen; formerly used as equivalent to subspecific epithet *q.v.*

subspontaneous Used of a plant that has been introduced into a new area and has successfully established itself in the new habitat.

substitute name *Nomen novum q.v.*

substitution Geographical replacement of one plant species by another, often closely related, species.

substitution, Law of That any extrinsic factor can be substituted by another so long as its effect on an organism is the same.

substitutional load The cost to a population of replacing one allele by another during evolutionary change; transitional load; *cf.* genetic load.

substrate 1: The substance acted upon by an enzyme, or that utilized by a microorganism as a food source. 2: Substratum *q.v.*

substrate race A local race having a characteristic coloration resembling that of its substrate.

substratohygrophilous Thriving on moist substrates; **substratohygrophile**, **substratohygrophily**.

substratoid Pertaining to epiphytes with a wide climatic and narrow edaphic amplitude; *cf.* climatoid.

substratum (substrata) The sediment, surface, or medium to which an organism is attached or upon which it grows; substrate.

subterranean Living or occurring beneath the surface of the earth; subterrestrial.

subtraction The loss of a hereditary factor.

subtribe A taxon at the rank of subtribe; a rank below tribe within the family-group.

subtroglophile A troglophile *q.v.* that does not reproduce in the cave habitat; *cf.* eutroglophile.

subtrogloxene An organism found only occasionally in underground caves and passages, and able to reproduce in the cave habitat but not fully tolerant of cave conditions; *cf.* eutrogloxene.

Subtropical convergence Antiboreal convergence *q.v.*

Subtropical zone The latitudinal zone between 23.5° and 34.0° in either hemisphere.

subvital A mutation that causes the death of between 10–50% of the mutants in the population; *cf.* quasi-normal, semilethal.

subxerophilous Thriving in moderately dry situations; **subxerophile, subxerophily**.

succession 1: Ecological succession *q.v.* 2: The chronological distribution of organisms within an area; the geological, ecological or seasonal sequence of species within a habitat or community.

successional speciation Phyletic speciation *q.v.*

successional species Successive morphospecies within a distinct evolutionary lineage representing the nodal points of a phyletic evolutionary trend that are sufficiently distinct to justify the assignment of species names.

succulency The condition of having specialized fleshy tissue in a plant root or stem for the conservation of water; **succulent**.

Sudanese Park Steppe region A subdivision of the African subkingdom of the Palaeotropical kingdom; see Appendix 4.

sudd A floating mass of plant material.

suffix The appended termination of a compound word; *cf.* prefix.

suffruticose chamaephyte A chamaephyte *q.v.* subtype in which the upper vegetative and flowering shoots die back leaving only the lower parts to survive unfavourable seasons.

sulphuretta The prokaryotic and protistan inhabitants of anaerobic sulphur-rich deposits; used in a more restricted sense than thiobios *q.v.*

sum of squares The sum of the squared deviations from the means, in an analysis of variance.

summer heat income The total amount of heat required to raise the temperature of a lake from 4° C to its highest observed summer heat content; *cf.* winter heat income.

super- Prefix meaning above, greater (in structure or significance); *cf.* sub-.

superdispersed Contagious distribution *q.v.*

superfamily name A scientific name of a taxon of superfamily rank.

superfetation Superfoetation *q.v.*

superfluent Used of an animal species in a community that is of equal significance to a subdominant plant species.

superfluous name *Nomen superfluum q.v.*

superfoetation The fertilization of a plant ovary by two or more kinds of pollen; hypercyesis; superfetation.

supergene A group of tightly linked genes usually transmitted as a single unit.

superior taxon A taxon of a higher rank than the one with which it is compared.

superlarvation Paedomorphosis *q.v.*

supernatant Floating at the surface.

supernumerary chromosome Any chromosome or chromosome fragment differing from the normal A-chromosomes *q.v.* in structure, genetic effectiveness and pairing behaviour; B-chromosome.

superoptimal Used of a stimulus that elicits a greater response than the natural stimulus.

superorganism Any colony possessing features of social organization analogous to the physiological properties of a single organism; any group of organisms acting as a single functional unit; supraorganism; epiorganism.

superovulation The production of an unusually large number of eggs at any one time.

superparasitism 1: Parasitism of a host that is itself a parasite; hyperparasitism; **superparasite**. 2: The multiple infection of a host by several parasites, usually of a similar kind.

superposition, law of That the lowest stratum of a geological formation was the first to be deposited and is therefore the oldest.

superspecies A monophyletic group of entirely or essentially allopatric species that are considered to be too distinct morphologically to be regarded as a single species; a cluster of incipient species (semispecies); in taxonomy the superspecies name is enclosed in square brackets between the two words of the binomen; collective species.

superterranean Occurring at or above ground level.

supertramp A species having a wide geographical range and very efficient means of dispersal, but which tends to be excluded by competition from species-rich faunas.

supplementary type In taxonomy, a described or figured specimen used to provide supplementary information about a previously described species.

suppression 1: Failure to develop. 2: In taxonomy, the rejection of an available name under the plenary powers of the ICZN, rendering it invalid for the purposes of priority but leaving it available for the purposes of homonymy.

suppressor gene A gene that suppresses the

phenotypic expression of a gene at a different locus.

supra- Prefix meaning above; *cf.* infra-.

supra citato (*supra cit.*) Latin, meaning cited above.

suprabenthic Hyperbenthic *q.v.*; **suprabenthos**.

suprafolious Growing on leaves.

suprageneric Above the rank of genus.

supralithion Those aquatic organisms swimming above rock, but dependent upon it as a food source.

supralittoral The region of the shore immediately above the highest water level and subject to wetting by spray or wave splash; splash zone; spray zone; supratidal; see Appendix 7; *cf.* mesic-supralittoral, submesic-supralittoral, xeric-supralittoral.

supraneuston Epineuston *q.v.*

supraorganism Superorganism *q.v.*

suprapelos Those aquatic organisms swimming above soft mud that are dependent upon the substratum as a food source.

suprapsammon Those aquatic organisms swimming above sand that are dependent upon the substratum as a food source.

supraspecific Above the rank of species.

supraterraneous Occurring at or above ground level.

supratidal The zone on the shore above mean high tide level; supralittoral.

surf The waves breaking on the shore.

surface pheromone A pheromone that has a very limited active space *q.v.* such that the organisms must be in contact or near contact in order to perceive it.

survival of the fittest The differential and greater success of the best adapted genotypes.

survival value The degree of effectiveness of a character in conferring fitness *q.v.* on an individual.

survivorship The proportion of individuals from a given cohort *q.v.* surviving at a given time; typically presented as a survivorship curve.

survivorship curve A graph of the number of surviving individuals of a given cohort plotted against age; three main forms of survivorship curves have been recognized – Type 1 in which the probability of survival decreases with age; Type 2 in which the probability of survival is constant with age; Type 3 in which the probability of survival increases with age.

suscept A plant or animal susceptible to, or harbouring, a disease organism.

susceptible Prone to influence by a stimulus, or to infection by parasites or pathogens; **susceptibility**.

suspended animation The temporary suspension of life processes.

suspension A dispersion of fine insoluble particulate matter in a fluid; **suspended**.

suspension feeder Any organism that feeds on particulate organic matter suspended in water.

suture zone A zone of secondary intergradation *q.v.*

Sverdrup A measure of ocean current flow equal to 1 million cubic metres per second.

swamp Wet spongy ground, saturated or intermittently inundated by standing water, typically dominated by woody plants but without an accumulation of surface peat.

swarm spore Any independently motile spore.

swarming Colony fission *q.v.* in bees.

sweepstakes bridge Chance crossings or migrations across a water barrier or other major biogeographic obstacle by rafting or other means of transport; the dispersal path is referred to as a sweepstakes route; sweepstake dispersal.

swell 1: A deep-water, wind-generated, wave system of long periodicity. 2: An area that has subsided less than the surrounding area and is covered by a thinner sequence of sedimentary deposits.

switch gene A gene that causes the epigenotype *q.v.* to switch to a different development pathway.

sycophagous Feeding on figs; **sycophage, sycophagy**.

sylvan Pertaining to woods.

sylvestral Pertaining to, or inhabiting, woods and shady hedgerows; sylvicolous; **sylvestrine**.

sylvicolous Inhabiting woodland; sylvestral; silvicolous; **sylvicole**.

symbiogenesis The evolutionary origins of symbiotic relationships between organisms.

symbiology The study of symbioses.

symbiont A participant in a symbiosis; a symbiotic organism; symbiote; **symbion**.

symbiosis The living together of two organisms; the relationship between two interacting organisms or populations, commonly used to describe all relationships between members of two different species, and also to include intraspecific associations; sometimes restricted to those associations that are mutually beneficial; symbioses; **symbiont, symbiote, symbiotic**.

symbiotrophic Used of an organism obtaining nourishment through a symbiotic relationship; **symbiotrophy**.

symmetric matrix A matrix with equal numbers of rows and columns which is unchanged when transposed; *cf.* matrix transpose.

sympaedium An assemblage of young animals playing together.

symparasite Any of several competing parasite species on a single host; **symparasitic**.

sympatric Used of populations, species or taxa occurring together in the same geographical area; the populations may occupy the same habitat (biotic sympatry) or different habitats (neighbouring sympatry) within the same geographical area; **sympatry**; *cf.* allopatric, dichopatric, parapatric.

sympatric hybridization The occasional production of hybrids between two well-defined sympatric species.

sympatric introgression The production of a new genotype by introgression *q.v.*, which then exists in the same populations as the original genotype; *cf.* allopatric introgression.

sympatric speciation The differentiation and attainment of reproductive isolation of populations that are not geographically separated and which overlap in their distributions.

symphily An amicable relationship between one organism (the symphile) and its host colony of social insects; **symphilic**.

sympleisiotypy The common possession of a pleisiotypic character state; *cf.* synapotypy.

symplesiomorphy The common possession of an ancestral (plesiomorph) character; *cf.* synapomorphy.

syn. Abbreviation of synonym or synonymy.

syn- Prefix meaning joined (in space or time).

synacme Homogamy *q.v.*

synanthropic Living close to human habitation; *cf.* eusynanthropic, exanthropic.

synapomorphy The common possession of a derived (apomorphic) homologous character; *cf.* symplesiomorphy.

synaposematic coloration Protective coloration in which the warning signal is shared with other species, as in Müllerian mimicry; synaposematism; *cf.* aposematic coloration.

synapotypy The common possession of a derived (apotypic) character state; *cf.* sympleisiotypy.

synapsis The pairing of homologous chromosomes during meiosis.

synaptospermy The clumping of seeds by means of cohering or interlocking surface structures.

synaptospory The clumping of spores by means of cohering or interlocking surface structures.

synchorology Study of the occurrence, distribution and classification of plant communities into the following regional units; region, province, sector, subsector, district and subdistrict.

synchronic Existing or occurring at the same time; contemporaneous; synchronous; *cf.* allochronic.

synchronic species Species occurring in the same geological time horizon; *cf.* allochronic species.

synchronogamic Having both male and female reproductive organs mature at the same time.

synchronopaedium A group of young animals of about the same age but of different parentage.

synchronous hermaphrodite An organism producing mature sperm and ova at the same time; simultaneous hermaphrodite; *cf.* consecutive hermaphrodite.

synclerobiosis Symbiosis between two species of social insects that usually inhabit separate colonies.

synclopia Lestobiosis *q.v.*

syncryptic Pertaining to the resemblance between unrelated organisms having similar cryptic coloration.

syncytium A multinucleate protoplasmic mass not differentiated into cells.

syndiacony A mutually beneficial relationship between ants and plants; **syndiaconic**.

syndynamics The study of successional changes in plant communities.

synecology Study of the ecology of organisms, populations, communities, or systems; ecological sociology; *cf.* autecology.

synecthry Commensalism in which the participants (synecthrans) display mutual dislike.

synergism 1: Cooperative action of two or more agencies such that the total is greater than the sum of the component actions; synergy; **synergic, synergetic, synergistic**; *cf.* antagonism. 2: Mutualism *q.v.* 3: The cooperative action of two microorganisms to effect a change that would not occur or would take place at a slower rate in axenic culture.

synethogametism Gametic compatibility; the ability of two gametes to fuse; *cf.* asynethogametism.

syngameon A group of sympatric or at least marginally sympatric semispecies *q.v.*

syngamete A diploid cell formed by the fusion of two haploid gametes; zygote.

syngamic sex determination Determination of sex as a result of gamete fusion and karyogamy *q.v.*

syngamodeme A biosystematic unit comprising all the coenogamodemes *q.v.* that are linked by the ability of some members to form viable but sterile hybrids; members are not capable of hybridizing with any other syngamodemes.

syngamy Sexual reproduction; fusion of male and female gametes; gametogamy; **syngametic**.

syngenesis 1: Sexual reproduction; **syngenetic**. 2: Descent from a common ancestor; coenogenesis.

syngenic Having the same set of genes; isogenic; syngeneic; *cf.* allogenic.

syngeographs Geographical equivalents *q.v.*

syngonic Producing male and female gametes in the same gonad; **syngony.**

synhesma A swarming society, or a swarm.

synhospitalic Used of two or more parasite species occurring on the same host species; *cf.* allohospitalic.

synisonym In taxonomy, one of two or more names having the same name bearing epithet (basionym).

synkaryon The zygote nucleus formed from the fusion of the gametic nuclei.

synkaryophyte Sporophyte *q.v.*

synkaryotic Diploid *q.v.*

synmorphology The study of floristic composition, minimal area and extent in time and space of a plant community.

synoecius Used of organisms that produce both male and female gametes; **synoecy.**

synoecy 1: A symbiosis between a colony of social insects and a tolerated guest organism (a synoekete). 2: The condition of a synoecius *q.v.* organism.

synoekete A guest organism tolerated within an insect colony; synoecete; synecete.

synonym One of two or more scientific names applied to the same taxon; **synonymous.**

synonym, commercial Commercial synonym *q.v.*

synonymum (*syn.*) Latin, meaning synonym.

synonymum novum (*syn. nov.*) Latin, meaning new synonym; used in synonymies to indicate that a name is being treated as a synonym for the first time.

synonymy The list of names considered by an author to apply to a given taxon, other than the name by which it should properly be known.

synopsis A summary of current knowledge; a brief description of the essential features of a taxon; **synoptic.**

synpiontology The study of ancient patterns of distribution and migration of plant communities.

synsystematics The classification of plant communities.

syntaxon A plant community of any rank; equivalent to taxon in idiotaxonomy *q.v.*

syntaxonomy The classification and nomenclature of plant communities; *cf.* idiotaxonomy.

syntechny Resemblance between different organisms as a result of adaptation to similar environments; convergent evolution.

synthetic evolution, theory of That evolution is best explained by a synthesis of many previous

theories, involving especially mutation and selection; *cf.* monistic evolution, theory of.

synthetic lethals Lethal chromosomes formed as a result of crossing over between normally viable chromosomes.

synthetic method Evolutionary method *q.v.*

synthetograph In taxonomy, a composite figure drawn from two or more individuals of the type series.

syntopic Pertaining to populations or species that occupy the same macrohabitat, are observable in close proximity and could thus interbreed; **syntopy.**

syntrophy The condition exhibited by two organisms mutually dependent for food; cross-feeding in which each organism derives one or more essential nutrients from the other; **syntrophic, syntrophism.**

syntype In taxonomy, any of two or more type specimens listed in the original description of a taxon when a holotype was not designated.

synusia A community of species with a similar life form and similar ecological requirements; a habitat of characteristic and uniform conditions; **synusial.**

synxenic Used of a culture comprising two or more organisms under controlled conditions.

synzoochorous Dispersed by the agency of animal species; **synzoochore.**

syringograde A form of locomotion in which the organism propels itself with a jet of water; jet propulsion.

syrtidad A dry sand-bar plant; syrtidophyte.

syrtidium A dry sand-bar community.

syrtidophilus Living on dry sand-bars; **syrtidophile, syrtidophily.**

syrtidophyte A dry sand-bar plant; syrtidad; **syrtidophyta.**

system 1: An assemblage of individual items formed into a single integrated whole. 2: The rocks formed during a given geological period or event.

systematic mutation A catastrophic mutation that would lead to the origin of an entirely new type of organism.

systematics The classification of living organisms into hierarchical series of groups emphasizing their phylogenetic interrelationships; often used as equivalent to taxonomy.

systematization The production of a natural classification based on phylogenetic relationships.

syzygy The time at which the sun and moon are in line, either in conjunction or in opposition, with respect to the Earth; *cf.* quadrature.

T

t. 1: Abbreviation of the Latin *teste* *q.v.*
2: Abbreviation of the Latin *tomus* *q.v.*

T Abbreviation of tera, denoting unit $\times 10^{12}$.

2,4,5-T 2,4,5-trichlorophenoxyacetic acid, a translocated hormone weedkiller used to control scrub and woody vegetation.

t.c. Abbreviation of the Latin *tomus citatus*, meaning in the volume cited.

tab. Abbreviation of the Latin *tabula*, meaning plate.

tacheion Actively moving aquatic organisms comprising both crawling (herpon) and free swimming forms.

tachy- Prefix meaning quick, fast.

tachyauxesis 1: Heterauxesis *q.v.* in which *a* from the equation $y = bx^a$ is greater than unity, so that a given structure is relatively larger in large individuals than in small ones. 2: Relatively rapid growth.

tachygen A structure that appears suddenly in phylogeny.

tachygenesis Accelerated ontogenetic or phylogenetic development, as in the omission of certain larval stages during development; *cf.* bradygenesis.

tachysporous Used of a plant that disperses its seeds quickly.

tachytelic Evolving at a relatively fast rate; used of a phase of rapid evolution occurring as populations shift from one major adaptive zone to another; *cf.* bradytelic, horotelic.

tactic 1: Pertaining to a taxis *q.v.* 2: A set of co-adapted traits, or a strategy.

tactile Pertaining to the sense of touch; tactual.

tactism Taxis *q.v.*

tagmosis The functional specialization occurring within metamerically segmented animals, producing a subdivision of the body into distinct regions (tagmata).

taiga Northern coniferous forest biome *q.v.*; the ecosystem adjacent to the arctic tundra, but used with varying scope to include only the arctic timberline ecotone through to the entire subarctic subalpine north temperate forest.

talandric Pertaining to rhythmical changes in the physiological processes of an organism.

tandem running Behaviour found in some social insects in which one individual follows closely behind another during exploration or recruitment.

tandem selection Selection for a second trait once the primary trait has reached a desired level of improvement.

tang line The highest continuous line on the shore along which a particular seaweed grows; has variously been applied to wracks, laminarians and other algal forms; *cf.* algal line.

tank epiphyte An epiphyte having roots but gathering water in the manner of a cistern epiphyte *q.v.*

tapestry A more or less continuous cover of trees on a steep slope.

taphocoenosis An assemblage of dead organisms brought together at or after the time of death, or of fossils that were buried together; thanatocoenosis; **taphocoenose**; *cf.* biocoenosis.

taphoglyph An imprint in a sediment of a dead or dying organism.

taphonomy Study of the environmental phenomena and processes that affect organic remains after death, including the processes of fossilization.

taphrad A ditch plant; taphrophyte.

taphrium A ditch community.

taphrophilus Thriving in ditches; taphrophilous; **taphrophile, taphrophily**.

taphrophyte A ditch plant; taphrad.

tautology Unnecessary repetition of a word or statement; **tautological, tautomer, tautomeric**.

tautomorphism The superficial resemblance of morphs belonging to two or more different polymorphic species; parallel polymorphism; **tautomorph**.

tautonymy In taxonomy, the identical spelling of the generic name and the epithet within a binomen or trinomen; tautonym; tautonymous name; *cf.* absolute tautonymy, virtual tautonymy.

tautotype The primary type of the type species of a genus established by virtue of tautonymy.

taximetrics The measurements of phenetic similarities between organisms and the incorporation of these data into systems of classification; numerical taxonomy; taxometrics, taxonometrics.

-taxis 1: Suffix meaning arrangement (of); **-taxy**. 2: Suffix meaning the directed response of a motile organism; aerotaxis, anemotaxis, aphototaxis, aphydrotaxis, apothermotaxis, argotaxis, barotaxis, biotaxis, chemotaxis, clinotaxis, cyclotaxis, diaphototaxis, electrotaxis, galvanotaxis, geotaxis, heliotaxis, hydrotaxis, hygrotaxis, klinotaxis, menotaxis, mnemotaxis, osmotaxis, oxygenotaxis, oxytaxis, pharotaxis, phobotaxis, photohoramotaxis, phototaxis, pneumotaxis, prophototaxis, prosaerotaxis, proschemotaxis, prosgalvanotaxis, proshydrotaxis, prososmotaxis, prosphototaxis, rheotaxis, scototaxis, seismotaxis, skototaxis, stereotaxis, strophotaxis, thermotaxis, thigmotaxis, tonotaxis, topochemotaxis, topogalvanotaxis, topophototaxis, topotaxis, traumatotaxis, trophotaxis, tropotaxis, vibrotaxis, zygotaxis.

taxis A directed reaction or orientation response of a motile organism towards (positive) or away from (negative) the source of the stimulus; tactism; **taxy, taxes, tactic;** *cf.* klinotaxis, telotaxis, tropotaxis.

taxocene A group of species belonging to a particular supraspecific taxon occurring together in the same association *q.v.*

taxometrics Taximetrics *q.v.*

taxon (taxa) A taxonomic group of any rank, including all the subordinate groups; any group of organisms, populations, or taxa considered to be sufficiently distinct from other such groups to be treated as a separate unit; taxonomic unit.

Taxon The official organ of the International Association of Plant Taxonomy.

taxon cycle Theory that a species spreads while adapted to one habitat, then becomes restricted in its range (often splitting into two or more species) while adapting to another habitat; in island systems the widespread low-elevation taxa on an island are the most recent colonists and the taxa restricted to montane rain forests are the older taxa on that island.

taxon vagum (***tax. vag.***) Latin, meaning uncertain taxon; occasionally used in names to indicate a taxon of uncertain rank.

taxonometrics Taximetrics *q.v.*

taxonomic distance An expression of the relationship between individuals or taxa in terms of multidimensional space, where each dimension represents a character, based on quantitative estimates of dissimilarity.

taxonomic group Any taxon, including all subordinate taxa.

taxonomic hierarchy A hierarchical system of taxonomic categories arranged in an ascending series of ranks; in Botany the following 12 main ranks are recognized – Kingdom, Division, Class, Order, Family, Tribe, Genus, Section, Series, Species, Variety, Form; in Zoology 7 main ranks are recognized – Kingdom, Phylum, Class, Order, Family, Genus, Species: additional categories can be introduced by the use of super- and sub- prefixes; see Appendix 10.

taxonomic position Rank; the position of a taxon in a hierarchy of classification.

taxonomic species Linnaean species *q.v.*

taxonomic synonyms Heterotypic synonyms *q.v.*

taxonomic system A hierarchy of classification.

taxonomy The theory and practice of describing, naming and classifying organisms; systematics; biosystematics; genonomy; **taxonomic**.

taxospecies A species based on overall similarity, determined by numerical taxonomic methods.

Taylor's power law That the variance of a population is proportional to a fractional power of the arithmetic mean; expressed as $s^2 = a\bar{x}^b$, where a is dependent on the size of the sampling unit and b is an index of dispersion varying from zero (for a uniform distribution) to infinity (for a highly contagious distribution).

Tchornozem Chernozem *q.v.*

tecnophagous Used of an individual that feeds on its own eggs; **technophage, technophagy**.

tectology The study of organisms as groups of morphological units or individuals.

tectonic Pertaining to the movement of the rigid plates that comprise the Earth's crust, and to deformation of the crustal plates.

tectonics Study of the structural features and processes of the Earth's crust.

tectonosphere The crustal zone of the Earth in which tectonic movement occurs.

tectoparatype A paratype used for microscopic examination to demonstrate generic or specific characters.

tectopleisiotype A specimen used for microscopic examination to demonstrate generic or specific characters, subsequent to the original description.

tectotype In taxonomy, any type comprising a microscope slide preparation or section.

tegulicolous Living on tiles, as of lichens; **tegulicole**.

tele- Prefix meaning beyond.

telegamic Pertaining to the process or behaviour of attracting a mate from a distance.

telegenesis Artificial insemination.

teleianthous Pertaining to a hermaphrodite flower.

telemorphosis A change in form in response to a remote stimulus.

teleology The doctrine that natural phenomena result from, or are shaped by, design or purpose; **teleological**.

teleonomy The doctrine that all structures and functions have evolved by the selective advantage conferred upon the organism.

teleorganic Used of processes or functions vital to the life of an organism.

teleosis Purposeful evolution or development; telic.

teleotypic character Any unique distinguishing character of a species.

teleplanic Used of pelagic larvae that have a protracted planktonic existence, with the capacity for very widespread dispersal.

telmatad A wet meadow plant; telmatophyte.

telmatium A wet meadow or marsh community; telmathium.

telmatology The study of wetlands, swamps and marshy areas.

telmatoplankton Planktonic organisms of freshwater bogs and marshes.

telmatophilous Thriving in wet meadows; **telmatophile, telmatophily**.

telmatophyte A wet meadow plant; telmatad; **telmatophyta**.

telmicolous Living in freshwater marshes; **telmicole**.

telmophagous Used of an organism, usually an insect, that feeds from a blood pool produced by tissue laceration; pool feeder; **telmophage**.

telocentric Used of a chromosome having the centromere located at one end; *cf*. metacentric.

telotaxis An orientation response of a motile organism to a single stimulus without reference to other sources of stimulation; goal orientation; *cf*. klinotaxis, tropotaxis.

TEM Transmission electron microscope.

temperate 1: Moderate; not excessive or extreme; used of climates with alternating long warm summers and short cold winters. 2: Pertaining to latitudes between the tropics and the polar circles in each hemisphere; *cf*. polar, tropical.

Temperate grassland biome Continental grassland regions such as pampas, prairies, steppes and veld, characterized by a rainfall intermediate between temperate forest and desert, a long dry season, seasonal extremes of temperature; dominated by grasses and large grazing mammals, with burrowing animals frequent; see Appendix 5.

Temperate forest biome Mixed forest of conifers and broad-leaf deciduous trees, or mixed conifer and broad-leaf evergreen trees, or entirely broad-leaf deciduous, or entirely broad-leaf evergreen trees, found in temperate regions across the world; characterized by high rainfall (temperate rain forests), warm summers, cold winters occasionally subzero, seasonality; typically with dense canopies, understorey saplings and tall shrubs, large animals, carnivores dominant, very rich in bird species; temperate rain forest biome; see Appendix 5.

Temperate rain forest biome Temperate forest biome *q.v.*

temperature coefficient Q_{10} A measure of the increase in rate of a reaction or process produced by an increase in temperature of 10° C.

temperature inversion A horizontal layer in a water column in which temperature increases with depth.

temperature scale 1: Fahrenheit scale *q.v.* 2: Centigrade scale *q.v.*, Celsius scale *q.v.*, Kelvin scale *q.v.*

temperature-salinity diagram A graph of temperature against salinity for a water column, from which different water masses can be identified; T-S curve.

temporal isolation The absence of interbreeding between populations that produce or release gametes at different times; a premating isolation *q.v.* mechanism.

temporal parallel communities Geographically separated fossil assemblages that occupy similar stratigraphic sequences and show similar taxonomic composition.

temporal polyethism Age polyethism *q.v.*

temporary parasite A parasite that makes contact with its host only for feeding.

teneral Immature.

tension zone Ecotone *q.v.*

tephra Clastic volcanic materials ejected from a volcano during an eruption and transported through the air.

tephrochronology Geological dating based on the layering of volcanic ash.

ter- Prefix meaning three, thrice, threefold.

tera- (T) Prefix used to denote unit × 10^{12}.

teratogenic Causing developmental malformations or monstrosities; **teratogenesis**.

tetatological specimen A deformed specimen; a monstrosity.

teratology The study of malformations and monstrosities; **teratologic**.

term A word used in taxonomy to designate the rank of a taxonomic category in a hierarchy of classification.

terminal addition The addition of new evolutionary features to the end of ancestral ontogenies so that formerly adult stages become the preadult stages of the descendants.

terminal extinction The loss of an entire species population without the production of any descendant lineage; *cf*. phyletic extinction.

terminal immigrant An organism transported outside its normal range to another area in which it can survive but in which it is unable to reproduce; expatriate.

termination In taxonomy, the end or ending of a name, determined by the gender of the noun to which it is attached.

termitarium A termites' nest.

termiticolous Inhabiting termite nests; **termiticole**.

termitology The study of termites.

termitophilic Having an affinity for termites; inhabiting termite nests; termitophilous; **termitophil, termitophile, termitophily**.

ternary Pertaining to a hybrid formed from crosses among three different genera.

ternary name Trinomen *q.v.*

terraneous Pertaining to, or living on, the land or ground surface; terrestrial.

terraqueous Comprising both land and water.

terrestrial Pertaining to, or living habitually on, the land or ground surface.

terricolous Living on or in soil; used of an organism that spends most of its active life on the ground; **terricole; terricoline**.

terrigenous Derived from the land.

terrigenous mud A marine sediment comprising at least 30% silt and sand derived from the land; *cf.* black mud, blue mud, green mud, red mud, white mud.

terriherbosa Herbaceous vegetation growing on dry land.

territoriality Behaviour related to the defence of a territory.

territory An area within the home range *q.v.* occupied more or less exclusively by an animal or group of animals of the same species, and held through overt defence, display or advertisement; **territorial**.

Tertiary A geological period of the Cenozoic era (*ca.* 65–1.6 million years B.P.), comprising the Palaeocene, Eocene, Oligocene, Miocene, and Pliocene; see Appendix 1.

tertiary parasite An organism parasitic on a hyperparasite *q.v.*

tertiary sex ratio The number of males per 100 females at sexual maturity; *cf.* primary sex ratio, secondary sex ratio.

test A statistical process used to determine whether to accept or reject a statistical hypothesis, or whether observed values differ significantly from expected values.

test statistic A statistic upon which a test is based.

teste (*t.*) Latin, meaning according to, on the evidence of; *fide*.

test-set reduction Character-set minimization *q.v.*

Tethys The epicontinental sea separating Laurasia from Gondwanaland following the break up of Pangaea in the Mesozoic; Tethys Sea; Tethyan Sea; Tethian Sea.

tetra- Prefix meaning four, fourfold.

tetraploid A polyploid having four sets of homologous chromosomes; **tetraploidy**.

tetrasomic Pertaining to an aneuploid *q.v.* having one or more homologous chromosomes represented four times.

texture The character or structure of a sediment or surface.

thalassad A marine plant; thalassophyte.

thalassic Pertaining to the seas or deep oceanic waters.

thalassium A marine plant community.

thalassoid Resembling or derived from ancestral marine forms.

thalasson The marine biotope; *cf.* crenon, stygon, troglon.

thalassophilus Thriving in the sea; **thalassophile, thalassophily**.

thalassophyte A marine plant; thalassad; **thalassophyta**.

thalassoplankton Marine planktonic organisms.

thamnobiontic Living in bushes and shrubs; **thamnobiont**.

thamnocolous Living on or in bushes and shrubs; **thamnocole**.

thamnophilic Thriving on, or living in, bushes or shrubs; thamnophilous; **thamnophile, thamnophily**.

thanatocoenosis An assemblage of organisms brought together after death; taphocoenosis; **thanatocenosis, thanatocenose, thanatocoenose**; *cf.* biocoenosis.

thanatogeography Study of the distributions of dead organisms; necrogeography.

thanatology Study of death and dead organisms.

thanatosis Death-feigning behaviour; death feint; hypnotic reflex; ascaphus reflex; tonic immobility; **thanatoid**.

thanatotope The total area in which the dead members of a taxon are deposited.

thegosis The process of tooth sharpening; tooth grinding behaviour.

thelygenic Producing offspring that are entirely, or predominantly, female; thelygenous; **thelygeny**; *cf.* amphogenic, arrhenogenic, monogenic, allelogenic.

thelytoky Obligatory parthenogenesis in which a female gives rise only to female offspring; thelyotoky; **thelytokous, thelytocous, thelyotocous**; *cf.* arrhenotoky.

theory A scientifically accepted general principle supported by a substantial body of evidence offered to provide an explanation of observed facts and as a basis for future discussion or investigation; **theoretical**.

theory of permanence of the continents Permanence of the continents, theory of *q.v.*

therium A plant succession caused by the activity of animals; therion.

therm- Prefix meaning heat; **thermo-**.

therm A secondary fps unit equal to 1×10^5 Btu or $1.055\ 06 \times 10^8$ joules; see Appendix 13.

thermad A plant of hot springs; thermophyte.

thermal layer Thermocline *q.v.*

Thermal Maximum Altithermal period *q.v.*

thermistor A temperature sensitive resistor; a bead of semi-conducting material in which the resistance rapidly decreases with increasing temperature that is used to calculate temperature; *cf.* thermocouple.

thermium A hot spring community.

thermo- Prefix meaning heat; **therm-**.

thermobiology Study of the effect of heat on living organisms and biological processes.

thermocleistogamy Self-pollination within flowers that fail to open because of unfavourable temperature conditions; **thermocleistogamic**.

thermocline 1: A boundary region in the sea between two layers of water of different temperature, in which temperature changes sharply with depth. 2: A horizontal temperature discontinuity layer in a lake in which the temperature falls by at least 1° C per metre depth; thermal layer.

thermocouple A temperature-sensitive bimetallic probe; *cf.* thermistor.

thermoduric Capable of surviving at a high temperature (70–80° C) which would be lethal to most vegetative organisms.

thermogenesis The production of heat; **thermogenic**.

thermohaline Pertaining to both temperature and salinity.

thermohaline circulation Oceanic circulation caused by differences in density between water masses, which is itself determined primarily by water temperature and salinity.

thermolabile Unstable or destroyed when exposed to temperatures greater than 55° C; *cf.* thermostable.

thermonasty A response to a non-directional temperature stimulus; **thermonastic**.

thermopegic Pertaining to hot-water springs.

thermoperiodic Pertaining to the response of an organism to periodic changes in temperature; **thermoperiodicity, thermoperiodism**.

thermophilic Thriving in warm environmental conditions; used of microorganisms having an optimum for growth above 45° C; thermophilous; thermophilus; **thermophil, thermophile, thermophily**.

thermophobic Intolerant of high temperatures.

thermophylactic Heat-resistant; tolerant of high temperatures.

thermophyte 1: A plant tolerant of, or thriving at, high temperatures. 2: A hot-spring plant; thermad.

thermosequence A group of related soils that differ in certain properties as a result of the effect of temperature as a soil forming factor; *cf.* catena.

thermosis Any change in an organism caused by heat.

thermostable Relatively stable or resistant to heat; *cf.* thermolabile.

thermotaxis 1: The directed response of a motile organism towards (positive) or away from (negative) a temperature stimulus; **thermotactic**. 2: Regulation of body temperature.

thermotoxy Death or injury caused by heat; **thermotoxic**.

thermotropism An orientation response to a temperature stimulus; **thermotropic**.

therodrymium A leafy-forest community.

theromegatherm A plant favouring a high summer temperature of 20° C; *cf.* theromesotherm.

theromesotherm A plant favouring a moderate summer temperature between 12 and 20° C; *cf.* theromegatherm.

therophyllous Deciduous; used of plants that have leaves only during the warm season.

therophyte An annual plant; a plant that completes its life cycle in a single season, passing unfavourable seasons as a seed; *cf.* Raunkiaerian life forms.

Thienemann's rule In phytogeography, that the most environmentally sensitive stages of the life cycle of a plant (the flowers and seedlings) set the basic limits for the maintenance and spread of the species.

thigmokinesis A change in the velocity of linear or angular movement of an organism in response to a contact stimulus; stereokinesis; **thigmokinetic**.

thigmomorphosis A change in form due to contact; **thigmomorphic**.

thigmotaxis A directed response of a motile organism to a continuous contact with a solid surface; stereotaxis; **thigmotactic**.

thigmothermic Used of an animal that draws heat into its body from contact with a warmed object in the environment; **thigmotherm**.

thigmotropism An orientation response to touch or a contact stimulus; stereotropism; haptotropism; **thigmotropic**.

thinad A sand-dune plant; thinophyte.

thinic Pertaining to sand dunes.

thinicolous Living on sand dunes; **thinicole**.

thinium A sand-dune community.

thinophilous Thriving on sand dunes; **thinophile, thinophily**.

thinophyte A sand-dune plant; thinad; **thinophyta**.

thiobios Those organisms inhabiting anaerobic sulphur-rich deposits; **thiobiotic**.

thiogenic Sulphur-producing.

thiophilic Thriving in sulphur-rich habitats; **thiophil, thiophile, thiophily**.

thixotropic Used of a colloidal sediment that becomes liquid as a result of agitation or pressure; **thixotropy**; *cf*. dilatency.

thremmatology Study of animal and plant breeding in domestic situations.

threpsology The study of nutrition; **threpsological**.

threshold The minimum level or value of a stimulus necessary to elicit a response; limen; *cf*. subliminal.

thryptophyte A symbiotic plant that produces a sublethal modification of the host tissue.

tidal current The alternating horizontal movement of water associated with the rise and fall of the tide.

tidal current cycle The period from one high water to the next; *cf*. tide cycle.

tidal datum Chart datum *q.v.*

tidal day The period between two consecutive high waters at a given place, averaging 24 h 51 min (24.84 h).

tidal flat An extensive flat tract of land alternately covered and uncovered by the tide, and comprising mostly unconsolidated mud and sand; tide flat, tideflat.

tidal marsh A low elevation marshy coastal area formed of mud and the root mat of halophytic plants, regularly inundated during high tides.

tidal zone Intertidal zone *q.v.*

tide The periodic rise and fall of the ocean water masses and atmosphere, produced by gravitational effects of the moon and sun on the Earth.

tide cycle The duration of a given tidal sequence, as for example a lunar month or a tidal day.

tide range The difference in height between consecutive high and low waters.

tideland The coastal area of land that is regularly covered and uncovered by the rise and fall of a normal daily tide.

till Material deposited directly by ice.

tilth The physical condition of a soil in relation to its fitness for plant growth.

timber line Tree line *q.v.*

time series Data or samples taken at given time intervals.

tiphad A pond plant; tiphophyte.

tiphic Pertaining to ponds.

tiphicolous Living in ponds; **tiphicole**.

tiphium A pond or pool community.

tiphophilous Thriving in ponds; **tiphophile, tiphophily**.

tiphophyte A pond or pool plant; tiphad; **tiphophyta**.

tocotropism An orientation response of an organism exhibited when giving birth to young.

token word In botanical nomenclature, a word or words used by an author to refer to a taxon but not intended as the name of that taxon.

tokogenetic relationship The genetic relationship between individuals.

tolerance 1: The ability of an organism to endure extreme conditions. 2: The range of an environmental factor within which an organism or population can survive.

tolerance, law of That the distribution of an organism is limited by its tolerance to the fluctuations of a single factor.

tombolo A sandbar connecting an island or group of islands with the mainland.

tomiparous Reproducing by fission; **tomiparity**.

tomus (*tom.; t.*) Latin, meaning volume.

tomus citatus (*tom. cit.*) Latin, meaning in the volume cited; used to avoid repetition of a bibliographic reference.

ton The secondary fps unit of mass equal to 2240 pounds or $1.016\ 047 \times 10^3$ kg; see Appendix 13.

tonic immobility Thanatosis *q.v.*

tonicity A measure of osmotic tension; defined by the osmotic response of cells immersed in a solution; *cf*. hypertonic, hypotonic, isotonic.

tonne (t) A derived unit of mass equal to 1000 kg, or 2205 pounds.

tonotaxis The directed response of a motile organism to a change in osmotic pressure or to an osmotic stimulus; osmotaxis; **tonotactic**.

tonotropism An orientation response to an osmotic stimulus; osmotropism; **tonotropic**.

tophomeotype In taxonomy, a specimen from the type locality identified by a recognized authority.

topical Occurring locally; confined to a small area; having a limited distribution.

topochemotaxis A directed response of a motile organism towards the source of a chemical stimulus; **topochemotactic**.

topocline A graded series of forms occurring through the geographical range of a taxon, but not necessarily correlated with an ecological gradient; *cf*. cline.

topodeme A local interbreeding population occurring in a particular geographical area.

topogalvanotaxis The directed response of a motile organism towards the source of an electrical stimulus; **topogalvanotactic**.

topogamodeme A relatively isolated, naturally interbreeding, population (gamodeme) occupying a particular area.

topographic desert A low rainfall desert region located in the middle of a continental land mass.

topography All natural and man-made surface features of a geographical area; **topographical**.

topomorph A geographical form, variety or subspecies.

toponym A scientific name based on the name of a locality, place or area; or the place-name itself.

topophototaxis The directed response of a motile organism towards the source of a light stimulus; **topophototactic**.

toposequence A group of related soils that differ in certain properties as the result of variations in topography; *cf.* catena.

topotaxis The directed response of a motile organism to spatial differences in the intensity of a stimulus, especially with reference to the source of the stimulus; **topotactic**.

topotropism An orientation response towards the source of a stimulus; **topotropic**.

topotype 1: In taxonomy, a specimen from the type-locality but not necessarily part of the type series; locotype. 2: A population differentiated in response to local geographical factors.

topsoil The upper soil layer or layers containing some organic matter.

torfaceous Pertaining to bogs; turfaceous.

torpid Dormant; lacking vigour; insensitive; stagnant; **torpidity**.

torpor A dormant state.

torr A unit of pressure defined as that pressure equal to a 1 mm column of mercury, or 1.33×10^{-3} bar (1.33×10^2 Nm^{-2}).

torrenticolous Living in river torrents; torrential.

total biomass growth (G) In ecological energetics, the total amount of organic matter accumulated by a population or trophic unit per unit time, per unit area or volume, including both elimination *q.v.* and weight loss *q.v.*

total illumination The illumination received by a point recorder from all angles at a given depth.

total range The entire area over which an individual moves during its lifetime.

total standing crop (TSC) The total organic matter of a given area, volume or ecosystem.

total sum of squares In an analysis of variance, the sum of the squared deviations of all observations from their means.

totipotency The potentiality of a cell isolated from a plant, of generating an entirely new plant.

totipotent 1: Bisexual. 2: Having the potential for various specializations during development, in response to external or internal factors; *cf.* unipotent.

toxic Poisonous.

toxicity The virulence of a poisonous substance (the toxicant).

toxicology The study of poisonous substances and their effects.

toxigenic Producing toxin; toxinogenic; toxicogenic.

toxin A biogenic poison, usually proteinaceous; toxicant.

toxiphoby, index of A qualitative scale ranking species according to their tolerance of polluted habitats.

trace Any distinct biogenic structure either fossil or recent, such as a track, trail, burrow, boring, faecal cast.

trace element An element essential for normal growth and development of an organism, but required only in minute quantities.

trace fossil A sedimentary structure resulting from the activity of a living animal, such as a track, burrow or tube; trace; ichnofossil; vestigiofossil; lebensspur.

trade wind Uniform tropical oceanic wind system blowing towards the equator from the north-east in the northern hemisphere and from the south-east in the southern hemisphere.

trail pheromone A chemical substance laid down as a trail by one individual and followed by another member of the same species; trail substance.

trait Any character or property of an organism.

trait group A group of individuals which show competitive, aggressive, sexual, defensive or other interaction with respect to a particular behavioural trait.

tramp A species having a wide geographical range and an efficient means of dispersal; tramp species.

trans- Prefix meaning across.

trans. nov. Abbreviation of the Latin *translatio nova*, meaning new transfer.

transad A species or pair of related species occurring on both sides of a geographical barrier, indicating a former continuity in the distribution of the species or of the ancestral species.

transalpine Situated on the north side of the Alps; *cf.* cisalpine.

transcription Synthesis of messenger RNA from a DNA molecule acting as a template; *cf.* translation.

transduction The transfer of genetic material from donor to recipient microorganism, as from one bacterial cell to another by a bacteriophage; *cf.* transformation.

transect A line or narrow belt used to survey the distributions of organisms across a given area.

transfer In taxonomy, a change in the position of a taxon, with or without a change in name.

transfer host Paratenic host *q.v.*

transfer RNA A relatively small RNA molecule that transfers a given amino acid to a polypeptide chain during translation *q.v.* of messenger RNA at the site of protein synthesis in a ribosome; tRNA.

transformation 1: A change of form; metamorphosis. 2: A change resulting from the acquisition of foreign genetic material. 3: In statistics, a change of variable; used to simplify calculations or to transform the distributions of random variables to a normal distribution.

transformation series The sequenced series of homologous character states representing an evolutionary trend in a character, typically from pleisiotypic (ancestral) to apotypic (derived) state, although the polarity *q.v.* is not necessarily determined; evolutionary transformation of a homologous character.

transgenation Point mutation *q.v.*

transgenosis The transfer of genes from one species to an unrelated species, in which they are maintained and manifested phenotypically.

transgeotropism Diageotropism *q.v.*

transgression The spread of the sea over a land area; *cf.* regression.

transgressive character species A plant species that is characteristic of two or more vegetation units of different syntaxonomic rank.

transgressive stratum The intermediate layer of a forest community comprising those species that are growing in order to replace the overstorey and understorey.

transheliotropism Diaheliotropism *q.v.*

transhydrogenation The sequential transfer of electrons in cellular oxidation cycles.

transient 1: Of short duration; transitory. 2: A stage in the phylogeny of a species.

transient polymorphism Genetic polymorphism that occurs while one form is in the process of replacing another, under the influence of natural selection; *cf.* polymorphism.

transilience Speciation by the rapid shift of a population across an adaptive valley from its original adaptive peak to a new one.

transitional adaptive zone An evolutionary pathway across a nonadaptive zone between major adaptive zones, in which the strength of selection and rate of evolution are not significantly greater than in the adaptive zones.

transitional load Substitutional load *q.v.*

transitional soil Intrazonal soil *q.v.*

translatio nova (*trans.nov.*) Latin, meaning new transfer; used to indicate that a taxon has been changed in position either by horizontal transfer to another genus or vertical transfer to a different taxonomic rank.

translation The sequential reading of the base sequence of messenger RNA and its translation into a specific amino acid chain during protein synthesis in ribosomes; *cf.* transcription.

transliterate To change the characters of one alphabet to the corresponding characters of another; **transliteration**.

translocated herbicide A herbicide that is dispersed throughout the plant tissue following absorption through the leaves or roots.

translocation 1: The transport of material within a plant. 2: Movement of a segment of a chromosome to another part of the same chromosome or to a different chromosome.

transmutation The theory that one species can evolve into another.

transpecific evolution Evolutionary events and processes above the species level, such as the origin of new higher taxa and new grades of organization; macroevolution.

transpiration Loss of water vapour from an organism through a membrane or through pores.

transpiration capacity The specific intensity of stomatal and cuticular water loss from leaves or other plant surfaces under defined conditions of evaporation.

transpiration efficiency The ratio of dry organic matter produced by a plant to water loss by transpiration, measured in grams dry matter per litre of water lost, usually over a period of several weeks; transpiration coefficient.

transpiration ratio The amount of water (in litres) consumed by a plant or stand of vegetation during the growing season, per kilogram of dry matter produced.

transplant Transfer of tissue from one part of an organism to another, or to another organism.

transport host Paratenic host *q.v.*

traplining A feeding strategy during which an animal visits a sequence of widely dispersed food sources, obtaining a small part of its daily food requirement at each feeding station.

trauma Injury or stress caused by an extrinsic agent; **traumatic, traumatism**.

traumatonasty A growth response to a traumatic stimulus; **traumatonastic**.

traumatotaxis Movement of organisms, cells or organelles in response to injury; traumotaxis; **traumatotactic**.

traumatotropism An orientation response to injury or stress; traumatropism; **traumatotropic**.

traumatropism Traumatotropism *q.v.*; **traumatropic**.

traumotaxis Traumatotaxis *q.v.*; **traumotactic**.

treatment effect In statistics, the change in response produced by a particular treatment.

treatment sum of squares In an analysis of variance, that part of the total sum of squares attributed to differences between treatments.

tree line The line which marks the northerly, southerly or upper altitudinal limit of tree cover; timber line.

tree-ring chronology Study of annual growth rings of trees for the purposes of dating; dendrochronology.

trellis diagram Matrix q.v.

tri- Prefix meaning three, thrice, threefold.

trial and error learning The development of an association between a stimulus and an independent activity as the result of reinforcement q.v. during appetitive behaviour.

Trias Triassic q.v.

Triassic A geological period of the Mesozoic era (ca. 245–210 million years B.P.); Trias; see Appendix 1.

tribe A rank within a hierarchy of classification; the principal category between family and genus; in botanical nomenclature takes the ending -eae; see Appendix 10.

tribium An ecological succession on eroded soil; a badland community.

trigeneric Pertaining to a hybrid derived from three different genera.

trigenetic Pertaining to a symbiont requiring three different hosts during the life cycle; cf. digenetic, monogenetic.

trigenic Used of characters or traits controlled by the integrated action of three genes; cf. monogenic, digenic, oligogenic, polygenic.

trigenomic Containing three independently derived genomes.

trigoneutic Producing three broods in a single season; trivoltine; **trigoneutism**; cf. monogoneutic, digoneutic, polygoneutic.

trihybrid An individual heterozygous in respect of three pairs of alleles.

trillion A million million, 10^{12}.

trimonoecious Used of an individual plant bearing male, female and hermaphrodite flowers; coenomonoecious.

trimorphic Used of organisms having three distinct forms in the life cycle, or in a population; **trimorphism**.

trinomen (trinomina) The scientific name of a subspecies, comprising a generic name, a specific epithet and a subspecific epithet; ternary name; trinominal name; trinomial name; **trinomial**.

trinomial name Trinomen q.v.

trinominal Comprising three words, as in the name of a subspecies.

trinominal name Trinomen q.v.

trioecious Used of a plant species having male, female and hermaphrodite flowers on different individuals; trioikous; **trioecy**; cf. dioecious, monoecious.

trioikous Trioecious q.v.

triplo- Prefix meaning threefold, treble.

triploid A polyploid having three sets of homologous chromosomes; **triploidy**.

tripton Non-living particulate matter suspended in water; a component of seston q.v.; abioseston; trypton.

trisomic Used of an aneuploid q.v. having one or more homologous chromosomes represented three times; **trisomy**.

trivalent An association of three chromosomes synapsed during the first meiotic division.

trivial name The specific name or epithet.

trivoltine Having three generations or broods per year; trigoneutic; cf. voltine.

trixenic Used of a mixed culture of one organism with three other species.

trixenous Used of a parasite utilizing three host species during its life cycle; **trixeny**; cf. dixenous, heteroxenous, monoxenous, oligoxenous.

tRNA Transfer RNA q.v.

troglobia The true cavernicolous fauna.

troglobie An obligate cavernicole; an organism found only in caves or subterranean passages; stygobie; **troglobite, troglobiont**; cf. troglophile, trogloxene.

troglodyte A subterranean organism; cave dweller; **troglodytic**.

troglon The biotope comprising subterranean water bodies in caves and subterranean passages; cf. crenon, stygon, thalasson.

troglophile An animal frequently found in underground caves or passages but not confined to them; eucaval; eutroglobiont; cf. troglobie, trogloxene.

troglophilic Thriving in caves and subterranean passages; stygophilic; **troglophile, troglophily**.

trogloxene An organism only occasionally found in underground caves or passages; tychocaval; xenocaval; cf. troglophile, troglobie.

trogloxenous Found only occasionally in caves or subterranean passages; stygoxenous; **trogloxene**.

troop A group of primates; a flock or assemblage of birds or mammals.

trophallaxis Mutual or unilateral exchange of food between colony members; oecotrophobiosis; reciprocal feeding; **trophallactic**.

-trophic Suffix meaning nourishment; -troph, -trophe; used to indicate the mode of nutrition of an organism or the nutrient utilized; cf. acidotrophic, allotrophic, amphitrophic, autotrophic, biotrophic, chemoautotrophic, chemolithotrophic, chemoorganotrophic, dermatotrophic, ectendotrophic, ectotrophic, haemotrophic, heterotrophic, holotrophic, hypotrophic, lecithotrophic, lithotrophic, mesotrophic, metatrophic, minerotrophic, mixotrophic, monotrophic, mycotrophic, myrmecotrophic, myxotrophic, oligotrophic, ombrotrophic, organotrophic, osmotrophic, paratrophic, phagotrophic, photoautotrophic, photolithotrophic, phototrophic, planktotrophic, polytrophic, prototrophic, pseudotrophic, rheotrophic, saprotrophic, symbiotrophic, syntrophic, zootrophic.

trophic Pertaining to nutrition; **trophism**.

trophic egg An egg, usually degenerate and non-viable, used as food by other members of a colony.

trophic group Trophic unit *q.v.*

trophic levels 1: The sequence of steps in a food chain or food pyramid *q.v.*, from producer to primary, secondary or tertiary consumer. 2: The nutrient status of a body of fresh water; categories include eutrophic, mesotrophic, oligotrophic, dystrophic.

trophic parasitism The invasion of a colony or social group by another organism in search of food.

trophic pathway The route of energy flow through a food pyramid *q.v.*; food chain.

trophic unit A group of individuals of one or more species occupying the same level of a trophic pyramid, that is a similar relative position in the sequence of steps in a food chain; trophic group.

trophobiosis Symbiosis in insects in which food is obtained from one species (the trophobiont) by another in return for protection.

trophodynamics Study of the energy relationships of feeding strategies and food webs.

trophogenic 1: Produced by, or resulting from, food or feeding behaviour. 2: Used of the illuminated surface layer of a lake in which photosynthesis occurs; *cf.* tropholytic.

trophology The study of nutrition.

tropholytic Pertaining to the deeper part of a lake below the trophogenic *q.v.* zone, in which organic matter is utilized as the energy source.

trophosphere Troposphere *q.v.*

trophotaxis The directed response of a motile organism towards (positive) or away from (negative) a food stimulus; **trophotactic**.

trophotropism An orientation response to food; **trophotropic**.

tropic Pertaining to a tropism *q.v.*

tropic tide The tide occurring twice monthly when

the moon reaches its greatest northerly or southerly declination; *cf.* solstitial tide.

tropical 1: Pertaining to the zone between the Tropic of Cancer (23° 27′ N) and the Tropic of Capricorn (23° 27′ S); *cf.* polar, temperate. 2: Used of a climate characterized by high temperature, humidity and rainfall, and having very rare light frosts at night.

tropical deserts Hot dry deserts situated close to the Tropics of Cancer and Capricorn, where subtropical high pressure meteorological conditions result in very low sporadic rainfall.

Tropical grassland biome The circumtropical savannah belt transitional between equatorial forest and grassland, characterized by constant high temperature, a long dry season, low fertility soil, frequent fires, dominated by grasses and scattered trees and large grazing mammals; tropical savannah biome; see Appendix 5.

tropical lake A lake having a surface temperature that never falls below 4° C at any time of year.

Tropical North American subregion Mexican subregion of the Neotropical region; see Appendix 3.

Tropical rain forest biome A circumtropical forest region characterized by constant high humidity and temperature, no frost, little seasonal fluctuation of climate except precipitation, typically with a rich variety of trees, a dense canopy, lianas and epiphytes, little undergrowth, an abundance of birds and arboreal vertebrates and the greatest diversity of animal life in any terrestrial biome; see Appendix 5.

Tropical savannah biome Tropical grassland biome *q.v.*

Tropical South American subregion Brazilian subregion of the Neotropical region; see Appendix 3.

tropical zone The latitudinal zone between 15.0° and 23.5° in either hemisphere.

tropicopolitan Cosmopolitan within the tropics.

-tropism Suffix meaning to turn; acrotropism, aerotropism, aitiotropism, allotropism, anaclinotropism, anatropism, anemotropism, anthotropism, antitropism, aphaptotropism, apheliotropism, aphercotropism, aphototropism, apogeotropism, autoorthotropism, autoskoliotropism, autotropism, barotropism, caloritropism, camptotropism, chemotropism, chromatotropism, chronotropism, clinotropism, cryotropism, diageotropism, diaheliotropism, diaphototropism, diatropism, dromotropism, edaphotropism, electrotropism, epigeotropism, epitropism, euphototropism, eurytropism, eutropism, exotropism, galvanotropism, gametotropism, gamotropism, geodiatropism, geonyctitropism, geoparallotropism, geoplagiotropism, geotropism, haptotropism, helcotropism, heliotropism, homolotropism, homotropism, hydrotropism, hygrotropism, katatropism, klinogeotropism, magnetotropism, mechanotropism, mesotropism, monotropism, narcotropism, nyctitropism, ombratropism, orthoheliotropism, orthophototropism, orthotropism, osmotropism, oxygenotropism, oxytropism, pantropism, paraheliotropism, parallelogeotropism, parallelotropism, paraphototropism, parathermotropism, phobophototropism, photoplagiotropism, phototropism, piezotropism, plagiogeotropism, plagiotropism, pneumotropism, polotropism, progeotropism, prohydrotropism, prophototropism, prosgeotropism, prosheliotropism, radiotropism, rheotropism, rugotropism, scototropism, seismotropism, selenotropism, siotropism, sitotropism, skototropism, somatotropism, stereotropism, thermotropism, thigmotropism, tocotropism, tonotropism, topotropism, transgeotropism, transheliotropism, traumatotropism, trophotropism, xerotropism, zenogeotropism, zenotropism.

tropism An orientation or growth response of a non-motile organism or one of its parts towards (positive) or away from (negative) a stimulus; occasionally used for motile organisms in which case it may be indistinguishable from a taxis *q.v.*; **tropic**.

tropodrymium A savannah forest community.

tropoparasite An organism that lives as an obligate parasite for only part of its life cycle.

tropopause The upper limit of the troposphere *q.v.*

tropophilous Thriving in an environment that undergoes marked periodic fluctuations of light, temperature and moisture; tropophilic; **tropophil, tropophily**.

tropophyte 1: A tropical plant. 2: A plant that thrives under mesic conditions for part of the year and xeric conditions at other times.

troposphere 1: The warm upper layer of oceanic water at middle and low latitudes, typically having strong currents; ocean waters above the thermocline *q.v.*; trophosphere. 2: The layer of the atmosphere below the stratosphere extending from ground level to 10–15 km above the Earth's surface, in which temperature falls rapidly with increasing altitude, typically with active convection currents; *cf.* mesosphere, stratosphere.

tropotaxis An orientation response of a motile organism in which the stimulus is compared

simultaneously on each side of the midline by bilateral sense organs, such that no deviations from the line of movement are required to compare the intensities of the stimuli; *cf.* klinotaxis, telotaxis.

truncated soil Soil which has had material removed from the upper layers by erosion.

tryphobiontic Inhabiting bogs.

trypton Tripton *q.v.*

T-S curve Temperature-salinity diagram *q.v.*

TSC Total standing crop *q.v.*

Tschernosem Chernozem *q.v.*

Tschernosion Chernozem *q.v.*

tsunami A large ocean wave caused by submarine volcanic or earthquake activity.

t-test A statistical method used for determining the significance of the difference between two means when the samples are small and drawn from a normally distributed population of which the standard deviation is unknown; measured in units of the standard error of the means; Student's t-test.

tubicolous Tube-dwelling; **tubicole**.

tumulus Pertaining to dunes.

Tundra biome A barren treeless region north of the Arctic Circle (arctic tundra), also found above the tree line of high mountains (alpine tundra) and on some subantarctic islands; characterized by very low winter temperatures, short cool summers, permafrost below a surface layer subject to summer melt; dominated by lichens, mosses, sedges and low shrubs, vast seasonal swarms of insects, and migratory animals; see Appendix 5.

Tundra soil A zonal soil with abundant litter and duff, a dark brown peaty A-horizon rich in organic material over a greyish B-horizon; formed under conditions of extreme cold, high humidity and poor drainage; characteristic of the Tundra biome *q.v.*

turbid Cloudy; opaque with suspended matter.

turbidite A marine sediment deposited at the base of a submarine slope by a turbidity current.

turbidity current A rapid undercurrent of relatively high density flowing down a submarine slope and transporting suspended sediment, often in great quantities; a gravity flow resulting from a mass of unstable sediment sliding down a submarine slope.

turbulence Irregular motion or agitation of liquids or gases; *cf.* laminar flow.

turfaceous Pertaining to bogs; torfaceous.

turfophilous Thriving in bogs; **turfophile, turfophily**.

turgescence The process of cellular or histological

distension due to internal pressure; turgor; turgidity.

turgor pressure Turgescence *q.v.*

turnover 1: In ecological energetics, the ratio of productive energy flow to standing crop biomass in a community or ecosystem; production turnover. 2: That fraction of a population or specified group which is exchanged (lost by mortality and emigration, replaced by recruitment) per unit time; individual turnover. 3: The extinction of species within an area and their replacement by other species. 4: Water circulation occurring in deep temperate lakes in spring and autumn, tending to equalize temperature throughout the water column.

turnover time 1: The time taken for the completion of a cycle within a system. 2: The time span from birth to death of an organism.

twin species A pair of sibling species *q.v.*

two tailed test In statistics, a test in which the critical region includes values on both sides of the mean of the sampling distribution (both left and right-hand tails); two sided test; *cf.* one tailed test.

two-state character Binary character *q.v.*

ty Thousand years.

tycho- Prefix meaning chance.

tychocaval Trogloxene *q.v.*

tychocoen Those members of a plant community that can thrive in a variety of habitat conditions and community types; *cf.* eucoen.

tychohyponeuston Organisms occurring briefly or by chance in the hyponeuston *q.v.* as a result of occasional or periodic vertical migrations; **tychohyponeustont, tychohyponeustonic.**

tycholimnetic Used of lacustrine organisms that are normally benthic, but occasionally free floating in the water column; tychopelagic.

tychoparthenogenesis Parthenogenesis occurring in normally sexually reproducing species.

tychopelagic Used of organisms that are normally benthic, but which have been carried up into the water column by chance factors; tycholimnetic.

tychoplankton Organisms occasionally carried into the plankton by chance factors such as turbulence; accidental plankton; pseudoplankton; **tychoplanktont, tychoplanktonic.**

tychopotamic Pertaining to aquatic organisms thriving in the still backwaters of rivers and streams; *cf.* autopotamic, eupotamic.

tychotroglobionte An organism rarely found in underground passages or caves.

typ. Abbreviation of the Latin *typus*, meaning type.

type In taxonomy, the single element of a taxon to which its name is permanently attached, and

on which the descriptive matter that satisfies the conditions of availability or valid publication is based.

-type Suffix meaning image, type.

type I error In statistics, the rejection of a null hypothesis when it is true and should be accepted.

type II error In statistics, the acceptance of a null hypothesis when it should be rejected.

type concept In taxonomy, a method of nomenclature that attaches a name permanently to a type specimen, a name which is retained by the type throughout all subsequent taxonomic acts; type method.

type culture A culture derived from a holotype, lectotype or neotype; sometimes used incorrectly in microbiology to signify a strain designated as the type of a new taxon.

type fixation Fixation *q.v.*

type genus The genus that is the type of a family-group taxon.

type horizon The geological stratum in which the type of a species or subspecies was located.

type host The host species from which the type of a parasite species or subspecies was collected.

type locality The exact geographical site at which the type of a species or subspecies was collected.

type method Type concept *q.v.*

type number The most commonly occurring chromosome number in a given taxonomic group.

type series In taxonomy, all the specimens on which the original description of a species or subspecies was based; type material.

type species The species designated as the type of a genus or subgenus.

type specimen A designated specimen or individual (holotype, lectotype, neotype) or one of a series of specimens (syntype) that is the type of a species or subspecies.

typical Pertaining to, or conforming to, the type *q.v.*

typification Fixation *q.v.*; the designation of a type.

typogenesis A period of explosive evolution in which new forms are produced rapidly; *cf.* typolysis, typostasis.

typographical error An error in a printed text for which the printer and not the author is deemed responsible; **typographic error.**

typological species A species defined on the characters of the type specimen; nomenspecies.

typological species concept The concept of a species as a group of organisms conforming to a common morphological plan, emphasizing the species as an essentially static, non-variable, assemblage; *cf.* multidimensional species concept, nondimensional species concept.

typology A method of classification or study of natural groups based on the assumption that all members of a taxonomic unit conform to a given morphological plan without significant variation.

typolysis A period of evolutionary history preceding the extinction of a form; *cf.* typogenesis, typostasis.

typonym An objective synonym; a later name given to a type specimen; **typonymic**.

typonymous homonym A typonym *q.v.* that is also a homonym *q.v.*; homonymous typonym.

typostasis A relatively static phase in evolutionary history in which few new forms are produced; *cf.* typogenesis, typolysis.

typotype In taxonomy, the type of a type; used of the specimen from which an illustration or description that constitutes a type is prepared.

typus Bauplan *q.v.*

typus conservandus (*typ. cons.*) Latin, meaning type to be conserved; used of the type of a conserved generic name which is different from that which would have to be adopted according to the Code but whose retention has been authorized by an International Botanical Congress.

U

ubiquitous Having a world-wide distribution, effect or influence; widespread; cosmopolitan; ecumenical; pandemic; **ubiquist, ubiquitist**.

Udden grade scale A scale of sediment particle size categories based on a doubling or halving, above or below respectively, the fixed reference point of 1 mm; see Appendix 11.

uliginous Inhabiting swampy soil or wet muddy habitats; helobius; paludal; palustrine; **uliginose**.

ulmification The process of peat formation.

ultimate factor The environmental factor to which the timing of a biological event is ultimately associated; ultimate causation; *cf*. proximate factor.

ultra- Prefix meaning beyond.

ultra-abyssal zone Hadal zone *q.v.*

ultra-abyssopelagic Pelagic in ocean depths greater than 6000 m; hadopelagic.

ultradian rhythm A biological rhythm with a period exceeding 24 h; *cf*. infradian rhythm.

ultrahaline Used of sea water having a salinity greater than 30 parts per thousand.

ultrananoplankton Planktonic organisms less than 2 μm in diameter, comprising picoplankton and femtoplankton; see Appendix 9.

ultraplankton Planktonic organisms less than 5 μm in length or diameter.

ultrastructure Cellular structure studied with the aid of a transmission electron microscope; fine structure.

ultraviolet radiation (UV) That part of the solar radiation spectrum below a wavelength of 380 nm.

umbraticolous Living in shaded habitats; skiophilous; **umbraticole**; *cf*. lucicolous.

umbrophilic Thriving in shaded habitats; heliophobous; skiophilic; **umbrophile, umbrophily**.

umwelt The total sensory input of an individual or species.

unavailable In taxonomy, used of a name or work that does not comply with the provisions of the Code or that has been annulled by the ICZN.

unavailable work In taxonomy, a work published prior to the starting point of the group, or one that does not conform to the provisions of the Code, or that has been annulled by the ICZN.

unbalanced hermaphrodite population One composed of a majority of hermaphrodites with varying degrees of maleness or femaleness and by a minority of pure male and female individuals; *cf*. balanced hermaphrodite population.

underdispersed distribution Uniform distribution *q.v.*; some confusion exists over the use of this term as a few authors have employed it with the opposite meaning, that is to describe contagious distributions *q.v.*

understorey The vegetation layer between the overstorey or canopy and the ground-storey of a forest community, formed by shade tolerant trees of moderate height; three component layers may be recognized, the upperunderstorey, middlestorey and the lowerunderstorey; understorey stratum.

unguligrade Walking or running on hooves or the modified tips of one or more digits.

uni- Prefix meaning one, single.

unicolonial Used of a population of social insects in which there are no behavioural colony boundaries so that the entire local population consists of a single colony.

unifactorial Pertaining to phenotypic characters controlled by a single gene; monogenic; monofactorial; *cf*. multifactorial.

uniform convergence hypothesis That convergent evolution in all parts of a phylogenetic tree takes place at a uniform rate.

uniform distribution A spatial distribution pattern in which values, observations or individuals are more regularly spaced out than in a random distribution, indicating that the presence of one individual decreases the probability of another individual occurring nearby; regular distribution; *cf*. contagious distribution, random distribution.

uniform rate hypothesis That any two lineages in a phylogenetic tree diverge at a constant rate with respect to each other.

uniformitarianism The doctrine or principle that natural geological processes affecting the Earth are still operating at essentially the same rate and intensity as they have throughout geological time; actualism; principle of uniformity; *cf*. catastrophism.

unigenesis Monogenesis *q.v.*

unimodal Used of a population or frequency distribution having a single mode or peak; *cf*. modality.

uninomen A scientific name comprising a single word; used for taxa at and above the rank of genus; uninominal name; monomial; uninomial; uninomial name.

uninomial Uninomen *q.v.*

uninominal name Uninomen *q.v.*

union A ranked category in the classification of vegetation; a stratal plant community comprising one or more species of similar physiognomy and life form.

uniovular Monozygotic *q.v.*

uniparental reproduction Reproduction from a single parent, as in vegetative reproduction, thelytoky and self-fertilizing hermaphroditism.

uniparous 1: Producing a single offspring in each brood; *cf.* biparous, multiparous. 2: Having only one brood during the life cycle.

unipotent Capable of developing into only one type of cell; *cf.* totipotent.

unisexual 1: Used of a population or generation composed of individuals of one sex only. 2: Used of an individual having either male or female reproductive organs and producing only male or female gametes; gonochoristic; dioecious; diclinous. 3: Used of a flower possessing only male or female reproductive organs; imperfect flower; *cf.* bisexual.

unistratal Pertaining to vegetation lacking distinguishable horizontal layering; *cf.* multistratal.

unit character 1: In taxonomy, a single character which cannot logically or conveniently be subdivided. 2: A character transmitted as a unit in heredity, controlled by one pair of alleles.

unitary Comprising a single unit or word; uninominal.

unite In taxonomy, to combine or join two or more taxa.

univalent A single chromosome, unpaired during the first meiotic division.

univariate analysis The analysis of one character or variable at a time.

universe In statistics, the entire statistical population.

univoltine Having one brood or generation per year; monovoltine; monogoneutic; *cf.* voltine.

univorous Feeding on only one type of food, or parasitizing a single host species; monophagous; **univore, univory**.

unstable Not constant; liable to change.

unused name In taxonomy, an available senior synonym that is not known to have been used as a valid name during the preceding 50 years.

unweighted pair group method A method of clustering OTUs which joins the smallest branches of the dendrogram first and the two largest branches last; each step groups together that pair of OTUs or pair of OTU-groups showing the greatest similarity to each other.

upperunderstorey Understorey *q.v.*

upwelling An upward movement of cold nutrient-rich water from ocean depths.

urkaryote One of the three primary kingdoms (urkingdoms *q.v.*) of living organisms, comprising all eukaryote organisms; *cf.* eubacteria, archaebacteria.

urkingdoms The three primary kingdoms proposed recently as a basic tripartite scheme for grouping all living organisms, as an alternative to the widely used five-kingdom *q.v.* scheme of Whittaker; *cf.* eubacteria, archaebacteria, urkaryote.

urophilic Thriving in habitats rich in ammonia; **urophile, urophily**.

US gallon A unit of volume equal to 0.8327 Imperial gallon *q.v.* ($3.785412 \ dm^3$).

usage In taxonomy, the sense in which a name has been applied, whether correctly or incorrectly according to the Code.

use and disuse, theory of Lamarckism *q.v.*

utilization time The time interval between a stimulus and the response; reaction time.

V

v. 1: Abbreviation of the Latin *vide*, meaning see; *visum*, seen. 2: Abbreviation of **volume**; **vol.** 3: Abbreviation of the Latin *varietas*, meaning variety; *var*.

v.et. Abbreviation of the Latin *vide etiam*, meaning see also.

vadal Floating close to the shore.

vadose Pertaining to water in the Earth's crust above the level of permanent ground water.

vagile Wandering; freely motile; mobile.

vagility The tendency of an organism or population to change its location or distribution with time; mobility; **vagile**.

vaginicolous Living in secreted sheaths or cases; **vaginicole**.

vagrant Unattached; wandering; used of unattached wind-blown organisms, and of organisms that move about by their own activity.

Valdian glaciation The most recent glaciation of the Quaternary Ice Age *q.v.* in Russia, with an estimated duration of about 70 thousand years.

valid In taxonomy, used of a name or nomenclatural act that is correct according to the provisions of the Code.

valid name The correct name for a particular taxon in zoology.

validated name A formerly invalid or unavailable name that has been made valid or available by the ICZN *q.v.*, by annulment or suppression of senior homonyms or synonyms.

validation To make valid or available a formerly invalid or unavailable name or invalid nomenclatural act; **validate**.

valvate dehiscence Spontaneous release of the contents of a ripe fruit through a valve, flap or other large aperture.

Van Valen's law Law of constant extinction *q.v.*

var. (v) Abbreviation of variety.

Varangian glaciation A widespread glaciation during the Precambrian era, about 670 million years B.P.

variability The property of being variable in form or quality.

variability, coefficient of In statistics, the standard deviation s, expressed as a percentage of the mean ($\bar{x}$); $CV = (s \times 100)/\bar{x}$; used to compare the amount of variation in populations having different means.

variable 1: Not conforming to type. 2: A property with respect to which individuals within a sample differ in some discernible way. 3: Any symbol or term to which a number of different numerical values may be assigned.

variable character A character that is present or absent, or that exhibits many different degrees of expression in individuals of the same population or species, or in some specified group.

variance In statistics, the square of the standard deviation, providing a measure of the dispersion of values about a mean; denoted by σ^2 and the sample estimate s^2; calculated as
$$s^2 = \Sigma(x - \bar{x})^2/n - 1.$$

variance ratio The ratio of two independent estimates of population variance, $F = S_1^2/S_2^2$.

variance to mean ratio The simple ratio of variance to mean, $s^2/\bar{x}$; used as an approximate test for agreement with the Poisson distribution, when it equals unity; also used as an approximate comparative index of dispersion with values ranging from 0 (uniform distribution) through 1 (random distribution) to Σx (highly contagious distribution).

variance-ratio test F-test *q.v.*

variant Any individual or group showing marked deviation from type, in form, quality or behaviour.

variant spelling In taxonomy, different spellings of epithets that are deemed to be homonyms.

variation 1: The divergence of structural and functional characteristics within a group, in particular those not attributable to differences in age, sex or life history stage. 2: The spread or dispersion of values about the mean.

variation, coefficient of Coefficient of variation *q.v.*

variegation The occurrence within an individual of a mosaic phenotype, typically with respect to coloration; partial albinism in plants.

variety 1: A rank in the hierarchy of botanical classification; the principal category between species and form; see Appendix 10. 2: An infrasubspecific taxon, such as varietas or cultivar; an ambiguous term often used for any variant group within a species; a group that differs from other varieties of the same subspecies including non-genetic variants and microgeographic races.

varve A layer of sediment deposited in a lake during the course of a single year; comprising a coarse layer deposited in spring and summer and a fine layer deposited in autumn and winter.

varve analysis A method of geological or archaeological dating based on the layering of fine clay sediments (varved clay) deposited annually in meltwater lakes of periglacial areas; varve chronology.

vector 1: In statistics, a quantity whose complete description requires two or more numbers. 2: A matrix with only one row or column; its elements are referred to as components. 3: An organism that carries or transmits a pathogenic agent; *cf.* carrier. 4: Any agency responsible for the introduction or dispersal of an animal or plant species.

vegetation The total plant life or cover in an area; also used as a general term for plant life; the assemblage of plant species in a given area; *cf.* faunation.

vegetative Pertaining to assimilative, somatic and trophic growth processes rather than to reproduction.

vegetative eunuch A plant which completes its growth and dies without producing seeds, thereby consuming resources but leaving no descendants.

vegetative reproduction Reproduction by asexual processes such as budding or fragmentation; vegetative division; vegetative propagation.

veld The open temperate grassland areas of southern Africa, typically with scattered shrubs or trees; veldt.

velocity Speed of motion.

Venezuela and Guianan region A subdivision of the Neotropical kingdom; see Appendix 4.

venomous Poisonous; noxious; venom-producing.

Verhulst logistic equation Logistic equation *q.v.*

vernacular name The common name of a species or group other than the formal Latin or latinized scientific name.

vernal Pertaining to the spring; *cf.* aestival, hibernal.

vernal pool A temporary pool formed during spring from meltwater or flood water.

vernalization A process of thermal induction in plants, in which growth and flowering are promoted by exposure to low temperatures; jarovization, yarovization.

versicoloured Exhibiting colour change, or different colours; variegated.

vertical classification A classification which groups ancestral and descendant stages of a phyletic line into a single taxon; *cf.* horizontal classification.

vertical illumination The illumination incident upon a horizontal plane at a given depth in a water body.

vespertine Pertaining to the evening; crepuscular; vesperal.

vestigial Degenerate or imperfectly developed; used of structures or functions that have become

diminished or reduced during the course of evolution or ontogeny; **vestige**.

vestigiofossil Trace fossil *q.v.*

viability A relative measure of the number of surviving individuals of any given phenotypic or genotypic class.

viable Having the capacity to live, grow, germinate or develop; **viability**.

viatical Growing on the roadside or beside paths.

vibrotaxis Seismotaxis *q.v.*; **vibrotactic**.

vicariads Vicars *q.v.*

vicariance The existence of closely related taxa or biota in different geographical areas, which have been separated by the formation of a natural barrier (a vicariance event).

vicariance event The splitting or division of a biota or taxon through the development of a natural biogeographical barrier.

vicariance model That the components of major biota evolve together and are subject to parallel effects of geographical and climatic fluctuations, and that modern biogeographical distributions result from the division of ancestral biotas by the formation of natural barriers; panbiogeographic model; *cf.* dispersalist model.

vicariants Closely related taxa isolated geographically from one another by a vicariance event *q.v.*

vicarious species Vicars *q.v.*

vicarism The existence of closely related and ecologically equivalent forms (vicars) that replace each other geographically; vicariation; *cf.* pseudovicarism.

vicarius Latin, meaning a substitute, substituting.

vicars Closely related and ecologically equivalent forms that tend to be mutually exclusive, occupying disjunct geographical areas; vicariads; vicarious species.

vice-counties Biogeographical subdivisions of former administrative counties in Britain, used as unit areas in distributional studies.

vicine Used of organisms invading or entering from adjacent areas or communities.

vicinism Variation in a population or individual that results from the close proximity of other organisms.

vicinus Latin, meaning neighbour; *vicini*.

vide etiam (*v. et.*) Latin, meaning see also.

videlicet (*viz.*) Latin, meaning namely.

vigour The intensity of growth or general metabolic activity of an organism, population or community.

virgin 1: Pertaining to a native habitat, fauna or flora that is essentially unaffected by the activities of man. 2: A female animal that has not copulated.

virginiparous Used of organisms reproducing only by parthenogenesis.

-viridae The ending of a name of a family in virological nomenclature.

-virinae The ending of a name of a subfamily in virological nomenclature.

virology The study of viruses; virusology.

virtual tautonymy In nomenclature, the nearly identical spelling, or same origin or meaning, of a generic name and of the epithet of one of the species or subspecies originally included in the genus; *cf*. absolute tautonymy.

virulence The capacity of a pathogen to invade host tissue and reproduce; the degree of pathogenicity.

viruliferous Virus carrying; used of organisms harbouring viruses.

-virus The ending of a name of a genus in virological nomenclature.

virusology The study of viruses; virology.

viscosity 1: Internal friction in fluids. 2: The relatively slow rate of dispersal of individuals in a population, and consequently of gene flow; population viscosity.

visible light That part of the solar radiation spectrum between the wavelengths 380–780 nm.

visible production Nett production *q.v.*

vitalism The view that phenomena exhibited by living organisms are the result of supernatural forces, as distinct from chemical and physical forces, that exert directional effects on variation and evolution.

vitality 1: The property of life peculiar to living organisms. 2: The property of an organism that enables it to survive unfavourable environmental fluctuations.

vitellogenous Yolk-producing.

viticolous Growing on vines; **viticole**.

viviparous 1: Producing live offspring from within the body of the parent; zoogonous; **viviparity**, **vivipary**; *cf*. larvipary, oviparous, ovoviviparous. 2: Germinating while still attached to the parent plant.

vixigregarious Sparsely distributed; occurring in small poorly defined groups.

viz. Abbreviation of the Latin *videlicet q.v.*

vocalization Production of songs, calls and other vocal sounds by animals.

vol. Abbreviation of volume.

volant Adapted for flying or gliding.

volcanism Volcanic activity; vulcanism.

volcanogenic Used of a sediment produced as a result of volcanic activity.

voltine Pertaining to the number of broods or generations per year or per season; usually quantified with a prefix as in univoltine, bivoltine, trivoltine, quadrivoltine, multivoltine.

voltinism A behavioural polymorphism in insects in which some members of a population enter diapause and others do not.

von Baer's law That development proceeds from the generalized (homogeneous) condition to the specialized (heterogeneous) condition, so that the earliest embryonic stages of related organisms are identical and distinguishing features develop later in ontogeny.

-vorous Suffix meaning feeding upon; **-vore**, **-vory**; ambivorous, aphidivorous, apivorous, calcivorous, carnivorous, detritivorous, erucivorous, exudativorous, folivorous, forbivorous, frugivorous, fucivorous, fungivorous, gallivorous, graminivorous, granivorous, gumivorous, herbivorous, insectivorous, larvivorous, lignivorous, limivorous, mellivorous, merdivorous, microbivorous, mucivorous, nectarivorous, nucivorous, omnivorous, phytosuccivorous, piscivorous, plurivorous, pupivorous, radicivorous, ranivorous, sanguivorous, univorous, zoosuccivorous.

voucher specimen Any specimen identified by a recognized authority for the purposes of forming a reference collection.

vulcanism Volcanic activity; volcanism.

W

Wallace effect The process of development and reinforcement of reproductive isolation between sexual populations that have reached the level of elementary biological species, by selection against hybrids that have relatively lower fitness; Wallace's hypothesis.

Wallace's line An imaginary line separating the Oriental and Australian zoogeographical regions drawn to the west of Weber's line *q.v.* and passing between the Philippines and Maluku (Moluccas) in the north, then south-west between Sulawesi and Borneo, and continuing south between Lombok and Bali.

Wallace's realms The major zoogeographical divisions – Neotropical, Nearctic, Palaearctic, Oriental, Australian and Ethiopian; see Appendix 3.

Wallacea The transition zone between the Oriental and Australian zoogeographical regions, bounded to the east by Weber's line *q.v.* and to the west by Wallace's line *q.v.*, and comprising Sulawesi, Lombok, Flores and Timor.

warm blooded Homoiothermic *q.v.*

warm monomictic Used of a lake with a winter overturn in which the water temperature never falls below 4° C; *cf.* cold monomictic.

warm temperate zone The latitudinal zone between $34.0°$ and $45.0°$ in either hemisphere.

warning coloration Conspicuous coloration used to advertise the noxious, unpalatable, or otherwise harmful properties of an organism to a potential predator; proaposematic coloration.

Warthe glaciation A glaciation of the Quaternary Ice Age in northern Germany and Poland with an estimated duration (together with the Saale glaciation) of 100 thousand years; see Appendix 2.

water balance The difference between the rate of water intake (absorption) by a plant and water loss (transpiration).

water capacity The amount of water that a soil can retain against gravity; usually expressed as a percentage per volume of soil.

water cycle The global biogeochemical cycle of water involving exchange between the hydrosphere, atmosphere, lithosphere and living organisms.

water humus Humus formed from organic material in aquatic environments.

water mass A body of water within an ocean characterized by its physicochemical properties of temperature, salinity, depth and movement.

water parasite A parasite that derives only water from its host.

water potential The thermodynamic state of the water within a plant cell, equal to the difference in free energy per unit volume between matrically bound, pressurized, or osmotically constrained, water and that of pure water.

water saturation deficit (WSD) A measure of the quantity of water lacking from a tissue as compared to complete saturation; calculated as $WSD = (W_s - W_{act})/(W_s) \times 100$ where W_s is the water content under conditions of saturation and W_{act} the actual water content.

water table The horizontal plane defining the upper limit of the ground layer fully saturated with water.

water vapour deficit Saturation deficit *q.v.*

watershed An elevated boundary area separating tributaries draining into different river systems; sometimes used to mean a drainage basin.

water-use efficiency of photosynthesis The ratio of photosynthesis to transpiration; calculated as $W_F = $ Photosynthesis/Amount of water transpired.

water-use efficiency of productivity The ratio of dry matter produced to the amount of water consumed; calculated as $W_P = $ Dry matter production/Water consumption.

watt (W) A derived SI unit of power, defined as the power equivalent to 1 joule per second; see Appendix 13.

wave An undulation or ridge on the surface of a fluid; a disturbance that moves through or over the surface of a medium.

W-chromosome The sex chromosome which in cases of female heterogamety is present in the female only; *cf.* X-chromosome, Y-chromosome, Z-chromosome.

weather Local, short term atmospheric conditions; *cf.* climate.

weathering Physical, chemical and biological changes resulting from exposure to the atmosphere, that accompany soil formation from parent rock; *cf.* biological weathering, chemical weathering, mechanical weathering.

Weber's line An imaginary line separating the Oriental and Australian zoogeographical regions drawn to the east of Wallace's line *q.v.* and passing between Maluku (the Moluccas) and Sulawesi to the north, and between Timor and the Kei Islands to the south.

weed Any plant growing where it is not wanted.

Weichselian glaciation The most recent glaciation of the Quaternary Ice Age in northern Germany and Poland with an estimated duration of about 70 thousand years; subdivided into Pomeranian, Frankfurt and Brandenburg glaciations; see Appendix 2.

weight loss (L) In ecological energetics, the biomass used for the metabolic processes of an individual, population or trophic unit, per unit time per unit area or volume, when respiration exceeds assimilation; expressed as weight loss (L) = total biomass produced (G) − production (P).

weighting A method of attaching different importance values to different characters.

Weismannism The concept that acquired characters are not inherited and that only changes in the germ plasm are transmitted from generation to generation.

Wentworth grade scale An extension of the Udden grade scale $q.v.$ with descriptive class terms for sediment particles; see Appendix 11.

West African rain forest region A subdivision of the African subkingdom of the Palaeotropical kingdom; see Appendix 4.

West African subregion A subdivision of the Ethiopian region; see Appendix 3.

West Australia Current A cold surface ocean current that flows north off the west coast of Australia, fed from the Antarctic Circumpolar Current and forming the eastern limb of the South Indian Gyre; see Appendix 6.

West Greenland Current A cold surface ocean current that flows north off the west coast of Greenland, giving rise to the cold Labrador Current; see Appendix 6.

West Indian Islands subregion Antillean subregion of the Neotropical region; see Appendix 3.

West Wind Drift A major cold surface ocean current that flows east through the Southern Ocean producing a circumglobal antarctic circulation; Antarctic Circumpolar Current; see Appendix 6.

Western and Central Asiatic region A subdivision of the Boreal kingdom; see Appendix 4.

Western subregion Californian subregion of the Nearctic region; see Appendix 3.

wetland An area of low lying land, submerged or inundated periodically by fresh or saline water.

white mud A terrigenous marine sediment derived from coral reef debris.

Wilcoxon two-sample test A non-parametric method in which sample counts are ranked in a single sequence, used to determine whether two independent random samples are drawn from populations having the same distribution.

wild type The natural or typical form of an organism, strain or gene, arbitrarily designated as standard or normal for comparison with mutant or aberrant individuals or alleles.

wilting Loss of turgidity in plants caused by an excess of water loss by transpiration over water uptake by absorption from the soil.

wilting coefficient The amount of water remaining in a soil when a plant is in a state of permanent wilting $q.v.$

wilting point The time at which the water content of a soil is reduced to a level at which permanent wilting occurs.

wind speed Beaufort wind scale $q.v.$ see Appendix 21.

windchill The effect of the wind in causing excessive cold penetration into plants.

windkill Plant death resulting from the effects of wind, often through windchill $q.v.$ or physiological drought $q.v.$

wing rule Rensch's rule $q.v.$

winnowing Grading and separation of the fine fractions of particulate matter by wind or water currents.

winter heat income The total amount of heat required to raise the temperature of a lake from its lowest heat content up to 4° C; $cf.$ summer heat income.

Wisconsin glaciation The most recent glaciation of the Quaternary Ice Age in North America, with an estimated duration (together with the Iowan glaciation) of about 70 thousand years; see Appendix 2.

Wolstonian glaciation A glaciation of the Quaternary Ice Age in the British Isles with an estimated duration of 100 thousand years; Gipping glaciation; see Appendix 2.

woodland A circumscribed area of vegetation dominated by a more or less closed stand of short trees.

woody plant A perennial plant having a secondarily thickened lignified stem.

work 1: Any structure or impression resulting from the activity of an animal, including tubes, burrows, nests and tracks. 2: In taxonomy, a scientific publication.

working hypothesis A hypothesis that serves as the basis for future experimentation.

Wright's inbreeding coefficient Relationship, coefficient of $q.v.$

Würm glaciation The most recent glaciation of the Quaternary Ice Age in the Alpine area with an estimated duration of about 70 thousand years; subdivided into Würm 3, Würm 2 and Würm 1 glaciations; see Appendix 2.

X

x-axis The horizontal axis of a graph; abscissa; *cf.* *y*-axis.

X-chromosome The sex chromosome which in cases of male heterogamety is present in both sexes; *cf.* W-chromosome, Y-chromosome, Z-chromosome.

xenautogamous Used of organisms in which cross fertilization normally occurs, but in which self-fertilization is possible; xenautogamy.

-xene Suffix meaning guest, visitor.

xenia Phenotypic changes in the maternal tissue of a fruit (endosperm) resulting from the direct influence of the pollen genotype.

xenic Pertaining to a culture containing one or more unidentified organisms.

xenobiosis Symbiosis in which one species lives freely within the colony of another species whilst maintaining broods separately; xeniobiosis.

xenobiotic A foreign (allochthonous) organic chemical; used of environmental pollutants such as pesticides in runoff water.

xenocaval Trogloxene *q.v.*

xenodeme A local interbreeding population of a parasite that differs from other demes in its host specificity.

xenodiagnosis A method of diagnosing a disease or detecting a parasite by infecting a test organism or suspected intermediate host.

xenoecic Inhabiting the empty domicile or shell of another organism; xenoecy.

xenogamy Cross fertilization; fertilization between flowers on different plants; *cf.* geitonogamy.

xenogeneic Originating in a member of another species.

xenogenesis 1: Spontaneous generation *q.v.*; abiogenesis. 2: Any unusual method of reproduction.

xenogenous Originating from outside the organism or system; exogenous; allochthonous.

xenograft Heterograft *q.v.*

xenology The study of host–parasite relationships.

xenomixis The fusion of gametes from different sources; exomixis.

xenomorphosis Heteromorphosis *q.v.*

xenoparasite A parasite infesting an organism that is not its normal host; xenoparasitism.

xerad Xerophyte *q.v.*

xerantic Becoming parched or dried up; withering.

xerarch succession An ecological succession beginning in a dry habitat; xerosere; *cf.* hydrarch succession, mesarch succession.

xerasium An ecological succession following drought or drainage.

xeric Having very little moisture; tolerating or adapted to dry conditions.

xeric-supralittoral The driest part of the supralittoral *q.v.* zone of the shore, wetted only by wind-borne spray; spray zone.

xero- Prefix meaning dry.

xerochastic Used of a fruit in which dehiscence is induced by desiccation; xerochasis, xerochasy; *cf.* hygrochastic.

xerochore That region of the Earth's surface covered by dry desert.

xerocleistogamy Self-pollination within flowers that remain unopened because of inadequate moisture; xerocleistogamic.

xerocolous Living under dry conditions; xerocole.

xerodrymium A dry thicket community.

xerogeophyte A plant which enters a resting stage during periods of drought.

xerohylad A dry-forest plant; xerohylophyte.

xerohylium A dry-forest community.

xerohylophilous Thriving in dry forests; xerohylophile, xerohylophily.

xerohylophyte A dry-forest plant; xerohylad; xerohylophyta.

xeromorphic Pertaining to plants having structural or functional adaptations to prevent water loss by evaporation, plants not necessarily confined to dry habitats; xeromorphism, xeromorphy.

xerophilous Thriving in dry habitats; xerophil, xerophile, xerophily.

xerophobous Intolerant of dry conditions; xerophobic; xerophobe, xerophoby.

xerophyte A plant living in a dry habitat, typically showing xeromorphic or succulent adaptations; a plant able to tolerate long periods of drought; zerophyte; xerophil; xerad; xerophytic; *cf.* hydrophyte, hygrophyte, mesophyte.

xerophyton Vegetation of dry habitats; xerophyta.

xeropoad A heath plant; xeropoophyte.

xeropoium A heathland community.

xeropoophilous Thriving in heathland; xeropoophile, xeropoophily.

xeropoophyte A heath plant; xeropoad; xeropoophyta.

xerosere 1: A xerarch succession *q.v.* 2: An ecological succession commencing on a dry rock surface.

xerosium An ecological succession on drained and dried soils.

xerostatic Used of an ecological succession completed under xeric conditions.

xerotaxis The condition of a plant succession that remains unaffected by drought.

xerothamnium A spiny-shrub community.

xerotherm A plant adapted to hot dry conditions.

Xerothermal period A postglacial interval of warmer and drier climate; approximately equivalent to the Altithermal period *q.v.* or the Subboreal period *q.v.*; Long Drought; *cf.* Hypsithermal period.

xerothermic Used of organisms tolerating or thriving in hot and dry environments; **xerotherm**.

xerotherous Used of organisms adapted to dry summer conditions.

xerotropism An orientation response of plants or plant structures to desiccation; **xerotropic**.

xylium A woodland community.

xylophagous Feeding on wood; dendrophagous; hylophagous; lignivorous; **xylophage, xylophagy**.

xylophilous Thriving on or in wood; **xylophile, xylophily**.

xylophyte 1: A woody plant; **xylophyta**. 2: A plant living in or on wood.

xylotomous Used of organisms able to cut or bore into wood.

Y

yard A derived fps unit of length equal to 0.9144 m.

Yarmouthian interglacial An interglacial period in the middle of the Quaternary Ice Age in North America; see Appendix 2.

yarovization Vernalization *q.v.*

y-axis The vertical axis of a graph; ordinate; *cf. x*-axis.

Y-chromosome The sex chromosome which in cases of male heterogamety is present in the male only; *cf.* W-chromosome, X-chromosome, Z-chromosome.

yield (Y) That part of production utilized by a consumer species or by a group belonging to a higher trophic level, or by man.

Yoda's 3/2 power law That a relationship exists between the dry weight (W) of surviving plants in a given area and the density (P) of the plants in that area; expressed as $W = C \times P^{-3/2}$ where C is a constant related to the growth architecture of the plant species.

yoked key Indented key *q.v.*

younger synonym Junior synonym *q.v.*

Z

Z-chromosome The sex chromosome which in cases of female heterogamety is present in both sexes; *cf.* W-chromosome, X-chromosome, Y-chromosome.

zeitgeber Any external stimulus that acts to trigger or phase a biological rhythm; synchronizer.

zelotypic Asexual *q.v.*

zenith The point on the celestial sphere directly above the observer, or above a given point on the surface; *cf.* nadir.

zenogeotropism Negative geotropism *q.v.*; also used as a synonym of diageotropism *q.v.*

zenotropism Negative tropism *q.v.*

zerophyte Xerophyte *q.v.*

zeta karyology In the study of chromosomes, the mapping of bands, chromosome puffs, nucleolar organizers and other landmarks on the basis of polytene chromosome analysis; *cf.* karyology.

zincophyte A plant adapted to, or tolerating, high zinc levels in the soil; frequently used as an indicator of this particular soil type.

zodiophilous Zoidiophilous *q.v.*

zoetic Pertaining to life *q.v.*

-zoic Suffix meaning animal.

zoidiogamy Fertilization by a motile male gamete; **zoidiogamic.**

zoidiophilous Pollinated by animals; zoophilous; zodiophilous; **zoidiophily.**

-zoite Suffix meaning animal.

zonal soil A deep soil having moderate internal and surface drainage with a well defined profile comprising several distinct horizons; used of mature soils with well developed characteristics that reflect the influence of vegetation and climate as soil forming factors; normal soil; *cf.* azonal soil, intrazonal soil.

zonality The zonal distribution of organisms; the zonal property of a habitat or area.

zonation The distribution of organisms in distinctive areas, layers or zones.

zone 1: An area, or subdivision of a biogeographical region that has a characteristic biota. 2: A stratum or series of strata distinguished by characteristic fossils; **zonal, zonate.**

zone of accumulation The B-horizon *q.v.* of a soil profile.

zonolimnetic Pertaining to a particular depth zone within a lake.

zooapocrisis The strategy or response of animals to environmental conditions.

zoobenthos Those animals living in or on the sea bed or lake floor.

zoobiotic Used of organisms that live as parasites on other animals.

zoocenosis Zoocoenosis *q.v.*

zoochorous Dispersed by the agency of animals; zoochoric; **zoochore, zoochory.**

zoocoenosis An animal community; zoocenosis.

zoodomatia Plant structures acting as shelters for animals.

zoodynamics Animal physiology; zoonomy.

zooecology The study of relationships between animals and their environment; animal ecology; *cf.* phytoecology.

zoogamete A motile gamete; planogamete; zoozygogamete.

zoogamous 1: Used of animals that reproduce sexually; **zoogamy.** 2: Used of plants having motile gametes.

zoogene The environment and associated deposits produced by abundant lime-secreting organisms.

zoogenesis 1: The origin of animal life on Earth; zoogeogenesis. 2: The evolution and development of an animal species; **zoogeny.**

zoogenetics Animal genetics.

zoogenic Produced by or associated with the activity of animals; zoogenous.

zoogeogenesis Zoogenesis *q.v.*

zoogeographical region Any of the major geographical divisions of the Earth, characterized by a particular faunal composition; the six original regions were Australian, Ethiopian, Nearctic, Neotropical, Oriental and Palaearctic; see Appendix 3.

zoogeography Study of the geographical distribution of animals and animal communities.

zoogonous Viviparous *q.v.*

zoolith An animal fossil; zoolite.

zoology The study of animals; **zoological.**

zoome An animal community considered as an ecological unit.

zoometry The application of statistical methodology to the study of animals; **zoometrics.**

zoomorphic 1: Pertaining to or produced by the activity of animals; **zoomorphosis.** 2: Having the form of an animal.

zooneuston The animals of the neuston; **zooneustonic, zooneustont.**

zoonomy Animal physiology; zoodynamics.

zoonosis A disease transmitted from animals to man under natural conditions; also used more loosely for any disease of animals; **zoonoses**.

zooparasite A parasitic animal.

zooparasitic Used of a parasite having an animal host.

zoophagous Feeding on animals or animal matter; **zoophage, zoophagy**.

zoophilous Pollinated by animals; having an affinity for animals; zoidiophilous; zodiophilous; **zoophile, zoophily**.

zoophobic Intolerant of animals; shunned or avoided by animals; **zoophobe, zoophoby**.

zoophyte Any animal that resembles a plant in morphology or mode of life; **zoophytic**.

zooplankton The animals of the plankton; **zooplankter, zooplanktont**; *cf.* phytoplankton.

zoosaprophagous Feeding on decaying animal matter; **zoosaprophage, zoosaprophagy**.

zoosemiotics The study of animal communication.

zoosis Any disease caused by an animal.

zoospore An independently motile spore in protistans, some fungi and algae.

zoosuccivorous Used of organisms that feed on liquid secretions of animals, or on decaying animal matter; **zoosuccivore, zoosuccivory**.

zootaxy Animal taxonomy.

zootechnics The science of breeding and rearing domestic animals; animal husbandry; **zootechny**.

zootic climax Any stable climax community dependent for its structure and composition upon continued grazing, or other stress from animal use.

zootoxin A poison produced by an animal.

zootrophic Heterotrophic *q.v.*

zoozygosphere A motile gamete; planogamete; zoogamete.

zugunruhe Migratory restlessness, such as directional hopping movements in birds that are unable to migrate.

zygogamy Isogamy *q.v.*

zygogenesis Reproduction during which male and female nuclei fuse; **zygogenetic**.

zygogenic Produced by fertilization; zygogenetic.

zygoid parthenogenesis Asexual reproduction from an egg that remains diploid or undergoes diploidization during its development.

zygophase The diploid phase of a life cycle; diplophase; *cf.* gamophase.

zygophyte A plant produced by sexual reproduction.

zygosis Union of gametes; conjugation; **zygotic**.

zygotaxis The mutual attraction between male and female gametes; **zygotactism, zygotactic**.

zygote A fertilized gamete; a diploid cell formed by the fusion of two haploid gametes; syngamete.

zymogenic Causing fermentation; **zymogenous**.

Appendix 1. GEOLOGICAL TIME SCALE

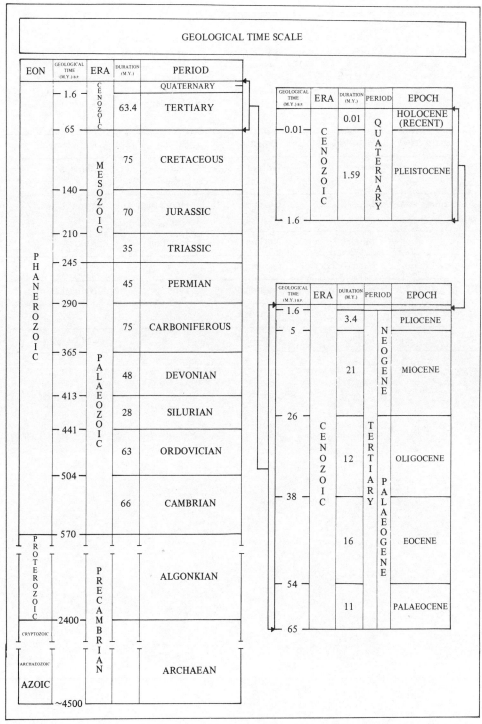

Appendix 2. CHRONOLOGY OF THE QUATERNARY ICE AGE

Estimated duration of Glacial Period (1000 years)	The Alpine area	British Isles	North Germany and Poland	North America
70	Würm 3 Würm 2 Würm 1	Devensian *(Weichselian)*	Pomeranian Frankfurt Brandenburgh	Wisconsin Iowan
Interglacial Period	Riss-Würm	Ipswichian	Eemian	Sangamonian
100	Riss	Wolstonian	Warthe Saale	Illinoian
'Great' Interglacial Period	Mindel-Riss	Hoxnian	Holsteinian	Yarmouthian
90	Mindel	Anglian	Elsterian	Kansan
Interglacial Period	Günz-Mindel	Cromerian		Aftonian
100	Günz	Beestonian	Elbe	Nebraskan
Interglacial Period		Pastonian		
260	Donau	Baventian		Pre-Nebraskan

Appendix 3. WALLACE'S ZOOGEOGRAPHICAL REGIONS

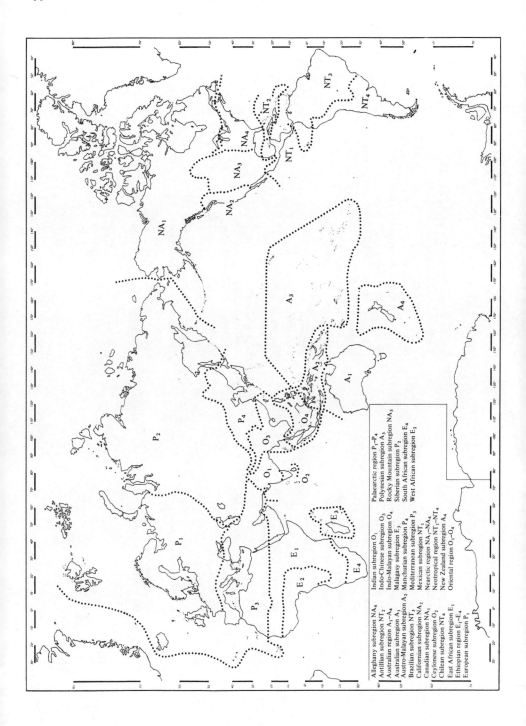

Alleghany subregion NA₄
Antillian subregion NT₂
Australian region A₁–A₄
Australian subregion A₁
Austro-Malayan subregion A₂
Brazilian subregion NT₃
Californian subregion NA₂
Canadian subregion NA₁
Ceylonese subregion O₂
Chilean subregion NT₄
East African subregion E₁
Ethiopian region E₁–E₄
European subregion P₁

Indian subregion O₁
Indo-Chinese subregion O₃
Indo-Malayan subregion O₄
Malagasy subregion E₃
Manchurian subregion P₄
Mediterranean subregion P₃
Mexican subregion NT₁
Nearctic region NA₁–NA₄
Neotropical region NT₁–NT₄
New Zealand subregion A₄
Oriental region O₁–O₄

Palaearctic region P₁–P₄
Polynesian subregion A₃
Rocky Mountain subregion NA₃
Siberian subregion P₂
South African subregion E₄
West African subregion E₂

Appendix 4. INDEX OF PHYTOGEOGRAPHICAL REGIONS

Appendix 4. PHYTOGEOGRAPHICAL REGIONS*

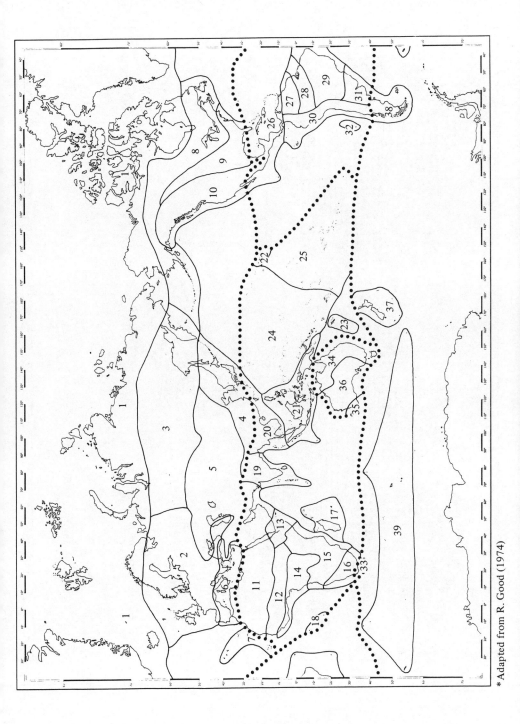

* Adapted from R. Good (1974)

274

Appendix 5. TERRESTRIAL BIOMES

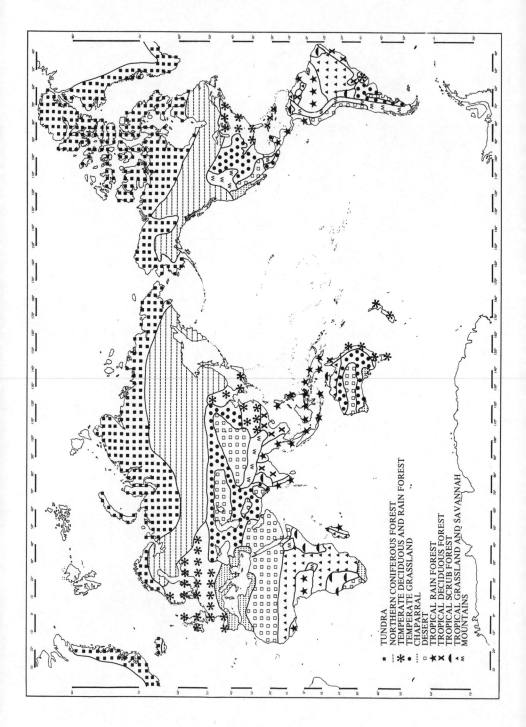

- TUNDRA
- ✳ NORTHERN CONIFEROUS FOREST
- ● TEMPERATE DECIDUOUS AND RAIN FOREST
- ∶ TEMPERATE GRASSLAND
- ☆ CHAPARRAL
- □ DESERT
- ✶ TROPICAL RAIN FOREST
- ✕ TROPICAL DECIDUOUS FOREST
- ◖ TROPICAL SCRUB FOREST
- ◣ TROPICAL GRASSLAND AND SAVANNAH
- ⋈ MOUNTAINS

Appendix 6. MAJOR OCEANIC SURFACE CURRENTS

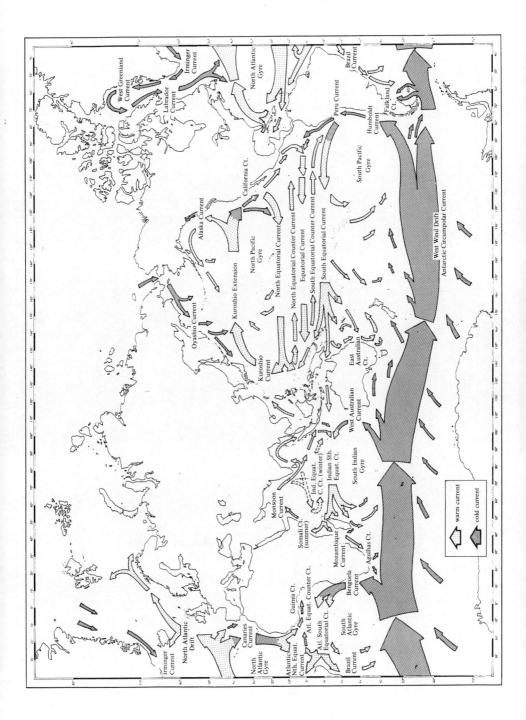

Appendix 7. MARINE DEPTH ZONES

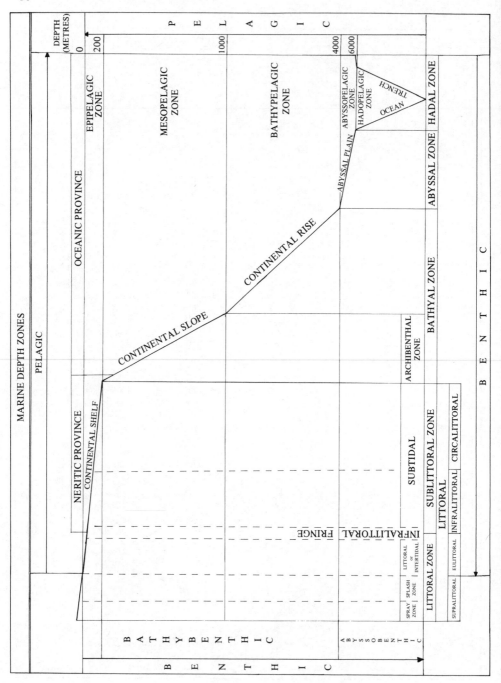

Appendix 8. LACUSTRINE DEPTH ZONES

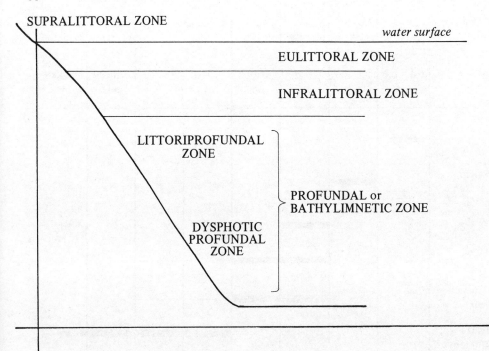

Appendix 9. PLANKTON SIZE CATEGORIES

PLANKTON	Femto-plankton (Ultrananoplankton)	Pico-plankton	Nano-plankton	Micro-plankton	Meso-plankton	Macro-plankton	Mega-plankton
Size . . .	0.02–0.2 μm	0.2–2.0 μm	2.0–20 μm	20–200 μm	0.2–20 mm	20–200 mm	0.2–2.0 m

Virio-plankton

Bacterio-plankton

Myco-plankton

Phyto-plankton

Protozoo-plankton

Metazoo-plankton

Nekton

SIZE (m) 10^{-8} 10^{-7} 10^{-6} 10^{-5} 10^{-4} 10^{-3} 10^{-2} 10^{-1} 10^{0}

Appendix 10. TAXONOMIC HIERARCHY: with recommended endings of certain categories

ZOOLOGY	ending	BOTANY	ending	VIROLOGY	ending (*algae*)	ending (*fungi*)	BACTERIOLOGY	ending
Kingdom		Kingdom						
Subkingdom								
Superphylum					(*algae*)	(*fungi*)		
Phylum		Division	-phyta	-phyta		-mycota	Division	
Subphylum		Subdivision	-phytina	-phytina		-mycotina	Subdivision	
Branch								
Superclass								
Class		Class	-opsida	-phyceae		-mycetes	Class	
Subclass		Subclass	-idae	-phycidae		-mycetidae	Subclass	
Infraclass								
Cohort								
Superorder		Superorder						
Order		Order	-ales				Order	-ales
Suborder		Suborder	-ineae				Suborder	-ineae
Infraorder								
Superfamily	-oidea							
Family	-idae	Family	-aceae	Family	-viridae		Family	-aceae
Subfamily	-inae	Subfamily	-oideae	Subfamily	-virinae		Subfamily	-oideae
Supertribe								
Tribe	-ini	Tribe	-eae				Tribe	-eae
Subtribe		Subtribe	-inae				Subtribe	-inae
Genus		Genus		Genus	-virus		Genus	
Subgenus		Subgenus					Subgenus	
		Section						
		Subsection						
		Series						
		Subseries						
Species		Species					Species	
Subspecies		Subspecies					Subspecies	(=Variety)
		Variety						
		Subvariety						
		Form						

Appendix 11. SEDIMENT PARTICLE SIZE CATEGORIES

PARTICLE GRADE SCALES		
UDDEN (mm)	PHI VALUE	WENTWORTH

UDDEN (mm)	PHI VALUE	WENTWORTH	
		BOULDER	
— 256.0	— −8.0		
		COBBLE	G R A V E L
— 64.0	— −6.0		
		PEBBLE	
— 4.0	— −2.0		
		GRANULE	
— 2.0	— −1.0		
		VERY COARSE	S A N D
— 1.0	— 0.0		
		COARSE	
— 0.5	— 1.0		
		MEDIUM	
— 0.25	— 2.0		
		FINE	
— 0.125	— 3.0		
		VERY FINE	
— 0.0625	— 4.0		
		SILT	M U D
— 0.0039	— 8.0		
		CLAY	
— 0.00024	— 12.0		
		COLLOID	

Appendix 12. SOIL PROFILE AND HORIZON NOMENCLATURE

HORIZON			DESCRIPTION	
O	Ol or L	A_{00}	Litter layer; loose fragmented organic matter in which original plant structures are still readily discernible	Horizons of maximum biological activity and eluviation
	Of or F		Fermentation layer; partially decomposed organic matter	
	Oh or H	A_0	Humified layer; heavily transformed organic matter with no discernible macroscopic plant remains	
A		A_1	Dark-coloured horizon, with high organic content intermixed with minerals. Ap denotes ploughed layer	
E		A_2 or E_a	Light-coloured horizon, with low organic content due to high eluviation	
		A_3	Transitional layer, sometimes absent	
B		B_1	Transitional layer, sometimes absent	Horizons of illuviation; accumulations of clay, humus or iron oxides leached or translocated from the upper layers
		B_2	A dark layer of accumulation of transported silicate, clay, minerals, iron and organic matter showing the maximum development of blocky and/or prismatic structure	
		B/C or B_3	Transitional layer	
C			Weathered parent material, comprising mineral substrate with little or no structure; containing a gleyed layer or layers of accumulated calcium carbonate or sulphate in some soils	
R		0	Underlying rock	

Appendix 13. SI UNITS

Base units of the International System (SI)

Physical Quantity	Name of SI Unit	SI Unit symbol
length	metre	m
mass	kilogram	kg
time	second	s
electric current	ampere	A
thermodynamic temperature	kelvin	K
luminous intensity	candela	cd
amount of substance	mole	mol

Derived units of the International System

Quantity	Name of SI Unit	SI Unit Symbol or abbreviation	Definition of SI unit
area	square metre		m^2
volume	cubic metre		m^3
frequency	hertz	Hz	s^{-1}
density	kilogram per cubic metre		kg/m^3
velocity	metre per second		m/s
angular velocity	radian per second		rad/s
acceleration	metre per second squared		m/s^2
angular acceleration	radian per second squared		rad/s^2
volumetric flow rate	cubic metre per second		m^3/s
force	newton	N	$kg.m/s^2$
surface tension	newton per metre, joule per square metre	N/m, J/m^2	kg/s^2
pressure	newton per square metre, pascal	N/m^2, Pa	$kg/m.s^2$
viscosity, dynamic	newton-second per square metre, pascal-second	$N.s/m^2$, Pa.s	$kg/m.s$
viscosity, kinematic	metre squared per second		m^2/s
work, torque, energy, quantity of heat	joule, newton-metre, watt-second	J, N.m, W.s	$kg.m^2/s^2$
power, heat flux	watt, joule per second	W, J/s	$kg.m^2/s^3$
heat flux density	watt per square metre	W/m^2	kg/s^3
volumetric heat release rate	watt per cubic metre	W/m^3	$kg/m.s^3$
heat transfer coefficient	watt per square metre kelvin	$W/m^2.K$	$kg/s^3.K$
heat capacity	joule per kilogram kelvin	$J/kg.K$	$m^2/s^2.K$
capacity rate	watt per kelvin	W/K	$kg.m^2/s^3.K$
thermal conductivity	watt per metre kelvin	$W/m.K$, $J.m/s.m^2.K$	$kg.m/s^3.K$
quantity of electricity	coulomb	C	A.s
electromotive force	volt	V, W/A	$kg.m^2/A.s^3$
electric field strength	volt per metre		V/m
electric resistance	ohm	Ω, V/A	$kg.m^2/A^2.s^3$
electric conductivity	ampere per volt metre	A/V.m	$A^2.s^3/kg.m^3$

Appendix 13 – *continued*
Derived units of the International System—*continued*

Quantity	Name of SI Unit	SI Unit Symbol or abbreviation	Definition of SI unit
electric capacitance	farad	F, A.s/V	$A^3.s^4/kg.m^2$
magnetic flux	weber	Wb, V.s	$kg.m^2/A.s^2$
magnetic field strength	ampere per metre		A/m
magnetomotive force	ampere		A
luminous flux	lumen	lm	cd.sr
luminance	candela per square metre		cd/m^2
illumination	lux, lumen per square metre	lx, lm/m^2	$cd.sr/m^2$

Table of SI – fps conversion

Unit/Quantity	Abbreviation	Metric	Imperial
Length		metre	foot
metre	m	–	3.280 839
foot	ft	0.3048	–
micron	μm	1×10^{-6}	$3.280\ 839 \times 10^{-6}$
angstrom	Å	1×10^{-10}	$3.280\ 839 \times 10^{-10}$
inch	in	2.54×10^{-2}	8.333×10^{-2}
yard	yd	0.9144	3.0
chain		20.1168	66.0
furlong		201.168	660.0
mile	mi	1609.344	5280.0
UK nautical mile		1853.184	6080.0
International nautical mile		1852.0	6076.114
fathom		1.8288	6.0
Area		square metre	square foot
square metre	m^2	–	10.763 904
square foot	ft^2	$9.290\ 304 \times 10^{-2}$	–
square inch	in^2	6.452×10^{-4}	6.944×10^{-3}
square yard	yd^2	0.836 127	9.0
acre		4046.856	43 560
hectare	ha	10 000	107 631
square mile		$2.589\ 988 \times 10^6$	$2.787\ 840 \times 10^7$
Volume		cubic metre	cubic foot
cubic metre	m^3	–	35.31467
cubic foot	ft^3	2.832×10^{-2}	–
cubic inch	in^3	$1.638\ 706 \times 10^{-5}$	$5.787\ 036 \times 10^{-4}$
cubic yard	yd^3	0.764 555	27.0
litre	l	1.0×10^{-3}	$3.531\ 467 \times 10^{-2}$
fluid ounce	fl.oz	$2.841\ 306 \times 10^{-5}$	$1.003\ 398 \times 10^{-3}$
pint	pt	$5.682\ 613 \times 10^{-4}$	$2.006\ 796 \times 10^{-2}$
bushel	bu	3.6369×10^{-2}	1.285 313
Imperial gallon	gal	4.54609×10^{-3}	$1.605\ 437 \times 10^{-1}$
US gallon		$3.785\ 412 \times 10^{-3}$	$1.336\ 806 \times 10^{-1}$
quart	qt	$1.136\ 523 \times 10^{-3}$	4.01359×10^{-2}

continued on p. 284

Appendix 13 – *continued*

Unit/Quantity	Abbreviation	Metric	Imperial
Mass		kilogram	pound
kilogram	kg	–	2.204 623
pound	lb	0.453 592	–
ounce	oz	$2.834\ 952 \times 10^{-2}$	6.25×10^{-2}
stone	st	6.350 586	14
quarter	qr	12.700 586	28
hundredweight	cwt	50.802 344	112
ton	t	1016.047	2240
tonne	t	1000	2204.623

SI prefixes and multiplication factors

Multiplication factor		Prefix	Symbol
1 000 000 000 000 000 000	10^{18}	exa	E
1 000 000 000 000 000	10^{15}	peta	P
1 000 000 000 000	10^{12}	tera	T
1 000 000 000	10^{9}	giga	G
1 000 000	10^{6}	mega	M
10 000	10^{4}	myria	
1 000	10^{3}	kilo	k
100	10^{2}	hecto	h
10	10^{1}	deca	da
0.1	10^{-1}	deci	d
0.01	10^{-2}	centi	c
0.001	10^{-3}	milli	m
0.000 001	10^{-6}	micro	μ
0.000 000 001	10^{-9}	nano	n
0.000 000 000 001	10^{-12}	pico	p
0.000 000 000 000 001	10^{-15}	femto	f
0.000 000 000 000 000 001	10^{-18}	atto	a

Appendix 14. TEMPERATURE CONVERSION CHART

CONVERSION CHART FAHRENHEIT ⇌ CELSIUS											
°F	◄ ►	°C	°F	◄ ►	°C	°F	◄ ►	°C	°F	◄ ►	°C
-22	-30	-34.4	91.4	33	0.6	204.8	96	35.6	318.2	159	70.6
-20.2	-29	-33.9	93.2	34	1.1	206.6	97	36.1	320	160	71.1
-18.4	-28	-33.3	95	35	1.7	208.4	98	36.7	321.8	161	71.7
-16.6	-27	-32.8	96.8	36	2.2	210.2	99	37.2	323.6	162	72.2
-14.8	-26	-32.2	98.6	37	2.8	212	100	37.8	325.4	163	72.8
-13	-25	-31.7	100.4	38	3.3	213.8	101	38.3	327.2	164	73.3
-11.2	-24	-31.1	102.2	39	3.9	215.6	102	38.9	329	165	73.9
-9.4	-23	-30.6	104	40	4.4	217.4	103	39.4	330.8	166	74.4
-7.6	-22	-30	105.8	41	5	219.2	104	40	332.6	167	75
-5.8	-21	-29.4	107.6	42	5.6	221	105	40.6	334.4	168	75.6
-4	-20	-28.9	109.4	43	6.1	222.8	106	41.1	336.2	169	76.1
-2.2	-19	-28.3	111.2	44	6.7	224.6	107	41.7	338	170	76.7
-0.4	-18	-27.8	113	45	7.2	226.4	108	42.2	339.8	171	77.2
1.4	-17	-27.2	114.8	46	7.8	228.2	109	42.8	341.6	172	77.8
3.2	-16	-26.7	116.6	47	8.3	230	110	43.3	343.4	173	78.3
5	-15	-26.1	118.4	48	8.9	231.8	111	43.9	345.2	174	78.9
6.8	-14	-25.6	120.2	49	9.4	233.6	112	44.4	347	175	79.4
8.6	-13	-25	122	50	10	235.4	113	45	348.8	176	80
10.4	-12	-24.4	123.8	51	10.6	237.2	114	45.6	350.6	177	80.6
12.2	-11	-23.9	125.6	52	11.1	239	115	46.1	352.4	178	81.1
14	-10	-23.3	127.4	53	11.7	240.8	116	46.7	354.2	179	81.7
15.8	-9	-22.8	129.2	54	12.2	242.6	117	47.2	356	180	82.2
17.6	-8	-22.2	131	55	12.8	244.4	118	47.8	357.8	181	82.8
19.4	-7	-21.7	132.8	56	13.3	246.2	119	48.3	359.6	182	83.3
21.2	-6	-21.1	134.6	57	13.9	248	120	48.9	361.4	183	83.9
23	-5	-20.6	136.4	58	14.4	249.8	121	49.4	363.2	184	84.4
24.8	-4	-20	138.2	59	15	251.6	122	50	365	185	85
26.6	-3	-19.4	140	60	15.6	253.4	123	50.6	366.8	186	85.6
28.4	-2	-18.9	141.8	61	16.1	255.2	124	51.1	368.6	187	86.1
30.2	-1	-18.3	143.6	62	16.7	257	125	51.7	370.4	188	86.7
32	0	-17.8	145.4	63	17.2	258.8	126	52.2	372.2	189	87.2
33.8	1	-17.2	147.2	64	17.8	260.6	127	52.8	374	190	87.8
35.6	2	-16.7	149	65	18.3	262.4	128	53.3	375.8	191	88.3
37.4	3	-16.1	150.8	66	18.9	264.2	129	53.9	377.6	192	88.9
39.2	4	-15.6	152.6	67	19.4	266	130	54.4	379.4	193	89.4
41	5	-15	154.4	68	20	267.8	131	55	381.2	194	90
42.8	6	-14.4	156.2	69	20.6	269.6	132	55.6	383	195	90.6
44.6	7	-13.9	158	70	21.1	271.4	133	56.1	384.8	196	91.1
46.4	8	-13.3	159.8	71	21.7	273.2	134	56.7	386.6	197	91.7
48.2	9	-12.8	161.6	72	22.2	275	135	57.2	388.4	198	92.2
50	10	-12.2	163.4	73	22.8	276.8	136	57.8	390.2	199	92.8
51.8	11	-11.7	165.2	74	23.3	278.6	137	58.3	392	200	93.3
53.6	12	-11.1	167	75	23.9	280.4	138	58.9	393.8	201	93.9
55.4	13	-10.6	168.8	76	24.4	282.2	139	59.4	395.6	202	94.4
57.2	14	-10	170.6	77	25	284	140	60	397.4	203	95
59	15	-9.4	172.4	78	25.6	285.8	141	60.6	399.2	204	95.6
60.8	16	-8.9	174.2	79	26.1	287.6	142	61.1	401	205	96.1
62.6	17	-8.3	176	80	26.7	289.4	143	61.7	402.8	206	96.7
64.4	18	-7.8	177.8	81	27.2	291.2	144	62.2	404.6	207	97.2
66.2	19	-7.2	179.6	82	27.8	293	145	62.8	406.4	208	97.8
68	20	-6.7	181.4	83	28.3	294.8	146	63.3	408.2	209	98.3
69.8	21	-6.1	183.2	84	28.9	296.6	147	63.9	410	210	98.9
71.6	22	-5.6	185	85	29.4	298.4	148	64.4	411.8	211	99.4
73.4	23	-5	186.8	86	30	300.2	149	65	413.6	212	100
75.2	24	-4.4	188.6	87	30.6	302	150	65.6	415.4	213	100.6
77	25	-3.9	190.4	88	31.1	303.8	151	66.1	417.2	214	101.1
78.8	26	-3.3	192.2	89	31.7	305.6	152	66.7	419	215	101.7
80.6	27	-2.8	194	90	32.2	307.4	153	67.2	420.8	216	102.2
82.4	28	-2.2	195.8	91	32.8	309.2	154	67.8	422.6	217	102.8
84.2	29	-1.7	197.6	92	33.3	311	155	68.3	424.4	218	103.3
86	30	-1.1	199.4	93	33.9	312.8	156	68.9	426.2	219	103.9
87.8	31	-0.6	201.2	94	34.4	314.6	157	69.4	428	220	104.8
89.6	32	0	203	95	35	316.4	158	70	437	225	107.2

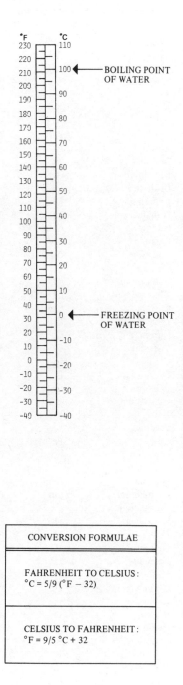

BOILING POINT OF WATER

FREEZING POINT OF WATER

CONVERSION FORMULAE
FAHRENHEIT TO CELSIUS: $°C = 5/9\ (°F - 32)$
CELSIUS TO FAHRENHEIT: $°F = 9/5\ °C + 32$

Appendix 15. CHEMICAL ELEMENTS

Name	Symbol	Atomic No.	Name	Symbol	Atomic No.
Actinium	Ac	89	Manganese	Mn	25
Silver	Ag	47	Molybdenum	Mo	42
Aluminium	Al	13	Nitrogen	N	7
Americium	Am	95	Sodium	Na	11
Argon	Ar	18	Niobium	Nb	41
Arsenic	As	33	Neodymium	Nd	60
Astatine	At	85	Neon	Ne	10
Gold	Au	79	Nickel	Ni	28
Boron	B	5	Nobelium	No	102
Barium	Ba	56	Neptunium	Np	93
Beryllium	Be	4	Oxygen	O	8
Bismuth	Bi	83	Osmium	Os	76
Berkelium	Bk	97	Phosphorus	P	15
Bromine	Br	35	Protactinium	Pa	91
Carbon	C	6	Lead	Pb	82
Calcium	Ca	20	Palladium	Pd	46
Cadmium	Cd	48	Promethium	Pm	61
Cerium	Ce	58	Polonium	Po	84
Californium	Cf	98	Praseodymium	Pr	59
Chlorine	Cl	17	Platinum	Pt	78
Curium	Cm	96	Plutonium	Pu	94
Cobalt	Co	27	Radium	Ra	88
Chromium	Cr	24	Rubidium	Rb	37
Caesium	Cs	55	Rhenium	Re	75
Copper	Cu	29	Rutherfordium	Rf	104
Dysprosium	Dy	66	Rhodium	Rh	45
Erbium	Er	68	Radon	Rn	86
Einsteinium	Es	99	Ruthenium	Ru	44
Europium	Eu	63	Sulphur	S	16
Flourine	F	9	Antimony	Sb	51
Iron	Fe	26	Scandium	Sc	21
Fermium	Fm	100	Selenium	Se	34
Francium	Fr	87	Silicon	Si	14
Gallium	Ga	31	Samarium	Sm	62
Gadolinium	Gd	64	Tin	Sn	50
Germanium	Ge	32	Strontium	Sr	38
Hydrogen	H	1	Tantalum	Ta	73
Hahnium	Ha	105	Terbium	Tb	65
Helium	He	2	Technetium	Tc	43
Hafnium	Hf	72	Tellurium	Te	52
Mercury	Hg	80	Thorium	Th	90
Holmium	Ho	67	Titanium	Ti	22
Iodine	I	53	Thallium	Tl	81
Indium	In	49	Thulium	Tm	69
Iridium	Ir	77	Uranium	U	92
Potassium	K	19	Vanadium	V	23
Krypton	Kr	36	Tungsten	W	74
Lanthanum	La	57	Xenon	Xe	54
Lithium	Li	3	Yttrium	Y	39
Lutetium	Lu	71	Ytterbium	Yb	70
Lawrencium	Lr	103	Zinc	Zn	30
Mendelevium	Md	101	Zirconium	Zr	40
Magnesium	Mg	12			

Appendix 16. SIGNS AND SYMBOLS

+   more; present; longer than; north; indicates a doubtful plant; wild type; graft hybrid; strain; derived character state

—   less; absent; shorter than; south; not utilized; ancestral character state

±   plus or minus; more or less; denotes uncertainty in quantitative measurement

∞   infinity; numerous; polyhybrid

×   hybrid cross; degree of magnification

:   ratio of; is to

∴   therefore

∵   because

!   factorial; placed after a species name indicates that identification was confirmed by comparison with type

①   annual

⊛   annual

②   biennial

⊙⊙   biennial

⊛⊛   biennial

⊙ ⊙   biennial

♂   male

♀   female

☿   neuter; hermaphrodite

⚥   hermaphrodite

⊖   egg

∝   female; very large number; numerous

♂   male

β   male

⚨   monoecious

⚨:♀   dioecious

♀♀♂   polygamous

◇   sex unknown

□   male individual

○   female individual

⧄   death of male

⦸   death of female

□—○   mating

□=○   consanguineous mating

mating, with male offspring

mating, with female offspring

dizygotic male twins

dizygotic female twins

monozygotic male twins

monozygotic female twins

₵   clone

†   poisonous

‡   extremely poisonous

⁎̲   northern hemisphere

‾⁎   southern hemisphere

|⁎   Old World

⁎| New World

continued on p. 288

288

Appendix 16 – *continued*

∅ phi, denotes polymorph when appended to a name

π pi

χ^2 chi-squared

μ arithmetic mean of a population

$\bar{x}$ arithmetic mean of sample of variable x

Σ summation; sum

σ standard deviation of population

σ^2 variance

σt density of sea water

⋁ magnification

/ per

e base of natural (Napierian) logarithm, 2.7183

= equal to; synonym

≡ identical; equivalent to

∼ equivalent to

≠ not equal to; not synonymous

< smaller than

> larger than

≤ equal to or smaller than

≥ equal to or larger than

≏ ≈ approximately equal to

∝ proportional to

§ section; division of a taxon

′ foot; minute

″ inch; second

‴ line

° degree (temperature, latitude, longitude)

† obscure species; imperfectly known; place before a personal name indicates decease destroyed, when used of a specimen; foss when applied to a taxon

* subspecies, when placed before a binomial, between the second and third word of a trin mial; footnote; used of a reference to indica a source of a good description or figure; infr subspecific rank

( ) parenthesis; round brackets; used to enclose subgenus; used to enclose the name of original author placed after an epithet that not in the original combination

[ ] brackets; square brackets; used to enclose superspecies group name; indicates a mi identification when used in a synonymy; use to enclose the name of the author of a n menclatural act first published anonymous

≘ corresponds to

% per cent

→ approaches

number; subsample

‖ merged into; primary homonym

secondary homonym

Appendix 17. ACRONYMS AND ABBREVIATIONS

a	are; atto	E	exa
A	assimilation; ampere	ed.	Editor
Å	ångstrom	edn	edition
abs	absolute	ECI	efficiency of consumer ingestion
ADP	adenosine diphosphate	EDP	electronic data processing
AFDW	ash-free dry weight	EEZ	exclusive economic zone
agg.	aggregate	f	femto
ancova	analysis of covariance	F	filial; egesta
anova	analysis of variance	$F_1, F_2, . F_n$	first, second, etc. filial generations
ATP	adenosine triphosphate	FAD	flavin adenine dinucleotide
AU	astronomical unit	fl.oz.	fluid ounce
av.	average	FMN	flavin mononucleotide
b.y.	billion years	fps	foot pound second
BOD	biological oxygen demand	ft	foot
BP	British Pharmacopoeia	FU	rejecta
B.P.	before present	g	gram
BTU	British thermal unit	G	giga
c	centi	gal	gallon
C	consumption	GMT	Greenwich Mean Time
°C	degrees Celsius; Centigrade	GPP	gross primary production
cal	calorie	GSI	gonosomatic index
Cal	Calorie	h	hecto; hour
CAM	crassulacean acid metabolism	H^+	hydrogen ion
CCD	carbonate compensation depth	H_2S	hydrogen sulphide
cd	candela	ha	hectare
c.g.	centre of gravity	HHW	higher high water
CGR	crop growth rate	HLW	higher low water
cgs	centimetre gram second	ht	height
cm	centimetre	HW	high water
CO	carbon monoxide	Hz	hertz
CO_2	carbon dioxide	I	inbreeding
conc.	concentration	$I_1, I_2, . I_n$	first, second, etc. inbreeding generations
cov.	covariance		
COV	crossover value	IAMS	International Association of Microbiological Societies
crit.	critical		
cu.	cubic	IAPT	International Association of Plant Taxonomy
CV	cultivar		
cwt	hundredweight	ICBN	International Code of Botanical Nomenclature
d	deci; day		
2,4-D	2, 4-dichlorophenoxyacetic acid	ICNB	International Committee on Nomenclature of Bacteria
d.f.	degrees of freedom		
da	deca	ICNV	International Committee on Nomenclature of Viruses
DBH	diameter at breast height		
DBHOB	diameter at breast height over bark	ICSB	International Committee on Systematic Bacteriology
DBHUB	diameter at breast height under bark		
DDT	dichlorodiphenyltrichloroethane	ICZN	International Commission on Zoological Nomenclature
DDVP	2, 2-dichlorovinyl dimethyl phosphate (dichlorvos)	ICTV	International Committee on Taxonomy of Viruses
DL	drought lethality		
DM	dry matter	IJSB	International Journal of Systemic Bacteriology
DNA	deoxyribonucleic acid		
DOC	dissolved organic matter (carbon)	IMS	Industrial Methylated Spirits
DPD	diffusion pressure deficit	in	inch
DSL	deep scattering layer	IQ	intelligence quotient
dyn	dyne	IR	infrared

continued on p. 290

290

Appendix 17. – *continued*

IRM	innate releasing mechanism	O_2	molecular oxygen
ITZN	International Trust for Zoological Nomenclature	O_3	ozone
		obs.	observed
J	joule	OTU	operational taxonomic unit
k	kilo	oz	ounce
K	carrying capacity of environment; kelvin	p.	page
		P	production; peta
Ky	thousand years	Pa	pascal
l	litre	PAL	present atmospheric level
LAD	leaf area duration	PAN	peroxi-acetyl-nitrate
LAI	leaf area index	PBB	polybrominated biphenyl
LAR	leaf area ratio	PCB	polychlorinated biphenyl
LAT	lowest astronomical tide	PhAR	photosynthetically active radiation
lb	pound	POC	particulate organic matter
LHW	lower high water	pp.	pages
LLW	lower low water	PP	community production
lm	lumen	PPA	percentage plant aeration
LW	low water	ppb	parts per billion
lx	lux	ppm	parts per million
m	metre; million	PPR	community production rate
M	mega	ppt	parts per thousand
m.y.	million years	PS	percentage similarity
max.	maximum	psi	pounds per square inch
MCD	mean character difference	pt	pint
mg	milligram	PWP	permanent wilting percentage
MHHW	mean higher high water	qt	quart
MHW	mean high water	r	intrinsic rate of increase
MHWN	mean high water neap	R	respiration
MHWS	mean high water spring	R_a	autotrophic respiration
mi	mile	RAI	radiational aridity index
min.	minimum	RCC	relative crowding coefficient
ml	millilitre	RDI	relative drought index
MLD	minimum lethal dose	Re	Reynolds number
MLLW	mean lower low water	R_e	ecosystem respiration
MLW	mean low water	rem	roentgen equivalent for man
MLWN	mean low water neap	RGR	relative growth rate
MLWS	mean low water spring	RH	relative humidity
mm	millimetre	R_h	heterotrophic respiration
mol	mole	RNA	ribonucleic acid
m-RNA	messenger RNA	RQ	respiratory quotient
MS	manuscript	r-RNA	ribosomal RNA
MSL	mean sea level	RWC	relative water content
MSS	manuscripts	RYT	relative yield total
MTL	mean tide level	s	second
MWL	mean water level	S	selfing
N	population size; haploid chromosome number; newton	S_1	first selfing generation
		s.d.	standard deviation
NAD	nicotinamide adenine dinucleotide	s.e.	standard error
NADP	nicotinamide adenine dinucleotide phosphate	SEM	scanning electron microscope
		SI unit	Système International d'Unités
NAR	nett assimilation rate	SLA	specific leaf area
NEC	nett ecosystem production	SO_2	sulphur dioxide
NH_3	ammonia	SSSI	Site of Special Scientific Interest
NPP	nett primary production	str.	strain
NS	not significant	t	tonne

Appendix 17. – *continued*

T	temperature; tera	UT	universal time
$T_{\frac{1}{2}}$	half life	UV	ultraviolet
2,4,5-T	2,4,5-trichlorophenoxyacetic acid	V	volt
TCDD	tetrachlorodibenzo-para-dioxin	vol.	volume
TEM	transmission electron microscope	W	watt
t-RNA	transfer RNA	WSD	water saturation deficit
TSC	total standing crop	wt	weight
U	excreta		

Appendix 18. LATIN ABBREVIATIONS

a.	*anno*	in the year
ab.	*aberratio*	aberrant
ad int.	*ad interim*	provisionally
adv.	*advena*	alien; introduced
aff.	*affinis*	related; akin to; adjacent
al.	*alii*	others
al.	*aliorum*	of others
ap.	*apud*	with; in the publication of
ascr.	*ascriptum*	ascribed to
auct.	*auctorum*	of authors
c.	*cum*	with
ca.	*circa*	about
cet.	*cetera*	the remainder
cf.; cfr.	*confer*	compare
cit.	*citatus*	cited
comb.	*combinatio*	combination
comb. nov.	*combinatio nova*	new combination
cons.	*conservandus*	to be conserved; to be kept
corr.	*correxit*	he, she or it corrected
cult.	*cultus*	cultivated
cv.	*cultivarietas*	cultivar
dat.	*datus*	given
ded.	*dedit*	he, she or it gave
descr.	*descriptio*	description
det.	*determinavit*	he, she or it identified
e. descr.	*ex descriptione*	from the description
e num.	*e numero*	from the number
e.g.	*exempli gratia*	for example
e.p.	*ex parte*	partly; in part
emend.	*emendatus*	emended
err. typogr.	*errore typographico*	by a typographical error
excl.	*exclusus*	excluded
exs.; exsic	*exsiccatus*	dried
f.	*fide*	according to
f.	*forma*	form
f. sp.	*forma specialis*	special form
fil.	*filius*	son
gen.	*genus*	genus
gen. et sp. nov.	*genus et species nova*	new genus and new species
gen. nov.	*genus novum*	new genus
gr.	*grupo*	the group of
h.	*hortus*	garden
hb.; herb.	*herbarium*	herbarium
hort.	*hortorum*	of gardens
hort.	*hortulanorum*	of gardeners
i.e.	*id est*	that is
ib.; ibid.	*ibidem*	in the same place
ic.	*icon*	illustration
id.	*idem*	the same
in adnot.	*in adnotatione*	in a note or annotation
in litt.	*in litteris*	in correspondence
in sched.	*in schedula*	on a label; on a herbarium label
incl.	*inclusus*	included
ined.	*ineditus*	unpublished

Appendix 18. – *continued*

inq.; inquil.	*inquilinus*	naturalized
l.c.; loc. cit.	*loco citato*	at the place cited
lat.	*latus*	broad, wide
m.	*mihi*	to me; of me
masc.	*masculus*	male
min. parte	*pro minore parte*	for the smaller part
MS; MSS	*manuscriptum; manuscripta*	manuscript; manuscripts
mus.	*museum*	museum
mut. char.	*mutatis characteribus*	with the characters changed
n.	*nobis*	to us, of us
n.	*nomen*	name
n.	*novus*	new
n. n.	*nomen novum*	new name
n. n.	*nomen nudum*	naked name
n. nov.	*nomen novum*	new name
n. sp.	*nova species*	new species
n.v.	*non visus*	not seen
nm.	*nothomorphus*	nothomorph
no.	*numero*	number
nob.	*nobis*	to us
nom.	*nomen*	name
nom. abort.	*nomen abortivum*	name contrary to Code
nom. alt.	*nomen alternativum*	alternative name
nom. ambig.	*nomen ambiguum*	ambiguous name
nom. anam.	*nomen anamorphosis*	name based on imperfect type
nom. conf.	*nomen confusum*	confused name
nom. cons.	*nomen conservandum*	conserved name
nom. dub.	*nomen dubium*	dubious name
nom. hybr.	*nomen hybridum*	hybrid name
nom. illeg.	*nomen illegitimum*	illegitimate name
nom. inval.	*nomen invalidum*	invalid name
nom. legit.	*nomen legitimatum*	legitimate name
nom. monstr.	*nomen monstrositatum*	name based on monstrosity
nom. nov.	*nomen novum*	new name
nom. nud.	*nomen nudum*	naked name
nom. oblit.	*nomen oblitum*	forgotten name
nom. obsc.	*nomen obscurum*	obscure name
nom. provis.	*nomen provisorium*	provisional name
nom. rejic.	*nomen rejiciendum*	rejected name
nom. superfl.	*nomen superfluum*	superflous name
non al.	*non aliorum*	not of other authors
nov.	*novus*	new
nov. n.	*novum nomen*	new name
nov. sp.	*nova species*	new species
op. cit.	*opere citato*	in the work cited
ordo nat.	*ordo naturalis*	natural order
orth. mut.	*orthographia mutata*	an altered spelling
p.	*pagina*	page
p.p.	*pro parte*	in part; partly
part.	*partim*	part
prop.	*propositus*	proposed
prov.	*provisorius*	provisional
q.e.	*quod est*	which is
q.v.	*quod vide*	which see

continued on p. 294

294

Appendix 18 – *continued*

r.; rr.	*rarus; rarissimus*	rare; very rare
recent.	*recentiorum*	of recent authors
s. ampl.	*sensu amplificato*	in an enlarged sense
s.l.	*sensu lato*	in a broad sense
s.n.	*sine numero*	unnumbered; without a number
s.s.; s. str.	*sensu stricto*	in the strict, or narrow, sense
sc.	*scilicet*	namely
sched.	*scheda*	label
sec.	*secundum*	according to
seq.	*sequens*	following
ser.	*series*	series
s-g.; subgen.	*subgenus*	subgenus
s-gg.	*subgenera*	subgenera
sp.	*species*	species
spp.	*species*	species (plural)
sp. ind.; sp. indet.	*species indeterminata*	indeterminate species
sp. n.; sp. nov.	*species nova*	new species
spec.	*specimen*	specimen
sphalm.	*sphalmata*	in error; by mistake
ssp.; subsp.	*subspecies*	subspecies
st.; stat.	*status*	rank
stat. nov.	*status novus*	new rank
supra cit.	*supra citato*	cited above
syn.	*synonymon, synonymia*	synonym, synonymy
syn. nov.	*synonymum novum*	new synonym
t.; tab.	*tabula*	plate
t.	*teste*	on the evidence of
t.; tom.	*tomus*	volume
tax. vag.	*taxum vagum*	uncertain taxon
trans. nov.	*translatio nova*	new transfer
typ.	*typus*	type
typ. cons.	*typus conservandus*	conserved type
v.; var.	*varietas*	variety
v.	*vide*	see
v.	*visum*	seen
v.; vol.	*volumen*	volume
v. et.	*vide etiam*	see also
viz.	*videlicet*	namely

Appendix 19. GREEK AND RUSSIAN ALPHABETS

GREEK			RUSSIAN			ROMAN NUMERALS	
Letter	Name	Transliteration	Letter	Name	Transliteration	I	1
						II	2
						III	3
Α α	alpha	a	А а	*ah*	a	IV	4
			Б б	*beh*	b	V	5
Β β	beta	b	В в	*veh*	v	VI	6
						VII	7
Γ γ	gamma	g	Г г	*geh*	g		
Δ δ	delta	d	Д д	*deh*	d	VIII	8
						IX	9
Ε ε	epsilon	e	Е е	*yeh*	ye	X	10
			Ж ж	*zheh*	zh	XX	20
Ζ ζ	zeta	z	З э	*zeh*	z	XXX	30
						XL	40
Η η	eta	e (or ē)	И и	*ee*	i	L	50
			Й й	*ee s krátkoi*	y		
Θ θ	theta	th	К к	*kah*	k	LX	60
						LXX	70
Ι ι	iota	i	Л л	*el*	l	LXXX	80
			М м	*em*	m	XC	90
Κ κ	kappa	k	Н н	*en*	n	C	100
						CC	200
Λ λ	lambda	l	О о	*oh*	o	CCC	300
Μ μ	mu	m	П п	*peh*	p	CD	400
			Р р	*err*	r	D	500
Ν ν	nu	n	С с	*ess*	s	DC	600
						DCC	700
Ξ ξ	xi	x	Т т	*teh*	t	DCCC	800
			У у	*ooh*	oo	CM	900
Ο ο	omicron	o	Ф ф	*eff*	f	M	1000
Π π	pi	p	Х х	*khan*	kh		
Ρ ϱ	rho	r	Ц ц	*tseh*	ts		
Σ σ,ς[1]	sigma	s	Ч ч	*cheh*	ch		
			Ш ш	*shah*	sh		
Τ τ	tau	t	Щ щ	*shchah*	shch		
Υ υ	upsilon	y	Ъ ъ	*tvyórdy znak*			
			Ы ы	*yery*	y		
Φ φ	phi	ph	Ь ь	*myakhki znak*			
Χ χ	chi	ch, kh	Э э	*eh oborótnoye*	eh		
			Ю ю	*yoo*	yu		
Ψ ψ	psi	ps	Я я	*yah*	ya		
Ω ω	omega	o (or ō)					

[1] At end of word.

Appendix 20. PROOF CORRECTION MARKS

No.	Meaning	Corresponding mark in text	Marginal mark	No.	Meaning	Corresponding mark in text	Marginal mark
1	Sign to show that marginal mark is concluded		/	33	Move to the right	enclosing matter to be moved to the right	
2	Delete (take out)	/		34	Move lines to right	at left side of group to be moved	
3	Delete and close-up	above and below letters to be taken out.		35	Move lines to left	enclosing matter to be moved to the left	
4	Delete and leave space	/	#	36	Move portion of matter so that it comes within the position indicated	at limits of required position	[ ]
5	Leave as printed	under letters or words to remain		37	Take letter or word from end of one line to beginning of next		
6	Change to capital letters	under letters or words to be altered		38	Take letter or word from beginning of one line to end of preceding line		
7	Change to small capitals	under letters or words to be altered		39	Raise lines	over lines to be moved	
8	Use capital letters for initial letters and small capitals for rest of words	under initial letters and under the rest of the words		40	Lower lines	under lines to be moved	
9	Change to lower case	Encircle letters to be altered		41	Correct the vertical alignment	‖	‖
10	Change to bold type	under letters or words to be altered		42	Straighten lines	through lines to be straightened	
11	Change to italics	under letters or words to be altered		43	Push down space	Encircle space affected	
12	Underline word or words			44	Begin a new paragraph		
13	Change to roman type	Encircle words to be altered		45	No fresh paragraph here	Between paragraphs	
14	(wrong fount) Replace by letter of correct fount	Encircle letters to be altered		46	The abbreviation or figure to be spelt out in full		spell out
15	Invert type	Encircle letter to be altered		47	Insert omitted portion of copy		out see copy
16	Replace by similar but undamaged character	Encircle letter to be altered	×	48	(Caret mark). Insert matter indicated in margin		
17	Substituted letters or signs under which this is placed to be 'superior'	/		49	Insert comma		
18	Inserted letters or signs under which this is placed to be 'superior'			50	Substitute comma	/	
19	Substituted letters or signs over which this is placed to be 'inferior'	/		51	Insert semi-colon		
20	Inserted letters or signs over which this is placed to be 'inferior'			52	Substitute semi-colon	/	
21	Use ligature (e.g. ffi) or diphong (e.g. œ)	enclosing letters to be altered		53	Insert full-stop		
22	Substitute separate letters for ligature or diphong		write out separate letters	54	Substitute full-stop	/	
23	Close-up—delete space between letters	linking words or letters		55	Insert colon		
24	Insert space		#	56	Substitute colon	/	
25	Space between lines or paragraphs		# >	57	Insert interrogation mark		
26	Make spacing equal	between words		58	Substitute interrogation mark	/	?/
27	Reduce space	between words		59	Insert exclamation mark		
28	Transpose	between letters or words numbered when necessary		60	Substitute exclamation mark	/	!/
29	Place in centre of line	Indicate position with	[ ]	61	Insert parentheses		
30	Indent one em			62	Insert (square) brackets		
31	Indent two ems			63	Insert hyphen		
32	Move to the left	enclosing matter to be moved to the left		64	Insert en (half-em) rule		en/
				65	Insert one-em rule		em/
				66	Insert two-em rule		2em/
				67	Insert apostrophe		
				68	Insert single quotation marks		
				69	Insert double quotation marks		
				70	Insert ellipsis		...
				71	Insert leader		
				72	Insert shilling stroke		
				73	Refer to appropriate authority anything the accuracy or suitability of which is doubted	Encircle words, etc., affected	(?)

Appendix 20. – *continued*

International paper sizes

Code	x (mm)	y (mm)	Code	x (mm)	y (mm)	Code	x (mm)	y (mm)
A0	841	1189	B0	1000	1414	C0	917	1297
A1	594	841	B1	707	1000	C1	648	917
A2	420	594	B2	500	707	C2	458	648
A3	297	420	B3	353	500	C3	324	458
A4	210	297	B4	250	353	C4	229	324
A5	148	210	B5	176	250	C5	162	229
A6	105	148	B6	125	176	C6	114	162

Appendix 21. THE BEAUFORT WIND SCALE

Beaufort number	Wind	Wind speed				Effects caused by the wind	
		knots	miles/h	m/s		at sea	on land
0	calm	<1	<1	<0.2		water surface smooth, mirror-like	air still; smoke rises vertically
1	light air	1–3	1–3	0.3–1.5		only ripples or small wavelets, no foam crests	smoke drifts downwind but vanes remain stationary
2	light breeze	4–6	4–7	1.6–3.3		distinct wavelets, small and short but not breaking	wind felt on face; leaves rustle, vanes moved
3	gentle breeze	7–10	8–12	3.4–5.3		large wavelets beginning to break, glassy foam, occasional white horses	leaves and twigs move constantly; wind extends small flag
4	moderate breeze	11–16	13–18	5.5–7.9		small waves somewhat longer win quite frequent white horses	wind raises dust, loose paper and moves small twigs and branches
5	fresh breeze	17–21	19–24	8.0–10.7		pronounced waves, distinctly elongated with many white horses, perhaps isolated spray	small trees in leaf begin to sway; crested wavelets form on inland waters
6	strong breeze	22–27	25–31	10.8–13.8		large breaking waves with extensive white foam crests; spray probable	large branches move, telegraph wires whistle, umbrellas difficult to control
7	strong wind	28–33	32–38	13.9–17.1		sea heaps up; streaks of white foam begin to be blown downwind	whole trees move; some wind resistance offered to walkers
8	fresh gale	34–40	39–46	17.2–20.7		moderately high waves with crests of considerable length, streaks of white foam well marked; spray blown from crests	twigs break off trees; greatly impedes progress of walkers
9	strong gale	41–47	47–54	20.8–24.4		high waves, rolling sea and dense streaks of white foam spray begin to reduce visibility	slight structural damage; chimney pots and roof tiles removed
10	whole gale	48–55	55–63	24.5–28.4		heavy rolling sea with very high waves; overhanging crests; great patches and dense streaks of foam	much structural damage; trees uprooted; seldom experienced inland
11	storm	56–65	64–75	28.5–32.6		extraordinarily high waves and deep troughs; sea surface with streaky foam; strong spray impedes visibility	widespread damage; very rarely experienced inland
12	hurricane	>65	>75	32.7–36.9		sea surface entirely white; air full of foam and spray	—